REVIEWS IN MINERALOGY

AND GEOCHEMISTRY

Volume 65   2007

KT-503-092

FLUID-FLUID INTERACTIONS

EDITORS:

Axel Liebscher   *Technical University of Berlin*
Berlin, Germany
and
GeoForschungsZentrum Potsdam
Potsdam, Germany

Christoph A. Heinrich   *ETH Zürich*
Zürich, Switzerland

COVER FIGURE CAPTION: Fluid inclusion trail in quartz from Stronghold Granite, Arizona. Separately trapped inclusions of co-existing vapor and hypersaline liquid (brine) demonstrate phase separation at near-magmatic temperature. The high-density brine inclusions are salt-saturated after cooling to ambient conditions, containing halite and sylvite daughter crystals; the vapor inclusions contain a large bubble and a little aqueous liquid; the largest inclusion has trapped a random mixture of the two fluids and thus shows intermediate phase proportions after cooling (photomicrograph courtesy of Andreas Audétat, Bayreuth); H_2O-NaCl phase diagram is redrawn after Heinrich CA, Driesner T, Stefánsson A, Seward TM (2004, Magmatic vapor contraction and the transport of gold from the porphyry environment to epithermal ore deposits. Geology 32:761-764).

*Series Editor: **Jodi J. Rosso***

MINERALOGICAL SOCIETY OF AMERICA
GEOCHEMICAL SOCIETY

SHORT COURSE SERIES DEDICATION

Dr. William C. Luth has had a long and distinguished career in research, education and in the government. He was a leader in experimental petrology and in training graduate students at Stanford University. His efforts at Sandia National Laboratory and at the Department of Energy's headquarters resulted in the initiation and long-term support of many of the cutting edge research projects whose results form the foundations of these short courses. Bill's broad interest in understanding fundamental geochemical processes and their applications to national problems is a continuous thread through both his university and government career. He retired in 1996, but his efforts to foster excellent basic research, and to promote the development of advanced analytical capabilities gave a unique focus to the basic research portfolio in Geosciences at the Department of Energy. He has been, and continues to be, a friend and mentor to many of us. It is appropriate to celebrate his career in education and government service with this series of courses.

Reviews in Mineralogy and Geochemistry, Volume 65

Fluid-Fluid Interactions

ISSN 1529-6466
ISBN 978-0-939950-77-5

COPYRIGHT 2007

THE MINERALOGICAL SOCIETY OF AMERICA
3635 CONCORDE PARKWAY, SUITE 500
CHANTILLY, VIRGINIA, 20151-1125, U.S.A.
WWW.MINSOCAM.ORG

FLUID-FLUID INTERACTIONS

65   *Reviews in Mineralogy and Geochemistry*   **65**

FROM THE EDITORS

The review chapters in this volume were the basis for a two day short course on *Fluid-Fluid Equilibria in the Crust* held at the Institute of Geology and Mineralogy, University of Cologne, Germany (August 16-17, 2007) prior to the Goldschmidt Conference 2007 in Cologne, Germany. This meeting and volume were sponsored by the Mineralogical Society of America, Geochemical Society, and the United States Department of Energy.

Any supplemental material and errata (if any) can be found at the MSA website *www.minsocam.org*.

Jodi J. Rosso, Series Editor
West Richland, Washington
June 2007

We thank the Mineralogical Society of America and the Geochemical Society for giving us the opportunity to edit this volume. We are grateful to all authors, who contributed state-of-the-art chapters on the various topics of fluid-fluid interactions. However, editing this volume would not have been possible without the assistance and editorial work of Jodi J. Rosso, which we especially acknowledge. We also thank all the reviewers for the individual chapters: A. Audétat, A. Belonoshko, L. Cathles, J. Cleverly, J. Hedenquist, S. Ingebritsen, H. Keppler, S. Kesler, P. Nabelek, R. Sadus, S. Simmons, A. Stavland, D. Vielzeuf, B.W.D. Yardley, who have done an excellent job and helped us to complete the editorial work in time.

Axel Liebscher   **Christoph A. Heinrich**
Berlin & Potsdam, Germany   *Zürich, Switzerland*

1529-6466/07/0065-0000$05.00   DOI: 10.2138/rmg.2007.65.0

FLUID-FLUID INTERACTIONS

65   *Reviews in Mineralogy and Geochemistry*   **65**

TABLE OF CONTENTS

3   Equations of State for Complex Fluids

Matthias Gottschalk

4   Liquid Immiscibility in Silicate Melts and Related Systems

Alan B. Thompson, Maarten Aerts, Alistair C. Hack

5   Phase Relations Involving Hydrous Silicate Melts, Aqueous Fluids, and Minerals

Alistair C. Hack, Alan B. Thompson, Maarten Aerts

6   Numerical Simulation of Multiphase Fluid Flow in Hydrothermal Systems

Thomas Driesner, Sebastian Geiger

7   Fluid Phase Separation Processes in Submarine Hydrothermal Systems

Dionysis I. Foustoukos, William E. Seyfried, Jr.

8   Fluids in Hydrocarbon Basins

Karen S. Pedersen, Peter L. Christensen

9   Fluid-Fluid Interactions in Geothermal Systems

Stefán Arnórsson, Andri Stefánsson, Jón Örn Bjarnason

10   Fluid Immiscibility in Volcanic Environments

James D. Webster, Charles W. Mandeville

11   Fluid-Fluid Interactions in Magmatic-Hydrothermal Ore Formation

Christoph A. Heinrich

12   Fluid Immiscibility in Metamorphic Rocks

Wilhelm Heinrich

Reviews in Mineralogy & Geochemistry
Vol. 65, pp. 1-13, 2007
Copyright © Mineralogical Society of America

1

Fluid–Fluid Interactions in the Earth's Lithosphere

Axel Liebscher

Institute for Applied Geosciences	*Department 4, Chemistry of the Earth*
Technical University of Berlin	*GeoForschungsZentrum Potsdam*
D-13355 Berlin, Germany	*Telegrafenberg*
	D-14473 Potsdam, Germany

axel.liebscher@tu-berlin.de

Christoph A. Heinrich

Isotope Geochemistry and Mineral Resources
Departement Erdwissenschaften NO
ETH Zentrum
CH-8092 Zurich, Switzerland
christoph.heinrich@erdw.ethz.ch

INTRODUCTION

Fluids rich in water, carbon and sulfur species and a variety of dissolved salts are a ubiquitous transport medium for heat and matter in the Earth's interior. Fluid transport through the upper mantle and crust controls the origin of magmatism above subduction zones and results in natural risks of explosive volcanism. Fluids passing through rocks affect the chemical and heat budget of the global oceans, and can be utilized as a source of geothermal energy on land. Fluid transport is a key to the formation and the practical utilization of natural resources, from the origin of hydrothermal mineral deposits, through the exploitation of gaseous and liquid hydrocarbons as sources of energy and essential raw materials, to the subsurface storage of waste materials such as CO_2.

Different sources of fluids and variable paths of recycling volatile components from the hydrosphere and atmosphere through the solid interior of the Earth lead to a broad range of fluid compositions, from aqueous liquids and gases through water-rich silicate or salt melts to carbon-rich endmember compositions. Different rock regimes in the crust and mantle generate characteristic ranges of fluid composition, which depending on pressure, temperature and composition are miscible to greatly variable degrees. For example, aqueous liquids and vapors are increasingly miscible at elevated pressure and temperature. The degree of this miscibility is, however, greatly influenced by the presence of additional carbonic or salt components. A wide range of fluid–fluid interactions results from this partial miscibility of crustal fluids. Vastly different chemical and physical properties of variably miscible fluids, combined with fluid flow from one pressure – temperature regime to another, therefore have major consequences for the chemical and physical evolution of the crust and mantle.

Several recent textbooks (e.g., Walther and Wood 1986; Carrol and Holloway 1994; Shmulovich et al. 1995; Jamtveit and Yardley 1997) and review articles (e.g., Brimhall and Crerar 1987; Eugster and Baumgartner 1987; Ferry and Baumgartner 1987; Ague 2003; Frape et al. 2003; German and Von Damm 2003; Kharaka and Hanor 2003; Schmidt and Poli 2003; Bodnar 2005; Kesler 2005; De Vivo et al. 2005; Green and Jung 2005) have addressed the role and diverse aspects of fluids in crustal processes. However, immiscibility of fluids and the associated phenomena of multiphase fluid flow are generally dealt with only in subsections

1529-6466/07/0065-0001$05.00   DOI: 10.2138/rmg.2007.65.1

with respect to specific environments and aspects of fluid mediated processes. This volume of *Reviews in Mineralogy and Geochemistry* attempts to fill this gap and to explicitly focus on the role that co-existing fluids play in the diverse geologic environments. It brings together the previously somewhat detached literature on fluid–fluid interactions in continental, volcanic, submarine and subduction zone environments. It emphasizes that fluid mixing and unmixing are widespread processes that may occur in all geologic environments of the entire crust and upper mantle. Despite different *P-T* conditions, the fundamental processes are analogous in the different settings.

The next two chapters of the book summarize the state of knowledge of the physical and chemical properties of coexisting crustal and upper-mantle fluids and their description with equations of state. Even though the book primarily focuses on fluids rich in volatile components and salts, some of the general principles of phase topology were first derived from theory and experiments on immiscible melt systems with or without water, which are therefore included as separate Chapters 4 and 5. After a chapter on numerical modeling of two-phase fluid flow, the remaining chapters review the processes of fluid–fluid interaction in different geological regimes, invariably emphasizing the close link between fluid properties and dynamic processes—many of which are of great practical relevance for assessing natural risks and for the supply of natural resources.

GEOLOGIC ENVIRONMENTS OF TWO-PHASE FLUIDS

In this introductory chapter we briefly address the different geologic environments in which crustal fluids may evolve (Fig. 1). We will then define some of the terms that are commonly used to describe different fluid phases and their interactions and introduce the principal thermodynamic phase relationships that control the compositional variation and other properties of crustal fluids (Figs. 2 to 5). For details, full referencing and in-depth discussion of these topics, the reader is referred to the individual chapters of this volume.

From a global perspective, water-bearing fluids link the different spheres of the Earth in very distinct environments (Fig. 1). In all of these settings at least two, and sometimes three fluid phases can stably coexist and interact with each other, as demonstrated by experimental data (Liebscher 2007, this volume), fluid inclusion observations and other geological evidence reviewed in the chapters of this book.

Submarine hydrothermal systems (Fig. 1b) chemically link the oceans with the oceanic crust, which itself has been extracted from the mantle (Foustoukos and Seyfried 2007, this volume). The oceanic crust becomes chemically modified by reaction with thermally convecting seawater circulating through it, but in turn modifies the chemical and isotopic composition of the seawater itself. Such hydrothermal systems develop primarily along mid ocean ridges (MOR) but also along back-arc spreading centers as well as in oceanic rift basins and around the submersed parts of intraplate volcanoes. The heat source of these dominantly dry basaltic systems is magmatic, but the overwhelming fluid source is seawater, infiltrating the oceanic crust and leading to spectacular black and white smokers on the seafloor. Fluid–fluid separation is primarily witnessed by large salinity variations within vent fluids not only on a global scale but also between neighboring vents from individual vent fields.

Interaction between heated seawater and rocks on the ocean floor extracts material from the oceanic crust and supplies it to the oceans, and also hydrates and metasomatizes the initially dry oceanic crust. *Metamorphic fluid systems* (Figs. 1c,d) are created by devolatilization reactions resulting when metasomatized oceanic crust and overlying sediments are subducted at convergent plate boundaries, or where continental collision leads to crustal thickening and regional metamorphism. Prograde metamorphic reactions generally lead to relatively low-

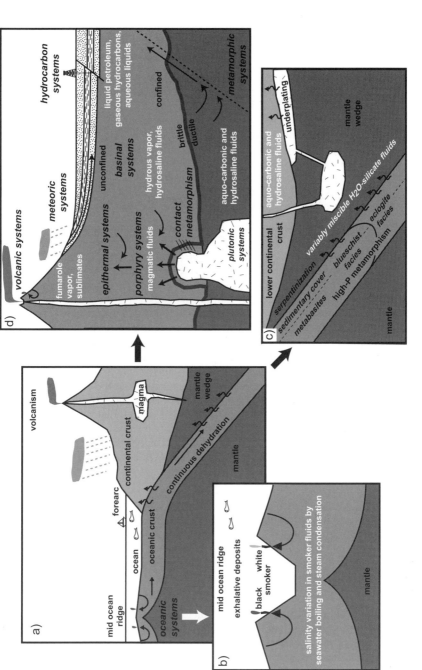

Figure 1. Schematic drawings of the different fluid systems and the corresponding fluid phase characteristics. (a) Global endogenic fluid cycle from mid ocean ridges through subduction zones to supra-subduction zone volcanism. (b) Hydrothermal system at mid ocean ridges. (c) High-pressure fluid/melt systems within subduction zones and principal fluid flow paths. (d) Hydrothermal fluid systems within continental crust. [(d) re-drawn and modified from Kesler (2005)].

salinity metamorphic fluids, but increasing contributions of CO_2 by carbonate-consuming reactions at high pressures and temperatures may generate aquo-carbonic fluids plus a stably coexisting fluid phase enriched in the salt components (Heinrich 2007b, this volume). A key conclusion of the chapter by W. Heinrich is that metamorphic fluids are more commonly in the two-phase state than previously thought, and that high-salinity fluids with reduced water activity can have a profound effect on metamorphic mineral stability relations.

When crustal material is subducted to the *deep slab and mantle environment* (Fig. 1c), further heating and compression produces fluids by high-pressure dehydration to eclogite-facies mineral assemblages. These fluids are expelled from the slab and can rise into the hotter overriding mantle, where they become an essential ingredient for the generation of calcalkaline magmas. At these conditions, the solubility of silicates in the water-rich fluid may increase to the point that aqueous solution and hydrous silicate melt become indistinguishable, as the two distinct coexisting mobile phases give way to a single-phase homogeneous mixture of water and silicate components in any proportion. This is discussed in the chapters by Hack et al. (2007, this volume) and Thompson et al. (2007, this volume), as an illustration of the principles of critical phase behavior in variably hydrous fluid–melt systems.

Slab-derived high-pressure fluids and mantle-derived silicate melts enriched in volatile components migrate upwards and ultimately recycle their volatile load to the surface and the atmosphere in *subduction-related magmatic-hydrothermal* environments (Fig. 1d). During ascent, they contribute to the evolution of calcalkaline magmas. From these, fluids of greatly variable composition density and viscosity are expelled, depending primarily on depth and temperature of emplacement of the magma and its crystallization in the crust. Thus, extremely saline fluids approaching hydrous salt melts can be generated, as well as partly coexisting vapor-like fluids of variable density and salinity. Resulting processes are reviewed in the chapters of Webster and Mandeville (2007, this volume) focusing primarily on volcanic processes including fumaroles and explosive eruptions, and by C. Heinrich (2007a, this volume) emphasizing fluid–fluid interactions in the subvolcanic to plutonic realm. Here, important types of mineral deposits are formed as a direct consequence of fluid mixing and phase separation, as further illustrated by numerical modeling of heat and mass transfer by multiple-phase fluid flow (Driesner and Geiger 2007, this volume). Fluids from plutons also contribute important components to *contact metamorphic fluid systems* (Fig. 1d) Here, aquo-carbonic fluids plus a stably coexisting fluid phase enriched in the salt components may form, which are compositionally similar to fluids in other metamorphic systems (Heinrich 2007b, this volume).

Fluid processes in *continental geothermal systems* (Fig. 1d), discussed by Arnórsson et al. (2007, this volume), are commonly linked to magmatism as well. They involve aqueous fluids of variable salinity, ranging from essentially salt-free meteoric water through shallow-circulating seawater to saline basin brines in continental rifts. The aqueous fluids are predominantly derived from the hydrosphere but may contain notable quantities of magmatic fluids. The generally dominant aqueous liquid commonly coexist with low-density vapor phase. Fluid mixing and fluid-phase separation are first-order processes governing the chemical and physical evolution of fluids in geothermal systems (Arnórsson et al. 2007)

Variably saline aqueous fluids coexist with liquid and gaseous *hydrocarbons in sedimentary basins* (Fig. 1d), which are discussed by Pedersen and Christensen (2007). These basinal fluid systems are developed in the upper parts of the continental crust above the brittle-ductile transition, and are therefore mostly unconfined at the top, but can become confined by compaction and cementation in their deeper parts. Aqueous fluids are meteoric and/or diagenetic in origin, locally with a deep crustal or even mantle component to their gas budget as indicated by helium and its isotopes. Up to three fluid phases can stably coexist and interact with each other and with the sedimentary rocks: variably saline aqueous liquid, a liquid petroleum phase, and a gas phase ranging from low-density water vapor to compressed mixtures of methane, carbon dioxide and other gases.

PRINCIPAL PHASE RELATIONS IN TWO-PHASE FLUID SYSTEMS

As a basis for further discussion throughout this volume, some general principles of fluid phase stability relations are illustrated using selected unary and binary systems. These are later followed up by a detailed review of experimental studies in geologically important model systems (Liebscher 2007), and a critical compilation of equations of state that were developed to mathematically describing phase properties as well as limiting miscibility surfaces (Gottschalk 2007).

One-component systems

In any one-component system (e.g., H_2O, CO_2, but also N_2, NaCl, SiO_2), all three phase states, i.e., solid, liquid and vapor, have identical composition. Their P-T stability fields are divariant and limited by the univariant solid–liquid, solid–vapor, and liquid–vapor equilibria, which meet at the invariant triple point where solid, liquid, and vapor coexist (Figs. 2a,b). In one-component systems, coexistence of liquid and vapor is restricted to the liquid-vapor equilibrium, which generally shifts to higher pressure with increasing temperature. While coexisting liquid and vapor along the liquid–vapor equilibrium curve have identical composition in one-component systems, they differ in physical properties like density (Fig. 2c) or viscosity. With increasing P and T along the equilibrium curve, however, any difference in physical properties between liquid and vapor continuously decreases and completely vanishes at the invariant critical point with P_c, T_c, and critical density ρ_c. At $P > P_c$ and $T > T_c$, liquid and vapor are physically indistinguishable and only one single phase fluid exists.

Binary fluid systems

The liquid–vapor phase topology of a one-component system changes drastically when a second component is added to the system. This additional component partitions either in favor of the liquid or in favor of the vapor. In two-component systems, coexisting liquid and vapor therefore not only differ in physical properties but also in composition. With composition as additional variable, the invariant critical point of the one-component system turns into the univariant critical curve of the two-component system (Fig. 2d), along which liquid and vapor attain identical physical and chemical properties. The orientation of the resulting critical curve in P-T space with respect to the critical isochore (i.e., line of constant critical density ρ_c) of the one-component system depends on the relative partitioning of the additional component between liquid and vapor: if the additional component has a preference for the liquid (e.g., NaCl in H_2O) the critical curve will be at lower P; if the additional component has a preference for the vapor (e.g., CO_2 or CH_4 in H_2O) it will be at higher P than the critical isochore of the one-component system (Fig. 2d). With respect to crustal hydrothermal systems, the most important two-component systems are water–salt systems with highly soluble salts, like H_2O-NaCl or H_2O-$CaCl_2$, and water–gas systems such as H_2O-CO_2, which have dramatically different properties.

In most water–salt systems (Fig. 3a,b), the critical curve is continuous over the entire P-T-x range and connects the critical points of the pure H_2O and salt system, respectively. At pressure below the critical curve, liquid and vapor of different composition and different physical properties coexist and form a two-fluid phase "tunnel" in P-T-x space. For any given temperature, the differences in physical and chemical properties of vapor and liquid increase along the tunnel surface with decreasing pressure. The two-fluid phase tunnel is limited towards lower pressure conditions by its intersection with the liquid–salt solubility surface. This intersection is called the solubility curve or vapor + liquid + solid salt equilibrium.

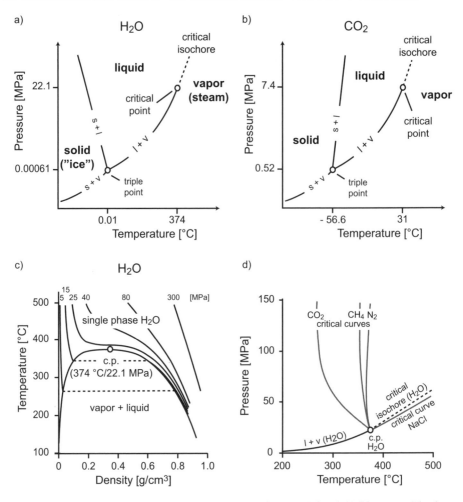

Figure 2. Pressure-temperature solid-liquid-vapor phase relations; (a) H_2O and (b) CO_2 system. Divariant curves s + v, s + l, and l + v denote two-phase coexistence of solid (s), vapor (v) and/or liquid (l). In one-component or unary systems, coexistence of vapor and liquid is restricted to the l + v equilibrium. The l + v equilibrium terminates (or vanishes) at the critical point. "Critical isochore" denotes the line of constant, critical density ($\rho_{crit.}$) in *P-T* space. The critical isochore may be used to distinguish between liquid ($\rho > \rho_{crit.}$) and vapor ($\rho < \rho_{crit.}$) even in the single phase fluid region. (c) Temperature – density relations in the H_2O system. Solid lines are isobars with pressure given at the top. Below the critical point, low-density vapor coexists with higher density liquid. With increasing temperature, the density contrast between vapor and liquid increases along the l + v equilibrium. (d) Effect of additional components on the critical behavior of H_2O dominated fluids. In two-component systems, the critical point turns into a critical curve, the principal orientation of which depends on the relative preference of the additional component for vapor (e.g., CO_2, CH_4, N_2) or liquid (e.g., NaCl).

Along this equilibrium, vapor, liquid, and solid salt coexist (i.e., both vapor and liquid are salt saturated), whereas at pressure conditions below the solubility curve, solid salt coexists with salt-saturated vapor only. The solubility curve terminates at the triple point of the pure salt system, usually at very high temperature but low pressure.

Two-component systems between H_2O and highly soluble salts, exemplified by the H_2O-NaCl model system, show continuous critical and solubility curves over the entire *P-T-x* space.

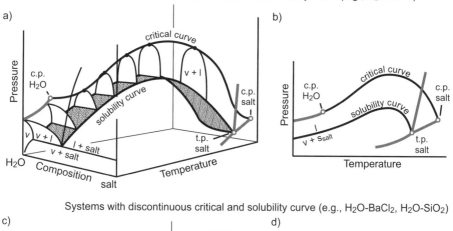

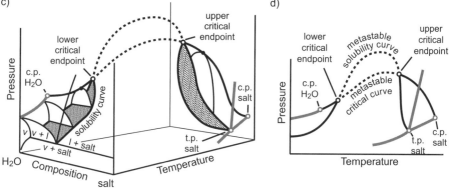

Figure 3. Principal fluid phase relations in two-component systems. Depending on the continuity of critical and solubility curve, two types of systems can be distinguished: Systems with continuous critical and solubility curves over the entire *P-T-x* space (a, b) and systems with discontinuous critical and solubility curves resulting in two critical endpoints, at which critical and solubility curve terminate (c, d). The first type of systems is typical for highly soluble components like H_2O-NaC whereas the second type is typical for only sparingly soluble components like H_2O-silicates (see text for more details); (a) redrawn and modified from Skippen (1988).

Alternatively, if the salt (e.g., $BaCl_2$) or oxide component (e.g., SiO_2) is only slightly soluble in H_2O (and vice versa), the solubility curve shifts to notably higher pressure conditions, thereby intersecting the critical curve (Fig. 3c,d). In such systems, neither the critical curve, nor the solubility curve, nor the two-fluid phase tunnel are continuous over the entire *P-T-x* space. Instead, two separate two-fluid phase volumes occur, which are separated by a single phase fluid + salt or oxide component field. The intersections of the solubility curve with the critical curve are termed lower and upper critical endpoints and are invariant in *P-T-x* space. In systems with silicates as the component in addition to water, the high-temperature region describes the wet melting behavior of the system, and the solubility curve is called the wet solidus. Further discussion of fluid–solid–melt phase relations can be found in the chapters of Thompson et al. (2007) and Hack et al. (2007).

Binary systems of water and a non-polar gas component, exemplified by the H_2O-CO_2 system, have notably different fluid properties than the typical H_2O-salt systems. The *P-T* dependence of their two-phase surface differs significantly from that of the typical H_2O-salt systems, reflecting the highly volatile and compressible nature of the second component. The critical curve shows a minimum as a function of pressure near 170 MPa, 265 °C and $x_{CO_2} \sim 0.3$, and the limiting two-phase immiscibility surface becomes narrower not only with increasing pressure, but also with increasing temperature. As a result, binary H_2O–CO_2 fluids exist as completely miscible single phase already at relatively shallow depths in the Earths interior, at all temperatures in excess of 350 to 400 °C. On the other hand, CO_2 and NaCl are almost immiscible at all geologically relevant conditions, and this is the reason why ternary fluids, containing water plus salts as well as CO_2, can show two-phase fluid coexistence throughout the entire range of *P-T* conditions in the crust (Liebscher 2007; Heinrich 2007b).

TERMINOLOGY OF FLUID PHASES AND PROCESSES

Due to the complex topology of phase stability fields in the H_2O–CO_2–NaCl fluid system ($\pm CH_4 \pm$ other salts), different *P-T* evolutions along the flow path of fluids through the Earth's lithosphere can lead to very different processes of fluid–fluid interaction. Fluids change their density and may mix or un-mix and the resulting phases may segregate from each other due to their different physical properties. To describe such processes with reference to experimental phase diagrams, a consistent terminology for phases and fluid interactions is desirable. The following terms (Heinrich 2005; Williams-Jones and Heinrich 2005) approximate the common usage in fluid-inclusion research, where the first evidence of multiple coexisting fluids in the earths interior originated (Roedder 1984; Samson et al. 2003). These widely used fluid terms can be consistently defined in H_2O – non-polar gas – salt systems of geological relevance, but they are not theoretically rigorous for fluid phase topologies in all other chemical systems. This is exemplified in the chapters by Thompson et al. (2007) and Hack et al. (2007), illustrating the stability relationships in melt–melt and melt–aqueous fluid systems at upper-mantle pressures.

Terms to describe fluid phases

We suggest that *fluid* is used as the overarching term for a physically mobile phase, irrespective of the proportions of salt and volatile components. It can be further specified as a *single-phase fluid* if emphasis is placed on a phase state away from a limiting two-phase boundary. The term "supercritical" is avoided for crustal fluids because it cannot always be defined unambiguously (see also below). A fluid can be a *liquid*, if it has a density that is greater than the critical density of the fluid composition in question, or a *vapor* if its density is lower than its critical density (see Figs. 2a,b for critical isochores in one-component systems). This definition divides—arbitrarily but unambiguously—the physically continuous single-phase stability field into two regions in pressure–temperature space. The separating surface is not a phase boundary but comprises the locus of all critical isochores, which are lines of constant fluid density emerging from all points along the critical curve of the system. Single-phase fluids of near-critical density are loosely but sometimes conveniently described as *intermediate-density fluids*. This definition of liquid, intermediate-density and vapor is consistent with the common description of isochoric homogenization of fluid inclusions, by bubble-point ("homogenization into the liquid"), near-critical ("meniscus fading") and dew-point transition ("homogenization to vapor"), respectively (Roedder 1984). *Hypersaline liquid* (loosely called "brine") is commonly used for a dense water-bearing liquid whose bulk salinity is high enough to saturate a salt crystal when cooled isochorically to room temperature (> 26 wt% salt in the model system H_2O–NaCl). The term *steam* should only be used for H_2O vapor and thus contains chemical information. *Gaseous* and *volatile*, although often used as nouns, are adjectives and should be exclusively used as such, to describe fluid components or species.

Terms describing fluid processes involving one or two phases

We use the term *fluid mixing* for any process of physical confluence and equilibration of two fluids that were initially out of equilibrium with each other, even if their interaction leads to phase separation into two coexisting fluids of different composition. For example, mixing between a homogeneous hypersaline H_2O–$NaCl$ liquid and a homogeneous H_2O–CO_2 fluid may lead to phase separation into an aqueous liquid of intermediate salinity containing minor CO_2, plus a stably coexisting vapor phase enriched in CO_2 (see Fig. 5).

Phase separation is used interchangeably with fluid un-mixing, as the overarching term of a single homogeneous fluid phase splitting into two coexisting fluid phases. Phase separation can either occur by *boiling*, if vapor bubbles nucleate from the liquid, or by *condensation*, which implies the formation of liquid droplets from the vapor. The distinction thus implies a *P-T* path to a point on the two-phase coexistence surface. In a one-component system, *boiling* implies crossing the liquid-vapor equilibrium towards higher temperature and/or lower pressure, whereas *condensation* implies the crossing of the liquid-vapor equilibrium towards lower temperature and/or higher pressure (Fig. 4a). In two- or multicomponent systems, where the separating phases have different compositions, *boiling* occurs when the *P-T* evolution of an initially homogeneous fluid intersects the limiting two-phase surface on the liquid side of the critical curve of the system, whereas *condensation* occurs upon intersection of the two-phase curve on the vapor side (Fig. 4b). Irrespective of composition, *boiling* and *condensation* imply changes in phase state and capture an important distinction between two types of phase separation.

In contrast to these terms for phase separation, the words *expansion* and *contraction* are used exclusively for homogeneous changes in density of a single fluid phase of fixed composition (Figs. 4a,b). *Fluid phase segregation* refers to physical separation of previously coexisting fluids. This may occur due to different buoyancy, viscosity, and/or wetting behavior of the two fluids. This terminology implies that any phase-separated fluid system that does not undergo phase segregation will not change its bulk composition; under new pressure–temperature conditions it may re-homogenize again to the initial single-phase fluid.

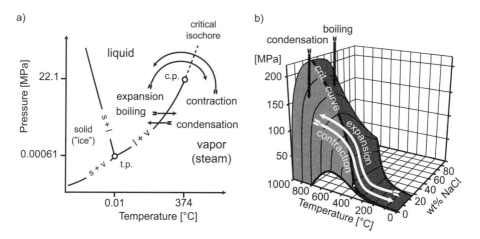

Figure 4. Schematic drawings showing the terminology for fluid phases and processes in (a) one-component systems as exemplified by the H_2O system and (b) two-component systems as exemplified by the H_2O-$NaCl$ system; (b) based on data by Driesner and Heinrich (2007).

A note on the term "supercritical"

The term "supercritical" is widely used in the literature on fluid-fluid equilibria. However, it is not applied consistently but used to describe rather different phenomena within the different systems. In some cases, its use is misleading or even meaningless. Except for specific cases outlined below, this term therefore should be avoided. In one-component systems, the term "supercritical" refers to P-T conditions above the critical point of the system, i.e., $P > P_c$ and $T > T_c$ (see Figs. 2a,b). Because in these systems P and T of the critical point are uniquely defined, the "supercritical" P-T field is likewise uniquely defined; even here, some authors use the term "supercritical" for high-temperature fluids at any pressure, because the fluid is single phase and changes in P or T do not induce phase separation. In two- or multi-component systems, critical behavior occurs along critical curves and is no longer uniquely defined in terms of P and T. "Supercritical," as implying $P > P_c$ and $T > T_c$, therefore becomes meaningless. This holds all the more if one considers that in two- or higher component systems the vapor-liquid two-phase field may open with increasing temperature as, for example, in the H_2O-NaCl system. In these systems, raising $T > T_c$ (at a given pressure) implies fluid phase separation instead of homogenization, contrary to what is intended by the term "supercritical." This picture may change if one uses "supercritical" with respect to critical endpoints in melt-fluid systems. For a given system, such a critical endpoint, i.e., where P and T of the critical curve and P and T of the wet solidus are identical and the wet solidus vanishes, is again uniquely defined with respect to P and T and so is the "supercritical" field. However, even in these cases, it will depend on the dP/dT slope of the critical curve, whether $T > T_{\text{critical endpoint}}$ implies single phase fluid (as intended; negative dP/dT slope) or fluid phase separation (positive dP/dT slope). For a more detailed and thorough discussion of critical endpoints and associated phenomena the reader is referred to Hack et al. (2007).

A completely different use of "supercritical" is found in the literature on oceanic hydrothermal systems (see also Foustoukos and Seyfried 2007). In these systems, the almost exclusive fluid source is seawater with a fixed salinity of ~ 3.2 wt% NaCl. P-T conditions for onset of fluid phase separation are therefore approximated by the liquid-vapor two-phase boundary of the 3.2 wt% NaCl isosalinity P-T section of the H_2O-NaCl system. This P-T section intersects the critical curve of the H_2O-NaCl system at ~ 30 MPa/404 °C, which is the critical point of seawater (Bischoff and Rosenbauer 1984). The terms "subcritical" and "supercritical" fluid phase separation then refer to this critical point of seawater and indicate fluid phase separation at P-T conditions below and above ~ 30 MPa/404 °C, respectively. At "subcritical" conditions, seawater has a higher salinity than the conjugate fluid, intersects the limiting liquid-vapor two-phase surface on the liquid side of the critical curve and the system starts to *boil*. At "supercritical" conditions, seawater has a lower salinity than the conjugate fluid, intersects the limiting liquid-vapor two-phase surface on the vapor side of the critical curve and the system starts to *condense*. The term "supercritical," as used in the literature of oceanic hydrothermal systems, is far from indicating a homogeneous single phase fluid state, but on the contrary indicates fluid-phase separation by condensation in the more general terminology proposed above. To avoid misunderstandings and for the comfort of non-specialists, we strongly recommend to avoid the inherently contradictory terms "subcritical phase separation" and "supercritical phase separation" in oceanic hydrothermal systems, but to use the terminology suggested above.

COMPOSITION AND PHASE STATE OF COMMON CRUSTAL FLUIDS

Crustal fluids are generally water-rich, but they are rarely pure H_2O and usually contain significant quantities of dissolved components. Dissolved salts such as NaCl, KCl, and $CaCl_2$, and polar gas components including CO_2, CH_4 and different nitrogen and sulfur species are the most important, which may even exceed the concentration of water in some crustal fluids. The principal compositional variations of water-bearing crustal fluids are broadly related to different geological environments outlined above (see Fig. 1). Consequences for fluid–fluid equilibria can be rationalized with respect to the simplified model system H_2O–NaCl–CO_2. This system approximates the properties of the more complex fluids, if NaCl is taken as representative for different salts (NaCl equivalent, $NaCl_{eq}$), and CO_2 is considered to represent other non-polar gas species.

As a first albeit simplified approximation, temperature and salinity may be used as physical and compositional discriminants between fluids from basinal, oceanic, metamorphic, and magmatic systems (Kesler 2005; his Fig. 4). The salinity of basinal fluids is extremely variable over approximately five orders of magnitude, between a few ppm in shallow meteoric regimes and ~60 wt% $NaCl_{eq}$ in fluids from evaporite-rich basins (Kharaka and Hanor 2003). The temperature of basinal fluids rarely exceeds 200 °C. Fluids from oceanic systems display a more restricted salinity range (~0.18 to 7.3 wt% $NaCl_{eq}$; German and Von Damm 2003; Foustoukos and Seyfried 2007) but typically reach higher temperatures up to ~400 °C. These temperatures refer to those measured in vent fluids, but oceanic system fluids may be considerably hotter in deeper parts so that fluid-fluid equilibria in these systems are not restricted to $T < 400$ °C. Metamorphic fluids generally display only moderate salinities (< ~5 wt% $NaCl_{eq}$); higher salinities may be encountered in evaporitic metasediments while extreme levels may be reached in some granulites (Crawford and Hollister 1986; Markl et al. 1998; W. Heinrich 2007). The temperature range of metamorphic fluids in mesothermal ore deposits as given by Kesler (2005; his Fig. 4) is ~200 to 450 °C, but metamorphic fluids in general will cover the whole T range of the metamorphic realm, i.e., up to > 700 °C in the granulite facies. Magmatic fluids display the largest range in salinity, up to 80 wt% $NaCl_{eq}$, and typically have temperatures > 400 °C (Heinrich 2007a).

This simplified temperature – salinity classification dramatically changes if CO_2 or additional polar gas components in the fluids are considered (Fig. 5). Oceanic fluids are restricted to H_2O-NaCl rich compositions and generally have low CO_2 contents (< 0.9 wt%, German and Von Damm 2003); here, CO_2 has no significant influence on fluid phase relations. Fluids from basinal systems may contain higher CO_2 concentrations than those from oceanic systems. CO_2 in sedimentary basins, and CH_4 to even greater extent, influence fluid stability relations and cause fluid phase separation into aqueous liquid and a gas phase. Contrary to fluids from oceanic and basinal systems, fluids from magmatic and metamorphic systems are truly ternary mixtures, and may contain considerable amounts of CO_2 as well as a major salinity component. The combined effects of salt and CO_2, which are mutually almost immiscible (Fig. 5), extends the possibility of two-phase fluid behavior to very hot magmatic, as well as very deep lower-crustal metamorphic environments.

In summary, Figure 5 shows the typical compositional ranges of major crustal fluid environments, superimposed on the limits of ternary miscibility in the H_2O-NaCl-CO_2 model system at widely differing geological *P-T* conditions. This overlay clearly shows that two-phase fluid coexistence must be widespread throughout the earths crust — even without considering hydrous silicate melts or liquid hydrocarbons, which may add a third fluid phase in magmatic or basin settings, respectively. This summary implies that fluid mixing and unmixing are expected to be common geological processes, wherever compositionally diverse crustal fluids flow through rocks along typical gradients in temperature and pressure.

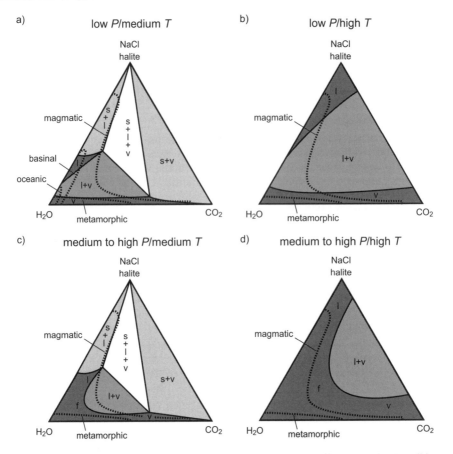

Figure 5. Principal phase relations in the ternary system H_2O-CO_2-NaCl at different crustal *P-T* conditions: a) and b) at pressure below and c) and d) at pressure above the critical curve of the sub-system H_2O-NaCl; a) and c) at temperature below and b) and d) at temperature above the melting curve of halite. Superimposed are schematic fields of fluid composition for basinal, oceanic, metamorphic, and magmatic fluids as given by Kesler (2005); ternaries redrawn and modified from W. Heinrich (2007).

ACKNOWLEDGMENTS

This introduction benefited from careful reading by K. Drüppel and G. Franz. Final editorial handling by J. Rosso is gratefully acknowledged.

REFERENCES

Ague JJ (2003) Fluid flow in the deep crust. Treatise Geochem 3:195-228
Arnórsson S, Stefánsson A, Bjarnason JÖ (2007) Fluid-fluid interactions in geothermal systems. Rev Mineral Geochem 65:259-312
Bischoff JL, Rosenbauer RJ (1984) The critical point and two-phase boundary of seawater, 200-500°C. Earth Planet Sci Lett 68:172-180
Bodnar RJB (2005) Fluids in planetary systems. Elements 1:9-12
Brimhall GH, Crerar DA (1987) Ore fluids: Magmatic to supergene. Rev Mineral 17:235-321

Carroll MR, Holloway JR (1994) Volatiles in Magmas. Rev Mineral Vol. 30, Mineralogical Society of America

Crawford ML, Hollister LS (1986) Metamorphic fluids, the evidence from fluid inclusions. *In*: Fluid-rock interactions during metamorphism. Walther JV, Wood BJ (eds) Springer, p 1-35

De Vivo B, Lima A, Webster JD (2005) Volatiles in magmatic-volcanic systems. Elements 1:19-24

Driesner T, Geiger S (2007) Numerical simulation of multiphase fluid flow in hydrothermal systems. Rev Mineral Geochem 65:187-212

Driesner T, Heinrich CA (2007) The System H_2O-NaCl. I. Correlation formulae for phase relations in pressure-temperature-composition space from 0 to 1000°C, 0 to 5000 bar, and 0 to 1 X_{NaCl}. Geochim Cosmochim Acta (accepted)

Eugster HP, Baumgartner L (1987) Mineral solubilities and speciation in supercritical metamorphic fluids. Rev Mineral 17:367-403

Ferry JM, Baumgartner L (1987) Thermodynamic models of molecular fluids at the elevated pressures and temperatures of crustal metamorphism. Rev Mineral 17:323-365

Foustoukos DI, Seyfried Jr. WE (2007) Fluid phase separation processes in submarine hydrothermal systems. Rev Mineral Geochem 65:213-239

Frape SK, Blyth A, Blomqvist R, McNutt RH, Gascoyne M (2003) Deep fluids in the continents: II. Crystalline rocks. Treatise Geochem 5:541-580

German CR, Von Damm KL (2003) Hydrothermal processes. Treatise Geochem 6:181-222

Gottschalk M (2007) Equations of state for complex fluids. Rev Mineral Geochem 65:49-97

Green II HW, Jung H (2005) Fluids, faulting, and flow. Elements 1:31-37

Hack AC, Thompson AB, Aerts M (2007) Phase relations involving hydrous silicate melts, aqueous fluids, and minerals. Rev Mineral Geochem 65:129-185

Heinrich CA (2005) The physical and chemical evolution of low to medium-salinity magmatic fluids at the porphyry to epithermal transition: a thermodynamic study. Mineral Deposita 39:864-889

Heinrich CA (2007a) Fluid – fluid interactions in magmatic-hydrothermal ore formation. Rev Mineral Geochem 65:363-387

Heinrich W (2007b) Fluid immiscibility in metamorphic rocks. Rev Mineral Geochem 65:389-430

Jamtveit B, Yardley BWD (1997) Fluid Flow and Transport in Rocks: Mechanisms and Effects. Chapman & Hall

Kesler SE (2005) Ore-forming fluids. Elements 1:13-18

Kharaka YK, Hanor JS (2003) Deep fluids in the continents: I. Sedimentary basins. Treatise Geochem 5:499-540

Liebscher A (2007) Experimental studies in model fluid systems. Rev Mineral Geochem 65:15-47

Markl G, Ferry J, Bucher K (1998) Formation of saline brines and salt in the lower crust by hydration reactions in partially retrogressed granulites from the Lofoten Islands, Norway. Am J Sci 298:705-757

Pedersen KS, Christensen P (2007) Fluids in Hydrocarbon Basins. Rev Mineral Geochem 65:241-258

Roedder E (1984) Fluid Inclusions. Rev Mineral Vol. 12, Mineralogical Society of America

Samson I, Anderson A, Marshall D (2003) Fluid Inclusions – Analysis and Interpretation. Mineral Assoc Can Short Course 32

Schmidt MW, Poli S (2003) Generation of mobile components during subduction of oceanic crust. Treatise Geochem 3:567-591

Shmulovich KI, Yardley BWD, Gonchar GG (1995) Fluids in the Crust. Chapman & Hall

Skippen G (1988) Phase relations in model fluid systems. Rendiconti Soc Ital Mineral Petrol 43:7-14

Thompson AB, Aerts M, Hack AC (2007) Liquid immiscibility in silicate melts and related systems. Rev Mineral Geochem 65:99-127

Walther JV, Wood BJ (1986) Fluid –Rock Interactions During Metamorphism. Springer

Webster J, Mandeville C (2007) Fluid immiscibility in volcanic environments. Rev Mineral Geochem 65:313-362

Williams-Jones AE, Heinrich CA (2005) Vapor transport of metals and the formation of magmatic-hydrothermal ore deposits. 100th Anniversary special paper, Econ Geol 100:1287-1312

Reviews in Mineralogy & Geochemistry
Vol. 65, pp. 15-47, 2007
Copyright © Mineralogical Society of America

2

Experimental Studies in Model Fluid Systems

Axel Liebscher

Institute for Applied Geosciences
Technical University of Berlin
D-13355 Berlin, Germany

Department 4, Chemistry of the Earth
GeoForschungsZentrum Potsdam
Telegrafenberg
D-14473 Potsdam, Germany

axel.liebscher@tu-berlin.de

INTRODUCTION

Numerous observations provide strong evidence that fluid-fluid coexistence is a widespread phenomenon in many geological processes: i) Oceanic hydrothermal fluids display a wide range in salinity, which clearly indicates fluid phase separation of the circulating seawater (see Foustoukos and Seyfried 2007a); ii) Selective enrichment of certain elements and coeval entrapped inclusions of different, co-existing fluids highlight the role immiscible fluids play in ore formation (see Heinrich 2007a); iii) Trace element and stable isotope characteristic as well as equilibrium mineral assemblages in contact aureoles prove immiscible fluids to be a ubiquitous feature in most contact metamorphism (see Heinrich 2007b); iv) Small scale fluid heterogeneity in eclogitic rocks indicate fluid immiscibility to occur in metamorphic rocks up to high or even ultra-high pressure (see Heinrich 2007b). Unfortunately, fluids are principally non-quenchable phases and direct quantitative information on fluid properties at P and T is in most cases not available. Experimental studies are thus central in studying geologic fluids in general and co-existing fluids in particular. They are still the primary source of quantitative information on physicochemical properties of co-existing crustal fluids: they allow, e.g., determine i) fluid phase relations as function of P, T, and x; ii) physical fluid properties like density and viscosity, and iii) trace element and isotope fractionation between co-existing fluids but also between co-existing fluids and solids. The experimental data can then be used as input in forward modeling approaches, the results and/or predictions of which may then be tested in natural systems (see Driesner and Geiger 2007).

Experimental data on co-existing fluids in systems of geological relevance already date back to the first half of last century (e.g., Keevil 1942). Since the 1960ies, numerous systematic experimental studies on fluid-fluid phase relations at crustal P and T conditions were performed (e.g., Sourirajan and Kennedy 1962; Ellis and Golding 1963; Tödheide and Franck 1963; Takenouchi and Kennedy 1964, 1965; Welsch 1973; Bischoff and Rosenbauer 1984, 1988; Bodnar et al. 1985; Japas and Franck 1985; Krader 1985; Rosenbauer and Bischoff 1987; Zhang and Frantz 1989; Hovey et al. 1990; Kotel'nikov and Kotel'nikova 1991; Shmulovich and Plyasunova 1993; Shmulovich et al. 1995; Lamb et al. 1996, 2002; Schmidt and Bodnar 2000). Several experimental studies have addressed trace element and stable isotope fractionation between co-existing fluids (e.g., Truesdell 1974; Bischoff and Rosenbauer 1987; Berndt and Seyfried 1990, 1997; Horita et al. 1995; Shmulovich et al. 1999, 2002; Pokrovski et al. 2002, 2005; Liebscher et al. 2005, 2006a,b, 2007). Although a lot of experimental studies at different P-T conditions, in different model fluid systems, and with different experimental techniques have been performed to address the different geological questions pertinent to fluid-fluid research, this review will show that the experimental data base is still insufficient to fully address the complexity of natural co-existing fluids. Compared to the rock-forming minerals, many pieces of the fluid-fluid puzzle are still lacking.

1529-6466/07/0065-0002$05.00   DOI: 10.2138/rmg.2007.65.2

After a short description of how the experimental data are presented, this review very briefly describes the principal experimental techniques applied to fluid-fluid research. Then the experimental results on the fundamental fluid-fluid phase relations in binary and ternary subsystems of the general H_2O-CO_2-CH_4-N_2-HCl-sulfur species-alkaline/earth alkaline salts-alkaline/earth alkaline hydroxides model system are reviewed. This general system is sufficient to describe most crustal fluids except hydrocarbons and silicate melts/silicate-rich high-pressure fluids. Describing the experimental studies on hydrocarbon systems and silicate melt/silicate-rich high-pressure fluids systems is beyond the scope of this review. For a review on fluid-fluid equilibria in these systems, the reader is referred to Hack et al. (2007), Pedersen and Christensen (2007), Thompson et al. (2007), and Webster and Mandeville (2007). Based on the information about fluid-fluid phase relations in the binary and ternary model systems, the available experimental data on trace element and stable isotope fractionation between co-existing fluids are reviewed. Finally, experimental data on fluid-mineral and fluid-rock interactions under two-fluid phase conditions are briefly addressed and a short outlook on potential future research topics is given.

DATA COMPILATION AND PRESENTATION

No uniform usage of units exists in the literature for the presentation of *P-V-T-x* data in model fluid systems. Pressure is presented in the different studies as (k)bar, (M)Pa, or atm and temperature as K or °C. These units are easily converted into each other and all pressure and temperature data are re-calculated and presented as MPa and °C throughout this review. Presentation of compositional data is even more diverse in the different studies. Commonly used units are wt%, mol%, mole fraction, (m)mol/kg, and (m)mol/l. Different units are not only used in different studies but may also be mixed in individual studies and even in individual diagrams: e.g., some studies use "wt% relative to H_2O" to express salt concentrations in ternary diagrams and "wt% relative to H_2O + salt" or mole fraction to express corresponding CO_2 concentrations. In addition, solute concentrations in co-existing fluids are considerably high at *P-T* conditions addressed by this review and any weight-based concentration value largely depends on the solutes molar weight: e.g., an H_2O-salt solution with $n_{salt}/(n_{H_2O} + n_{salt}) = 0.1$ corresponds to 26.5 wt% NaCl but 50.3 wt% $CaCl_2$. Thus, any weight-based concentration values prohibit direct comparison between results obtained in the different model fluid systems. To allow for direct comparison between the different studies and between different systems, all compositional data for the presentation of *P-V-T-x* relations are therefore re-calculated throughout this review on a mole fraction basis with $x_{component\ A} = n_{component\ A}/\Sigma n_{system}$. The reader should be aware that this re-calculation leads to some changes in the shape of the here presented phase topologies when compared with the original references.

One of the key parameters in binary H_2O-salt systems is the location of the critical curve, as this defines the upper pressure limit for fluid phase separation at any given temperature (see below). However, not all studies on *P-V-T-x* relations in binary H_2O-salt systems present data for the location of the critical curve. In binary H_2O-salt systems at isothermal conditions, an albeit only purely empirically derived equation relates critical pressure P_c, sampling pressure P, and composition x_{salt}^{vapor} and x_{salt}^{liquid} of co-existing vapor and liquid:

$$x_{salt}^{liquid} - x_{salt}^{vapor} = \text{const.}\ (P_c - P)^b$$

with $x_{salt} = n_{salt}/(n_{salt} + n_{H_2O})$ and $b = 0.5$ (Pitzer et al. 1987; Bischoff and Rosenbauer 1988; see also discussion between Harvey and Levelt Sengers 1989 and Pitzer and Tanger 1989). Thus, plotting the squared mole fraction difference $(x_{salt}^{liquid} - x_{salt}^{vapor})^2$ of co-existing vapor and liquid versus sampling pressure *P* yields the critical pressure P_c for $(x_{salt}^{liquid} - x_{salt}^{vapor})^2 = 0$. According to the law of rectilinear diameters, the average mole fractions $(x_{salt}^{liquid} + x_{salt}^{vapor})/2$ of co-existing vapor and liquid should define a linear relationship with sampling pressure, which intersects the critical pressure P_c at the critical composition x_c (Kay and Rambosek 1953; Rowlinson and Swinton 1982; Dubois et al. 1994). For those studies on *P-V-T-x* relations in binary H_2O-salt systems that lack data for the critical curve, critical pressure P_c and critical composition x_c for the respective isotherms were calculated for this review according to the above mentioned relation between $(x_{salt}^{liquid} - x_{salt}^{vapor})^2$, $(x_{salt}^{liquid} + x_{salt}^{vapor})/2$, *P*, P_c, and x_c. But the reader has to be aware that this relation is only empirical without theoretical foundation and that especially $b = 0.5$ may not apply to all systems. However, depending on amount and quality of the underlying x_{salt}^{liquid}, x_{salt}^{vapor}, and *P* data, uncertainties of the derived P_c and x_c are probably below ± 0.1 MPa for P_c and ± 0.005 for x_c (Bischoff and Rosenbauer 1988; Dubois et al. 1994).

Trace element and stable isotope fractionation between co-existing fluids depends on a complex interplay between pressure, temperature, composition, and physicochemical properties like density and pH of the co-existing fluids. Some of these factors may simultaneously influence trace element and stable isotope fractionation in opposite directions: e.g., increasing temperature principally results in less fractionation but concomitantly implies (at isobaric conditions in H_2O-salt systems) increasing opening of the liquid-vapor two-phase field and therefore increasing compositional and density differences between the co-existing fluids, which favors higher fractionation (e.g., Shmulovich et al. 2002). Unfortunately, no generally applicable theory for trace element and stable isotope fractionation between co-existing fluids is yet available. One of the prime factors that govern trace element fractionation between co-existing fluids is the density difference (e.g., Pokrovski et al. 2005). However, systematic density data as function of pressure and temperature are only available for the H_2O-NaCl system (Khaibullin and Borisov 1966; Urusova 1975; Bischoff and Rosenbauer 1988; Bischoff 1991). Trace element fractionation data in this system are therefore presented in this review as function of the density difference between co-existing fluids. In case the original reference lack density data, the densities of co-existing fluids were calculated based on given *P-T-x* data and the density data of Bischoff (1991). All other trace element and stable isotope fractionation data are presented as function of $P_c - P$, i.e., pressure distance to the critical curve for any particular isotherm. This procedure is at variance with data presentation in some of the original references (e.g., $^{18}O/^{16}O$ and D/H stable isotope fractionation by Shmulovich et al. 1999; see below) and is confronted with the problem that in some systems P_c is not that well constrained, resulting in comparable large uncertainty of $P_c - P$. However, it allows for direct comparison of results obtained at different *P-T* conditions and in different systems and by that to draw some general conclusions.

EXPERIMENTAL TECHNIQUES APPLIED TO FLUID-FLUID STUDIES

Several different experimental techniques have been applied to studies on co-existing fluids. This is due to i) the diverse questions in fluid-fluid studies (e.g., *P-T-x* relations, densities of co-existing fluids, vapor-liquid fractionation of trace elements and stable isotopes, mineral solubility under two-phase fluid conditions, speciation of the different fluid components), ii) the diverse chemical systems investigated (e.g., simple binary H_2O-non polar gas and H_2O-salt systems, more complicated ternary or even quaternary systems, systems in which physicochemical parameters like pH, f_{O_2}, f_S, f_{H_2}, f_{H_2S} ideally have to be controlled or at least recorded), and iii) the wide range of *P* and *T* conditions (e.g., from below the critical point of pure H_2O to high temperature of up to 1000 °C and to pressure of up to 500 MPa or even higher). This review will not present and discuss in detail all different techniques used; for more details the reader is referred to the original references given in the respective sections below. Instead, some general aspects are outlined here.

Somewhat simplified and generalized speaking, fluids are non-quenchable phases, in contrast to most solid phases and higher viscosity melts. This creates the main and principle problem in experimental studies on fluid-fluid systems. To derive information on the co-existing fluids at P and T, the systems have either to be studied *in situ* or the co-existing fluid phases have to be physically separated from each other at P and T to be then analyzed separately at ambient conditions. *In situ* studies have been applied by several workers to study *P-V-T-x* relations (e.g., Tödheide and Franck 1963; Welsch 1973; Blencoe et al. 2001). In these studies, a precisely known amount of fluid is filled into an autoclave of known volume (with $V = f(T)$); i.e., m_{bulk}, V_{bulk}, x_{bulk}, and ρ_{bulk} are well constrained. The system is then heated up and some system parameter is recorded on-line as $f(T)$. At the transition from an immiscible system to a single-phase fluid, $\partial_{parameter}/\partial_T$ will change. Because m_{bulk}, V_{bulk}, x_{bulk}, and ρ_{bulk} are either constant (m_{bulk}, x_{bulk}) or well constrained (V_{bulk}, ρ_{bulk}) during the experiment, the phase transition from a multiphase to a single-phase fluid is then precisely defined in terms of P, T, x, and ρ.

Experimental studies, which physically separate the co-existing fluids for later analysis, can be subdivided into those that use synthetic fluid inclusions (SynFlinc) and those that use different types of hydrothermal autoclaves. For SynFlinc studies, fluid and (mostly) quartz crystals are loaded into common capsules and brought to the desired *P-T* conditions. Here, the quartz crystals trap the co-existing fluids as synthetic fluid inclusions (for a detailed description of this method see Bodnar and Sterner 1987; Sterner and Bodnar 1991). The SynFlincs are then analyzed by classical microthermometric techniques (for an introduction into principles of fluid inclusion research see Roeder 1984) for salinity (e.g., Bodnar et al. 1985; Zhang and Frantz 1989; Dubois et al. 1994) or, only most recently, by Laser Ablation-Inductively Coupled Plasma-Mass Spectrometry (LA-ICP-MS) for trace element composition to determine trace element fractionation. While the SynFlinc technique is a very elegant way to sample fluids at the *P-T* conditions of interest without any change in P and T introduced by the sampling procedure itself, the technique has some shortcomings: i) SynFlincs may not sample pure endmember fluids but some mixture of the co-existing fluids (see, e.g., data by Dubois et al. (1994) for the H_2O-LiCl system presented below); ii) SynFlincs may change during the cooling path of the experimental samples; iii) Melting temperature and homogenization temperature may be difficult determined in low-density vapor inclusions; iv) Determination of salinity depends on the knowledge of the ambient pressure phase relations of the system of interest. However, compared to the autoclave techniques, the SynFlinc technique is much less restricted in accessible *P-T-x* conditions.

Studies based on autoclave techniques are performed with externally heated autoclaves that allow for recovering fluid samples at P and T through some kind of sampling line (see Seyfried et al. 1987). Autoclaves and sampling line(s) are designed such that they allow for stratification of the lower-density fluid from the higher-density fluid and consequently sampling of pure endmember fluids. This is achieved via sampling in an up-right position of the (generally rotatable) autoclave with the lower-density fluid at the top and the higher-density fluid at the bottom (this sampling technique therefore also provide information on the relative densities of co-existing fluids; see below for H_2O-CO_2 example by Takenouchi and Kennedy 1964). The great advantage of the autoclave technique is that it allows for analyzing the recovered fluid samples for major and trace element composition by standard techniques like ion chromatography, ICP-MS, and ICP-OES (optical emission spectrometry) and for isotopic composition by mass spectrometry. However, like the SynFlinc technique, the autoclave technique has some shortcomings: i) Extraction of a fluid sample decrease the bulk density of the system and by that the pressure. Fluid-fluid equilibria are very pressure sensitive, especially near critical conditions, and the sampling procedure itself therefore shifts the equilibrium composition of the co-existing fluids. Two different approaches are used to overcome this problem: flexible reaction cells, which allow for external pressure control and large-volume autoclaves, for which the sampling volume is very small compared to the autoclave volume and pressure drop during sampling is consequently very small; ii) Fluid samples are immediately quenched from run to ambient *P-T* conditions during sampling. This quenching may result in precipitation of some solute load within the sampling line(s); iii) To avoid precipitation of the major salt(s) during sampling, experiments are normally only preformed up to x_{salt}^{liquid} (at run conditions) $< x_{salt}^{saturation}$ (at ambient conditions); iv) *P-T* conditions for use of most of the autoclaves are restricted to medium *T* (< 700 °C) and *P* (< 150 MPa).

P-V-T-x RELATIONS IN BINARY MODEL FLUID SYSTEMS

Based on the second component (e.g., non polar gases, electrolytes) in addition to H_2O, binary model fluid systems that are addressed in this review may be chemically grouped and subdivided into i) H_2O-non polar gas systems, ii) H_2O-salt systems, and iii) additional and geologically less important systems. Although binary fluid systems in general are a simplified representation of natural crustal fluids (oceanic hydrothermal fluids may be an exception, for which the binary H_2O-NaCl systems is a reasonable and probably valid approximation; see Foustoukos and Seyfried 2007a), experimental studies in these systems provide the necessary robust framework for the more realistic ternary, quaternary or even higher component systems, for which the binary systems form the boundary conditions.

H_2O-non polar gas systems

These systems include the geological relevant systems H_2O-CO_2 (Fig. 1a), H_2O-CH_4 (Fig. 1b), and H_2O-N_2 (Fig. 1c), of which the H_2O-CO_2 system has attracted the most attention. Fluid-fluid equilibria in the system H_2O-CO_2 has been studied, among others, by Tödheide and Franck (1963), Takenouchi and Kennedy (1964), Sterner and Bodnar (1991), Seitz and Blencoe (1997), and Blencoe et al. (2001). Welsch (1973) and Shmonov et al. (1993) studied the system H_2O-CH_4, and Japas and Franck (1985) the system H_2O-N_2. Experimental data for the system H_2O-Xe are given in Welsch (1973) and Franck et al. (1974). Despite some differences in actual *P-T-x* conditions, the principal phase relations in the three systems H_2O-CO_2, H_2O-CH_4, and H_2O-N_2 resemble each other: The available data indicate that immiscibility is generally restricted to temperatures below about 400 °C. Under most crustal *P-T* conditions these systems are therefore characterized by a homogeneous single-phase fluid. Nevertheless, fluid immiscibility at *T* < 400 °C in these systems can be of economic interest. Their critical curves (Fig. 1d) consistently exhibit temperature minima. These minima occur at 60 to 70 MPa/ ~ 365 °C in the H_2O-N_2 system (Japas and Franck 1985), ~ 100 MPa/353 °C in the H_2O-CH_4 system (Welsch 1973), and 155 to 190 MPa/~ 265 °C in the H_2O-CO_2 system (Takenouchi and Kennedy 1964). With increasing pressure, the critical curves then shift to slightly higher temperature; however, the data indicate that up to 300 MPa critical temperatures will not exceed ~ 275 °C in the H_2O-CO_2 system, ~ 380 °C in the H_2O-CH_4 system, and ~ 390 °C in the H_2O-N_2 system. *T-x* sections at 100 and 200 MPa indicate that in this *P*-range pressure has only a negligible effect on the location of the critical curves. They also show that the vapor-liquid two-phase fields in H_2O-non polar gas systems close with increasing temperature but widen with increasing pressure (Fig. 1e). Pressure and temperature therefore have the opposite effect on extend of immiscibility in H_2O-non polar gas systems than in H_2O-salt systems where the vapor-liquid two-phase fields widen with increasing temperature but close with increasing pressure (see below).

Takenouchi and Kennedy (1964) studied the H_2O-CO_2 system by means of the autoclave technique. Their data therefore provide information on the relative densities of the co-existing H_2O- respectively CO_2-rich fluids (Fig. 1f). Up to about 210 MPa at 264 °C and 230 MPa at 260 °C, the CO_2-rich fluid was sampled from the upper part of the autoclave indicating lower relative density of this fluid up to these conditions. Contrary, at higher pressure conditions, the H_2O-rich fluid was sampled from the upper part of the autoclave, suggesting a density inversion between H_2O- and CO_2-rich fluids in this *P-T* range (Takenouchi and Kennedy 1964). The indicated *P-T* range of the density inversion in the H_2O-CO_2 system agrees fairly

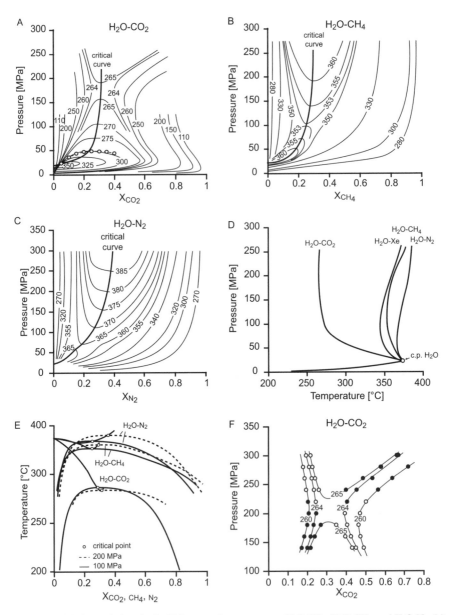

Figure 1. Fluid phase relations in the H$_2$O-non polar gas systems H$_2$O-CO$_2$, H$_2$O-CH$_4$, and H$_2$O-N$_2$. (a), (b), and (c) Pressure-composition relations in the above systems; thin lines are isotherms with temperature in °C given by numbers; thick lines represent the critical curves. (d) Pressure-temperature relations of the critical curves in the above systems. (e) Isobaric representation of the vapor-liquid two-phase fields at 100 and 200 MPa, open circles represent the respective critical points. (f) Enlargement of (a) showing sampling conditions: open circles represent samples recovered from the top part of the autoclave, filled circles represent samples recovered from the bottom part of the autoclave. The data indicate a density inversion in the system H$_2$O-CO$_2$, see text for further details. [data for H$_2$O-CO$_2$ from Takenouchi and Kennedy 1964, open circles in (a) from Blencoe et al. 2001; for H$_2$O-CH$_4$ from Welsch 1973; for H$_2$O-N$_2$ from Japas and Franck 1985]

well with those *P-T* conditions, at which the pure end member systems H_2O and CO_2 display a density inversion (Takenouchi and Kennedy 1964).

H_2O-salt systems

Besides H_2O-CO_2, H_2O-salt systems are the geologically most important binary systems, as most fluids at least contain some amount of chlorine and therefore some amount of salt. Contrary to H_2O-non polar gas systems, which are (in general) completely miscible at temperature above about 400 °C, H_2O-salt systems may show fluid immiscibility up to > 250 MPa at 800 °C (see below). This *P-T* range includes pressure-temperature conditions as they occur in geothermal systems (see Arnórsson and Stefánsson 2007), oceanic hydrothermal systems (see Foustoukos and Seyfried 2007a), magmatic to hydrothermal ore deposits (see Heinrich 2007a), volcanic and sub volcanic hydrothermal systems (see Webster and Mandeville 2007), and contact metamorphism (see Heinrich 2007b). Experimental studies in H_2O-salt systems are therefore directly relevant to important geologic processes, although some of the above mentioned systems are more adequately described by ternary or quaternary systems (see below).

Sodium chloride NaCl, $CaCl_2$, and KCl are the most abundant and important salts in crustal fluids. The concentration of other salts like LiCl, CsCl, $SrCl_2$, or $MgCl_2$ in crustal fluids is normally at the minor or even trace element level and these salts have therefore only negligible effects on the fluid phase relations of the respective system. Experimental studies on these H_2O-salt systems nevertheless provide important information on general systematics in H_2O-salt systems. These studies are therefore also reviewed here.

H_2O-NaCl. H_2O-NaCl is the most important binary H_2O-salt system and by far the most experimental studies on fluid-fluid equilibria have focused on this system. This system is therefore used here to illustrate principal features of H_2O-salt systems and its phase topology is summarized in Figure 2 together with results of some of the fundamental experimental studies. Early studies include Schröer (1927), Ölander and Liander (1950), Khaibullin and Borisov (1966) and Marshall and Jones (1974). Keevil (1942) studied the vapor pressure of H_2O-NaCl solutions at 183 to 646 °C and determined the vapor + liquid + solid three phase curve, while Urusova and Ravich (1971) studied the vapor pressure of H_2O-NaCl solutions at 350 and 400 °C up to salt saturation. Sourirajan and Kennedy (1962) presented a systematic study on the vapor-liquid two-phase surface at 350 to 700 °C/11.4 to 124 MPa. First density data along the vapor-liquid two-phase surface were presented by Khaibullin and Borisov (1966) and Urusova (1975). Bischoff and Rosenbauer (1984) studied the vapor-liquid phase relations between 200 and 500 °C for a fluid with seawater-like composition of 3.2 wt% NaCl (i.e., $x_{NaCl} = 0.01$). Bodnar et al. (1985) extended the experimental data set to 1000 °C and 130 MPa and studied in detail the 100 MPa isobar; their liquid data were then re-calculated and slightly modified by Chou (1985). Chou also calculated the densities of co-existing vapor and liquid along the three phase curve based on the *P-T* data by Keevil (1942). Bischoff et al. (1986) studied vapor-liquid relations near the critical temperature of H_2O and the vapor + liquid + solid three phase curve from 300 to 500 °C. Rosenbauer and Bischoff (1987) and Bischoff and Rosenbauer (1988) presented detailed studies of the 380, 400, 405, 415, 450, 475, and 500 °C isotherms with special emphasis given to the determination of the critical pressure and critical composition. Combining and summarizing all available experimental data, Bischoff and Pitzer (1989) presented a consistent description of the *P-T-x* vapor-liquid two-phase surface from 300 to 500 °C and pressure between the three phase curve and the critical curve. Bischoff (1991) extended this summary by adding also density data for vapor and liquid. Shmulovich et al. (1995) then derived additional data for the 400, 500, 600, 650, and 700 °C isotherm. The most recent representation of the complete phase topology in the H_2O-NaCl system up to 1000 °C/220 MPa, and $x_{NaCl} = 0$ to 1.0 is given by Driesner and Heinrich (2007), based on a re-evaluation of all available experimental data.

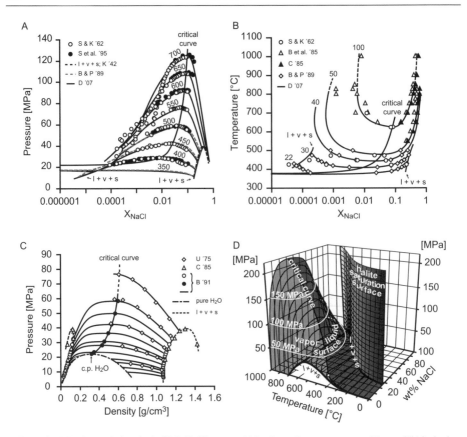

Figure 2. Fluid phase relations in the H₂O-NaCl system. (a) Isothermal pressure-composition and (b) isobaric temperature-composition relations, (c) density relations between vapor and liquid, and (d) three dimensional *P-T-x* diagram from Driesner and Heinrich (2007). [data sources: K '42: Keevil 1942; S & K '62: Sourirajan and Kennedy 1962; U '75: Urusova 1975; B et al. '85: Bodnar et al. 1985; C '85: Chou 1985; B & P '89: Bischoff and Pitzer 1989; S et al. '95: Shmulovich et al. 1995; D '07: Driesner and Heinrich 2007]

Most experiments on the *P-T-x* relations in H₂O-salt systems are performed along individual isotherms, and Figure 2a therefore shows an isothermal representation of the H₂O-NaCl system based on the data by Keevil (1942), Sourirajan and Kennedy (1962), Bischoff and Pitzer (1989) Shmulovich et al. (1995), and Driesner and Heinrich (2007). At isothermal conditions above the critical point of H₂O, the vapor-liquid two-phase field has a parabola-form shape, which has its crest at high pressure and opens towards decreasing pressure. This crest defines the critical point of each respective isotherm, at which any physicochemical difference between vapor and liquid vanishes. The critical curve starts at the critical point of H₂O (374 °C/22.1 MPa), connects all the individual critical points, and defines the highest pressure at which for a given temperature fluid immiscibility can occur. At pressure conditions above the critical curve, only a single-phase fluid is stable. Within the range of reasonable crustal *P-T* conditions, the critical curve in the H₂O-NaCl system monotonously extends from the critical point of pure H₂O towards higher pressure and higher salinity with increasing temperature. This is a characteristic feature of almost all H₂O-salt systems (an exception being the system H₂O-BaCl₂, which shows discontinuous critical and solubility curves; Valyashko et al. 1987) and unlike to the H₂O-non polar gas systems presented above. At $T < T_{\text{triple NaCl}}$ (= 800.7 °C), each isotherm, and therefore also the vapor-liquid two-phase surface, terminates at the vapor + liquid + solid three phase curve, which is defined by the intersection of the liquid surface with the salt saturation surface. At pressure conditions below the vapor + liquid + solid three phase curve no liquid is stable and vapor co-exists with solid salt. The numerous experimental studies in the H_2O-NaCl system agree fairly well within their experimental and analytical uncertainties. Only the NaCl concentrations in the vapor at 700 °C as determined by Sourirajan and Kennedy (1962) tend to be slightly higher than indicated by the other studies. The critical curve of the H_2O-NaCl system is well established and indicates that the critical pressure increases from the critical point of pure H_2O at 22.1 MPa/374 °C to 124,6 MPa/ 700 °C (Driesner and Heinrich 2007).

Figure 2b shows an isobaric representation of the *P-T-x* relations in the H_2O-NaCl system based on data by Sourirajan and Kennedy (1962), Bodnar et al. (1985), Chou (1985), Bischoff and Pitzer (1989), and Driesner and Heinrich (2007). Contrary to isothermal conditions, the vapor-liquid two-phase field at isobaric conditions has its critical point at lowest temperature and opens with increasing temperature. In binary H_2O-salt systems, low pressure and high temperature therefore favor fluid immiscibility, and isothermal decompression or isobaric heating may lead to fluid phase separation. This behavior of H_2O-salt systems is at variance with the H_2O-non polar gas systems and the behavior of solid solution series, in which solvi generally close with increasing temperature but open with increasing pressure.

Like x_c, the critical density ρ_c increases with increasing temperature and pressure along the critical curve (Fig. 2c; Urusova 1975; Chou 1985; Knight and Bodnar 1989; Bischoff 1991). At 550 °C/76 MPa and $x_{NaCl} = 0.054$, the critical density in the H_2O-NaCl system is already $\rho_c = 0.61$ g/cm^3 compared to $\rho_c = 0.322$ g/cm^3 in the pure H_2O system. This shows that for crustal *P-T* conditions the vapor in H_2O-salt systems may be considerably dense with peculiar physicochemical properties that are significantly different from those at low *P-T* conditions and in the pure H_2O system (see Heinrich 2007a for the role of vapor in ore-forming processes).

H_2O-KCl; -CaCl$_2$; -MgCl$_2$; -LiCl; -CsCl. The *system H_2O-KCl* (Fig. 3a) has been studied at temperature above the critical point of pure H_2O by Hovey et al. (1990) along the 380 and 410 °C isotherms, Dubois et al. (1994) along the 500 and 600 °C isotherms, and Shmulovich et al. (1995) along the 400, 500, 600, 650, and 700 °C isotherms. Keevil (1942) reported vapor pressures of liquids with KCl concentrations up to salt saturation. The results from the different studies agree well, except for some vapor data at 500 °C by Dubois et al. (1994), and the phase relations in the H_2O-KCl system are well established. Critical pressures as determined based on the data by Shmulovich et al. (1995) are 28.2 MPa at 400 °C, 58.8 MPa at 500 °C, 91.2 MPa at 600 °C, 106.6 MPa at 650 °C, and 121.2 MPa at 700 °C. These data fairly well agree with the corresponding critical pressures in the H_2O-NaCl system and the differences between the H_2O-KCl and H_2O-NaCl systems are only minor, at least in the *P-T* range studied. In natural samples, the behavior of KCl bearing fluids may therefore reasonably well be approximated by the *P-T-x* behavior of the much better studied H_2O-NaCl system.

The *system H_2O-CaCl$_2$* (Fig. 3b) has been studied at temperatures above the critical point of pure H_2O by Marshall and Jones (1974) up to 400 °C and 1.8 molal CaCl$_2$, Oakes et al. (1994) up to 550 °C and 3 molal CaCl$_2$, Tkachenko and Shmulovich (1992) and Shmulovich et al. (1995) at 400, 500, and 600 °C, and Bischoff et al. (1996) at 380, 400, 430, and 500 °C. The data by Zhang and Frantz (1989) provide additional constraints on compositional limits of the vapor limb at 600 and 700 °C. Vapor pressures of liquids with CaCl$_2$ concentrations up to salt saturation at 250 to 400 °C are reported by Ketsko et al. (1984). At temperatures below the critical point of pure H_2O, *P-V-T-x* data are presented by Zarembo et al. (1980) and Wood et al. (1984) at 150 to 350 °C, Valyashko et al. (1987) at 300 °C and Crovetto et al. (1993) at 350 and 370 °C. The overall vapor-liquid phase topology of the H_2O-CaCl$_2$ system is well established, although the data sets by Shmulovich et al. (1995) and Bischoff et al. (1996) show some discrepancies. The data by Shmulovich et al. (1995) indicate a slightly narrower

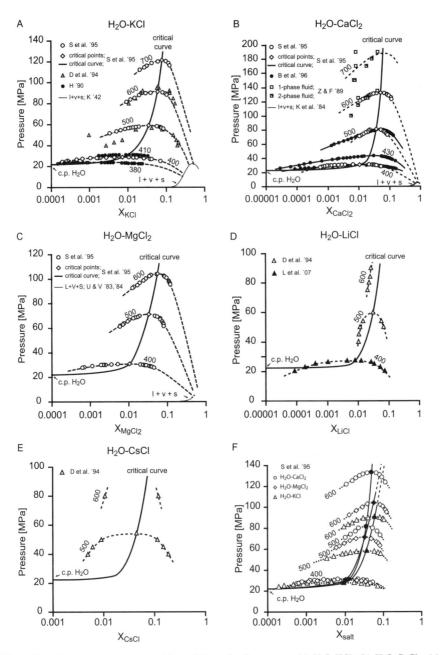

Figure 3. Isothermal pressure-composition relations for the systems (a) H₂O-KCl, (b) H₂O-CaCl₂, (c) H₂O-MgCl₂, (d) H₂O-LiCl, (e) H₂O-CsCl, and (f) comparison of the systems H₂O-KCl, H₂O-CaCl₂, and H₂O-MgCl₂. [data sources: K '42: Keevil 1942; U & V '83, '84: Urusova and Valyashko 1983, 1984; K et al. '84: Ketsko et al. 1984; Z & F '89: Zhang and Frantz 1989; H et al. '90: Hovey et al. 1990; D et al. '94: Dubois et al. 1994; B et al. '96: Bischoff et al. 1996; S et al. '95: Shmulovich et al. 1995; L et al. '07: Liebscher et al. 2007]

vapor-liquid two-phase field than the data by Bischoff et al. (1996). Additionally, calculated critical points at 400 (31.4 MPa/x_{CaCl_2} = 0.012) and 500 °C (81.6 MPa/x_{CaCl_2} = 0.036) based on the data by Shmulovich et al. (1995) suggest lower salt concentrations at comparable pressure than reported by Bischoff et al. (1996) for these conditions (30.0 MPa/x_{CaCl_2} = 0.018 at 400 °C; 80.6 MPa/x_{CaCl_2} = 0.053 at 500 °C). Combining the results of the different studies, the data, nevertheless, clearly show that critical pressures are significantly higher in the H_2O-$CaCl_2$ system than in the H_2O-NaCl and H_2O-KCl systems. Extrapolating the critical curve as defined by the data of Shmulovich et al. (1995) in combination with the results by Zhang and Frantz (1989) at 700 °C indicates a critical pressure of ~ 190 MPa at 700 °C compared to 124.6 MPa/ 700 °C in the H_2O-NaCl system (Driesner and Heinrich 2007) and 121.2 MPa/700 °C in the H_2O-KCl system (Shmulovich et al. 1995).

The *system H_2O-$MgCl_2$* (Fig. 3c) has been studied by Shmulovich et al. (1995) at 400, 500, and 600 °C. Vapor pressures of liquids with $MgCl_2$ concentrations up to salt saturation at 250 to 350 °C are reported by Urusova and Valyashko (1983, 1984; additional data can be found in Valyashko et al. 1988). Critical points at 400, 500, and 600 °C, as determined based on the data by Shmulovich et al. (1995), are at 30.8, 71.6, and 104.2 MPa, notably higher pressures than in the H_2O-alkaline salt systems. This tendency to higher critical pressures in the H_2O-$MgCl_2$ system is consistent with results for the H_2O-$CaCl_2$ system.

The *system H_2O-LiCl* (Fig. 3d) has been studied by Dubois et al. (1994) at 500 (vapor and liquid data) and 600 °C (only vapor data) and by Liebscher et al. (2007) at 400 °C. The data by Dubois et al. (1994) suggest an extremely narrow vapor-liquid two-phase field at 500 and 600 °C, which is not indicated by the data of Liebscher et al. (2007) at 400 °C and which is at variance with all other H_2O-salt systems studied so far (see also below). Dubois et al. (1994) derived their data by synthetic fluid inclusions and the extremely narrow vapor-liquid two-phase field may indicate that they did not analyze pure vapor and liquid inclusions but some mixed fluid inclusions. If this is the case, then the derived critical pressure of 59.8 MPa at 500 °C is most probably likewise erroneous and slightly to high. At 400 °C the critical pressure is 27.6 MPa (Liebscher et al. 2007), slightly lower than in the H_2O-NaCl and H_2O-KCl systems.

To the author's knowledge, the *system H_2O-CsCl* (Fig. 3e) has so far only been studied by Dubois et al. (1994) at 500 and 600 °C. The available data are restricted to three vapor-liquid data at 500 and only one vapor-liquid datum at 600 °C. The critical point at 500 °C is at 54.5 MPa, a slightly lower pressure than in the systems H_2O-NaCl and H_2O-KCl.

The experimental results by Shmulovich et al. (1995) provide an internally consistent set of data, at least from the experimental point of view, for the H_2O-KCl, -$MgCl_2$, and -$CaCl_2$ systems and allow for comparison between these different systems (Fig. 3f). The data indicate that the general shape (in terms of x_{salt}) of the vapor-liquid two-phase field in the different systems is comparable; they give no hint to any significant difference in the opening of the vapor-liquid two-phase fields with increasing distance to the critical curves in the different systems. This clearly suggests that the data by Dubois et al. (1994) on the H_2O-LiCl systems, which indicate an extremely narrow two-phase field, are most probably erroneous. The data by Shmulovich et al. (1995) also nicely show that for a given temperature the critical curve shifts to higher pressure with $P_c^{H_2O\text{-}KCl} < P_c^{H_2O\text{-}MgCl_2} < P_c^{H_2O\text{-}CaCl_2}$.

Limits of fluid immiscibility in H_2O-salt systems. At any given temperature, the field of fluid immiscibility in binary H_2O-salt systems is bounded by the three phase curve (vapor + liquid + solid) towards lower pressure and by the critical curve towards higher pressure. A lot of *P-T-x* data on the three phase curve and the critical curve in the different systems are available and allow addressing the *P-T-x* conditions, at which fluid phase separation occurs in geological environments (Fig. 4).

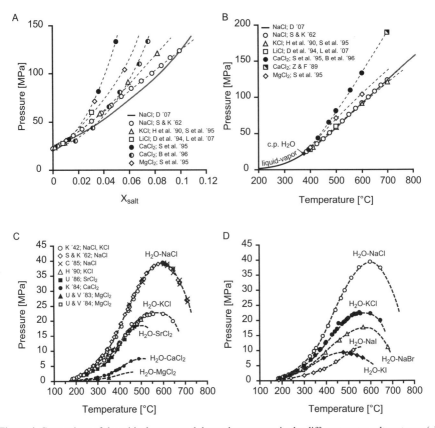

Figure 4. Comparison of the critical curves and three-phase curves in the different water-salt systems: (a) Pressure-composition and (b) pressure-temperature relations of the critical curves; (c) Three-phase curves for chloride systems and (d) influence of different anions on the location of the three-phase curve. [data sources: K '42: Keevil 1942 (also all data in (d)); S & K '62: Sourirajan and Kennedy 1962; U & V '83, '84: Urusova and Valyashko 1983, 1984; K et al. '84: Ketsko et al. 1984; C '85: Chou 1985; U '86: Urusova 1986; Z & F '89: Zhang and Frantz 1989; H et al. '90: Hovey et al. 1990; D et al. '94: Dubois et al. 1994; S et al. '95: Shmulovich et al. 1995; B et al. '96: Bischoff et al. 1996; D '07: Driesner and Heinrich 2007; L et al. '07: Liebscher et al. 2007]

The experimentally determined P_c-x_c relations slightly scatter and display some notable inconsistencies between different data sets (Fig. 4a): In the H_2O-NaCl system, the critical curve by Sourirajan and Kennedy (1962) indicates slightly higher P_c for $x_c < 0.1$ but lower P_c for $x_c > 0.1$ when compared to the most recent compilation by Driesner and Heinrich (2007); in the H_2O-$CaCl_2$ system significant differences appear between the results by Shmulovich et al. (1995) and Bischoff et al. (1996), e.g., for $x_{CaCl_2} = 0.05$ the data by Bischoff et al. (1996) indicate $P_c \sim 75$ MPa whereas the data by Shmulovich et al. (1995) indicate $P_c \sim 142$ MPa. In general, however, for a given x_{salt} the available data indicate $P_c^{NaCl} < P_c^{KCl} < P_c^{MgCl_2} \leq P_c^{CaCl_2}$. Contrary to the P_c-x_c relations, the P_c-T_c relations are very consistent between the different studies (Fig. 4b). Within analytical and experimental uncertainties, the critical curves in the alkaline salt systems H_2O-LiCl, H_2O-NaCl, and H_2O-KCl coincide. In all three systems P_c is an almost linear function of T_c and increases with increasing temperature from the critical point of pure H_2O (22.1 MPa/374 °C) to about 120 MPa at 700 °C. The critical curves in the earth alkaline salt systems H_2O-$MgCl_2$ and H_2O-$CaCl_2$ are likewise almost linear functions of T_c, but are shifted to higher P_c when compared to the alkaline salt systems. Additionally, there appears a notable pressure difference between the critical curves in the H_2O-$MgCl_2$ and H_2O-$CaCl_2$ systems, the later being at higher pressure. Overall, the data indicate that for a crustal temperature of 800 °C, fluid immiscibility in binary H_2O-alkaline salt systems is restricted to $P < 150$ MPa whereas in earth alkaline salt dominated systems fluid immiscibility may occur at pressure of up to ~ 250 MPa.

The location of the three phase curve vapor + liquid + solid is quite different in the different systems although its general shape is identical (Fig. 4c,d). Data for the systems H_2O-$NaCl$, H_2O-KCl, H_2O-$MgCl_2$, H_2O-$CaCl_2$, and H_2O-$SrCl_2$ indicate that the three phase curve in earth alkaline salt systems is at lower pressure than in alkaline salt systems (Fig. 4c): The highest pressure attained by the three phase curve is at ~ 40 MPa in the H_2O-$NaCl$ system but only at ~ 8 MPa in the H_2O-$CaCl_2$ system. The limits of fluid immiscibility in earth alkaline salt dominated systems are therefore not only expanded to higher P along the critical curve but also to lower P along the three phase curve when compared to alkaline salt dominated systems. The data for the three phase curve also indicate that increasing atomic weight of the cation decreases the pressure of the three phase curve and thus increases the field of fluid immiscibility in the alkaline salt systems, but the opposite effect occurs in earth alkaline salt systems: increasing pressure of the three phase curve and decreasing field of fluid immiscibility with increasing atomic weight of the cation. The effect of different anions on the location of the three phase curve has been studied by Keevil (1942) for the alkaline salt systems H_2O-$NaCl$, H_2O-$NaBr$, H_2O-NaI, H_2O-KCl and H_2O-KI (Fig. 4d). These data consistently indicate decreasing pressure of the three phase curve with increasing atomic weight of the anion.

Hydrolysis in H_2O-salt systems. Chlorine is one of the most important ligands in hydrothermal fluids and its availability largely controls solubility and mobility of several metal ions. In addition, the solubility of certain elements strongly depends on the acid, basic and/or alkaline character of hydrothermal fluids. In H_2O-salt systems hydrolysis of the general form

$$x \, H_2O + M^{x+}Cl^-_x = x \, HCl + M^{x+}(OH)^-_x$$

occurs, providing a source for HCl and hydroxides. At ambient conditions this equilibrium is strongly shifted to the left hand side and H_2O-salt mixtures are neutral with negligible amounts of HCl (e.g., Shmulovich et al. 1995). At higher crustal pressure and temperature conditions, however, the equilibrium may shift to the right hand side resulting in considerable amounts of HCl and $M^{x+}(OH)^-_x$. In case of fluid phase separation, HCl fractionates into the vapor whereas $M^{x+}(OH)^-_x$ fractionates into the liquid (Shmulovich and Kotova 1982; Petrenko et al. 1989; Shmulovich et al. 1995, 2002; Bischoff et al. 1996). This gives rise to an HCl-enriched and $M^{x+}(OH)^-_x$-depleted, potentially acidic vapor and a co-existing HCl-depleted and $M^{x+}(OH)^-_x$-enriched, potentially basic or alkaline liquid. The actual concentration of H_3O^+, however, and thus the acidic or basic/alkaline nature of vapor and liquid depends on the dissociation of HCl and $M^{x+}(OH)^-_x$ at the specific *P-T* conditions.

Despite its obvious geochemical importance, hydrolysis of the above form has only rarely been studied experimentally. Bischoff et al. (1996) studied the formation of HCl and the onset of hydrolysis in the system H_2O-$CaCl_2$ between 380 and 500 °C. Indirect evidence for hydrolysis and the concomitant opposite vapor-liquid fractionation of HCl and $M^{x+}(OH)^-_x$ comes from quench-pH measurements on experimentally phase separated fluids by Vakulenko et al. (1989) and Shmulovich et al. (2002). Bischoff et al. (1996) performed experiments at varying pressure along the 380, 400, 430, and 500 °C isotherms (see Fig. 5a for 400 and 500 °C data). Vapor samples recovered below 23.0 MPa at 380 °C, 25.0 MPa at 400 °C, 26.0 MPa at 430 °C, and 58.0 MPa at 500 °C showed scatter and reversals of trend in Cl concentration. Vapor samples from the 400 and 500 °C isotherms were additionally analyzed for both Ca and Cl. With decreasing pressure along both isotherms, i.e., increasing opening of the vapor-liquid two-phase field, Cl concentration progressively and notably exceeds Ca concentration. As the Ca data roughly follow the $CaCl_2$ trend of the vapor limb at higher pressure, Ca concentrations

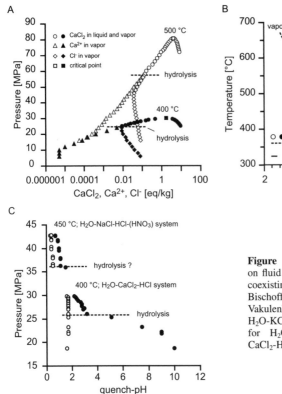

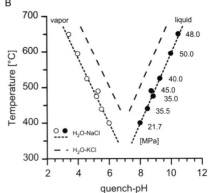

Figure 5. Effect of hydrolysis reaction on fluid phase relations and pH values in coexisting vapor and liquid. Data from (a) Bischoff et al. (1996) for H_2O-$CaCl_2$, (b) Vakulenko et al. (1989) for H_2O-$NaCl$ and H_2O-KCl, and (c) Shmulovich et al. (2002) for H_2O-$NaCl$-HCl-(HNO_3) and H_2O-$CaCl_2$-HCl, see text for details.

most probably reflect simple $CaCl_2$ solubility in vapor, whereas the marked increase in Cl indicates formation of HCl (Bischoff et al. 1996). The data therefore suggest onset of the hydrolysis reaction

$$2H_2O + CaCl_2 = 2HCl + Ca(OH)_2$$

at 23.0 MPa/380 °C, 25.0 MPa/400 °C, 26.0 MPa/430 °C, and 58.0 MPa/500 °C. The amount of HCl produced by this reaction is remarkably and may reach 0.1 mol/kg in the vapor. The increasing HCl concentration in the vapor is also mirrored by semiquantitative quench-pH measurements, which yielded pH values between 1 and 3 (Bischoff et al. 1996). The effect of hydrolysis on HCl concentration in vapor and $M^{x+}(OH)^-_x$ concentration in co-existing liquid is also shown by quench-pH measurements by Vakulenko et al. (1989; Fig. 5b) and Shmulovich et al. (2002; Fig. 5c). Vakulenko et al. (1989) studied the H_2O-$NaCl$ system at 21.7 to 50 MPa/~ 400 to 650 °C. Their start solution had a neutral pH. After quenching, pH values in the vapor were as low as ~ 3.5 whereas pH values in co-existing liquid reached up to ~ 10.5. Shmulovich et al. (2002) performed experiments in the systems H_2O-$NaCl$-HCl-HNO_3 at 450 °C ($pH_{start} = 0.5$) and H_2O-$CaCl_2$-HCl at 400 °C ($pH_{start} = 1.8$), originally designed to study vapor-liquid fractionation of the REE (see below). The quench-pH of liquid samples was found to be generally higher than the quench-pH of co-existing vapor. At 400 °C in the system H_2O-$CaCl_2$-HCl, Shmulovich et al. (2002) found a marked increase in quench-pH of the liquid samples recovered below about 25.5 MPa, which they attributed to the onset of notable hydrolysis reaction. The 25.5 MPa/400 °C observed by Shmulovich et al. (2002) as onset of hydrolysis in the system H_2O-$CaCl_2$-HCl correspond well with the 25.0 MPa/400 °C observed by Bischoff et al. (1996) for the pure H_2O-$CaCl_2$ system

Other binary system

Fluid immiscibility has been studied in several other binary systems, especially at comparable low P and T conditions, but these systems are mostly of only minor importance for crustal processes: e.g., Keevil (1942) studied the systems H_2O-Na_2CO_3 and H_2O-Na_2SO_4 up to ~ 370 °C/21 MPa and Urusova (1974) studied the system H_2O-NaOH at 350 to 550 °C/11 to 90 MPa. Of these, only the data by Urusova (1974) on the H_2O-NaOH system are discussed here in greater detail (Fig. 6). Urusova (1974) focused on the composition of the liquid but also presented some vapor data, which allows comparison with the H_2O-NaCl system. The data indicate that at constant P and T x_{NaOH}^{vapor} and x_{NaOH}^{liquid} are generally higher than corresponding x_{NaCl}^{vapor} and x_{NaCl}^{liquid} in the H_2O-NaCl system. As a consequence, the critical curve in the H_2O-NaOH system is

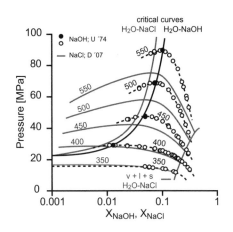

Figure 6. Fluid phase relations in the system H_2O-NaOH compared to the system H_2O-NaCl. [data sources: U '74: Urusova 1974; D '07: Driesner and Heinrich 2007]

shifted to higher concentration and notable higher pressure: At 550 °C, the critical point in the H_2O-NaOH system is at x_{NaOH} = 0.1 and 90 MPa compared to x_{NaCl} =0.06 and 75 MPa in the H_2O-NaCl system (NaCl data from Driesner and Heinrich 2007). Although Urusova (1974) did not study the vapor + liquid + solid three phase curve, her data at 400 and 450 °C clearly show that the three phase curve is at lower P but higher x than in the H_2O-NaCl system. For natural systems, this means that the field of fluid immiscibility in highly basic and alkaline fluids is expanded and fluid behavior may no longer be approximated by the H_2O-NaCl system.

P-V-T-x RELATIONS IN TERNARY MODEL FLUID SYSTEMS

Among the experimentally studied model fluid systems, fluid-fluid equilibria in ternary H_2O-salt-non polar gas systems are those with the highest geological relevance. Fluid phase relations in these ternary systems are controlled by the fluid phase relations within and between the two binary boundary systems H_2O-non polar gas and H_2O-salt. The third binary boundary system salt-non polar gas can be taken as immiscible over the entire crustal P-T range. Despite differences in detailed fluid phase relations between the different ternary H_2O-salt-non polar gas systems, some general results of the different experimental studies may be outlined: i) Adding salt to a H_2O-non polar gas system decreases the solubility of the non polar gas in the liquid at given P and T (salting out effect); ii) The solubility of salt in the H_2O-non polar gas mixture is generally low; iii) In ternary H_2O-salt-non polar gas systems the field of fluid immiscibility is significantly expanded in terms of P, T, and x when compared to the binary boundary systems H_2O- non polar gas and H_2O-salt. In these ternary fluid systems, fluid immiscibility may therefore prevail over the entire crustal P-T range (see Heinrich 2007b). Below, experimental data in the ternary systems H_2O-NaCl-CO_2, H_2O-$CaCl_2$-CO_2, and H_2O-NaCl-CH_4 are presented in some detail whereas other systems are only briefly addressed.

H_2O-NaCl-CO_2 system

Like the two binary boundary systems H_2O-CO_2 and H_2O-NaCl, the H_2O-NaCl-CO_2 system is the best studied ternary model fluid system. A large number of experimental studies has investigated the fluid phase relations within the H_2O-NaCl-CO_2 ternary at crustal P-T

conditions (e.g., Ellis and Golding 1963; Takenouchi and Kennedy 1965; Naumov et al. 1974; Gehrig 1980; Kotel'nikov and Kotel'nikova 1990; Johnson 1991; Frantz et al. 1992; Joyce and Holloway 1993; Shmulovich and Plyasunova 1993; Schmidt et al. 1995; Gibert et al. 1998; Shmulovich and Graham 1999, 2004; Schmidt and Bodnar 2000; Anovitz et al. 2004).

The effect of NaCl on the solubility of CO_2 in H_2O liquid at temperature conditions below the critical curve of the H_2O-CO_2 binary has been studied by Takenouchi and Kennedy (1965). Their data at 150 and 250 °C show notably decreasing CO_2 concentrations in the liquid with increasing NaCl concentrations (Fig. 7): At 100 MPa, e.g., addition of only $x_{NaCl} = 0.0715$ [here $x_{NaCl} = n_{NaCl}/(n_{NaCl} + n_{H_2O})$] decreases the solubility of CO_2 in the liquid from $x_{CO_2} \sim 0.04$ (at 150 °C) and 0.12 (at 250 °C) in the pure H_2O-CO_2 binary system (Takenouchi and Kennedy 1964) to only $x_{CO_2} \sim 0.02$ (at 150 °C) and 0.03 (at 250 °C) in the H_2O-NaCl-CO_2 ternary. This salting out effect therefore significantly expands the field of immiscibility in the ternary compared to the binary system.

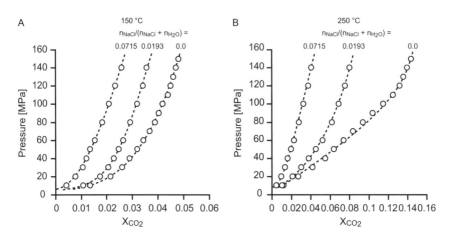

Figure 7. Effect of NaCl on the CO_2 solubility in H_2O liquid at (a) 150 and (b) 250 °C. [data for NaCl from Takenouchi and Kennedy 1965, data for pure H_2O-CO_2 from Takenouchi and Kennedy 1964]

Immiscibility in the ternary H_2O-NaCl-CO_2 system at temperature conditions above the critical curve of the H_2O-CO_2 binary has been studied, among others, by Kotel'nikov and Kotel'nikova (1990) at 400 to 800 °C/100 and 200 MPa, Shmulovich and Plyasunova (1993) at 500 °C/500 MPa, Joyce and Holloway (1993) at 200 to 500 MPa/700 to 850 °C, Gibert et al. (1998) at 900 °C/500 and 700 MPa, Schmidt and Bodnar (2000) at 275 to 650 °C/44 to 370 MPa, and Anovitz et al. (2004) at 500 °C/50 MPa. Figure 8 summarizes the experimental results at 500 °C/50 and 500 MPa and at 200 MPa/600 and 800 °C. The data show that immiscibility prevails over most portions of the ternary system at all studied conditions. Solubility of NaCl within H_2O-CO_2 mixtures and of CO_2 within H_2O-NaCl mixtures only very rarely exceeds x_{NaCl}, $x_{CO_2} > 0.1$. Only at H_2O rich compositions, i.e., near the H_2O apex, NaCl and CO_2 concentrations in ternary mixtures might be slightly higher. At 500 °C/50 MPa, the H_2O-NaCl binary still exhibits vapor-liquid immiscibility, and the data by Anovitz et al. (2004) nicely show the principal phase assemblages to be expected in ternary H_2O-salt-non polar gas systems (Fig. 8): i) At H_2O rich compositions, a NaCl poor H_2O-CO_2 vapor may coexists with a CO_2 poor H_2O-NaCl liquid, defining a vapor + liquid two phase field; ii) The stability field of the NaCl poor H_2O-CO_2 vapor extend to pure CO_2 compositions whereas the stability field of the CO_2 poor H_2O-NaCl liquid is limited towards higher NaCl concentration by NaCl saturation. For higher NaCl concentrations, CO_2 poor H_2O-NaCl liquid coexists with solid salt, defining

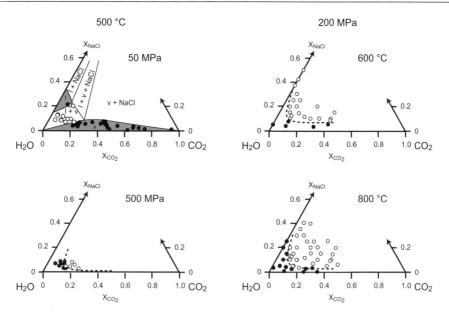

Figure 8. Fluid phase relations in the ternary system H_2O-NaCl-CO_2 at 500 °C/50 and 500 MPa and at 200 MPa/600 and 800 °C based on synthetic fluid inclusions. Open circles represent two coexisting fluid phases whereas filled circles represent single phase fluid conditions. [data sources: 500 °C/50 MPa: Anovitz et al. 2004; 500 °C/500 MPa: Shmulovich and Plyasunova 1993; 200 MPa/600 and 800 °C: Kotel'nikov and Kotel'nikova 1990]

a liquid + NaCl two phase field; iii) At intermediate compositions, NaCl poor H_2O-CO_2 vapor and CO_2 poor H_2O-NaCl liquid coexist with solid salt, defining a vapor + liquid + solid salt three phase field; iv) At higher CO_2 concentration, NaCl poor H_2O-CO_2 vapor coexists with solid salt, defining a vapor + solid salt two phase field (the evolution of these principal phase assemblages with P and T is shown in Fig. 3 of Heinrich 2007b; this volume).

At isothermal (Fig. 8 left) and isobaric (Fig. 8 right) conditions increasing pressure and temperature, respectively, increase the filed of immiscibility, especially the solubility of NaCl in H_2O-CO_2 vapor notably decreases: At 500 °C/50 MPa, maximum NaCl concentration in H_2O-CO_2 vapor is x_{NaCl} ~ 0.074 whereas at 500 °C/500 MPa the data indicate maximum NaCl concentration in H_2O-CO_2 vapor of only $x_{NaCl} < 0.025$; at 200 MPa/600 °C, maximum NaCl concentration in H_2O-CO_2 vapor is x_{NaCl} ~ 0.052 whereas at 200 MPa/800 °C the data indicate maximum NaCl concentration in H_2O-CO_2 vapor of only $x_{NaCl} < 0.032$ (additional presentations of the H_2O-NaCl-CO_2 system are given in Fig. 5 of Heinrich 2007b; this volume).

H_2O-$CaCl_2$-CO_2 system

The shape of the immiscibility field of the H_2O-$CaCl_2$-CO_2 ternary has been experimentally determined within the range 0.1 to 0.9 GPa and 500 to 800 °C by Zhang and Frantz (1989), Plyasunova and Shmulovich (1991), Shmulovich and Plyasunova (1993), and Shmulovich and Graham (2004) (Fig. 9). The principal phase topology resembles that of the H_2O-NaCl-CO_2 system: $CaCl_2$ strongly partitions into the H_2O rich liquid and even small amounts of $CaCl_2$ cause immiscibility over a wide P-T range. Compared to the H_2O-NaCl-CO_2 system, however, the $CaCl_2$ concentrations in H_2O-CO_2 vapor are extremely low and rarely exceed x_{CaCl_2} ~ 0.007. At isobaric conditions, the data consistently show widening of the immiscibility field with increasing temperature from 500 to 700 °C (in Fig. 9 from left to right). Additionally, at isothermal conditions of 500 and 700 °C (in Fig. 9 from top to bottom), the data consistently

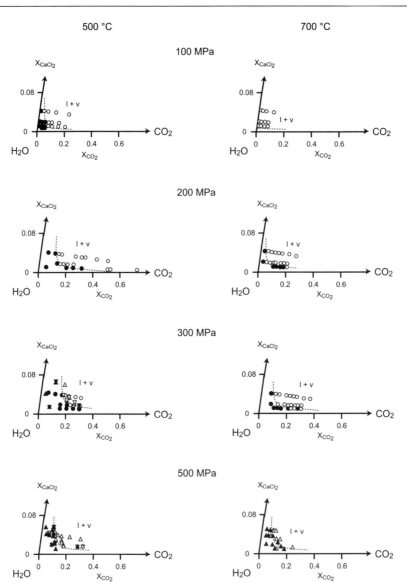

Figure 9. Fluid phase relations in the ternary system H_2O-$CaCl_2$-CO_2 at 500 and 700 °C/100 to 500 MPa based on synthetic fluid inclusions. Open symbols represent two coexisting fluid phases whereas filled symbols represent single phase fluid conditions. [data sources: circles: Zhang and Frantz 1989; upward triangles: Shmulovich and Plyasunova 1993; downward triangles: Plyasunova and Shmulovich 1991]

indicate that the field of immiscibility decreases with increasing pressure from 100 to 300 MPa but again notably increases with further increase in pressure to 500 MPa (additional presentations of the H_2O-$CaCl_2$-CO_2 system can be found in Figs. 4 and 5 of Heinrich 2007b; this volume).

H_2O-NaCl-CH_4

Krader (1985) studied the H_2O-NaCl-CH_4 system at 364 to 470 °C/40 to 263 MPa and Lamb et al. (1996, 2002) at 100 and 200 MPa/400 to 600 °C. The latter also presented tie-lines for co-existing vapor and liquid as well as compositional limits of the different one-, two- and three-phase fields (Fig. 10). The principal phase relations resemble those in the other ternary systems. Like in the H_2O-$CaCl_2$-CO_2 system, increasing pressure from 100 to 200 MPa at isothermal conditions decreases the field of immiscibility (in Fig. 10 from left to right). Unfortunately, no data are available at higher pressure for the H_2O-NaCl-CH_4 system. The effect of increasing temperature at isobaric conditions on the width and position of the vapor-liquid two-phase field is not that evident as in the other ternary systems, however, the data at 100 MPa suggest decreasing NaCl solubility in H_2O-CH_4 vapor with increasing temperature and thus a widening of the vapor-liquid two-phase field. But the data by Lamb et al. (1996, 2002) clearly show i) a steepening of the vapor-liquid tie-lines with increasing temperature and ii) a notable shrinking of the vapor + liquid + solid salt three phase field in favor of the increasing vapor + liquid and vapor + solid salt two phase fields (a presentation of the H_2O-NaCl-CH_4 phase relations in terms of wt% can be found in Fig. 6 of Heinrich 2007b; this volume).

Three-dimensional presentations of the H_2O-NaCl-CH_4 system showing the evolution of the phase relations as function of T and x at isobaric conditions have been presented by Krader (1985; here redrawn in Fig. 11 for 50 and 100 MPa). These presentations show the effects of

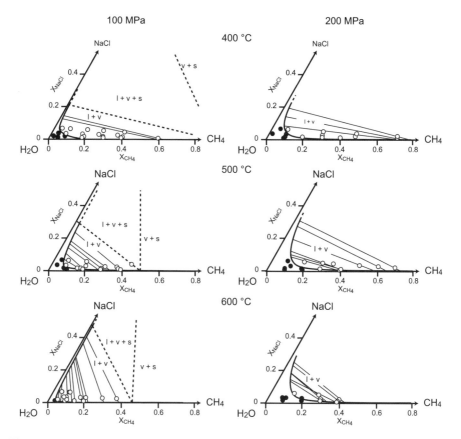

Figure 10. Fluid phase relations in the ternary system H_2O-NaCl-CH_4 at 100 and 200 MPa/400 to 600 °C based on synthetic fluid inclusions. Open circles represent two coexisting fluid phases whereas filled circles represent single phase fluid conditions. Solid lines connect coexisting vapor and liquid compositions, dashed lines limit the vapor + liquid + solid three phase assemblage. [data sources: 100 MPa: Lamb et al. 1996; 200 MPa: Lamb et al. 2002]

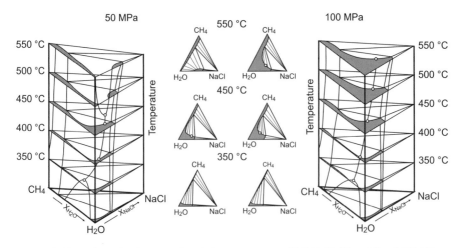

Figure 11. Three dimensional presentation of the ternary system H_2O-NaCl-CH_4 at 50 and 100 MPa up to 550 °C. Phase relations shown in isothermal diagrams are only schematic, composition of coexisting phases should not be read from these. [redrawn and modified after Krader 1985]

closing and opening, respectively, of the vapor-liquid two-phase fields in the binary H_2O-CH_4 and H_2O-NaCl boundary systems as well as the shift of the critical curve of the vapor-liquid two-phase field from the critical point of the pure H_2O-CH_4 system towards increasing NaCl concentrations (at 50 MPa terminating at the critical point of the pure H_2O-NaCl system). The principal phase relations shown in these diagrams can equally well, albeit only qualitatively, be translated into the other ternary H_2O-salt-non polar gas systems.

Limits of immiscibility in ternary H_2O-salt-non polar gas systems

The experimental data sets at 500 °C/100 MPa and 600 °C/200 MPa for the three ternary systems H_2O-NaCl-CO_2, H_2O-$CaCl_2$-CO_2, and H_2O-NaCl-CH_4 (Zhang and Frantz 1989; Kotel'nikov and Kotel'nikova 1990; Lamb et al. 1996, 2002) allow comparing the limits of vapor + liquid immiscibility in these systems (Fig. 12). The data clearly show that at these P-T conditions the compositional extend of the vapor-liquid two-phase field increases in the order H_2O-NaCl-CO_2 < H_2O-NaCl-CH_4 < H_2O-$CaCl_2$-CO_2. The solubility of NaCl in H_2O-CH_4 vapor and $CaCl_2$ in H_2O-CO_2 vapor is extremely low and for x_{CH4} and x_{CO_2} > 0.2 almost negligible. In these systems even smallest amounts of salt are sufficient to induce fluid phase separation. Likewise, the one phase field of ternary mixtures is very restricted and differs not only between the systems but also with changing pressure and temperature. The crest of the vapor-liquid two-phase field is at x_{H_2O} ~ 0.8 in the H_2O-NaCl-CO_2 system at both P-T conditions and in the H_2O-NaCl-CH_4 system at 600 °C/200 MPa. It is at x_{H_2O} ~ 0.9 in the H_2O-NaCl-CH_4 system at 500 °C/100 MPa and in the H_2O-$CaCl_2$-CO_2 system at 600 °C/200 MPa and at almost x_{H_2O} ~ 0.95 in the H_2O- $CaCl_2$-CO_2 system at 500 °C/100 MPa (a comparison between the H_2O-NaCl-CO_2 and H_2O-$CaCl_2$-CO_2 systems at 500 °C/500 MPa and 800 °C/900 MPa is given in Fig. 5 by Heinrich 2007b; this volume).

Other ternary systems

Experimental studies in ternary systems others than those described above are rare. As these systems are also of only minor geologic importance, the respective studies are only shortly summarized here. Zhang and Frantz (1992) studied the system H_2O-CO_2-CH_4 between 400 and 600 °C/100 and 300 MPa. The limits of fluid immiscibility in this ternary system were found to be between 300 and 350 °C depending on pressure and bulk composition. Under typical

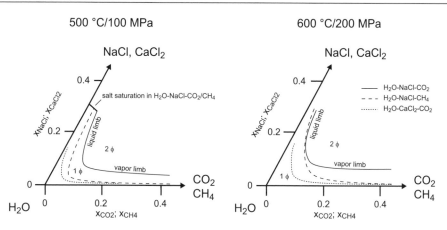

Figure 12. Limits of fluid miscibility in the ternary systems H_2O-NaCl-CO_2, H_2O-$CaCl_2$-CO_2, and H_2O-NaCl-CH_4 at 500 °C/100 MPa and 600 °C/200 MPa. 1Φ denotes single phase fluid conditions, 2Φ denotes two coexisting fluid phases.

crustal *P-T* conditions this system exhibit complete miscibility. Seitz et al. (1994) determined volumetric fluid properties in the ternary system CO_2-CH_4-N_2 at 200 °C/100 MPa.

TRACE ELEMENT AND STABLE ISOTOPE FRACTIONATION

Fluid phase separation is a potentially important and especially effective mechanism to fractionate and ultimately concentrate certain elements (for the role of fluid phase separation in ore formation see Heinrich 2007a). However, studies on trace element and stable isotope fractionation are comparable rare and restricted to simple model fluid systems: In seawater-like H_2O-NaCl model systems, Bischoff and Rosenbauer (1987) studied vapor-liquid fractionation of K, Ca, Mg, Fe, Mn, Zn, Cu, B, SO_4, CO_2, and H_2S at ~ 390 °C and Berndt and Seyfried (1990) that of Li, K, Ca, Sr, Ba, Br, and B at 425 to 450 °C; Shmulovich et al. (2002) then studied the fractionation of the rare earth elements (REE) at 350 to 450 °C in the systems H_2O-NaCl±HNO_3±HCl and H_2O-$CaCl_2$-HCl; Pokrovski et al. (2005) presented a set of systematic fractionation data in the H_2O-NaCl system for As, Sb, Fe, Zn, Cu, Ag, and Au at 350, 400, and 450 °C; the fractionation of B and Br in the H_2O-NaCl system was systematically studied by Liebscher et al. (2005, 2006) between 380 and 450 °C; fractionation data at notably higher *P-T* conditions include those for Cu by Williams et al. (1995) at 800 °C/100 MPa and 850 °C/50 MPa, for B by Schatz et al. (2004) at 800 °C/100 MPa, and for Fe and Au by Simon et al. (2005) at 800 °C/110 to 145 MPa.

Studies on the vapor-liquid fractionation of stable isotopes include those by Horita et al. (1995), Berndt et al. (1996), Shmulovich et al. (1999), and Driesner and Seward (2000) on D/H in the systems H_2O-NaCl and H_2O-KCl; by Horita et al. (1995), Shmulovich et al. (1999) and Driesner and Seward (2000) on $^{18}O/^{16}O$ in the systems H_2O-NaCl and H_2O-KCl; by Spivack et al. (1990) and Liebscher et al. (2005) on $^{11}B/^{10}B$ in the system H_2O-NaCl; by Magenheim (1995), Phillips (1999), and Liebscher et al. (2006) on $^{37}Cl/^{35}Cl$ in the system H_2O-NaCl; and by Foustoukos et al. (2004) and Liebscher et al. (2007) on $^7Li/^6Li$ in the systems H_2O-NaCl and H_2O-LiCl, respectively.

Trace element fractionation

In their experimental study, Pokrovski et al. (2005) showed that the density difference between co-existing vapor and liquid is the prime factor that controls trace element fractionation between co-existing vapor and liquid. The available experimental data on trace element fractionation below 500 °C in the binary H_2O-NaCl system (only for which systematic density data are available, see above) are therefore plotted versus the vapor-liquid density difference in Figures 13 and 14. For each element, the data define a (roughly) linear trend forced through $\log(c_{vapor}/c_{liquid}) = 0$ at $\log(\rho_{vapor}/\rho_{liquid}) = 0$; i.e., no fractionation at the critical curve. At $T <$ 500 °C, all studied elements fractionate in favor of the liquid. Depending on their fractionation behavior, however, different groups of elements can be distinguished. Gold, Sb, and especially As display an about 2 to 4 orders of magnitude weaker preference for the liquid than Fe, Cu, Zn, and Ag, which form a relative coherent group of elements (Fig. 13a,b). Pokrovski et al. (2005) observed two different sets of fractionation data for Sb, which they attributed to two different Sb species (Fig. 13a). Boron shows the smallest vapor-liquid fractionation of all elements studied, whereas Br vapor-liquid fractionation is comparable to that of Sb, Au, and the alkalines (Fig. 13c,d). The alkaline elements Li and K have a notable smaller vapor-liquid fractionation than the earth alkaline elements Ca, Sr, and Ba (Fig. 13d). However, alkaline as well as earth alkaline elements indicate increasing preference for the liquid with increasing atomic weight. The fractionation data for As (Pokrovski et al. 2002, 2005) and B (Liebscher et al. 2005) allow addressing the influence of temperature on the vapor-liquid fractionation (Fig. 14). At a given density difference, vapor-liquid fractionation gets smaller with increasing temperature. In case of B, the decreasing preference for the liquid with increasing temperature is also highlighted by the data of Schatz et al. (2004) at 800 °C/100 MPa, which indicate preferential fractionation of B into the vapor. Unfortunately, all the above data were derived in experiments free of sulfur. Natural data indicate that sulfur species may play an important role for the transport of certain elements, e.g., Cu, Au, As, and Ag (see Heinrich 2007a; this volume). First experimental data in sulfur bearing systems (Nagaseki and Hayashi 2006; Pokrovski et al. 2006) indicate that the vapor-liquid fractionation behavior of these elements may significantly change in the presence of sulfur (see also Fig. 3 of Heinrich 2007a; this volume).

The experimental results by Shmulovich et al. (2002) on the vapor-liquid fractionation of the rare earth elements (REE) at 350 to 450 °C in the systems H_2O-NaCl±HNO_3±HCl and H_2O-$CaCl_2$-HCl consistently showed a general strong preference of the REE for the liquid (Fig. 15). However, the data also show that this preference is strongest for the light REE and weakest for the heavy REE. Fluid phase separation is therefore expected to fractionate the light from the heavy REE and by this changing the REE patterns in the co-existing phases. Shmulovich et al. (2002) calculated the exchange coefficients K_D^{REE-La} [= $(REE/La)_{vapor}/(REE/La)_{liquid}$], which describe the relative vapor-liquid fractionation of the individual REE. They found a linear dependence of K_D^{REE-La} on the pressure difference to the critical curve. Additionally, for a given pressure difference to the critical curve K_D^{REE-La} values increase from light to heavy REE and are higher in the H_2O-$CaCl_2$-HCl system at 400 °C than in the H_2O-NaCl±HNO_3±HCl system at 450 °C. However, it is not clear whether this different behavior is due to the different temperature or due to the different chemical systems.

Stable isotope fractionation

Horita et al. (1995; up to 350 °C) and Driesner and Seward (2000; up to 413 °C) studied the effect of NaCl and KCl on the stable isotope fractionation of D/H and $^{18}O/^{16}O$ (Fig. 16). Above 200 °C, 1000 $\ln\alpha_{v-l}$(D) is generally positive and slightly increases with increasing temperature. Contrary, 1000 $\ln\alpha_{v-l}$(^{18}O) is generally negative but vapor-liquid fractionation decreases with increasing temperature. A different effect of NaCl versus KCl on D/H fractionation is not apparent in the data, while the data indicate a slightly smaller effect of KCl than of NaCl on $^{18}O/^{16}O$ fractionation at higher temperature. The authors then used their fractionation data to calculate the apparent isotope salt effect 1000 $\ln\Gamma_{app}$, i.e., the difference in isotope fractionation in the salt bearing systems compared to the pure H_2O system (the apparent isotope salt effect is therefore only defined below the critical point of the pure H_2O system). 1000 $\ln\Gamma_{app}$(D) is generally negative and the effect of adding salt to the system notably increases with increasing

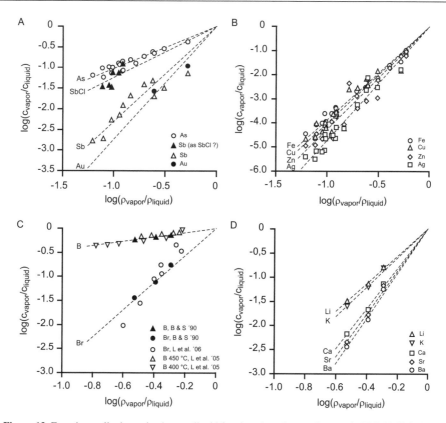

Figure 13. Experimentally determined vapor-liquid fractionation of trace elements in H_2O-NaCl fluids as function of the density difference between vapor and liquid. Dashed lines are linear fits forced through the origin. [data sources: (a) and (b): Pokrovski et al. 2005; B & S '90: Berndt and Seyfried 1990; L et al. '05: Liebscher et al. 2005; L et al. '06: Liebscher et al. 2006b; (d): Bischoff and Rosenbauer 1987]

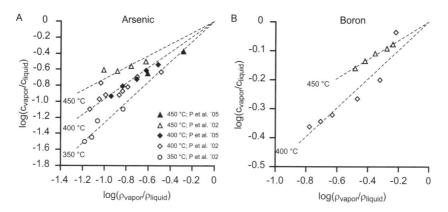

Figure 14. Temperature dependence of experimentally determined vapor-liquid fractionation of (a) As and (b) B in H_2O-NaCl fluids as function of the density difference between vapour and liquid. Dashed lines are linear fits forced through the origin. [data sources: P et al. '02: Pokrovski et al. 2002; P et al. '05: Pokrovski et al. 2005; (b): Liebscher et al. 2005]

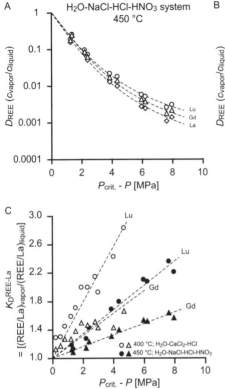

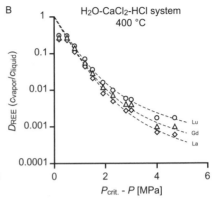

Figure 15. Experimentally determined vapor-liquid fractionation of the rare earth elements as function of the distance to the critical curve in the (a) H_2O-NaCl-HCl-HNO_3 and (b) H_2O-$CaCl_2$-HCl systems. (c) Calculated REE exchange coefficients K_D^{REE-La} between vapor and liquid. [data from Shmulovich et al. 2002]

temperature. $1000 \ln\Gamma_{app.}(^{18}O/^{16}O)$, on the other hand, is positive at $T > 150\ °C$ and the effect of adding salt to the system is much less pronounced than in the case of D/H fractionation.

Stable isotope vapor-liquid fractionation above the critical point of pure H_2O has been studied for D/H (Berndt et al. 1996; Shmulovich et al. 1999), $^{18}O/^{16}O$ (Shmulovich et al. 1999; Driesner and Seward 2000), $^{11}B/^{10}B$ (Spivack et al. 1990; Liebscher et al. 2005), $^7Li/$ ^6Li (Foustoukos et al. 2004; Liebscher et al. 2007), and $^{37}Cl/^{35}Cl$ (Magenheim 1995; Phillips 1999; Liebscher et al. 2006) (Fig. 17). No coherent vapor-liquid fractionation behavior of the different stable isotope systems is apparent. Within the systems D/H and $^{11}B/^{10}B$, the heavy isotope has a preference for the vapor whereas in the $^{18}O/^{16}O$ system the heavy isotope fractionates into the liquid. In the systems $^7Li/^6Li$ and $^{37}Cl/^{35}Cl$ no clear trend is visible at all, and the data suggest a lack of notable stable isotope fractionation in these systems, at least at the experimental P-T conditions. The fractionation data for D/H, $^{18}O/^{16}O$, $^{11}B/^{10}B$ have been fitted with a logarithmic function of the form

$$1000 \ln\alpha_{v\text{-}l} = a + b \ln[(P - P_{crit.}) + e^{(-a/b)}]$$

following Berndt et al. (1996). This allows for extrapolation of the stable isotope vapor-liquid fractionation data to salt saturated conditions, where fractionation is largest. Extrapolated maximum stable isotope fractionation is $1000 \ln\alpha_{v\text{-}l}(D) = 15.1\ ‰$ at $350\ °C$, $= 14.5\ ‰$ at $400\ °C$, and $= 13.6\ ‰$ at $450\ °C$; $1000 \ln\alpha_{v\text{-}l}(^{18}O) = -1.6\ ‰$ at $350\ °C$ and $= -1.13\ ‰$ at $450\ °C$; and $1000 \ln\alpha_{v\text{-}l}(^{11}B) = 0.94\ ‰$ at $400\ °C$ and $= 1.43\ ‰$ at $450\ °C$. The reader has to be aware that contrary to the extrapolation presented here, Shmulovich et al. (1999) interpreted their data for $1000 \ln\alpha_{v\text{-}l}(D)$ and $1000 \ln\alpha_{v\text{-}l}(^{18}O)$ as linearly dependent on NaCl concentration

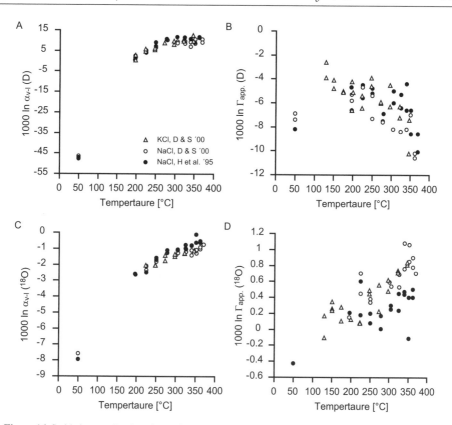

Figure 16. Stable isotope fractionation and apparent isotope salt effect for (a), (b) D/H and (c), (d) $^{18}O/^{16}O$ between vapor and liquid in the systems H_2O-NaCl and H_2O-KCl as function of temperature. [data sources: open triangles: Horita et al. 1995 for H_2O-NaCl; open circles: Driesner and Seward 2000 for H_2O-NaCl; filled circles: Driesner and Seward 2000 for H_2O-KCl]

in the liquid. Consequently, they fitted and extrapolated their data with a linear function of salt concentration in the liquid and arrived at notably higher maximum fractionation at salt saturation: Their data interpretation indicates maximum stable isotope fractionation of $1000 \ln\alpha_{v\text{-}l}(D) = 26$ to $28\ ‰$ at 350 to 600 °C and $1000 \ln\alpha_{v\text{-}l}(^{18}O) \sim -2\ ‰$ at 350 to 600 °C. Clearly, more experimental data at conditions further away from the critical curve are needed to adequately address and extrapolate stable isotope fractionation to salt saturated conditions.

FLUID-MINERAL AND FLUID-ROCK INTERACTIONS UNDER TWO-FLUID PHASE CONDITIONS

Only comparable few experimental studies so far have addressed fluid-mineral and fluid-rock interactions in the presence of co-existing fluid phases. The limited available studies are far from establishing a systematic experimental framework but rather spotlight various important effects that co-existing fluids may have on fluid-mineral and fluid-rock interactions. Some exemplary studies are reviewed here, for further information the reader is kindly referred to the references cited in these studies. Mineral solubility in vapor-liquid systems has been studied by Petrenko et al. (1990) and Foustoukos and Seyfried (2007b); Bischoff and Rosenbauer (1987) studied the effect of fluid phase separation on basalt-seawater interaction

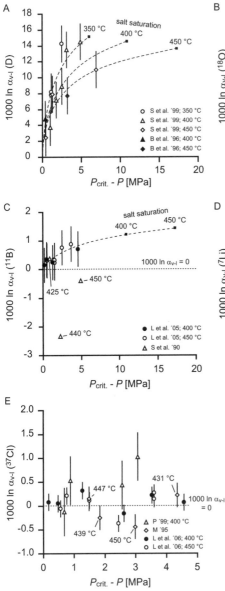

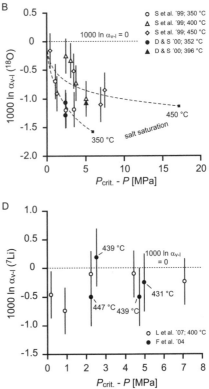

Figure 17. Stable isotope fractionation between vapor and liquid for (a) D/H, (b) $^{18}O/^{16}O$, (c) $^{11}B/^{10}B$, (d) $^{7}Li/^{6}Li$, and (e) $^{37}Cl/^{35}Cl$ as function of the distance to the critical curve. Data in (a), (b), and (c) have been fitted with a logarithmic function of the form $1000 \ln\alpha_{v-l} = a + b \ln[(P - P_{crit.}) + e^{(-a/b)}]$ (dashed lines), see text for details. [data sources: S et al. '90: Spivack et al. 1990; M '95: Magenheim 1995; B et al. '96: Berndt et al. 1996; P '99: Phillips 1999; S et al. '99: Shmulovich et al. 1999; D & S '00: Driesner and Seward 2000; F et al. '04: Foustoukos et al. 2004; L et al. '05, '06, '07: Liebscher et al. 2005, 2006a, 2007]

and corresponding metal transport; mineral equilibria in the presence of immiscible H_2O-CO_2-electrolyte solutions were studied by Shmulovich and Kotova (1982) while Kaszuba et al. (2005) evaluated fluid-rock reactions in the presence of H_2O-CO_2-NaCl fluids; Holness and Graham (1991) finally studied dihedral angles in the system H_2O-CO_2-NaCl-calcite and their implications for fluid flow during metamorphism.

Petrenko et al. (1990) studied the solubility of either synthetic galena, troilite or sphalerite in either aqueous NH_4Cl-, $MgCl_2$- or HCl-solutions. Pressure conditions during the experiments were not determined exactly but are assumed to not exceed 15 to 16 MPa, implying that the systems were immiscible in all runs. With decreasing pressure, i.e., decreasing autoclave filling and consequently increasing vapor fraction, the authors observed increasing mineral solubility, which was most pronounced in the H_2O-NH_4Cl system. They attributed this increase in solubility to increasing formation of HCl via the reaction

$$NH_4Cl = NH_3 + HCl$$

Interestingly, the authors speculate that NH_3 preferentially fractionates into the vapor whereas HCl fractionates into the liquid, by that decreasing the quench-pH of the liquid. This would be at variance to what is observed for hydrolysis in water-salt systems in which HCl preferentially fractionates into the vapor (see above).

Foustoukos and Seyfried (2007b) studied quartz solubility in aqueous NaCl ± KCl solutions at 365 to 430 °C/21.9 to 38.1 MPa, i.e., near and within the two-phase region. The observed results were identical in the H_2O-NaCl and H_2O-NaCl-KCl systems within experimental and analytical uncertainties. Runs within the two-phase field showed dissolved silica concentrations in the vapor that range from 2.81 to 14.6 mmolal. With increasing chloride concentration in the vapor, due to changing *P-T* conditions of the experiments, silica concentrations likewise increased. Unfortunately, the authors were unable to sample the co-existing liquid due to experimental constraints. Therefore, no information is available about dissolved silica concentrations in the liquid or about different net effects of single phase versus two-phase fluid conditions on quartz solubility. The authors then used the observed relation between pressure, temperature, chloride concentration and silica concentration to model the conditions of fluid phase separation in vent systems at 9°50′N East Pacific Rise (for more details see Foustoukos and Seyfried 2007a).

Bischoff and Rosenbauer (1987) performed leaching experiments between natural basalt and two-phase fluids of ordinary and evolved seawater-like composition along adiabatic ascent paths from 415 °C/33 MPa to ~ 390 °C/25 MPa. The evolved seawater-like composition was characterized by lower SO_4, Mg, Ca, Fe, Mn, B, SiO_2, and CO_2 when compared to ordinary, natural seawater. However, both fluids showed comparable effects upon experimental fluid-basalt interaction: i) During adiabatic ascent and expansion the quench-pH of the liquid decreases whereas that of the vapor increases, the opposite to what is expected from experimental hydrolysis in water-salt systems (see above); ii) heavy-metal concentrations increase in the liquid and are generally higher than in co-existing vapor, which itself is nevertheless characterized by notable contents of SiO_2, Fe and other heavy metals. The vapor thus constitutes an effective transport medium for these elements even at the low *P-T* conditions studied; iii) both fluids gained notable amounts of Zn from the basalt under two-fluid phase conditions when compared to single fluid phase conditions; and iv) fluid-phase separation increases the solubility of pyrite. The authors also concluded that density differences between vapor and liquid are only minor along the studied ascent path and that vapor-liquid mixtures most likely ascend together along the same conduit allowing for continuous equilibration.

Shmulovich and Kotova (1982) studied the carbonatization reaction of grossular to calcite, anorthite, wollastonite/quartz in the presence of aqueous $CaCl_2$ and KCl solutions at 500 to 700 °C/100 to 400 MPa. The *P,T,x* conditions of their experiments corresponded to either single phase or two-phase fluid conditions. While experiments under single phase fluid conditions were reproducible and internally consistent, the results from experiments under two-phase fluid conditions were not reproducible. The author speculate that in these experiments different reaction kinetics for vapor and liquid, fractionation of the hydrolysis products of $CaCl_2$ between vapor and liquid, and different solubility of calcite in vapor and liquid influenced the results in an unpredictable manner.

The effect of injecting CO_2 into a NaCl fluid-rock system was studied by Kaszuba et al. (2005) at 200 °C/20 MPa in order to evaluate the integrity of a geologic carbon repository.

A 5.5 molal NaCl fluid was reacted with a mixture of synthetic arkose (as representation of an aquifer) and natural argillaceous shale (as representation of an aquitard). After 32 days and reaching steady state conditions, the system was injected with CO_2 and allowed to react for an additional 45 days. Injection of CO_2 yielded a fluid with $x_{CO_2}^{bulk} \sim 0.19$, which under the experimental conditions un-mixes into a liquid with $x_{CO_2} \sim 0.02$ and a vapor with $x_{CO_2} \sim 0.82$ (Takenouchi and Kennedy 1964). A parallel experiment was run for 77 days without injection of CO_2. Upon CO_2 injection, the authors observed a decrease in liquid pH and a significant, abrupt increase in Ca (15%), Mg (800 %), SiO_2 (80 %), Fe (100 %), and Mn (500 %) concentrations. Compared to the CO_2-free, single phase fluid experiment, the results indicate accelerated fluid-silicate reaction rates and enhanced silica solubility under two-phase fluid conditions. Based on these findings, the authors suggest the change between single phase and two-phase fluid conditions as a potential mechanism to mobilize respectively precipitate silica, potentially leading to silica cement and quartz vein mineralization.

The permeability of a rock is controlled by the geometry of the fluid-filled pores in the rock, which itself depends on the wetting behavior or dihedral angle (Θ) of the fluid against the solids: For $\Theta < 60°$, a connected grain-edge network of fluid filled channels establishes, making the rock permeable to fluid flow; for $\Theta > 60°$, however, isolated, non-connected fluid filled pores form and the rock is impermeable to fluid flow. In two- or multi-phase fluid systems, differences between the dihedral angle of the co-existing fluids will therefore largely control whether the co-existing fluids are able to segregate from each other. Holness and Graham (1991) experimentally determined the dihedral angles between fluid and calcite in the system H_2O-CO_2-NaCl-calcite. Experiments were performed at 100 and 200 MPa/550 to 750 °C with binary H_2O-CO_2 and H_2O-NaCl and ternary H_2O-CO_2-NaCl fluids, resulting in single phase and two-phase fluid conditions depending on P,T and x. Under two-phase fluid conditions, the authors observed a clear bimodal distribution of Θ for H_2O-CO_2-NaCl fluids with $\Theta_1 \sim 80$ to 90° and $\Theta_2 \sim 40$ to 55°. By comparison with results from runs with binary, single phase H_2O-CO_2 and H_2O-NaCl fluids, they concluded that Θ_1 represents a vapor with about $x_{CO_2} = 0.9$ whereas Θ_2 represents a liquid with about 40 to 50 wt% NaCl. In runs with two-phase H_2O-NaCl fluids, a bimodal distribution of Θ was only observed for rapid cooling; slowly cooled runs showed a unimodal distribution of Θ. In summary, the authors conclude that i) in marbles a connected fluid phase will only form for fluids with either $x_{CO_2} \sim 0.5$ or > 30 wt% NaCl, ii) quartzites/psammites act as aquifers relative to limestones for water-rich or saline conditions, iii) compared to quartzites/psammites only limestones will permit pervasive fluid flow for fluids with $x_{CO_2} \sim 0.2$ to 0.6, and iv) both lithologies are permeable to highly saline fluids (examples of two-phase fluid flow under metamorphic conditions are given in Heinrich 2007b, this volume).

CONCLUDING REMARKS

The principal fluid phase relations in the basic binary water-salt and water-non polar gas and ternary water-salt-non polar gas systems are well established. However, there are still substantial gaps in knowledge: i) In binary water-salt systems, systematic density data as function of P,T,x are restricted to the H_2O-NaCl system. But density data are not only important for the understanding of trace element fractionation between co-existing fluids (see above) but are essential for any modeling of multiphase fluid flow in hydrothermal systems (see Driesner and Geiger 2007, this volume); ii) Experimental studies in the system H_2O-$SrCl_2$ and in iron chloride bearing systems are virtually absent. Compared to H_2O-NaCl, the experimental data bases for the systems H_2O-LiCl, H_2O-CsCl, and H_2O-$MgCl_2$ are insufficient; iii) In ternary water-salt-non polar gas systems, the exact position of vapor-liquid tie-lines as function of P and T are improperly known; iv) Experiments on fluid phase relations in quaternary system are still lacking; v) The effect of hydrolysis is only poorly constraint in binary water-salt systems and has not yet been explicitly studied in other systems.

Experimental studies on trace element and stable isotope fractionation between co-existing fluids focused on simple binary water-salt systems with chlorine as only or dominant anion. Only most recently have experiments addressed trace element fractionation between co-existing fluids in sulfur-bearing systems. Experiments on trace element and stable isotope fractionation in CO_2 bearing systems are to the author's knowledge lacking. Recent advantages in equations of state (see Gottschalk 2007, this volume) and numerical modeling (see Driesner and Geiger 2007, this volume) now provide the opportunity to model multiphase fluid flow. The next step is to combine these modeling approaches with trace element and stable isotope fractionation data. For this purpose, however, the experimental data base for trace element and stable isotope fractionation between co-existing fluids has to be notably enlarged in P, T, and x.

The few experimental studies on fluid-mineral and fluid-rock interactions under two-fluid phase conditions show that the physicochemical characteristics of fluid-mineral/rock interaction may significantly change under multiphase fluid conditions when compared to the simple single phase fluid case. As permeability, mineral solubility, fluid phase segregation (among others) are key parameters for the physical and chemical evolution of any hydrothermal system, experimental determination of the physicochemical characteristics of fluid-mineral/rock under multiphase fluid flow is definitely a challenging but also promising task for future research.

ACKNOWLEDGMENTS

This paper benefited from critical and thorough reading and comments by T. Driesner and C. A. Heinrich. Extensive and continuous discussions with W. Heinrich, M. Gottschalk, C. Schmidt and R. L. Romer over the last years significantly improved the author's knowledge (and hopefully understanding) of immiscible fluids but any error is his own. Final editorial handling by J. Rosso is gratefully acknowledged.

REFERENCES

Anovitz LM, Labotka TC, Blencoe JG, Horita J (2004) Experimental determination of the activity-composition relations and phase equilibria of H_2O-CO_2-NaCl fluids at 500 °C, 500 bars. Geochim Cosmochim Acta 68:3557-3567
Arnórsson S, Stefánsson A, Bjarnason JÖ (2007) Fluid-fluid interactions in geothermal systems. Rev Mineral Geochem 65:259-312
Berndt ME, Seal II RR, Shanks III WC, Seyfried Jr. WE (1996) Hydrogen isotope systematics of phase separation in submarine hydrothermal systems: Experimental calibration and theoretical methods. Geochim Cosmochim Acta 60:1595-1604
Berndt ME, Seyfried Jr WE (1990) Boron, bromine, and other trace elements as clues to the fate of chlorine in mid-ocean ridge vent fluids. Geochim Cosmochim Acta 54:2235-2245
Berndt ME, Seyfried Jr WE (1997) Calibration of Br/Cl fractionation during subcritical phase separation of seawater: Possible halite at 9 to 10°N East Pacific Rise. Geochim Cosmochim Acta 61:2849-2854
Bischoff JL (1991) Densities of liquids and vapors in boiling NaCl-H_2O solutions: A *PVTx* summary from 300° to 500 °C. Am J Sci 291:309-338
Bischoff JL, Pitzer KS (1985) Phase relations and adiabats in boiling seafloor geothermal systems. Earth Planet Sci Lett 75:327-338
Bischoff JL, Pitzer KS (1989) Liquid-vapor relations for the system NaCl-H_2O: Summary of the *P-T-x* surface from 300 to 500 °C. Am J Sci 289:217-248
Bischoff JL, Rosenbauer RJ (1984) The critical point and two-phase boundary of seawater, 200-500 °C. Earth Planet Sci Lett 68:172-180
Bischoff JL, Rosenbauer RJ (1987) Phase separation in seafloor geothermal systems: An experimental study of the effects on metal transport. Am J Sci 287:953-978
Bischoff JL, Rosenbauer RJ (1988) Liquid-vapor relations in the critical region of the system NaCl-H_2O from 380 to 415 °C: A refined determination of the critical point and two-phase boundary of seawater. Geochim Cosmochim Acta 52:2121-2126
Bischoff JL, Rosenbauer RJ, Fournier RO (1996) The generation of HCl in the system CaCl$_2$-H_2O: Vapor-liquid relations from 380-500 °C. Geochim Cosmochim Acta 60:7-16

Bischoff JL, Rosenbauer RJ, Pitzer KS (1986) The system NaCl-H$_2$O: relations of vapour-liquid near the critical temperature of water and of vapour-liquid-halite from 300 to 500 °C. Geochim Cosmochim Acta 50:1437-1444

Blencoe JG, Naney MT, Anovitz LM (2001) The CO$_2$-H$_2$O system: III. A new experimental method for determining liquid-vapor equilibria at high subcritical temperatures. Am Mineral 86:1100-1111

Bodnar RJ, Burnham CW, Sterner SM (1985) Synthetic fluid inclusions in natural quartz. III. Determination of phase equilibrium properties in the System H$_2$O-NaCl to 1000 °C and 1500 bars. Geochim Cosmochim Acta 49:1861-1873

Bodnar RJ, Sterner SM (1985) Synthetic fluid inclusions in natural quartz. II. Application to *PVT* studies. Geochim Cosmochim Acta 49:1855-1859

Bodnar RJ, Sterner SM (1987) Synthetic fluid inclusions. *In:* Hydrothermal Experimental Techniques. Ulmer GC, Barnes HL (eds) Wiley, p 423-457

Chou I-M (1985) Phase relations in the system NaCl-KCl-H$_2$O. III: Solubilities of halite in vapor-saturated liquids above 445 °C and redetermination of phase equilibrium properties in the system NaCl-H$_2$O to 1000 °C and 1500 bars. Geochim Cosmochim Acta 51:1965-1975

Crovetto R, Lvov SN, Wood RH (1993) Vapor pressures and densities of NaCl(aq) and KCl(aq) at temperature 623 K and CaCl$_2$(aq) at the temperatures of 623 K and 643 K. J Chem Thermodyn 25:127-138

Driesner T, Geiger S (2007) Numerical simulation of multiphase fluid flow in hydrothermal systems. Rev Mineral Geochem 65:187-212

Driesner T, Heinrich CA (2007) The System H$_2$O-NaCl. I. Correlation formulae for phase relations in pressure-temperature-composition space from 0 to 1000 °C, 0 to 5000 bar, and 0 to 1 X$_{NaCl}$. Geochim Cosmochim Acta (accepted)

Driesner T, Seward TM (2000) Experimental and simulation study of salt effects and pressure/density effects on oxygen and hydrogen stable isotope liquid-vapor fractionation for 4-5 molal aqueous NaCl and KCl solutions to 400 °C. Geochim Cosmochim Acta 64:1773-1784

Dubois M, Weisbrod A, Shtuka A (1994) Experimental determination of the two-phase (liquid and vapour) region in water-alkali chloride binary systems at 500° and 600 °C using synthetic fluid inclusions. Chem Geol 115:227-238

Ellis AJ, Golding RM (1963) The solubility of carbon dioxide above 100 °C in water and in sodium chloride solutions. Am J Sci 261:47-60

Foustoukos DI, James RH, Berndt ME, Seyfried Jr WE (2004) Lithium isotopic systematics of hydrothermal vent fluids at the Main Endeavour Field, Northern Juan de Fuca Ridge. Chem Geol 212:17-26

Foustoukos DI, Seyfried Jr. WE (2007) Fluid phase separation processes in submarine hydrothermal systems. Rev Mineral Geochem 65:213-239

Foustoukos DI, Seyfried Jr WE (2007b) Quartz solubility in the two-phase and critical region of the NaCl-KCl-H2O system: Implications for submarine hydrothermal vent systems at 9°50′N East Pacific Rise. Geochim Cosmochim Acta 71:186-201

Franck EU, Lentz H, Welsch H (1974) The system water-xenon at high pressures and temperatures. Z Phys Chem 93:95-108

Frantz JD, Popp RK, Hoering TC (1992) The compositional limits of fluid immiscibility in the system H2O-NaCl-CO2 as determined with the use of synthetic fluid inclusions in conjunction with mass spectrometry. Chem Geol 98:237-255

Gehrig M (1980) Phasengleichgewichte und *PVT*-Daten ternärer Mischungen aus Wasser, Kohlendioxid und Natriumchlorid bis 3 kbar und 550 °C. PhD Dissertation, University of Karlsruhe, Karlsruhe, Germany

Gibert F, Guillaume D, Laporte D (1998) Importance of fluid immiscibility in the H$_2$O-NaCl-CO$_2$ system and selective CO$_2$ entrament in granulites: experimental phase diagram at 5-7 kbar, 900 °C and wetting textures. Eur J Mineral 10:1097-1123

Gottschalk M (2007) Equations of state for complex fluids. Rev Mineral Geochem 65:49-97

Hack AC, Thompson AB, Aerts M (2007) Phase relations involving hydrous silicate melts, aqueous fluids, and minerals. Rev Mineral Geochem 65:129-185

Harvey JN, Levelt Sengers JMH (1989) On the NaCl-H$_2$O coexistence curve near the critical temperature of H$_2$O. Chem Phys Lett 156:415-417

Heinrich CA (2007a) Fluid – fluid interactions in magmatic-hydrothermal ore formation. Rev Mineral Geochem 65:363-387

Heinrich W (2007b) Fluid immiscibility in metamorphic rocks. Rev Mineral Geochem 65:389-430

Holness MB, Graham CM (1991) Equilibrium dihedral angles in the system H$_2$O-CO$_2$-NaCl-calcite and implications for fluid flow during metamorphism. Contrib Mineral Petrol 108:368-383

Horita J, Cole DR, Wesolowski DJ (1995) The activity-composition relationship of oxygen and hydrogen isotopes in aqueous salt solutions: III. Vapor-liquid water equilibration of NaCl solutions to 350 °C. Geochim Cosmochim Acta 59:1139-1151

Hovey JK, Pitzer KS, Tanger IV JC, Bischoff JL, Rosenbauer RJ (1990) Vapor-liquid phase equilibria of potassium chloride-water mixtures: Equation-of-state representation for KCl-H_2O and NaCl-H_2O. J Phys Chem 94:1175-1179

Japas ML, Franck EU (1985) High pressure phase equilibria and *PVT*-data of the water-nitrogen system to 673 K and 250 MPa. Ber Bunsenges Phys Chem 89:793-800

Johnson EL (1991) Experimentally determined limits for H_2O-CO_2-NaCl immiscibility in granulites. Geology 19:925-928

Joyce DB, Holloway JR (1993) An experimental determination of the thermodynamic properties of H_2O-CO_2-NaCl fluids at high pressures and temperatures. Geochim Cosmochim Acta 57:733-746

Kaszuba JP, Janecky DR, Snow MG (2005) Experimental evaluation of mixed fluid reactions between supercritical carbon dioxide and NaCl brine: Relevance to the integrity of a geologic carbon repository. Chem Geol 217:277-293

Kay WB, Rambosek GM (1953) Liquid-vapor equilibrium relations in binary systems. Ind Eng Chem 45:221-226

Keevil NB (1942) Vapor pressures of aqueous solutions at high temperatures. J Am Chem Soc 64:841-850

Ketsko VA, Urusova MA, Valyashko VM (1984) Solubility and vapour pressure of solutions in the $CaCl_2$-H_2O System at 250-400 °C. Russ J Inorg Chem 29:1398-1399, [translated from Zhurnal Neorganicheskoi Khimii 29:2443-2445, 1984]

Khaibullin K, Borisov NM (1966) Experimental investigation of the thermal properties of aqueous and vapour solutions of sodium and potassium chloride at phase equilibrium. Teplofizika Vysokikh Temperatur 4:518-523 [English translation 489-494]

Knight CL, Bodnar RJ (1989) Synthetic fluid inclusions: IX. Critical PVTX properties of NaCl-H_2O solutions. Geochim Cosmochim Acta 53:3-8

Kotel'nikov AR, Kotel'nikova ZA (1991) The phase state of the H_2O-CO_2-NaCl system examined from synthetic fluid inclusions in quartz. Geochem Intern 29:55-66, [translated from Geokhimiya 4:526-537, 1990]

Krader T (1985) Phasengleichgewichte und kritische Kurven des ternären Systems H_2O-CH_4-NaCl bis 250 MPa und 800 K. PhD Dissertation, University of Karlsruhe, Karlsruhe, Germany

Lamb WM, McShane CJ, Popp RK (2002) Phase relations in the CH_4-H_2O-NaCl system at 2 kbar, 300 to 600 °C as determined using synthetic fluid inclusions. Geochim Cosmochim Acta 66:3971-3986

Lamb WM, Popp RK, Boockoff LA (1996) The determination of phase relations in the CH_4-H_2O-NaCl system at 1 kbar, 400 to 600 °C using synthetic fluid inclusions. Geochim Cosmochim Acta 60:1885-1897

Liebscher A, Barnes J, Sharp Z (2006a) Liquid-vapor fractionation of the chlorine isotopes: Experimental calibration at 400 to 450 °C/23 to 42 MPa. Chem Geol 234:340-345

Liebscher A, Lüders V, Heinrich W, Schettler G (2006b) Br/Cl signature of hydrothermal fluids: Liquid-vapor fractionation of bromine revisited. Geofluids 6:113-121

Liebscher A, Meixner A, Romer RL, Heinrich W (2005) Liquid-vapor fractionation of boron and boron isotopes: Experimental calibration at 400 °C/23 MPa to 450 °C/42 MPa. Geochim Cosmochim Acta 69:5693-5704

Liebscher A, Meixner A, Romer RL, Heinrich W (2007) Experimental calibration of the vapour-liquid phase relations and lithium isotope fractionation in the system H2O-LiCl at 400 °C/20-28 MPa. Geofluids 7:1-7

Magenheim AJ (1995) Oceanic borehole fluid chemistry and analysis of chlorine stable isotopes in silicate rocks. PhD Dissertation, University of California, San Diego, California

Marshall WE, Jones EV (1974) Liquid-vapor critical temperatures of aqueous electrolyte solutions. J Inorg Nucl Chem 36:2313-2318

Nagaseki H, Hayashi K (2006) The effect of sulfur on the vapor-liquid distribution of Cu and Zn in boiling hydrothermal fluid by SRXRF microanalysis of fluid inclusions. International Mineralogical Association Conference, Kobe, Abstract K104-006

Naumov VB, Khakimov AKH, Khodakovskiy IL (1974) Solubility of carbon dioxide in concentrated chloride solutions at high temperatures and pressures. Geochem Int 11:31-41

Oakes CS, Bodnar RJ, Simonson JM, Pitzer KS (1994) Critical and supercritical properties for 0.3 to 3.0 mol*kg⁻¹ $CaCl_2$(aq). Geochim Cosmochim Acta 58:2421-2431

Ölander A, Liander H (1950) The phase diagram of sodium chloride and steam above the critical point. Acta Chem Scand 4:1437-1445

Pedersen KS, Christensen P (2007) Fluids in Hydrocarbon Basins. Rev Mineral Geochem 65:241-258

Petrenko GV, Malinin SD, Arutyunyan LA (1989) Solubility of heavy-metal sulfides in heterogenous chloride solution-vapor phase systems. Geochem Intern 28:101-105, [translated from Geokhimiya 6:882-887, 1989]

Phillips J (1999) Chlorine isotopic composition of hydrothermal vent fluids from the Juan De Fuca Ridge. Master Thesis, University of North Carolina, Wilmington, North Carolina

Pitzer KS, Bischoff JL, Rosenbauer RJ (1987) Critical behaviour of dilute NaCl in H_2O. Chem Phys Lett 134:60-63

Pitzer KS, Tanger JC (1989) Critical exponents for the coexistence curves for NaCl-H$_2$O near the critical temperature of H$_2$O. Reply to comment by A.H. Harvey and J.M.H. Levelt Sengers. Chem Phys Lett 156:418-419

Plyasunova NV, Shmulovich KI (1991) Phase equilibria in the system H$_2$O-CO$_2$-CaCl$_2$ at 500 °C. Transactions USSR Acad Sci 320:221-225

Pokrovski GS, Borisova Y, Harrichoury JC (2006) The effect of sulfur on vapor-liquid partitioning of metals in hydrothermal systems: An experimental batch-reactor study. Abstr. Goldschmidt Conference, Geochim Cosmochim Acta 33(8):657-660

Pokrovski GS, Roux J, Harrichoury JC (2005) Fluid density control on vapor-liquid partitioning metals in hydrothermal systems. Geology 33:657-660

Pokrovski GS, Zakirov IV, Roux J, Testemale D, Hazemann JL, Bychkov AY, Golikova GV (2002) Experimental study of arsenic speciation in vapor phase to 500 °C: Implications for As transport and fractionation in low-density crustal fluids and volcanic gases. Geochim Cosmochim Acta 66:3453-3480

Roedder E (1984) Fluid Inclusions. Rev Mineral Vol 12, Mineralogical Society of America

Rosenbauer RJ, Bischoff JL (1987) Pressure-composition relations for coexisting gases and liquids and the critical points in the system NaCl-H$_2$O at 450, 475, and 500 °C. Geochim Cosmochim Acta 51:2349-2354

Rowlinson JS, Swinton FL (1982) Liquids and Liquid Mixtures. Butterworth

Schatz OJ, Dolejš D, Stix J, Williams-Jones AE, Layne GD (2004) Partitioning of boron among melt, brine and vapor in the system haplogranite-H$_2$O-NaCl at 800 °C and 100 MPa. Chem Geol 210:135-147

Schmidt C, Bodnar RJ (2000) Synthetic fluid inclusions: XVI. *PVTX* properties in the system H$_2$O-NaCl-CO$_2$ at elevated temperatures, pressures, and salinities. Geochim Cosmochim Acta 64:3853-3869

Schmidt C, Rosso KM, Bodnar RJ (1995) Synthetic fluid inclusions: XIII. Experimental determination of PVT properties in the system H$_2$O + 40 wt.% NaCl + 5 mol.% CO$_2$ at elevated temperature and pressure. Geochim Cosmochim Acta 59:3953-3959

Schröer E (1927) Untersuchungen über den kritischen Zustand. I. Beitrag zur Kenntnis des kritischen Zustandes des Wassers und wässriger Lösungen. Z Phys Chem 129:79-110

Seitz JC, Blencoe JG (1997) Experimental determination of the volumetric properties and solvus relations of H$_2$O-CO$_2$ mixtures at 300-400 °C and 75-1000 bars. Proc Fifth Intern Sym Hydro Reactions 109-112. Oak Ridge National Laboratory, Oak Ridge, TN

Seitz JC, Blencoe JG, Joyce DB, Bodnar RJ (1994) Volumetric properties of CO$_2$-CH$_4$-N$_2$ fluids at 200 °C and 1000 bars: A comparison of equations of state and experimental data. Geochim Cosmochim Acta 58:1065-1071

Seyfried WE, Janecky DR, Berndt ME (1987) Rocking autoclaves for hydrothermal experiments: II. The flexible reaction-cell system. *In:* Hydrothermal Experimental Techniques. Ulmer GC, Barnes HL (eds) Wiley, p 216-239.

Shmonov VM, Sadus RJ, Franck EU (1993) High pressure and supercritical phase equilibria PVT-Data of the binary water + methane mixture to 723 K and 200 MPa. J Phys Chem 97:9054-9059

Shmulovich K, Heinrich W, Möller P, Dulski P (2002) Experimental determination of REE fractionation between liquid and vapour in the systems NaCl-H$_2$O and CaCl$_2$-H$_2$O up to 450 °C. Contrib Mineral Petrol 144:257-273

Shmulovich KI, Graham CM (1999) An experimental study of phase equilibria in the system H$_2$O-CO$_2$-NaCl at 800 °C and 9 kbar. Contrib Mineral Petrol 136:247-257

Shmulovich KI, Graham CM (2004) An experimental study of phase equilibria in the systems H$_2$O-CO$_2$-CaCl$_2$ and H$_2$O-CO$_2$-NaCl at high pressures and temperatures (500-800 °C, 0.5-0.9 GPa): geological and geophysical applications. Contrib Mineral Petrol 146:450-462

Shmulovich KI, Kotova PP (1982) Mineral equilibria in a hot H$_2$O-CO$_2$-electrolyte fluid. Geokhimiya 10:1440-1453

Shmulovich KI, Landwehr D, Simon K, Heinrich W (1999) Stable isotope fractionation between liquid and vapour in water-salt systems up to 600 °C. Chem Geol 157:343-354

Shmulovich KI, Plyasunova NV (1993) Phase equilibria in ternary systems formed by H$_2$O and CO$_2$ with CaCl$_2$ or NaCl at high *T* and *P*. Geochem Intern 30(12):53-71, [translated from Geokhimiya 5:666-684, 1993]

Shmulovich KI, Tkachenko SI, Plyasunova NV (1995) Phase equilibria in fluid systems at high pressures and temperatures. *In:* Fluids in the Crust: Equilibrium and transport properties. Shmulovich KI, Yardley BWD, Gonchar GG (eds) Chapman & Hall, London, p 193-214

Simon AC, Frank MR, Pettke T, Candela PA, Picolli PM, Heinrich CA (2005) Gold partitioning in melt-vapor-brine systems. Geochim Cosmochim Acta 69:3321-3335

Sourirajan S, Kennedy GC (1962) The system H$_2$O-NaCl at elevated temperatures and pressures. Am J Sci 260:115-141

Spivack AJ, Berndt ME, Seyfried Jr WE (1990) Boron isotope fractionation during supercritical phase separation. Geochim Cosmochim Acta 54:2337-2339

Sterner SM, Bodnar RJ (1991) Synthetic fluid inclusions. X: Experimental determination of P-V-T-X properties in the CO_2-H_2O system to 6 kb and 700 °C. Am J Sci 291:1-54

Takenouchi S, Kennedy GC (1964) The binary system H_2O-CO_2 at high temperatures and pressures. Am J Sci 262:1055-1074

Takenouchi S, Kennedy GC (1965) The solubility of carbon dioxide in NaCl solutions at high temperatures and pressures. Am J Sci 263:445-454

Thompson AB, Aerts M, Hack AC (2007) Liquid immiscibility in silicate melts and related systems. Rev Mineral Geochem 65:99-127

Tkachenko SI, Shmulovich K (1992) Liquid-vapor equilibria in water-salt systems (NaCl, KCl, $CaCl_2$, $MgCl_2$) at 400-600 °C. Dokl Akad SSSR 326:1055-1059 [in Russian]

Tödheide K, Franck EU (1963) Das Zweiphasengebiet und die kritische Kurve im System Kohlendioxid-Wasser bis zu Drucken von 3500 bar. Z Phys Chem Neue Folge 37:387-401

Truesdell AH (1974) Oxygen isotope activities and concentrations in aqueous salt solutions at elevated temperatures: Consequences for isotope geochemistry. Earth Planet Sci Lett 23:387-396

Urusova MA (1974) Phase equilibria in the sodium hydroxide-water and sodium chloride-water systems at 350-550 °C. Russ J Inorg Chem 19(3):450-454 [translated from Zhurnal Neorganicheskoi Khimii 19:828-833, 1974]

Urusova MA (1975) Volume properties of aqueous solutions of sodium chloride at elevated temperatures and pressures. Russ J Inorg Chem 20:1717-1721 [translated from Zhurnal Neorganicheskoi Khimii 20:3103-3110, 1975]

Urusova MA (1986) Vapour pressure in the $SrCl_2$-H_2O system at temperatures above 250 °C. Russ J Inorg Chem 31:1104-1105 [translated from Zhurnal Neorganicheskoi Khimii 31:1916-1917, 1986]

Urusova MA, Ravich MI (1971) Vapor pressure and solubility in the sodium chloride-water system at 350° and 400 °C. Russ J Inorg Chem 16 (10):1534

Urusova MA, Valyashko VM (1983) Solubility, vapour pressure, and thermodynamic properties of solutions in the $MgCl_2$-H_2O system at 300-350 °C. Russ J Inorg Chem 28:1045-1048 [translated from Zhurnal Neorganicheskoi Khimii 28:1845-1849, 1983]

Urusova MA, Valyashko VM (1984) Vapour pressure and thermodynamic properties of aqueous solutions of magnesium chloride at 250 °C. Russ J Inorg Chem 29:1395-1396 [translated from Zhurnal Neorganicheskoi Khimii 29:2437-2439, 1984]

Urusova MA, Valyashko VM (1987) The vapor pressure and the activity of water in concentrated aqueous solutions containing the chlorides of alkali metals (Li, K, Cs) and alkaline earth metals (Mg, Ca) at increased temperatures. Russ J Inorg Chem 32 (1):23-26 [translated from Zhurnal Neorganicheskoi Khimii 32:44-48, 1987]

Vakulenko AG, Alekhin YuV, Rasina MV (1989) Solubility and thermodynamic properties of alkali chlorides in steam. Proc. II Intern. Sympos. 'Properties of water and steam', Praque, pp. 395-401

Valayashko VM, Urusova MA, Fogt V [Voigt W], Emons GG (1988) Properties of solutions in the $MgCl_2$-H_2O system over wide temperature and concentration ranges. Russ J Inorg Chem 33:127-130 [translated from Zhurnal Neorganicheskoi Khimii 33:228-232, 1988]

Valyashko VM, Urusova MA, Ketsko VA, Kravchuk KG (1987) Phase equilibria and thermodynamic properties of solutions in the $MgCl_2$-H_2O, $CaCl_2$-H_2O, $SrCl_2$-H_2O, and $BaCl_2$-H_2O systems at elevated temperatures. Russ J Inorg Chem 32:1634-1639 [translated from Zhurnal Neorganicheskoi Khimii 32:2811-2819, 1987]

Webster J, Mandeville C (2007) Fluid immiscibility in volcanic environments. Rev Mineral Geochem 65:313-362

Welsch H (1973) Die Systeme Xenon-Wasser und Methan-Wasser bei hohen Drücken und Temperaturen. PhD Dissertation, University of Karlsruhe, Karlsruhe, Germany

Williams TJ, Candela PA, Piccoli PM (1995) The partitioning of copper between silicate melts and two-phase aqueous fluids: An experimental investigation at 1 kbar, 800 °C and 0.5 kbar, 850 °C. Contrib Mineral Petrol 121:388-399

Wood SA, Crerar DA, Brantley SL, Borcsik M (1984) Mean molal stoichiometric activity coefficients of alkali halides and related electrolytes in hydrothermal solutions. Am J Sci 284:668-705

Zarembo VI, Lvov SN, Matuzenko MYu (1980) Saturated vapor pressure of water and activity coefficients of calcium chloride in the $CaCl_2$-H_2O system at 423-623 K. Geochem Int 17:159-162

Zhang Y-G, Frantz JD (1989) Experimental determination of the compositional limits of immiscibility in the system $CaCl_2$-H_2O-CO_2 at high temperatures and pressures using synthetic fluid inclusions. Chem Geol 74:289-308

Zhang Y-G, Frantz JD (1992) Hydrothermal reactions involving equilibrium between minerals and mixed volatiles: 2. Investigations of fluid properties in the CO_2-CH_4-H_2O system using synthetic fluid inclusions. Chem Geol 100:51-72

Reviews in Mineralogy & Geochemistry
Vol. 65, pp. 49-97, 2007
Copyright © Mineralogical Society of America

3

Equations of State for Complex Fluids

Matthias Gottschalk

Department 4: Chemistry of the Earth
Section 4.1: Experimental Geochemistry and Mineral Physics
GeoForschungsZentrum
Telegrafenberg
14473 Potsdam, Germany
gottschalk@gfz-potsdam.de

INTRODUCTION

Compositions of coexisting fluids at elevated pressure-temperature conditions can be calculated by applying thermodynamics as in any other chemical equilibrium. This calculation requires the evaluation of the thermodynamic properties of the phase components i at the respective pressure P, temperature T, and the individual abundances x_i, i.e., composition. In comparison with solids, fluids normally have distinct PVT-behaviors. Furthermore, because of their disordered state and the relatively weak bonding between molecules, fluid species mix much more easily than solid-phase components. The evaluation of the thermodynamic properties of fluids requires, therefore, special treatment and procedures that differ significantly from those used for solids. For fluids, this treatment includes a distinct and different standard state than for solids and a special description of the volume V as a function of P, T (or P as a function of V and T) and x_i, i.e., an equation of state (EOS).

PRINCIPLES

For geological systems, which are usually defined by the variables P and T, the Gibbs free energy G is the function of choice to describe equilibria. At constant P and T the chemical equilibrium condition is defined by (the Table 1 lists the symbols used):

$$dG = 0 \tag{1}$$

The partial derivative of G with respect to P is the volume V

$$\left(\frac{\partial G}{\partial P} \right)_{T,n_i} = V \tag{2}$$

Therefore, the pressure-dependence of G is obtained by integration of Equation (2) with boundaries from the reference pressure P_r, i.e., ambient pressure of 0.1 MPa, to the required P

$$G_{(P,T)} = G_{(P_r,T)} + \int_{P_r}^{P} V dP \tag{3}$$

For the practical treatment of mixed phases, G has to be replaced by its partial derivative with respect to composition, the chemical potential μ_i of the phase component i[1]

---

[1] The term *phase component* is used as a synonym for a endmember of a solution, i.e., for fluids specific species or molecules, whereas the term *component* is used in the classic thermodynamical sense.

     DOI: 10.2138/rmg.2007.65.3

Table 1. Notation and symbols

EOS	equation of state

sub- and superscripts

o	pure phase
$o*$	hypothetical standard state of an ideal gas
m	mixture
r	reference conditions ($P_r = 0.1$ MPa, $T_r = 298.15$ K)
c	critical property
i	of phase component i

thermodynamic variables and properties

P	pressure
T	temperature
V, v	volume, molar volume
ρ	molar density
N_A	Avogadro´s number
A, a	Helmholtz free energy, molar Helmholtz free energy
G, g	Gibbs free energy, molar Gibbs free energy
H, h	enthalpy, molar enthalpy
S, s	entropy, molar entropy
U, u	internal energy, molar internal energy
R	gas constant
a	activity
f	fugacity
x	mole fraction
n	number of moles
n_T	total number of moles
z	compressibility factor
φ	fugacity coefficient
μ_i	chemical potential of phase component i
k, δ	binary interaction parameter

EOS specific properties

Γ	intermolecular potential
α	molecular polarizability
ε, σ	parameters of Lennard-Jones potential
μ	dipole moment
r	molecular distance
a, b	parameters of the van der Waals equation
B, C, D	parameters of the virial equation
y	hard-sphere parameter

$$\mu_i = \left(\frac{\partial G}{\partial n_i}\right)_{P,T,n_{j\neq i}} \tag{4}$$

For solids using the standard state of a pure phase component at reference conditions P_r and T_r, (i.e., 0.1 MPa, 298 K) integration over P leads to,

$$\mu_{i(P,T)} = h^o_{i(P_r,T)} - Ts^o_{i(P_r,T)} + \int_{P_r}^{P} v^o_i\,dP + RT\ln a_i \tag{5}$$

where h_i^o and s_i^o are the molar enthalpy and entropy at P_r and T, v_i^o the molar volume as a function of P at T of the pure phase component i, and a_i is its activity, which also incorporates in this case any volume effects of mixing.

However, for fluids the standard state is defined differently, and Equation (5) is not used in this form. The principal thermodynamic equations are rearranged in such a way that, at the standard state, the properties of a pure fluid are taken as if this fluid would behave like an ideal gas. For fluids, the standard state is, therefore, that of a *hypothetical ideal gas* at P_r at any required temperature T. The use of this standard state leads to the following expression,

$$\mu_{i(P,T)} = h^{o*}_{i(P_r,T)} - Ts^{o*}_{i(P_r,T)} + \int_{0}^{P}\left(\left(\frac{\partial V}{\partial n_i}\right)_{P,T,n_{j\neq i}} - \frac{RT}{P}\right)dP + RT\ln\frac{P}{P_r} + RT\ln x_i \tag{6}$$

in which the asterisk denotes the *hypothetical ideal gas* standard state properties. Details of the derivation of Equation (6), and Equation (11) in the following, can be found for example in Beattie (1955) or in Appendix A.

The integral term in Equation (6) provides the definition of the fugacity coefficient φ_i^m at P and T [2]

$$RT\ln\varphi^m_{i(P,T)} = \int_{0}^{P}\left(\left(\frac{\partial V}{\partial n_i}\right)_{P,T,n_{j\neq i}} - \frac{RT}{P}\right)dP \tag{7}$$

Thus Equation (6) can be written as

$$\mu_{i(P,T)} = h^{o*}_{i(P_r,T)} - Ts^{o*}_{i(P_r,T)} + RT\ln\frac{x_i\varphi^m_{i(P,T)}P}{P_r} \tag{8}$$

and with definition of the fugacity f_i^m

$$f^m_{i(P,T)} = x_i\varphi^m_{i(P,T)}P \tag{9}$$

Equation (9) becomes[3]:

$$\mu_{i(P,T)} = h^{o*}_{i(P_r,T)} - Ts^{o*}_{i(P_r,T)} + RT\ln\frac{f^m_{i(P,T)}}{P_r} \tag{10}$$

---

[2]   The lower integration boundary is strictly 0 MPa and not just some low pressure, so the fluid will behave like an ideal gas and that $f=p$ as often stated (e.g., Beattie 1955; Ferry and Baumgartner 1987; Anderson and Crerar 1993; Nordstrom and Munoz 1994; Anderson 2005). For the straightforward derivation, see Appendix A.

[3]   Note the term P_r in the denominator of Equation (10). In many textbooks this is term sloppily omitted, because the pressure unit employed is *bar* and the reference pressure is set to a value of 1 bar. If, however, SI units are used, P_r is 0.1 MPa and the symbol has to be carried along.

The difference between the thermodynamic properties of a real fluid and an ideal gas is then calculated by evaluating its fugacity. It must be emphasized that a real fluid has always a fugacity coefficient not equal to 1 ($\ln\varphi_i^m \neq 0$), also at P_r, but only at zero pressure and infinite volume.

The choice of the *hypothetical ideal gas* standard state, which is by no means an approximation but mathematically exact, has distinct advantages over the conventional standard state usually used for solids. First, the thermodynamic properties at standard conditions are those of an ideal gas, e.g., its heat capacity is that of an ideal gas with isolated atoms or molecules without any intermolecular interactions, which may be calculated for example by *ab initio* calculations and *molecular* simulations. Second, the evaluation of φ_i^m includes all excess mixing properties.

The calculation of φ_i^m using Equation (7) requires an EOS expression explicit in V, i.e., a function of P and T. As it will be shown below most EOS are explicit in P instead, i.e., a function of V and T, which often can not be transformed to V as a function of P and T. In that case, the following equation can be used instead (Appendix A):

$$\mu_{i(P,T)} = h^{o*}_{i(P_r,T)} - Ts^{o*}_{i(P_r,T)} + \int\limits_{V}^{\infty} \left(\left(\frac{\partial P}{\partial n_i} \right)_{V,T,n_{i\neq j}} - \frac{RT}{V} \right) dV - RT \ln \frac{Pv}{RT} + RT \ln \frac{x_i P}{P_r} \tag{11}$$

Equation (11) is normally written using the compressibility factor z, which is defined as

$$z = \frac{Pv}{RT} \tag{12}$$

where v is the molar volume ($v = V/n_T$). Inserting (12) into (11) leads to:

$$\mu_{i(P,T)} = h^{o*}_{i(P_r,T)} - Ts^{o*}_{i(P_r,T)} + \int\limits_{V}^{\infty} \left(\left(\frac{\partial P}{\partial n_i} \right)_{V,T,n_{i\neq j}} - \frac{RT}{V} \right) dV - RT \ln z + RT \ln \frac{x_i P}{P_r} \tag{13}$$

In this case the fugacity coefficient φ_i^m at V and T is defined as:

$$RT \ln \varphi_{i(V,T)}^m = \int\limits_{V}^{\infty} \left(\left(\frac{\partial P}{\partial n_i} \right)_{V,T,n_{i\neq j}} - \frac{RT}{V} \right) dV - RT \ln z \tag{14}$$

Using Equation (14), the simplified equations

$$\mu_{i(P,T)} = h^{o*}_{i(P_r,T)} - Ts^{o*}_{i(P_r,T)} + RT \ln \frac{x_i \varphi_{i(V,T)}^m P}{P_r} \tag{15}$$

and

$$\mu_{i(P,T)} = h^{o*}_{i(P_r,T)} - Ts^{o*}_{i(P_r,T)} + RT \ln \frac{f_{i(V,T)}^m}{P_r} \tag{16}$$

can be written if again the fugacity f_i is defined as:

$$f_{i(V,T)}^m = x_i \varphi_{i(V,T)}^m P \tag{17}$$

For the evaluation of the fluid properties, and hence also for any calculations of possible phase separations, the expressions (7) or (14) have to be calculated for every involved phase component. This calculation requires appropriate EOS for the respective fluids.

TYPES OF EOS

Many specific articles and reviews dealing with EOS exist in the chemical and physicochemical literature (e.g., the more recent reviews by Ferry and Baumgartner 1987, Anderko 1990, Sandler 1994, Prausnitz et al. 1999, Sengers et al. 2000, Wei and Sadus 2000). In the geological literature, knowledge about EOS and its state-of-the-art is much less widespread and, therefore, a brief overview is given here.

Ideal gas

The simplest EOS is that of an ideal gas,

$$V = \frac{n_T RT}{P} \tag{18}$$

or,

$$P = \frac{n_T RT}{V} \tag{19}$$

where R is the universal gas constant and n_T the total number of molecules. An ideal gas is characterized by the absence of any intermolecular forces and the particles themselves have zero volume. As such, the volume of an ideal gas is able to approach zero at a temperature of 0 K. Because of the non-existing intermolecular forces, an ideal gas never condenses, liquids can not be described and, as a consequence, fluid separation is not possible. Hence, the EOS assigns for each pressure only one specific volume (Fig. 1a). Furthermore because of the lack of intermolecular interaction, mixtures of ideal gases also mix ideally.

Virial equation

The ideal gas law can be expanded in a MacLaurien series in powers of volume,

$$P = \frac{n_T RT}{V}\left(1 + \frac{n_T B_{(T)}}{V} + \frac{n_T^2 C_{(T)}}{V^2} + \frac{n_T^3 D_{(T)}}{V^3} \cdots \right) \tag{20}$$

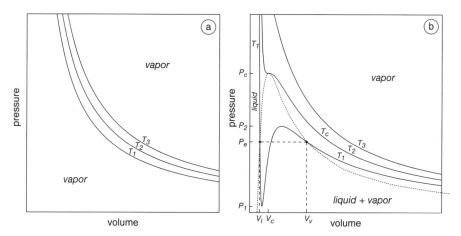

Figure 1. vT-relationships for a) an ideal gas and b) for a van der Waals fluid. For temperatures below the critical temperature in certain pressure regions, the van der Waals equation provides two valid solutions for v. The pressure P_e, at which liquid and vapor are in equilibrium can be calculated by considering Equation (26). The dashed curve designates the boundary of the field of two coexisting fluids.

in which B, C, D, $\cdots$ are volume-independent but temperature-dependent coefficients. The virial EOS has a rigorous theoretical foundation in statistical thermodynamics. The virial coefficients are directly related to interactions between isolated molecule clusters. The parameter B represents interaction between molecule pairs, C represents interaction of three molecules, D for four molecules, and so forth. The parameters for a fluid are either determined using experimental results or calculated by statistical mechanic methods. Whereas the virial equation is theoretically sound, its applicability is limited to fluids at conditions of low and medium densities.

Using the compressibility factor z (see Eqn. 12), Equation (20) can be expressed also as:

$$z = 1 + \frac{n_T B_{(T)}}{V} + \frac{n_T^2 C_{(T)}}{V^2} + \frac{n_T^3 D_{(T)}}{V^3} \cdots \tag{21}$$

Another form of the virial equation can be written if z is expressed as a power series of P

$$z = 1 + B'P + C'P^2 + D'P^3 + \cdots \tag{22}$$

The parameters in Equations (21) and (22) are related if they are determined from experimental results for z which were determined for V going to infinity and P to zero, respectively (see also Prausnitz et al. 1999, Appendix C).

In that case the following relationships hold:

$$B' = \frac{B}{RT}, \quad C' = \frac{C - B^2}{(RT)^2}, \quad D' = \frac{D - 3BC + 2B^3}{(RT)^3} \tag{23}$$

More details concerning the virial EOS are given for example in Trusler (2000).

Cubic EOS

The first equation that was able to reasonably describe the properties of both a vapor and a liquid phase was proposed by van der Waals (1873)

$$P = \frac{n_T RT}{V - n_T b} - \frac{n_T^2 a}{V^2} \tag{24}$$

In this equation, the volume available to the particles is reduced by a characteristic co-volume b, which is related to the volume required by the particles themselves. This parameter stands for repulsion forces between particles. Additionally, the pressure is decreased by intermolecular interaction, i.e., attractive forces, by the term $n_T^2 a / V^2$. In this case a is the characteristic interaction parameter for the respective fluid.

Equation (24) can be arranged to yield a cubic form in V,

$$-PV^3 + (n_T bP + n_T RT)V^2 - n_T^2 aV + abn_T^3 = 0 \tag{25}$$

therefore, an EOS, such as the van der Waals equation, is also called a cubic EOS. The cubic form of Equation (25) has the property that analytical solutions for the three roots in V exist. For certain regions in P and T, two of these roots are imaginary. Figure 1b illustrates this result. For temperatures above the critical temperature T_c, there always exists only one real root, and for each pressure, only one real volume. Below T_c there are pressure regions in which all three roots are real as, for example, in Figure 1b for pressures between P_1 and P_2. However, only one pressure in this interval, i.e., P_e, yields the valid saturated vapor pressure for the specific fluid at a chosen temperature. To calculate P_e, an additional condition has to be met. The volume work at a constant temperature T during transformation from the liquid to vapor state, i.e., evaporation,

$$w = \left(V_v - V_l \right) P_e \tag{26}$$

has to be equal to the same work calculated using the van der Waals EOS

$$w = \int_{V_l}^{V_v} \left(\frac{n_T RT}{V - n_T b} - \frac{n_T^2 a}{V^2} \right) dV \tag{27}$$

Then, for coexisting phases the lowest root corresponds to the volume of the liquid and the highest to that of the vapor. At T_c, all three roots, i.e., volumes, are identical.

If a is a special function of T, a further advantage of a cubic EOS is, that the parameters a and b for each pure fluid can be extracted from its critical data. The critical point (P_c and T_c) is defined by the following conditions:

$$\left(\frac{\partial P}{\partial v} \right)_{T_c} = \left(\frac{\partial^2 P}{\partial v^2} \right)_{T_c} = 0 \tag{28}$$

If Equation (28) is applied to a pure fluid the parameters a and b can be determined if P_c and T_c are known. In the case of the van der Waals equation a and b are:

$$a = \frac{27}{64} \frac{R^2 T_c^2}{P_c} \tag{29}$$

$$b = \frac{1}{8} \frac{RT_c}{P_c} \tag{30}$$

The van der Waals parameters a and b are also related to the second viral coefficient B:

$$B = \frac{bV}{V - n_T b} - \frac{a}{RT} \tag{31}$$

B, however, should only be a function of T. This can be accomplished if the relation $V \gg n_T b$ in Equation (31) is used, which is at least valid for low P and large V

$$B = b - \frac{a}{RT} \tag{32}$$

Whereas the van der Waal EOS is capable of predicting vapor-liquid equilibria qualitatively, it is not suitable for quantitatively evaluating phase equilibria. Therefore a wide range of modifications of the attractive part of the van der Waals EOS exists in the literature (e.g., Anderko 2000; Wei and Sadus 2000).

Redlich and Kwong (1949) introduced successfully a temperature dependence to the attractive interaction term

$$P = \frac{n_T RT}{V - n_T b} - \frac{n_T^2 a}{\sqrt{T} V (V + n_T b)} \tag{33}$$

Because of the success of the Redlich-Kwong equation, Soave (1972) proposed a more generalized form of the attractive term than used in Equation (33),

$$P = \frac{n_T RT}{V - n_T b} - \frac{n_T^2 a(T)}{V (V + n_T b)} \tag{34}$$

with

$$a(T) = 0.4275 \left(\frac{R^2 T_c^2}{P_c} \right) \left[1 + m \left(1 - \left(\frac{T}{T_c} \right)^{\frac{1}{2}} \right) \right]^2 \tag{35}$$

$$m = 0.480 + 1.574\omega - 0.175\omega^2 \tag{36}$$

$$b = 0.08664 \left(\frac{R T_c}{P_c} \right) \tag{37}$$

within which the constants in Equations (35) and (37) are derived from the critical parameters using Equation (28) as for the van der Waals equation, and ω is an empirical parameter describing the acentric factor of the molecules, i.e., deviation from a spherical form.

Experimentally observed z-values (Eqn. 12), at the critical point of pure fluids, i.e., z_c, with geological relevance, range between 0.229 for H_2O and 0.306 for H_2 (Poling et al. 2000). In contrast, the van der Waals EOS generally predicts a z_c-value of 0.375 and the Redlich-Kwong and Soave EOS a z_c of 0.333, which are too large. Peng and Robinson (1976) altered the attractive part of the van der Waals EOS in such a way that z_c becomes 0.307:

$$P = \frac{n_T R T}{V - n_T b} - \frac{n_T^2 a}{V(V - n_T b) + n_T b(V + n_T b)} \tag{38}$$

with:

$$a(T) = 0.45724 \left(\frac{R^2 T_c^2}{P_c} \right) \left[1 + m \left(1 - \left(\frac{T}{T_c} \right)^{\frac{1}{2}} \right) \right]^2 \tag{39}$$

$$m = 0.37464 + 1.5422\omega - 0.26922\omega^2 \tag{40}$$

$$b = 0.07780 \left(\frac{R T_c}{P_c} \right) \tag{41}$$

The Peng-Robinson EOS seems to be better suited for hydrocarbons than the Soave EOS down to intermediate values for the acentric factor.

Abbott (1979) showed that any cubic EOS, with the appropriate asymptotic behavior as V approaches b, can be written in a generic form

$$P = \frac{n_T R T}{V - n_T b} - \frac{n_T^2 \theta (V - n_T \eta)}{(V - n_T b)(V^2 + \delta n_T V + n_T^2 \varepsilon)} \tag{42}$$

in which the parameters b, δ, ε, θ, and η depend generally on T and composition and are individual for each fluid species. Equation (42) can be considered as a generalization of the van der Waals equation, and EOS such as (24), (33), (34), and (38) can be expressed by Equation (42) using a proper choice for the parameters b, δ, ε, θ, and η.

Further cubic EOS with three and more parameters are discussed in Anderko (2000).

Hard-sphere extension of the EOS

A notable and quite successful modification of the repulsive term is the accurate and rigorous hard-sphere extension by Carnahan and Starling (1969). The result of their computer simulations by treating the molecules as hard spheres yielded as a result the following term for the repulsive interaction:

$$P = \frac{n_T RT}{V}\left(\frac{1 + y + y^2 - y^3}{(1-y)^3}\right) \tag{43}$$

with:

$$y = \frac{n_T b}{4V} = \frac{b}{4v} \tag{44}$$

Equation (43) can be combined with the different attractive terms of the cubic EOS. In such a way a modified van der Waals EOS

$$P = \frac{n_T RT}{V}\left(\frac{1 + y + y^2 - y^3}{(1-y)^3}\right) - \frac{n_T^2 a}{V^2} \tag{45}$$

or modified Redlich-Kwong EOS

$$P = \frac{n_T RT}{V}\left(\frac{1 + y + y^2 - y^3}{(1-y)^3}\right) - \frac{n_T^2 a_{(T)}}{V(V + n_T b)} \tag{46}$$

can be formulated. Equations such as (45) or (46) are, however, no longer cubic equations in volume, but rather expressions to the power of 5. As a consequence, no analytical roots for V exist anymore, instead they have to be searched for numerically. However, with the computer power available today, this search is not a problem.

Integrated EOS

Another approach to derive an EOS is to consider the Helmholtz free energy A of a fluid directly. Thereby, the intermolecular potentials Γ are used in their integrated and averaged form. Figure 2 illustrates some simple potentials. The simplest potential for repulsion is that of hard spheres (Fig. 2a):

$$\Gamma_{(r)} = \begin{cases} \infty & \text{for} \quad r \leq \sigma \\ 0 & \text{for} \quad r > \sigma \end{cases} \tag{47}$$

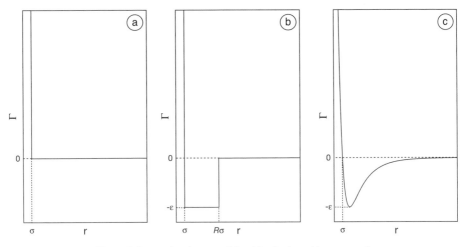

Figure 2. Intermolecular potentials: a) hard sphere, b) square well, and c) Lennard-Jones. For discussion see text.

This function just means that the potential is infinity for an intermolecular distance r below σ, which is the collision diameter. Carnahan and Starling (1969) have shown that, for this potential, an exact solution exists (see Eqn. 43).

A very simplified attraction term can be added to this potential, (Fig. 2b), which describes a potential with a minimum of ε between σ and $R\sigma$. Mathematically such a square-well potential can be described by:

$$\Gamma_{(r)} = \begin{cases} \infty & \text{for} & r \leq \sigma \\ -\varepsilon & \text{for} & \sigma < \varepsilon \leq R\sigma \\ 0 & \text{for} & r > R\sigma \end{cases} \tag{48}$$

More realistic is the Lennard-Jones potential (Fig. 2c):

$$\Gamma_{(r)} = 4\varepsilon \left(\left(\frac{\sigma}{r} \right)^{12} - \left(\frac{\sigma}{r} \right)^{6} \right) \tag{49}$$

In the Lennard-Jones potential, the repulsive wall is not vertical but goes with the power of 12 and allows some, but energetically costly, penetration of the two molecules. The attraction goes with the power of 6 and models the dispersion forces.

As for the cubic EOS, the individual contributions to A are added. For the simplest case A is just the sum of the contribution of the ideal gas A^{ideal} and some residual term $A^{residual}$

$$A = A^{ideal} + A^{residual} \tag{50}$$

Whereas the ideal contribution is calculated using the ideal standard state properties by integrating the ideal gas law, the residual term is split into well-known contributions, which are treated as the reference frame and a term which should describe the rest, the perturbation term,

$$A^{residual} = A^{reference} + A^{perturbation} \tag{51}$$

such that:

$$A = A^{ideal} + A^{reference} + A^{perturbation} \tag{52}$$

Possible terms for $A^{residual}$, besides others, are those from hard-sphere repulsion, dispersion, electrostatic interactions like dipole-dipole, induced dipole-dipole, quadrupole-quadrupole, dipole-quadrupole, and induced dipole-quadrupole interactions. Empirical terms can also be formulated

$$A^{residual} = A^{hs} + A^{dis} + A^{dip-dip} + A^{ind-dip} + A^{quad-quad} + A^{quad-dip} + A^{ind-quad} + \cdots + A^{empirical} \tag{53}$$

Perturbation theory is often used (e.g., Boublík 2000) in conjunction with integrated EOS from which the term $A^{perturbation}$ is named. In an oversimplified manner, the perturbation terms stem from series expansions of a well-known reference frame.

An example for an $A^{residual}$-term is the application of the hard-sphere potential from Equation (43). Subtraction of the ideal gas part and integration over the proper boundaries yields the expression A^{hs}

$$A^{hs} = n_T a^{hs} = \int_V^{\infty} \left(\frac{n_T RT}{V} \left(\frac{1 + y + y^2 - y^3}{(1-y)^3} \right) - \frac{n_T RT}{V} \right) dV = n_T RT \frac{4y - 3y^2}{(1-y)^2} \tag{54}$$

In conjunction with an integrated EOS, A is often expressed as a molar function a using the molar volume v or ρ like $a(\rho, T, x_i)$ or $a(v, T, x_i)$ as in Equation (54) instead of $A(V, T, n_i)$. In that case the fugacity coefficient φ_i^m can be calculated for use in Equation (15) with

$$\ln \varphi_{i(V,T,x_i)}^m = \left(\frac{\partial n_T a_{(\rho\, or\, v,T,x)}^{residual} \big/ RT}{\partial n_i} \right)_{\rho\, or\, v,T,n_{j \ne i}} - \ln z + z - 1 \tag{55}$$

instead of Equation (14). Details and derivation are in Appendix A.

MIXING AND COMBINING RULES

The subject of this volume is fluid mixtures and their possible segregation of separate phases. The phase rule predicts that, in a fluid consisting of two and more phase components, at constant P and T, coexistence of fluids with different compositions is in principle possible. As a result the thermodynamic evaluation of phase equilibria requires EOS of such mixtures. Such an EOS for fluid mixtures can be obtained for example if, for the required phase components, i.e., species or molecules, the respective EOS for the pure phase components are available, which have the same mathematical form. In this case, at least in principle, the necessary EOS parameters of each pure species can be combined in one way or another, and parameters for the respective mixtures may be obtained.

The question arises as to how the EOS parameters should be combined. The respective hints come from the theory behind the virial EOS (Eqn. 20) which is theoretically rigorous. As described above the parameters B, C, D, and so forth represent the 2-, 3-, 4-, and higher body interactions in a virial EOS. It is evident that in mixtures the number of these interactions depend on the quantities of the actual species. It can be shown that the virial interaction parameters in a mixture have to be calculated by summing over all products of the probabilities of the respective species combinations times its interaction parameter

$$B_{(T,x)} = \sum_i \sum_j x_i x_j B_{ij(T)} \tag{56}$$

$$C_{(T,x)} = \sum_i \sum_j \sum_k x_i x_j x_k C_{ijk(T)} \tag{57}$$

$$D_{(T,x)} = \sum_i \sum_j \sum_k \sum_l x_i x_j x_k x_l D_{ijkl(T)} \tag{58}$$

$$\vdots$$

Because Equations (56) to (58) are exact and the virial EOS has been proven to work well at least at middle and low densities, the such predicted behavior for mixtures should be considered at least as the low-pressure boundary condition for other types of EOS when expanded to the virial form (Sandler and Orbey 2000). However, these equations provide no information *per se* about the way in which the mixed species parameters such as B_{ij}, C_{ijk}, and D_{ijkl} are related to the respective parameters of the pure species, i.e., B_{ii}, B_{jj} or C_{iii}, C_{jjj}, C_{kkk} and so forth.

Mathematically it can be shown that the van der Waals equation can be expressed as a Taylor series:

$$\frac{PV}{n_T RT} = \sum_{m=0}^{\infty} \left(\frac{n_T b}{V} \right)^m - \frac{n_T a}{RTV} \tag{59}$$

As in Equation (32), Equation (59) shows the relationship between the virial equation (Eqn. 20) and the van der Waals equation (24). If Equation (59) is expanded up to $m = 1$ then:

$$B_{(T,x)} = \sum_i \sum_j x_i x_j B_{ij(T)} = b - \frac{a}{RT} = \sum_i \sum_j x_i x_j \left(b - \frac{a}{RT} \right)_{ij} \tag{60}$$

Equation (60) implies that:

$$a = \sum_i \sum_j x_i x_j a_{ij} \tag{61}$$

$$b = \sum_i \sum_j x_i x_j b_{ij} \tag{62}$$

However, if Equation (59) is expanded to higher order terms then:

$$C_{(T,x)} = \sum_i \sum_j \sum_k x_i x_j x_k C_{ijk(T)} = b^2 \tag{63}$$

$$D_{(T,x)} = \sum_i \sum_j \sum_k \sum_l x_i x_j x_k x_l D_{ijkl(T)} = b^3 \tag{64}$$

$$\vdots$$

It is obvious that there exists no compositional dependence for b that satisfies Equations (62), (63), (64) and so on, simultaneously. In other words, for the van der Waals EOS, the compositional dependence of Equations (63), (64) and all higher order components are not consistent with Equation (62).

As stated above, the virial equation provides no clue for combining rules to extract values for the parameters such as B_{ij}, C_{ijk}, and D_{ijkl}. As a consequence, there is no immediate hint for the calculation of a_{ij} and b_{ij} for the van der Waals equation or any other EOS for application to mixtures.

Practically, the most commonly used combining rules for the van der Waals and other EOS are,

$$a_{ij} = \left(a_i a_j\right)^{\frac{1}{2}} \left(1 - k_{ij}\right) \tag{65}$$

$$b_{ij} = \frac{1}{2}\left(b_i + b_j\right)\left(1 - \delta_{ij}\right) \tag{66}$$

where k_{ij} and δ_{ij} are binary-interaction parameters which are usually obtained by fitting the EOS to experimental results. For the case that k_{ij} and δ_{ij} are 0, Equations (61) and (62) become:

$$a = \sum_i \sum_j x_i x_j \left(a_i a_j\right)^{\frac{1}{2}} \tag{67}$$

$$b = \sum_i x_i b_i \tag{68}$$

Equations (67) and (68), the geometric mean and arithmetic mean, respectively, are the classical van der Waals mixing rules that are often also used for parameters occurring in cubic and other EOS. They are, however, only usable for moderate non-idealities.

The justification for Equations (65) and (66) is difficult. Whereas the mixed inter-molecular interaction parameter a_{ij} in Equation (65) is related to and may be deduced from intermolecular potential theory, Equation (66) describes the excluded-volume parameter b_{ij} as if the molecules are hard spheres and δ_{ij} is 0. Equation (66) is, however, not unequivocal and it can be argued that even for hard spheres, the diameter is of importance and not the volume (Sandler and Orbey 2000). In that case b_{ij} would be:

$$b_{ij} = \left(b_i^{1/3} + b_j^{1/3}\right)^3 \tag{69}$$

Also with the proper arguments, other combining rules are possible.

Boublík (1970) and Mansoori et al. (1971) extended the hard-sphere concept to mixtures and derived the following expression which replaces Equation (43) for mixtures

$$P = \frac{n_T RT}{V} \left(\frac{1 + \left(3\frac{DE}{F} - 2\right)y + \left(3\frac{E^3}{F^2} - 3\frac{DE}{F} + 1\right)y^2 - \left(\frac{E^3}{F^2}\right)y^3}{(1-y)^3} \right) \tag{70}$$

With y defined as in Equation (44) and:

$$D = \sum_{i=1}^{m} x_i \sigma_i, \quad E = \sum_{i=1}^{m} x_i \sigma_i^2, \text{ and } \quad F = \sum_{i=1}^{m} x_i \sigma_i^3 \tag{71}$$

Additionally:

$$b = \frac{2}{3}\pi N_a F \tag{72}$$

Here σ_i is the hard-sphere diameter. The integrated form of the hard-sphere repulsion term is then

$$A^{hs} = n_T RT \left(\frac{\frac{3DE}{F}y - \frac{E^3}{F^2}}{1-y} + \frac{\frac{E^3}{F^2}}{(1-y)^2} + \left(\frac{E^3}{F^2} - 1\right)\ln(1-y) \right) \tag{73}$$

with y defined as in Equation (44). For mixtures, Equation (73) can be used instead of Equation (54).

More detailed reviews dealing with EOS parameters in mixtures and combining rules can be found in Matteoli et al. (2000) and Sandler and Orbey (2000).

SPECIFIC EOS FOR MIXTURES APPLICABLE TO GEOLOGIC SETTINGS

For the calculation of fugacity coefficients, the integrals in the Equations (7) or (14) have to be solved or the respective Helmholtz free energy expressions for Equation (53) are required, both of which need appropriate EOS. However, not all EOS are equal. There exist EOS in the literature that are tuned for special purposes, e.g., EOS which reproduce precise *PVT*-relationships only in restricted regions, for example for fluid inclusion research or EOS for studies of critical behavior. Such EOS are ill-suited for use in Equations (7) or (14), because any appropriate EOS must cover the whole range from 0 to the required P.

This volume covers coexisting fluids, and as stated before, EOS that are able to deal with fluid mixtures are required for equilibrium calculations. But in any case, the necessary boundary conditions are that any such EOS must also equally well deal with the respective pure phase components. This requirement can easily be checked with accurate EOS for pure phase components (species) which are available in the literature. For example, for H_2O, CO_2, CH_4, O_2, N_2, and Ar, accurate EOS are available from Wagner and Pruß (2002), Span and Wagner (1996), Setzmann and Wagner (1991), Schmidt and Wagner (1984), Span et al. (2000), and Tegeler et al. (1999), respectively[4]. These dedicated EOS for distinct species cannot be

---

[4] For many discussed EOS, programs running with *Mathematica*® are provided in conjunction with this volume as electronic supplements.

used for mixtures, however, except if ideal mixing is assumed or the mixing properties are determined and available otherwise. Ideal mixing does not induce phase separation, however.

For geologic, geochemical, and petrologic applications useful EOS should cover the occurring species and be valid up to the relevant P and T conditions. In the following, the presented EOS[5] mostly cover combinations of the phase components H_2O, CO_2, CH_4, CO, H_2, O_2, N_2, Ar, He, Ne, NaCl, and KCl. EOS for hydrocarbons are handled in detail in Pedersen and Christensen (2007). The applicable maximum pressure should be at least 100 MPa or higher and maximum temperatures should be above 800 °C. In the following, emphasis is placed on the background and kind of formulation of the EOS. Quality and success of the EOS are hard to evaluate and may vary with the specific requirements of applications.

Holloway (1976), Holloway (1981)

Phase components: H_2O, CO_2, CH_4, CO, H_2, N_2, H_2S, SO_2.
Applicable range given by the author: 723-2073 K, up to 4 GPa.

Based on the work of De Santis et al. (1974), Holloway proposed an EOS using a modified Redlich-Kwong equation in which the a parameter is also a function of T

$$P = \frac{n_T RT}{V - n_T b} - \frac{n_T^2 a_{(T)}}{\sqrt{T} V (V + n_T b)} \tag{74}$$

The parameter a is split into a constant a_0 and polynomial a_1

$$a_{(T)} = a^0 + a_{(T)}^1 \tag{75}$$

The mixing and the combining rules are those from Equations (67) and (68), except for the subsystem H_2O-CO_2 for which a modified expression of the type of Equation (65) is used

$$a_{H_2O\text{-}CO_2} = \left(a_{H_2O}^0 a_{CO_2}^0 \right)^{\frac{1}{2}} + \frac{1}{2} R^2 T^{5/2} K \tag{76}$$

Here K is the equilibrium constant describing the observed complex formation between H_2O and CO_2. For combinations of H_2O or CO_2 with other species the cross term is defined as:

$$a_{H_2O\text{-}j} = \left(a_{H_2O}^0 a_j \right)^{\frac{1}{2}} \quad \text{or} \quad a_{CO_2\text{-}j} = \left(a_{CO_2}^0 a_j \right)^{\frac{1}{2}} \tag{77}$$

Using this EOS, considerable deviations from ideal mixing for the H_2O-CO_2 system have been calculated by Holloway (1976), but no unmixing seems to be predicted in the EOS validity range. Flowers (1979) pointed out that the equation used by Holloway (1976) to calculate the fugacity coefficients did not obey the Gibbs-Duhem relation and was in error. The corrected EOS predicts considerably more ideal fugacity coefficients than the uncorrected EOS.

Bowers and Helgeson (1983)

Phase components: H_2O, CO_2, NaCl.
Applicable range given by the authors: between 623 and 873 K, above 50 MPa and no particu-
lar maximum P given, maximum x_{NaCl} = 0.14 (35 wt%) relative to H_2O+NaCl.

This EOS is very similar to the approach taken by Holloway (1976). It uses a modified Redlich-Kwong equation like Equation (74). Differences are in the calculation of the cross terms with H_2O when applying the Equations (61) and (62). The a_{H_2O} parameter is split into a pure function a_{H_2O} and a cross term function $a^°_{H_2O}$, both of which depend on temperature. Together with the parameter b_{H_2O} these are also functions of w, where w corresponds to the

---

5   The available EOS are numerous and the following presentation can hardly be complete.

weight fraction of NaCl relative to $H_2O+NaCl$. The $a_{H_2O\,CO_2}$ term is calculated according Equation (76). The EOS does consider the phase component NaCl only implicitly using the parameter w, and does not use terms such as a_{NaCl}, $a^°_{NaCl}$, or b_{NaCl}. For this reason, the EOS is not a true ternary EOS. Whereas the EOS reproduces the molar volumes in its validity region well, it turns out that it can not be applied to calculate phase relations in the ternary. As a consequence, calculated values for φ_i^m are also doubtful. A FORTRAN program to calculate fluid properties using this EOS is provided by Bowers and Helgeson (1985). Bakker (1999) extended this EOS and added the phase components CH_4 and N_2.

Kerrick and Jacobs (1981), Jacobs and Kerrick (1981a,b)

Phase components: H_2O, CO_2, CH_4.
Applicable range given by the authors: up to 573-1323 K, at least 2 GPa.

This EOS uses a modified Redlich-Kwong equation including a hard-sphere term and the parameter a is a function of V and T

$$P = \frac{n_T RT}{V}\left(\frac{1+y+y^2-y^3}{(1-y)^3}\right) - \frac{n_T^2 a_{(V,T)}}{V(V+n_T b)} \tag{78}$$

with

$$a_{(V,T)} = c + \frac{n_T d}{V} + \frac{n_T^2 e}{V^2} \tag{79}$$

Here the parameters c, d, and e are cubic equations of T, and y is defined according to Equation (44). The mixing rules are claimed to be according to Equations (67) and (68). But from the Appendix in Kerrick and Jacobs (1981) and the program provided by Jacobs and Kerrick (1981a) it is evident that for the cross terms instead of Equation (67) relations of the type

$$c_{ij} = \left(c_i c_j\right)^{1/2}, \quad d_{ij}\left(d_i d_j\right)^{1/2}, \quad e_{ij}\left(e_i e_j\right)^{1/2} \tag{80}$$

are used. Thus polynomial cross terms were calculated instead. For mixtures this calculation produces a significantly different numerator in Equation (78) than using the appropriate van der Waals mixing rule. Furthermore, because Equation (79) is an arbitrary polynomial, and the attractive interaction a is always positive, some of the parameters c, d, and e might be negative. There is the inherent possibility, that as a side effect for mixtures, the cross terms in Equation (80) will result in a negative root and will be therefore imaginary. Furthermore, if for two phase components the specific parameters are both negative, according to Equation (80) inconsistent positive cross terms will result. Applying this cross terms to a pure phase with a negative c_{ii}, d_{ii}, or e_{ii} will result in a positive one. Kerrick and Jacobs (1981) took care that the problem did not arise by making sure that the parameters c, d, and e are positive in the validity region of the EOS. However, extrapolation and extension by adding additional species is error-prone besides the used rules are not the van der Waals mixing rules.

The calculated activity coefficients using this EOS show significant positive deviations from ideal mixing at elevated pressures, and unmixing is implied at temperatures <400 °C.

Rimbach and Chatterjee (1987)

Phase components: H_2O, H_2.
Applicable range given by the authors: above 273 K, up to 10 and 20 GPa for H_2 and H_2O,
  respectively.

This EOS uses a modified Redlich-Kwong in which the parameter a is a function of P and T and b only of P

$$P = \frac{n_T RT}{V - n_T b_{(P)}} - \frac{n_T^2 a_{(P,T)}}{\sqrt{T} V(V + n_T b_{(P)})} \tag{81}$$

The chosen mixing and combining rules are those of Equations (61), (62), (65), and (66) using fitted values for k_{ij} and δ_{ij}.

Grevel and Chatterjee (1992)

Phase components: H_2O, H_2.
Applicable range given by the authors: above 673 K and at least 3300 K, at least 10 GPa.

This EOS is based on the approach taken by Halbach and Chatterjee (1982) for pure H_2O and uses a modified Redlich-Kwong equation in which the parameters a and b are functions of P and T

$$P = \frac{n_T RT}{V - n_T b_{(P,T)}} - \frac{n_T^2 a_{(P,T)}}{\sqrt{T} V(V + n_T b_{(P,T)})} \tag{82}$$

For the phase component H_2O the PT-space is split into two regions. For temperatures below 400 °C the vapor saturation curve and for temperatures above a linear relationship is used as a divider. In each region different functions for a and b were added to the function derived by Halbach and Chatterjee (1982). For H_2 single functions for a and b were derived. Van der Waals mixing rules according to Equations (67) and (68) were applied for the mixtures.

The choice to make a and b a function of P, as in Rimbach and Chatterjee (1987) and Halbach and Chatterjee (1982), makes integration not straightforward, because the EOS can not be solved either for P or V and must be therefore performed numerically. As a result of the different PT-regions for a and b used for H_2O, discontinuities in V are created which are also not advantageous for the calculation of fluid properties such as fugacity coefficients, and introduces systematic errors for pure as well as mixture properties.

Grevel (1993)

Phase components: H_2O, CH_4.
Applicable range given by the author: above 150 K and at least 2273 K, up to 23 GPa.

Like the EOS of Grevel and Chatterjee (1992), Grevel (1993) uses a modified Redlich-Kwong equation in which both parameters a and b are functions of P and T

$$P = \frac{n_T RT}{V - n_T b_{(P,T)}} - \frac{n_T^2 a_{(P,T)}}{\sqrt{T} V(V + n_T b_{(P,T)})} \tag{83}$$

As for H_2O in Grevel and Chatterjee (1992), the PT-space is split into several regions for CH_4 with different functional dependencies for the parameter a. For higher P and T the parameter a_{CH_4} becomes negative. Therefore instead of the actual van der Waals mixing rules in Equations (67) and (68) for $a_{H_2O-CH_4}$, the relation

$$a_{H2O-CH4} = |a_{H2O} a_{CH4}|^{1/2} \tag{84}$$

is used. This choice is arbitrary and introduces discontinuities in isochore slopes. Furthermore, mixed and pure systems do not converge. The significance of the calculated φ_i^m-values is therefore questionable.

Spycher and Reed (1988)

Phase components: H_2O, CO_2, CH_4.
Applicable range given by the authors: up to 1273 K, up to 50 MPa for mixtures.

The EOS is based on a virial expansion in P

$$z = 1 + B'_{(T)}P + C'_{(T)}P^2 \tag{85}$$

in which the parameters B' and C' are polynomial functions of T. As mixing rules, Equations (56) and (57) were used and the required cross terms were fitted to experimental results as functions of T. It must be noted that the mixing rules (56) and (57) are sensu stricto valid only for virial equations of the type of Equation (21), and not for Equation (85) as used here.

Saxena and Fei (1987, 1988)

Phase components: H_2O, CO_2, CH_4, CO, H_2, O_2
Applicable range given by the authors: above 400 K for all but for H_2O (>673 K), no explicit pressure and temperature limit is provided but according to Saxena and Fei (1987), the EOS is valid for crustal conditions.

The properties of the pure phase components and those of the mixture are calculated separately. The pure components are treated with a virial EOS which uses the reduced pressure $P_r = P/P_c$ as the variable (Saxena and Fei 1987),

$$z = A'_{(T_r)} + B'_{(T_r)}P_r + C'_{(T_r)}P_r^2 \tag{86}$$

in which the parameters A', B', and C' are functions of the reduced temperature $T_r = T/T_c$. P_c and T_c are the critical pressure and temperature for each pure phase component.

For non-polar molecules the mixture properties are calculated using a Kihara potential,

$$\Gamma_{(r)} = \begin{cases} \infty & \text{for } r < 2a \\ 4\varepsilon\left(\left(\dfrac{\sigma - 2a}{r - 2a}\right)^{12} - \left(\dfrac{\sigma - 2a}{r - 2a}\right)^{6}\right) & \text{for } r \geq 2a \end{cases} \tag{87}$$

which is similar to the Lennard-Jones potential described previously, but does not theoretically allow complete penetration of the molecules. The parameter $2a$ is the spherical molecular core diameter which describes the maximum penetration depth. For interactions with involvement of polar molecules, a Stockmayer potential is used

$$\Gamma_{(r)} = 4\varepsilon\left(\left(\frac{\sigma}{r}\right)^{12} - \left(\frac{\sigma}{r}\right)^{6}\right) - \frac{\mu^2}{r^3} F(\theta_1, \theta_2, \theta_3) \tag{88}$$

The Stockmayer potential is a modified Lennard-Jones potential which considers the direction $(\theta_1, \theta_2, \theta_3)$ dependent dipole-dipole interactions and includes the dipole moment μ of the polar species.

The mixing rules for the pure non-polar and pure polar interactions are:

$$r_{ij} = \frac{1}{2}(r_i + r_j), \quad \sigma_{ij} = \frac{1}{2}(\sigma_i + \sigma_j), \quad \varepsilon_{ij} = \left(\varepsilon_i \varepsilon_j\right)^{1/2}, \quad \mu_{ij} = \left(\mu_i \mu_j\right)^{1/2} \tag{89}$$

For a pair of polar (i) and non-polar (j) molecules, the following relation is used:

$$\sigma_{ij} = \frac{\left(\sigma_i + \sigma_j\right)}{2\left(1 + \dfrac{\alpha_{jj}\mu_{ii}\left(\varepsilon_{ii}/\varepsilon_{jj}\right)^{1/2}}{4\sigma_{jj}^3\left(\varepsilon_{ii}\sigma_{ii}^3\right)^{1/2}}\right)^{1/6}} \tag{90}$$

in which α is the polarizability of the non-polar molecule.

The calculated mixing properties for each binary system are expressed by a van Laar equation using a binary interaction parameter W_{ij} and the molar volumes of the pure phase components v_k^o,

$$G_{ij}^{excess} = \frac{n_T W_{ij} x_i x_j v_i^o v_j^o}{\left(v_i^o + v_j^o\right)\left(x_i v_i^o + x_j v_j^o\right)} \tag{91}$$

which in turn can be combined to ternary and higher systems. Because the calculated energies are two-body interactions, ternary and higher parameters are neglected. Appropriate differentiation in respect to n_i yields the required expressions for the activity constants.

Shi and Saxena (1992)

Phase components: H_2O, CO_2, CH_4, CO, O_2, H_2, S_2, SO_2, COS, H_2S.
Applicable range given by the authors: up to 2500 K, up to 2 GPa.

The calculation of the pure and mixed fluid properties is, with slight modifications, identical to Saxena and Fei (1987, 1988) described above. For pure CO_2, CH_4, CO, O_2, H_2, S_2, SO_2, COS, and H_2S a virial equation of the type

$$z = A'_{(T_r)} + B'_{(T_r)} P_r + C'_{(T_r)} P_r^2 + D'_{(T_r)} P_r^3 \tag{92}$$

is used, in which the parameters A', B', C', and D' are functions of T_r. The chosen functional dependence for each parameter is

$$Q_{(T)} = Q_1 + Q_2 T_r^{-1} + Q_3 T_r^{-3/2} + Q_4 T_r^{-3} + Q_5 T_r^{-4} \tag{93}$$

for $P < 100$ MPa and

$$Q_{(T)} = Q_1 + Q_2 T_r + Q_3 T_r^{-1} + Q_4 T_r^2 + Q_5 T_r^{-2} + Q_6 T_r^3 + Q_7 T_r^{-3} + Q_8 \ln T_r \tag{94}$$

or $P \geq 100$ MPa. The properties of H_2O are calculated using the EOS of Saul and Wagner (1989).

The properties of the fluid mixtures are calculated with the approach given by Saxena and Fei (1988), as described above.

Belonoshko and Saxena (1992c), Belonoshko et al. (1992)

Phase components: H_2O, CO_2, CH_4, CO, O_2, H_2, S_2, SO_2, COS, H_2S, N_2, NH_3, Ar.
Applicable range given by the authors: up to 2273 K for 100 MPa, 3273 K for 500 MPa, 4000 K for up to 100 GPa, lower pressure limit 500 MPa.

This EOS is based on publications of Belonoshko and Saxena (1992a,b) and does not fit into the pattern outlined in the beginning of this chapter. It is based on intermolecular potentials such as the integrated EOS, but uses results from molecular dynamical simulations as input instead. The intermolecular potential applied is of the α-*Exp-6* type which is also sometimes referred to as a modified Buckingham potential,

$$\Gamma_{(r)} = \frac{\varepsilon}{\alpha - 6}\left(6\exp\left(\alpha\left(1 - \frac{r}{r_{min}}\right)\right) - \alpha\left(\frac{r}{r_{min}}\right)^6\right) \tag{95}$$

where ε is the minimum potential energy at the intermolecular separation r_{min}, and α determines the steepness of the repulsion wall. By applying this potential in molecular dynamical simulations large sets of *PVT* data were generated which were then in turn fitted to the following term which looks in part like a virial equation,

$$P_r = \left(\frac{A}{V_r - G} + \frac{B}{V_r^2} + \frac{F}{V_r^3} + \frac{C}{V_r^m} \right) \left(1 + D(\alpha - 12.75) + E(\alpha - 12.75)^2 \right) \tag{96}$$

where

$$P_r = \frac{P\sigma^3}{\varepsilon}, \quad V_r = \frac{V}{\sigma^3}, \quad T_r = \frac{kT}{\varepsilon}, \quad \sigma = \frac{r_{min}}{2^{1/6}} \tag{97}$$

and A, B, C, D, E, F, and G are either functions of only T_r or T_r and V_r, k is the Boltzmann constant and σ is the collision diameter, the distance at which the potential Γ is 0.

The applied mixing rules for ε and α are related to the Equations (61), (62), and (67), whereas the combining rule for the collision diameter σ is like Equation (69)

$$\sigma = \left(\sum_i \sum_j x_i x_j \sigma_{ij}^3 \right)^{1/3} \tag{98}$$

$$\varepsilon = \frac{\sum_i \sum_j x_i x_j \varepsilon_{ij} \sigma_{ij}^3}{\sum_i \sum_j x_i x_j \sigma_{ij}^3} \tag{99}$$

$$\alpha = \frac{\sum_i \sum_j x_i x_j \gamma_{ij} \varepsilon_{ij} \sigma_{ij}^3}{\sum_i \sum_j x_i x_j \varepsilon_{ij} \sigma_{ij}^3} \tag{100}$$

with:

$$\sigma_{ij} = \frac{1}{2}(\sigma_i + \sigma_j), \quad \varepsilon_{ij} = (\varepsilon_i \varepsilon_j)^{1/2}, \quad \alpha_{ij} = (\alpha_i \alpha_j)^{1/2} \tag{101}$$

According to the authors, the rather high lower boundary pressure of validity (500 MPa at 400 K) is related to the use of the α-*Exp-6* type potential. As a consequence for the evaluation of the fugacity coefficients using Equation (14), the integral must be split into a low and a high V part. For volumes corresponding to pressures below 500 MPa another but valid EOS has to be used to obtain satisfactory accuracies for φ_i^m.

For the application of this EOS and that of Shi and Saxena (1992), the authors provide a program called SUPERFLUID (Belonoshko et al. 1992).

Duan et al. (1992c, 1996)

Phase components: H_2O, CO_2, CH_4, CO, O_2, H_2, Cl_2, H_2S, N_2.
Applicable range given by the authors: up to 2800 K, up to 30 GPa.

In Duan et al. (1996) the EOS presented by Duan et al. (1992c) for pure phase components is extended to fluid mixtures. The EOS is a virial equation

$$z = \frac{P_{ref} V_{ref}}{n_T R T_{ref}} = 1 + \frac{B}{V_{ref}} + \frac{C}{V_{ref}^2} + \frac{D}{V_{ref}^4} + \frac{E}{V_{ref}^5} + \frac{F}{V_{ref}^2} \left(1 + \frac{\gamma}{V_{ref}^2} \right) \exp\left(-\frac{\gamma}{V_{ref}^2} \right) \tag{102}$$

in which P_{ref}, V_{ref}, and T_{ref} are defined as follows:

$$P_{ref} = \frac{\varepsilon_{ref} \sigma_i^3 P}{\varepsilon_i \sigma_{ref}^3}, \quad V_{ref} = \frac{\sigma_{ref}^3 V}{\sigma_i^3}, \quad T_{ref} = \frac{\varepsilon_{ref} T}{\varepsilon_i} \tag{103}$$

The parameters ε and σ are those of the Lennard-Jones potential. The index ref signifies the pure reference fluid CH_4 for which the parameters B, C, D, E, F, and γ were determined and the index i stands for any pure fluid for which the EOS is intended to be valid. Except for γ, these parameters are functions of T_{ref}. For mixtures, the following mixing and combining rules are used:

$$\varepsilon = \sum_i \sum_j x_i x_j k_{1,ij} \left(\varepsilon_i \varepsilon_j \right)^{1/2}, \quad \sigma = \sum_i \sum_j x_i x_j k_{2,ij} \left(\frac{\sigma_i + \sigma_j}{2} \right), \quad k_{1.ii} = 1, \quad k_{2.ii} = 1 \quad (104)$$

Here the mixing parameters $k_{1,ij}$ and $k_{2,ij}$ are treated as constants and have been fitted using experimental results.

Duan et al. (2000) checked the validity of this EOS for H_2O, CO_2, CH_4, N_2 mixtures by comparing the calculated properties to experimental results and to other EOS with favorable results.

Duan et al. (1992a,b)

Phase components: H_2O, CO_2, CH_4.
Applicable range given by the authors: up to 1273 K, up to 300 MPa.

The EOS used is very similar to the approach taken by Duan et al. (1992c, 1996)

$$z = 1 + \frac{BV_c}{V} + \frac{CV_c^2}{V^2} + \frac{DV_c^4}{V^4} + \frac{EV_c^5}{V^5} + \frac{FV_c^2}{V^2} \left(\beta + \frac{\gamma V_c^2}{V^2} \right) \exp\left(-\frac{\gamma V_c^2}{V^2} \right) \quad (105)$$

and differs in the additional parameter β, and the replacement of the Lennard-Jones parameters by the critical properties of the pure phase components and uses reduced properties instead:

$$P_r = \frac{P}{P_c}, \quad V_r = \frac{V}{V_c}, \quad T_r = \frac{T}{T_c} \quad (106)$$

Accordingly the parameters B, C, D, E, F are functions of the reduced temperature T_r and the parameters β and γ are constants. Different from the approach above is the treatment of the mixing and combining rules. Mixing requires the calculation of the cross terms for the parameters B, C, D, E, F, and γ and also one for the critical volume V_c. However, whereas V_c for a pure phase component has a physical meaning, for a mixture it does not. Mixing rules according to Equations (56) to (58) and cross terms for each term in Equation (105), are listed below:

$$B^{mix} V_c^{mix} = \sum_i \sum_j x_i x_j B_{ij} V_{c,ij} \quad (107)$$

$$B_{ij} = \left(\frac{B_i^{1/3} + B_j^{1/3}}{2} \right)^3 k_{1,ij}, \quad V_{c,ij} = \left(\frac{V_{c,i}^{1/3} + V_{c,j}^{1/3}}{2} \right)^3, \quad k_{1.ii} = 1 \quad (108)$$

$$C^{mix} \left(V_c^{mix} \right)^2 = \sum_i \sum_j \sum_k x_i x_j x_k C_{ijk} V_{c,ijk}^2 \quad (109)$$

$$C_{ijk} = \left(\frac{C_i^{1/3} + C_j^{1/3} + C_k^{1/3}}{3} \right)^3 k_{2,ijk}, \quad V_{c,ijk} = \left(\frac{V_{c,i}^{1/3} + V_{c,j}^{1/3} + V_{c,k}^{1/3}}{3} \right)^3 \quad (110)$$

$$k_{2.iii} = 1, \quad k_{2.ijj} = k_{2.iij}, \quad k_{2.ikk} = k_{2.iik}, \quad k_{2.jkk} = k_{2.jjk} \quad (111)$$

$$D^{mix} \left(V_c^{mix} \right)^4 = \sum_i \sum_j \sum_k \sum_l \sum_m x_i x_j x_k x_l x_m D_{ijklm} V_{c,ijklm}^4 \quad (112)$$

$$D_{ijklm} = \left(\frac{D_i^{1/3} + D_j^{1/3} + D_k^{1/3} + D_l^{1/3} + D_m^{1/3}}{5} \right)^3 \tag{113}$$

$$V_{c,ijklm} = \left(\frac{V_{c,i}^{1/3} + V_{c,j}^{1/3} + V_{c,k}^{1/3} + V_{c,l}^{1/3} + V_{c,m}^{1/3}}{5} \right)^3 \tag{114}$$

$$E^{mix} \left(V_c^{mix} \right)^5 = \sum_i \sum_j \sum_k \sum_l \sum_m \sum_n x_i x_j x_k x_l x_m x_n E_{ijklmn} V_{c,ijklmn}^5 \tag{115}$$

$$E_{ijklmn} = \left(\frac{E_i^{1/3} + E_j^{1/3} + E_k^{1/3} + E_l^{1/3} + E_m^{1/3} + E_n^{1/3}}{6} \right)^3 \tag{116}$$

$$V_{c,ijklmn} = \left(\frac{V_{c,i}^{1/3} + V_{c,j}^{1/3} + V_{c,k}^{1/3} + V_{c,l}^{1/3} + V_{c,m}^{1/3} + V_{c,n}^{1/3}}{6} \right)^3 \tag{117}$$

$$F^{mix} \left(V_c^{mix} \right)^2 = \sum_i \sum_j x_i x_j F_{ij} V_{c,ij}^2 \tag{118}$$

$$F_{ij} = \left(\frac{F_i^{1/3} + F_j^{1/3}}{2} \right)^3 k_{1,ij}, \quad V_{c,ij} = \left(\frac{V_{c,i}^{1/3} + V_{c,j}^{1/3}}{2} \right)^3 \tag{119}$$

$$\beta = \sum_i x_i \beta_i \tag{120}$$

$$\gamma^{mix} \left(V_c^{mix} \right)^2 = \sum_i \sum_j \sum_k x_i x_j x_k \gamma_{ijk} V_{c,ijk}^2 \tag{121}$$

$$\gamma_{ijk} = \left(\frac{\gamma_i^{1/3} + \gamma_j^{1/3} + \gamma_k^{1/3}}{3} \right)^3 k_{3,ijk} \tag{122}$$

$$k_{3.ijk} = 1 \text{ for } i = j = k, \quad k_{3.ijj} = k_{3.iij}, \quad k_{3.ikk} = k_{3.iik}, \quad k_{3.jkk} = k_{3.jjk} \tag{123}$$

A program to calculate the fluid properties using this EOS has been published by Nieva and Barragan (2003).

Duan and Zhang (2006)

Phase components: H_2O, CO_2.
Applicable range given by the authors: up to 2573 K, up to 10 GPa.

An identical EOS to the one described above (Duan et al. 1992a,b) is used and the parameters calibrated to experimental results and volumes derived from molecular simulations.

Anderko and Pitzer (1993a,b), Duan et al. (1995)

Phase components: H_2O, NaCl, KCl and H_2O, NaCl, CO_2.
Applicable range given by the authors: between 773 and up to 1200 K and below 500 MPa for the system containing H_2O, CO_2, and NaCl. If KCl is considered as a phase component, maximum temperature is 973 K.

Anderko and Pitzer (1993a) implemented a reasonable successful EOS for H_2O-NaCl mixtures. Anderko and Pitzer (1993b) added the phase component KCl, and later Duan et al. (1995) adapted CO_2 to the base system H_2O-$NaCl$[6]. NaCl as well as KCl are treated as undissociated species. The foundation of this EOS is an integrated formulation of the Helmholtz free energy and adds a hard-sphere, dipole-dipole interaction and a perturbation term to the ideal gas properties:

$$A = A^{ideal} + A^{hs} + A^{dip-dip} + A^{per} \tag{124}$$

On a molar basis, the hard-sphere contribution is calculated according to Equation (73) which is derived for mixtures:

$$\frac{a^{hs}}{RT} = \frac{A^{hs}}{n_T RT} = \frac{\frac{3DE}{F}y - \frac{E^3}{F^2}}{1-y} + \frac{\frac{E^3}{F^2}}{(1-y)^2} + \left(\frac{E^3}{F^2} - 1\right)\ln(1-y) \tag{125}$$

The definitions for y, D, E, and F are from Equations (44), (71), and (72). The dipole-dipole interaction term uses a [0,1] Padé approximation

$$\frac{a^{dip-dip}}{RT} = \frac{A^{dip-dip}}{n_T RT} = \frac{A_2}{1 - A_3/A_2} \tag{126}$$

The required terms in Equation (126) are calculated according to the following equations:

$$A_2 = -\frac{4}{3}\sum_{i=1}^{m}\sum_{j=1}^{m}x_i x_j \frac{b_i b_j}{b_{ij}} y_{ij}\tilde{\mu}_i^2\tilde{\mu}_j^2 I_{2(\eta)} \tag{127}$$

with:

$$y_{ij} = \frac{b_{ij}}{4v}, \quad b_{ij} = \left(\frac{b_i^{1/3} + b_j^{1/3}}{2}\right)^3 \tag{128}$$

and:

$$A_3 = \frac{10}{9}\sum_{i=1}^{m}\sum_{j=1}^{m}\sum_{k=1}^{m}x_i x_j x_k \frac{b_i b_j b_k}{b_{ij}b_{ik}b_{jk}} y_{ijk}^3\tilde{\mu}_i^2\tilde{\mu}_j^2\tilde{\mu}_k^2 I_{3(\eta)} \tag{129}$$

with:

$$y_{ijk} = \frac{b_{ijk}}{4v} = \frac{(b_{ij}b_{ik}b_{jk})^{1/3}}{4v} \tag{130}$$

In Equations (127) and (129) $\tilde{\mu}_i$ is the reduced dipole moment,

$$\tilde{\mu}_i = \left(\frac{\mu_i^2}{\sigma_i^3 kT}\right)^{1/2} \tag{131}$$

and the functions $I_{2(\eta)}$ and $I_{3(\eta)}$ are integrals over pair and triple distribution functions and depend on η:

$$\eta = \frac{1}{4v}\sum_{i=1}^{m}\sum_{j=1}^{m}x_i x_j b_{ij} \tag{132}$$

---

[6]   There seems to be some frustration in the community and practical implementations of this EOS are scarce. Because of a few typing errors in the required constants, but mainly resulting from some confusion in the presented derivatives many programming efforts failed. In Appendix B, a complete set of constants and required equations is provided.

The perturbation term is a virial expansion of the form:

$$\frac{a^{per}}{RT} = \frac{A^{per}}{n_T RT} = -\frac{1}{RT}\left(\frac{a}{v} + \frac{acb}{4v^2} + \frac{adb^2}{16v^3} + \frac{aeb^3}{64v^4}\right) \tag{133}$$

In this equation, the quantities a, acb, adb^2, and aeb^3 are related to the second, third, fourth, and fifth virial coefficients, respectively. Accordingly, the second virial coefficient is a quadratic function of composition, the third a cubic, and so on. Besides the co-volume b relevant for 2, 3, 4, and 5 body interactions, the respective quantities involve mixing parameters characteristic for each considered binary join, which in turn are functions of T.

Using this EOS binary and ternary fluid phase equilibria can be successfully calculated. Figure 3 shows an example that illustrates the phase relationships in the system H_2O- CO_2-NaCl as a function of P and T. In Heinrich (2007) further isothermal-isobaric phase equilibria are presented.

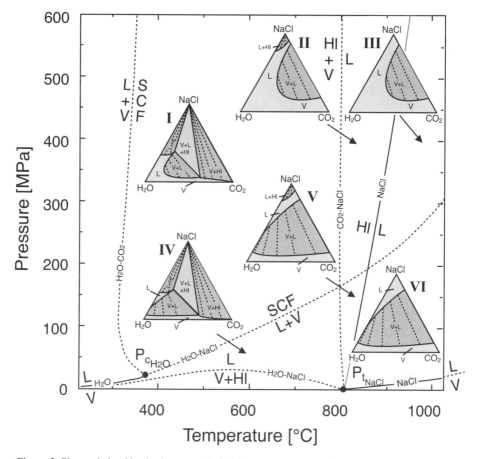

Figure 3. Phase relationships in the system H_2O-CO_2-NaCl as function of P and T. The ternary phase boundaries, including the tie lines connecting coexisting fluids, are calculated using the EOS by Duan et al. (1995). [Used by permission of Springer, from Heinrich et al. (2004), *Contributions to Mineralogy and Petrology*, Vol. 148, Fig. 1, p. 133.]

Duan et al. (2003)

Phase components: H_2O, CH_4, NaCl and H_2O, CO_2, CH_4, NaCl.
Applicable range given by the authors: 573 to 1300 K, up to 500 MPa.

The approach used by Anderko and Pitzer (1993a) and later taken up by Duan et al. (1995) has been augmented in this publication further by the phase component CH_4.

Jiang and Pitzer (1996), Duan et al. (2006)

Phase components: H_2O, $CaCl_2$ and H_2O, $MgCl_2$.
Applicable range given by the authors: 523 to 973 K, up to 150 MPa.

Jiang and Pitzer (1996) presented an EOS for the system H_2O-$CaCl_2$ and Duan et al. (2006) for the systems H_2O-$CaCl_2$ and H_2O-$MgCl_2$, which are based on the approach taken by Anderko and Pitzer (1993a), except that dipole-quadrupole and quadrupole-quadrupole interactions were additionally taken into account. The respective Helmholtz free energy is then:

$$A = A^{ideal} + A^{hs} + A^{dip-dip} + A^{dip-quad} + A^{quad-quad} + A^{per} \tag{134}$$

Churakov and Gottschalk (2003a,b)

Phase components: H_2O, CO_2, CH_4, CO, H_2, O_2, N_2, Ar, He, Ne and 88 other organic and inorganic phase components.
Applicable range given by the authors: at least 2000 K and up to 10 GPa.

This EOS is based on perturbation theory and uses the following expression for the Helmholtz free energy,

$$A = A^{ideal} + A^{LJ} + A^{dip-dip} + A^{ind-dip} \tag{135}$$

which considers expressions derived from intermolecular potentials according to the Lennard-Jones, dipole-dipole and induced dipole-dipole interaction. A^{LJ} is split into terms for the hard-sphere, high temperature and random phase approximation (see also Shmulovich et al. 1982)

$$A^{LJ} = A^{hs} + A^{hta} + A^{rpa} \tag{136}$$

As in Anderko and Pitzer (1993a), for $A^{dip-dip}$, a [0,1] Padé approximation is used.

$$A^{dip-dip} = \frac{A_2^{dip-dip}}{\left(1 - A_3^{dip-dip} / A_2^{dip-dip}\right)} \tag{137}$$

with:

$$A_2^{dip-dip} = -\frac{2\pi N_A^2 \rho}{3kT} \sum_{i=1}^{m} \sum_{j=1}^{m} x_i x_j \frac{\mu_i^2 \mu_j^2}{\sigma_{ij}^3} J_{ij} \tag{138}$$

$$A_3^{dip-dip} = \frac{32\pi^3 N_A^3 \rho^2}{135(kT)^2} \left(\frac{14\pi}{5}\right)^{1/2} \sum_{i=1}^{m} \sum_{j=1}^{m} \sum_{k=1}^{m} x_i x_j x_k \frac{\mu_i^2 \mu_j^2 \mu_k^2}{\sigma_{ij}\sigma_{ik}\sigma_{jk}} K_{ijk} \tag{139}$$

and:

$$A_2^{ind-dip} = -2\pi N_A^2 \rho \sum_{i=1}^{m} \sum_{j=1}^{m} x_i x_j \frac{\alpha_j \mu_i^2 + \alpha_i \mu_j^2}{\sigma_{ij}^3} J_{ij} \tag{140}$$

Here N_A is the Avogadro's number, μ_i the dipole moment and α_i the polarizability of the molecule i, k the Boltzmann constant, and J_{ij} and K_{ijk} are the integrals over pair and triple distribution functions of the reference Lennard-Jones fluid. For the Equations (138) to (140), the following combining rules are applied:

$$\sigma_{ij} = \frac{1}{2}\left(\sigma_i + \sigma_j\right), \quad \varepsilon_{ij} = \left(\varepsilon_i \varepsilon_j\right)^{1/2} \tag{141}$$

The mixing rules for the Lennard-Jones contributions are:

$$\sigma_{mix}^3 = \sum_{i=1}^{m}\sum_{j=1}^{m} x_i x_j \sigma_{ij}^3 \tag{142}$$

$$\varepsilon_{mix} = \frac{1}{\sigma_{mix}^3} \sum_{i=1}^{m}\sum_{j=1}^{m} x_i x_j \varepsilon_{ij} \sigma_{ij}^3 \tag{143}$$

Figure 4 shows calculated fluid-vapor phase equilibria using this EOS for the system H_2O-Xe, H_2O-CH_4 and H_2O-CO_2 in the range between 523-633 K and 0-300 MPa and compares them to interpolated experimental results. For the polar - non-polar systems H_2O-Xe and H_2O-CH_4, the calculated and the experimental results are in good agreement despite using parameters only for pure phase components. Larger deviations are observed for the system H_2O-CO_2, which results from the fact that the EOS does not treat dipole-quadrupole interactions, an important type of interaction in this system. In Figure 5, equilibria of coexisting fluids in the systems H_2O-N_2, H_2O-Ar, and H_2O- CH_4 are plotted for pressures up to 6 GPa and shows that considerable ranges of fluid unmixing is predicted in these systems. Whereas with this EOS, the quality of the calculated properties for the smaller molecules is very good, because of the involvement of only three constants (ε, σ, μ or α) for each phase component, for complicated molecules these have to be evaluated carefully before being applied.

CALCULATION OF PHASE EQUILIBRIA

For the compositional evaluation of coexisting fluids the respective phase equilibria have to be calculated. Generally the equilibrium conditions are, that for all phase components the chemical potentials have to be constant at P and T in all coexisting phases:

$$\mu_i^I = \mu_i^{II} \tag{144}$$

For binary systems, this is relatively easily accomplished by solving two equations simultaneously. For ternary and higher systems, numerical and practical problems become evident and equilibrium calculations are increasingly difficult.

A better approach and the method of choice is the minimization of the Gibbs free energy. This method is also able to handle saturated solids, i.e., change of fluid bulk compositions can be treated. The development of minimization algorithms is, however, an art itself. Such programs are mostly proprietary to specific scientific groups or available only commercially. Even if such programs are obtained, it is hard if not impossible to adapt them, so that the required EOS can be applied with it. However, a promising minimization approach is that of Karpov et al. (1997), which can be programmed, at least in principle, just using the equations provided in the publication[7].

CONCLUSIONS

Numerous equations of state are available for mixtures of non-polar and polar fluids, mainly for crustal P and T conditions. For the calculation of fluid equilibria in the mantle

---

[7] A free but compiled implementation can be found and downloaded from *http://les.web.psi.ch/Software/ GEMS-PSI/*.

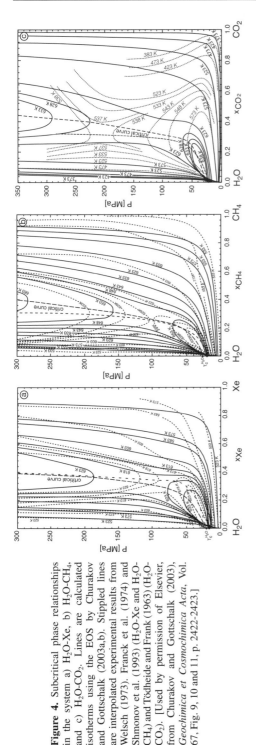

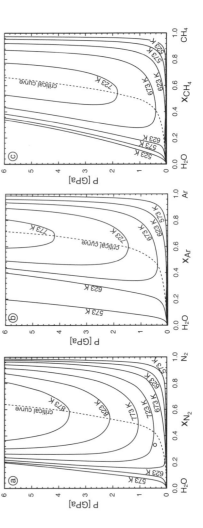

Figure 4. Subcritical phase relationships in the system a) H_2O-Xe, b) H_2O-CH_4, and c) H_2O-CO_2. Lines are calculated isotherms using the EOS by Churakov and Gottschalk (2003a,b). Stippled lines are interpolated experimental results from Welsch (1973). Franck et al. (1974) and Shmonov et al. (1993) (H_2O-Xe and H_2O-CH_4) and Tödheide and Frank (1963) (H_2O-CO_2). [Used by permission of Elsevier, from Churakov and Gottschalk (2003), *Geochimica et Cosmochimica Acta*, Vol. 67, Fig. 9, 10 and 11, p. 2422-2423.]

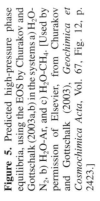

Figure 5. Predicted high-pressure phase equilibria, using the EOS by Churakov and Gottschalk (2003a,b) in the systems a) H_2O-N_2, b) H_2O-Ar, and c) H_2O-CH_4. [Used by permission of Elsevier, from Churakov and Gottschalk (2003), *Geochimica et Cosmochimica Acta*, Vol. 67, Fig. 12, p. 2423.]

even for relatively low T conditions in subducting slabs, the situation becomes much less comfortable. Missing are especially EOS at such conditions which treat salts such as NaCl, KCl, and $CaCl_2$ as phase components. Furthermore, all the EOS described above treat these salts as undissociated species, which very probably does not describe the real situation. EOS handling total or partially dissociated brines at high P and T do not exist to my knowledge.

At elevated conditions the differences between fluids and melts vanish (see Hack et al. 2007). Fluids and melts become indistinguishable. Thus all major elements become phase components in fluids. An ultimate EOS should handle these. Since the last review in this series by Ferry and Baumgartner (1987) concerning EOS for fluids, the focus has been shifted from deriving EOS by fitting parameters to experimental results to EOS which consider intermolecular interactions directly. The last approaches seems to be promising and with further increase in available computer power more EOS applicable to problems in geosciences can be expected.

ACKNOWLEDGMENTS

I would like to thank Anatoly Belonoshko, Christian Schmidt, and Barrie Clarke for their reviews and efforts to enhance the manuscript. I also like to thank Axel Liebscher for his patience and for not pushing too hard.

REFERENCES

Abbott MM (1979) Cubic equations of state: an interpretive review. Adv Chem Ser 182:47-70
Anderko A (1990) Equation of state methods for the modelling of phase equilibria. Fluid Phase Equilibria 61:145-225
Anderko A (2000) Cubic and generalized van der Waals equations. *In*: Equations of State for Fluids and Fluid Mixtures. Sengers JV, Kayser RF, Peters CJ, White HJ Jr (eds) Elsevier, p 75-126
Anderko A, Pitzer KS (1993a) Equation-of-state representation of phase equilibria and volumetric properties of the system NaCl-H_2O above 573 K. Geochim Cosmochim Acta 57:1657-1680
Anderko A, Pitzer KS (1993b) Phase equilibria and volumetric properties of the systems KCl-H_2O and NaCl-KCl-H_2O above 573 K: equation of state representation. Geochim Cosmochim Acta 57:4885-4897
Anderson G (2005) Thermodynamic of Natural Systems. Cambridge University Press
Anderson GM, Crerar DA (1993) Thermodynamics in Geochemistry. Oxford University Press
Bakker RJ (1999) Adaptation of the Bowers and Helgeson (1983) equation of state to the H_2O-CO_2-CH_4-N_2-NaCl system. Chem Geol 154:225-236
Beattie JA (1930) A rational basis for the thermodynamic treatment of real gases and mixtures of real gases. Phys Rev 36:132-143
Beattie JA (1949) The computation of the thermodynamic properties of real gases and mixtures of real gases. Phys Rev 44:141-191
Beattie JA (1955) Thermodynamic properties of real gases and mixtures of real gases. *In*: Thermodynamics and Physics of Matter. Vol 1. Rossini FD (ed) Princeton University Press, p 240-337
Beattie JA, Stockmayer WH (1951) The thermodynamics and statistical mechanics of real gases. *In*: A Treatise on Physical Chemistry. Vol 2. Taylor HS, Glasstone S (eds) D. van Nostrand, p 187-352
Belonoshko AB, Saxena S (1992a) A molecular dynamics study of the pressure-volume-temperature properties of supercritical fluids: I H_2O. Geochim Cosmochim Acta 55:381-387
Belonoshko AB, Saxena S (1992b) A molecular dynamics study of the pressure-volume-temperature properties of supercritical fluids: II CO_2, CH_4, CO, O_2, and H_2. Geochim Cosmochim Acta 55:3191-3208
Belonoshko AB, Saxena S (1992c) A unified equation of state for fluids of C-H-O-N-S-Ar composition and their mixtures up to high temperatures and pressures. Geochim Cosmochim Acta 56:3611-3636
Belonoshko AB, Shi P, Saxena SK (1992) SUPERFLUID: a FORTRAN-77 program for calculation of Gibbs free energy and volume of C-H-O-N-S-Ar mixtures. Computers Geosci 18:1267-1269
Boublík T (1970) Hard-sphere equation of state. J Chem Phys 53:471-472
Boublík T (2000) Perturbation theory. *In*: Equations of State for Fluids and Fluid Mixtures. Sengers JV, Kayser RF, Peters CJ, White HJ Jr (eds) Elsevier, p 127-168
Bowers TS, Helgeson HC (1983) Calculation of the thermodynamic and geochemical consequences of nonideal mixing in the system H_2O-CO_2-NaCl on phase relations in geologic systems: Equation of state for H_2O-CO_2-NaCl fluids at high pressures temperatures. Geochim Cosmochim Acta 47:1247-1275

Bowers TS, Helgeson HC (1985) Fortran programs for generating inclusion isochores and fugacity coefficients for the system H_2O-CO_2-NaCl at high pressures and temperatures. Computers Geosci 111:203-213

Carnahan NF, Starling KE (1969) Equation of state for non-attracting rigid spheres. J Chem Phys 51:635-636

Churakov SS, Gottschalk M (2003a) Perturbation theory based equation of state for polar molecular fluids: I. Pure fluids. Geochim Cosmochim Acta 67:2397-2414

Churakov SS, Gottschalk M (2003b) Perturbation theory based equation of state for polar molecular fluids: II. Fluid mixtures. Geochim Cosmochim Acta 67:2415-2425

De Santis R, Breedveld GJF, Prausnitz JM (1974) Thermodynamic properties of aqueous das mixtures at advanced pressures. Ind Eng Chem Proc Design Dev 13:374-377

Duan Z, Zhang Z (2006) Equation of state of the H_2O, CO_2, and H_2O-CO_2 systems up to 10 GPa and 2573.15 K: molecular simulations with ab initio potential surface. Geochim Cosmochim Acta 70:2311-2324

Duan Z, Møller N, Weare JH (1992a) An equation of state for the CH_4-CO_2-H_2O system: I. Pure systems from 0 to 1000 °C and 0 to 8000 bar. Geochim Cosmochim Acta 56:2605-2617

Duan Z, Møller N, Weare JH (1992b) An equation of state for the CH_4-CO_2-H_2O system: II. Mixtures from 50 to 1000 °C and 0 to 1000 bar. Geochim Cosmochim Acta 56:2619-2631

Duan Z, Møller N, Weare JH (1992c) Molecular dynamics simulation of PVT properties of geological fluids and a general equation of state of nonpolar and weakly polar gases up to 2000 K and 20000 bar. Geochim Cosmochim Acta 56:3839-3845

Duan Z, Møller N, Weare JH (1995) Equation of state for the NaCl-H_2O-CO_2 system: prediction of phase equilibria and volumetric properties. Geochim Cosmochim Acta 59:2869-2882

Duan Z, Møller N, Weare JH (1996) A general equation of state for supercritical fluid mixtures and molecular dynamics simulation of mixture PVTX properties. Geochim Cosmochim Acta 60:1209-1216

Duan Z, Møller N, Weare JH (2000) Accurate prediction of the thermodynamic properties of fluids in the system H_2O-CO_2-CH_4-N_2 up to 2000 K and 100 kbar from a corresponding states/one fluid equation of state. Geochim Cosmochim Acta 64:1069-1075

Duan Z, Møller N, Weare JH (2003) Equations of state for the NaCl-H_2O-CH_4 system and the NaCl-H_2O-CO_2-CH_4 system; phase equilibria and volumetric properties above 573 K. Geochim Cosmochim Acta 67:671-680

Duan Z, Møller N, Weare JH (2006) A high temperature equation of state for the H_2O-$CaCl_2$ and H_2O-$MgCl_2$ systems. Geochim Cosmochim Acta 70:3765-3777

Ferry JM, Baumgartner L (1987) Thermodynamic models of molecular fluids at the elevated pressures and temperatures of crustal metamorphism. Rev Mineral 17:323-366

Flowers GC (1979) Correction of Holloway's (1977) adaptation of the modified Redlich-Kwong equation of state for calculation of the fugacities of molecular species in supercritical fluids of geologic interest. Contrib Mineral Petrol 69:315-318

Franck EU, Lentz H, Welsch H (1974) The system water-xenon at high pressures and temperatures. Z Phys Chem N F 93:95-108

Gillespie LJ (1925) Equilibrium pressures of individual gases in mixtures and the mass-action law for gases. J Chem Soc 47:305-312

Gillespie LJ (1936) Methods for the thermodynamic correlation of high pressure gas equilibria with the properties of pure gases. Chem Rev 18:359-371

Grevel K-D (1993) Modified Redlich-Kwong equations of state for CH_4 and CH_4-H_2O fluid mixtures. N Jb Mineral M 1993:462-480

Grevel K-D, Chatterjee ND (1992) A modified Redlich-Kwong equation of state for H_2-H_2O fluid mixtures at high pressures and at temperatures above 400 °C. Eur J Mineral 4:1303-1310

Hack AC, Thompson AB, Aerts M (2007) Phase relations involving hydrous silicate melts, aqueous fluids, and minerals. Rev Mineral Geochem 65:129-185

Halbach H, Chatterjee ND (1982) An empirical Redlich-Kwong-type equation of state for water to 1000 °C and 200 kbar. Contrib Mineral Petrol 79:337-345

Heinrich W (2007) Fluid immiscibility in metamorphic rocks. Rev Mineral Geochem 65:389-430

Heinrich W, Churakov SS, Gottschalk M (2004) Mineral-fluid equilibria in the system CaO-MgO-SiO_2-H_2O-CO_2-NaCl and the record of reactive flow in contact metamorphic aureoles. Contrib Mineral Petrol 148:131-144

Holloway JR (1976) Fugacity and activity of molecular species in supercritical fluids. *In*: Thermodynamics in Geology. Fraser DG (ed) D. Reidel Publishing Company, p 161-181

Holloway JR (1981) Compositions and volumes of supercritical fluids in the Earth's crust. *In*: Short Course in Fluid Inclusions: Fluid Inclusions: Petrologic Applications. Vol 6. Hollister LS, Crawford ML (eds) Mineralogical Association of Canada, p 13-38

Jacobs GK, Kerrick DM (1981a) APL and FORTRAN programs for a new equation of state for H_2O, CO_2 and their mixtures at supercritical conditions. Computers Geosci 7:131-143

Jacobs GK, Kerrick DM (1981b) Methane: An equation of state with application to the ternary system H_2O-CO_2-CH_4. Geochim Cosmochim Acta 45:607-614

Jiang S, Pitzer KS (1996) Phase equilibria and volumetric properties of aqueous $CaCl_2$ by an equation of state. AIChE J 42:585-594

Karpov IK, Chudnenko KV, Kulik DA (1997) Modeling chemical mass transfer in geochemical processes: Thermodynamic relations, conditions of equilibria, and numeric algorithms. Am J Sci 297:767-806

Kerrick DM, Jacobs GK (1981) A modified Redlich-Kwong equation for H_2O, CO_2, and H_2O-CO_2 mixtures at elevated pressures and temperatures. Am J Sci 281:735-767

Mansoori GA, Carnahan NF, Starling KE, Leland TW (1971) Equilibrium thermodynamic properties of the mixture of hared spheres. J Chem Phys 54:1523-1525

Matteoli E, Hamad EZ, Mansoori GA (2000) Mixtures of dissimilar molecules. *In*: Equations of State for Fluids and Fluid Mixtures. Sengers JV, Kayser RF, Peters CJ, White HJ Jr (eds) Elsevier, p 359-380

Nieva D, Barragan RM (2003) HCO-TERNARY: A FORTRAN code for calculating P-V-T-X properties and liquid vapor equilibria of fluids in the system H_2O-CO_2-CH_4. Computers Geosci 29:469-485

Nordstrom DK, Munoz JL (1994) Geochemical Thermodynamics. Blackwell Scientific Publications

Pedersen KS, Christensen P (2007) Fluids in hydrocarbon basins. Rev Mineral Geochem 65:241-258

Peng DY, Robinson DB (1976) A new two-constant equation of state. Ind Eng Chem Fundam 15:59-64

Poling BE, Prausnitz JM, O'Connell JP (2000) Properties of Gases and Liquids. McGraw-Hill Professional

Prausnitz JM, Lichtenthaler RN, Gomes de Azevedo E (1999) Molecular Thermodynamics of Fluid-Phase Equilibria. Prentice-Hall

Redlich O, Kwong JNS (1949) On the thermodynamics of solutions. V. An equation of state. Fugacities of gaseous solutions. Chem Rev 44:233-244

Rimbach H, Chatterjee ND (1987) Equations of state for H_2, H_2O, and H_2-H_2O fluid mixtures at temperatures above 0.01 °C and at high pressures. Phys Chem Mineral 14:560-569

Sandler SI (1994) Models for Thermodynamic and Phase Equilibria Calculations. Chemical Industries. Vol. 52. Marcel Dekker Inc.

Sandler SI, Orbey H (2000) Mixing and combining rules. *In*: Equations of State for Fluids and Fluid Mixtures. Sengers JV, Kayser RF, Peters CJ, White HJ Jr (eds) Elsevier, p 321-357

Saul A, Wagner W (1989) A fundamental equation of state for water covering the range from the melting line to 1273 K a pressures up to 25,000 MPa. J Phys Chem Ref Data 18:1537-1564

Saxena SK, Fei Y (1987) Fluids at crustal pressures and temperatures. I. Pure species. Contrib Mineral Petrol 95:370-375

Saxena SK, Fei Y (1988) Fluid mixtures in the C-H-O system at high pressure and temperature. Geochim Cosmochim Acta 52:505-512

Schmidt RW, Wagner W (1984) A new form of the equation of state for pure substances and its application to oxygen. Fluid Phase Equilibria 19:175-200

Sengers JV, Kayser RF, Peters CJ, White HJ Jr (eds) (2000) Equations of State for Fluids and Fluid Mixtures. Experimental Thermodynamics. Elsevier

Setzmann U, Wagner W (1991) A new equation of state and tables of thermodynamic properties for methane covering the range from the melting line to 625 K at pressures up to 1000 MPa. J Phys Chem Ref Data 20:1061-1155

Shi P, Saxena, SK (1992) Thermodynamic modeling of the C-H-O-S fluid system. Am Mineral 77:1038-1049

Shmonov VM, Sadus RJ, Franck EU (1993) High pressure and supercritical phase equilibria PVT-data of the binary water + methane mixture to 723 K and 200 MPa. J Phys Chem 97:9054-9059

Shmulovich KI, Tereshchenko YN, Kalinichev AG (1982) The Equation of state and isochors for nonpolar gases up to 2000 K and 10 GPa. Trans Geokhim 11:1598-1613

Soave G (1972) Equilibrium constants from a modified Redlich-Kwong equation of state. Chem Eng Sci 27:1197-1203

Span R, Lemmon EW, Jacobsen RT, Wagner W, Yokozeki A (2000) A reference equation of state for the thermodynamic properties of nitrogen for temperatures from 63.151 to 1000 K and pressures to 2200 MPa. J Phys Chem Ref Data 29:1361-1433

Span RW, Wagner W(1996) A new equation of state for carbon dioxide covering the fluid region from the triple point temperature to 1100 K at pressures up to 800 MPa. J Phys Chem Ref Data 25:1509-1596

Spycher NF, Reed MH (1988) Fugacity coefficients of H_2, CO_2, CH_4, H_2O and of H_2O-CO_2-CH_4 mixtures: a virial equation treatment for moderate pressures and temperatures applicable to calculations of hydrothermal boiling. Geochim Cosmochim Acta 52:749-749

Tegeler C, Span R, Wagner W (1999) A new equation of state for argon covering the fluid region for temperatures from the melting line to 700 K at pressures up to 1000 MPa. J Phys Chem Ref Data 28:779-851

Tödheide K, Frank EU (1963) Das Zweiphasengebiet und die kritische Kurve im System Kohlendioxid-Wasser bis zu Drucken von 3500 bar. Z Phys Chem Neue Folge 37:387-401.

Trusler JPM (2000) The virial equation of state. *In*: Equations of State for Fluids and Fluid Mixtures. Sengers JV, Kayser RF, Peters CJ, White HJ Jr (eds) Elsevier, p 35-74

van der Waals JD (1873) Over de continuiteit van de gas- en vloeistoftoestand. Dissertation, Leyden

Wagner W, Pruß A (2002) The IAPWS formulation 1995 for the thermodynamic properties of ordinary water substance for general and scientific use. J Phys Chem Ref Data 31:387-535

Wei YS, Sadus RJ (2000) Equations of state for the calculations of fluid-phase equilibria. AIChE J 46:169-196
Welsch H (1973) Die Systeme Xenon-Wasser und Methan-Wasser bei hohen Drücken und Temperaturen.
    Dissertation, Karlsruhe

APPENDIX A

Derivations

The thermodynamic treatment of fluids at elevated pressures and temperatures has been laid out in numerous publications (e.g., Gillespie 1925, 1936; Beattie 1930, 1949, 1955, Beattie and Stockmayer 1951). For the derivation of the general accepted forms of the fugacity the so called "general limit method" is used, which avoids basically indefinite integrals[8] if real fluids are handled using ideal gases as a reference model. In most textbooks of physical chemistry the "general limit method" is also used to introduce the term fugacity. While the reasoning and the derived equations are correct, it seems that many students and some teachers misunderstand the logic behind "general limit method," which leads to errors in textbooks, publications, and lecture notes, at least in geosciences. This problem is especially true for the necessary reference and standard conditions of fluids, which requires a sufficient low pressure P^* such that a fluid behaves as an ideal gas. This again makes it sometimes difficult for students to get a grip on the term fugacity.

In my opinion much of the confusion caused by the "general limit method" can be avoided using very similar logic, but rearranging the necessary equations. In addition no assumptions have to be made, but the principal and simple properties of an ideal gas are used directly avoiding the limit considerations. These properties are that the particles have zero volume with no acting intermolecular forces. As the resulting equations defining the fugacity are identical to the "general limit method," some intermediate results may or may not enhance the understanding of the practically used reference and standard conditions. To improve the understanding of the derivation intermediate steps are also included.

Let us start from the definitions of Helmholtz and Gibbs free energies A and G using the inner energy U, the enthalpy H and the entropy S

$$A = U - TS \tag{A1}$$

$$G = H - TS \tag{A2}$$

In the following the partial derivatives of A and G in respect to P and T at constant composition n_i

$$\left(\frac{\partial A}{\partial V} \right)_{T,n_i} = -P \tag{A3}$$

$$\left(\frac{\partial G}{\partial P} \right)_{T,n_i} = V \tag{A4}$$

$$\left(\frac{\partial A}{\partial T} \right)_{V,n_i} = \left(\frac{\partial G}{\partial T} \right)_{P,n_i} = -S \tag{A5}$$

and the chemical potentials μ_i

---

[8] Integrals of the form $\int_0^P V dP$ are indefinite. The volume of a fluid is indefinite at a pressure of 0.

$$\mu_i = \left(\frac{\partial A}{\partial n_i}\right)_{V,T,n_{j\neq i}} = \left(\frac{\partial G}{\partial n_i}\right)_{P,T,n_{j\neq i}} \tag{A6}$$

are used.

P and T as state variables

First we consider P and T as the state variables. In that case G is the appropriate state function. G as a function of P at constant T is calculated by integrating Equation (A4). The integration constant in Equation (A7) is $G_{(P_r,T)}$ at the respective reference pressure P_r and any T. The usual choice of P_r is 0.1 MPa. For any real system we have:

$$G_{(P,T)} = G_{(P_r,T)} + \int_{P_r}^{P} V dP \tag{A7}$$

Pure fluids. For the moment, the standard state of a fluid is the real pure fluid phase component. Pure phase components are generally designated by ^{o}. We will change the standard state for the fluid later, however. For such a real pure fluid i in its standard state, Equation (A7) becomes:

$$G_{i(P,T)}^o = G_{i(P_r,T)}^o + \int_{P_r}^{P} V_i^o dP \tag{A8}$$

Now the first rearrangement of Equation (A8) uses the ideal gas law, where n_T is the total number of moles ($n_T = \Sigma n_i$),

$$V = \frac{n_T RT}{P} \tag{A9}$$

by expanding (e.g., Gillespie 1925) the argument in the integral of Equation (A8)

$$G_{i(P,T)}^o = G_{i(P_r,T)}^o + \int_{P_r}^{P}\left(V_i^o - \frac{n_T RT}{P}\right)dP + \int_{P_r}^{P}\frac{n_T RT}{P}dP \tag{A10}$$

Within the first integral, the volume of an ideal gas is subtracted, and it must be compensated by the second integral. Now the argument in the first integral describes the deviation of the volume of a real fluid from ideal gas behavior.

The second rearrangement concerns the lower integration boundary of the first integral in Equation (A10). Instead of P_r, the lower boundary is changed to 0, which has to be compensated again, whereas the second integral of Equation (A10) can be solved immediately

$$G_{i(P,T)}^o = G_{i(P_r,T)}^o - \int_{0}^{P_r}\left(V_i^o - \frac{n_T RT}{P}\right)dP + \int_{0}^{P}\left(V_i^o - \frac{n_T RT}{P}\right)dP + n_T RT\ln\frac{P}{P_r} \tag{A11}$$

For pure fluids we have

$$\mu_i^o = \left(\frac{\partial G_i^o}{\partial n_i}\right)_{P,T} = \frac{G_i^o}{n_T} = g_i^o \tag{A12}$$

and we get a molar quantity (molar quantities are designated here by small letters) which we can either designate as $\mu_{i(P,T)}^o$ or $g_{i(P,T)}^o$:

$$\mu_{i(P,T)}^o = g_{i(P,T)}^o = \mu_{i(P_r,T)}^o - \int_0^{P_r}\left(v_i^o - \frac{RT}{P}\right)dP + \int_0^P\left(v_i^o - \frac{RT}{P}\right)dP + RT\ln\frac{P}{P_r} \qquad (A13)$$

In Equation (A13), we have an infinite integral by the means of Gillespie (the lower integration boundary is 0), but which has a finite value and, therefore, is harmless and solvable if we have an EOS. Equation (A13) provides the definition of the fugacity coefficient for a pure fluid:

$$RT\ln\varphi_{i(P,T)}^o = \int_0^P\left(v_i^o - \frac{RT}{P}\right)dP \qquad (A14)$$

Equation (A14) defines the fugacity coefficient at any pressure dependent only on the upper integration boundary, either P or P_r in Equation (A13). It is important to note that for the derivation of Equation (A14), no assumptions were necessary and no general limits have to be considered but the ideal gas law *per se*.

As in Equation (A13) we can also write

$$\mu_{i(P_r,T)}^o = g_{i(P_r,T)}^o \qquad (A15)$$

which leads with Equation (A14) to:

$$\mu_{i(P,T)}^o = g_{i(P_r,T)}^o - RT\ln\varphi_{i(P_r,T)}^o + RT\ln\varphi_{i(P,T)}^o + RT\ln\frac{P}{P_r} \qquad (A16)$$

Now we come to the reconsideration of the standard state. The second term of the right-hand side of Equation (A16) is the term which includes the fugacity coefficient of the fluid at T and reference pressure P_r. At a pressure of 0, it is possible to treat any fluid like an ideal gas, because neither the particle volume nor intermolecular forces play a role as the particle distances are infinite. If we recapitulate that this term represents the sum (integral) of deviations from a pure real fluid and a pure ideal gas, then the difference of the first two terms in Equation (A16) are the thermodynamic properties of an ideal gas $g_{i(P_r,T)}^{o*}$. Ideal gas properties are designated by an asterisk *

$$g_{i(P_r,T)}^{o*} = g_{i(P_r,T)}^o - RT\ln\varphi_{i(P_r,T)}^o \qquad (A17)$$

In other words, if we know the thermodynamic ideal gas properties of a pure phase component and are able to calculate the fugacity coefficient at P_r (having an EOS) we know the properties of the real fluid at P_r or any other P. Most thermodynamic ideal gas properties can be obtained using, for example, *ab initio* calculations and molecular simulations. Because of the lack of intermolecular forces only single-particle properties and velocity distributions have to be considered to evaluate the inner energy u_i^{o*} and the heat capacity $c_{P,i}^{o*}$. The partitioning function or g_i^{o*} has to be calculated to obtain the entropy s_i^{o*}. For any fluid phase component, it is convenient to choose the ideal gas as its standard state. In addition, because $g_{i(P_r,T)}^{o*}$ in Equation (A17) is a corrected term and not a real quantity, this special choice is also called the standard state of an *hypothetical ideal gas*. As for solids the reference conditions are usually 0.1 MPa and 298.15 K.

Using Equation (A17) we can also write for Equation (A16)

$$\mu_{i(P,T)}^o = g_{i(P_r,T)}^{o*} + RT\ln\frac{\varphi_{i(P,T)}^o P}{P_r} \qquad (A18)$$

or with Equation (A2):

$$\mu^o_{i(P,T)} = h^{o*}_{i(P_r,T)} - Ts^{o*}_{i(P_r,T)} + RT \ln \frac{\varphi^o_{i(P,T)}P}{P_r} \tag{A19}$$

Knowing the heat capacity of the *hypothetical ideal gas* $c^{o*}_{P,i}$ as a function of T any chemical potential $\mu^o_{i(P,T)}$ can be calculated provided that we have an EOS

$$\mu^o_{i(P,T)} = h^{o*}_{i(P_r,T_r)} + \int_{T_r}^{T} c^{o*}_{P,i} dT - T \left(s^{o*}_{i(P_r,T_r)} + \int_{T_r}^{T} \frac{c^{o*}_{P,i}}{T} dT \right) + RT \ln \frac{\varphi^o_{i(P,T)}P}{P_r} \tag{A20}$$

Fluid mixtures. Instead of the molar quantity of pure fluids in Equation (A18) we can also write in general:

$$G^o_{i(P,T)} = G^{o*}_{i(P_r,T)} + \int_{0}^{P} \left(V^o_i - \frac{n_T RT}{P} \right) dP + n_T RT \ln \frac{P}{P_r} \tag{A21}$$

If we consider mixtures this equation alters to

$$G_{(P,T)} = G^*_{(P_r,T)} + \int_{0}^{P} \left(V - \frac{n_T RT}{P} \right) dP + n_T RT \ln \frac{P}{P_r} \tag{A22}$$

in which the pure are replaced by the mixed properties. If we recapitulate that $G^*_{i(P_r,T)}$ is then the property of a mixture of ideal gases we get

$$G_{(P,T)} = \sum_i n_i \left(g^{o*}_{i(P_r,T)} + RT \ln x_i \right) + \int_{0}^{P} \left(V - \frac{n_T RT}{P} \right) dP + n_T RT \ln \frac{P}{P_r} \tag{A23}$$

Because of the missing intermolecular forces, no volume of mixing occurs and we have to consider only the ideal entropy of mixing ($\sum n_i RT \ln x_i$). It is convenient to incorporate the entropy of mixing in to the last term of Equation (A23)

$$G_{(P,T)} = \sum_i n_i g^{o*}_{i(P_r,T)} + \int_{0}^{P} \left(V - \frac{n_T RT}{P} \right) dP + RT \sum_i n_i \ln \frac{x_i P}{P_r} \tag{A24}$$

Differentiation using Equation (A6) gets the respective μ_i for each component i:

$$\mu_{i(P,T)} = g^{o*}_{i(P_r,T)} + \int_{0}^{P} \left(\left(\frac{\partial V}{\partial n_i} \right)_{P,T,n_{j \neq i}} - \frac{RT}{P} \right) dP + RT \ln \frac{x_i P}{P_r} \tag{A25}$$

So the fugacity coefficient φ_i^m in the mixture is defined as:

$$RT \ln \varphi^m_{i(P,T)} = \int_{0}^{P} \left(\left(\frac{\partial V}{\partial n_i} \right)_{P,T,n_{j \neq i}} - \frac{RT}{P} \right) dP \tag{A26}$$

The evaluation of $\varphi^m_{i(P,T)}$ requires an EOS but now for fluid mixtures. Using Equation (A26) and we get the general expression:

$$\mu_{i(P,T)} = g^{o*}_{i(P_r,T)} + RT \ln \frac{x_i \varphi^m_{i(P,T)}P}{P_r} \tag{A27}$$

or:

$$\mu_{i(P,T)} = h_{i(P_r,T)}^{o*} - Ts_{i(P_r,T)}^{o*} + RT \ln \frac{x_i \varphi_{i(P,T)}^m P}{P_r} \qquad (A28)$$

With the definition of the fugacity

$$f_{i(P,T)}^m = x_i \varphi_{i(P,T)}^m P \qquad (A29)$$

we get

$$\mu_{i(P,T)} = h_{i(P_r,T)}^{o*} - Ts_{i(P_r,T)}^{o*} + RT \ln \frac{f_{i(P,T)}^m}{P_r} \qquad (A30)$$

As a result, the following relations also hold for the activity coefficient γ_i and activity a_i:

$$f_{i(P,T)}^o = \varphi_{i(P,T)}^o P \qquad (A31)$$

$$\varphi_{i(P,T)}^m = \gamma_i \varphi_{i(P,T)}^o \qquad (A32)$$

$$f_{i(P,T)}^m = a_i f_{i(P,T)}^o \qquad (A33)$$

Again, for the definition of the fugacity coefficient of a phase component in a fluid mixture no other prerequisite than the ideal gas definition is needed.

V and T as state variables

Because most EOS are explicit in P instead of V, the Helmholtz free energy has to be used to calculate μ_i. The line of arguments is, however, identical to the one given for the Gibbs free energy. The dependence of A on V is given by integrating (A3):

$$A_{(V,T)} = A_{(V_r,T)} - \int_{V_r}^{V} P dV \qquad (A34)$$

Pure fluids. Expanding Equation (A34) using the ideal gas law for a pure phase component we get:

$$A_{i(V,T)}^o = A_{i(V_r,T)}^o - \int_{V_r^o}^{V^o} \left(P - \frac{n_T RT}{V} \right) dV - \int_{V_r^o}^{V^o} \frac{n_T RT}{V} dV \qquad (A35)$$

Splitting the first integral by expanding the upper integration boundary to infinity[9] and solving the second we get:

$$A_{i(V,T)}^o = A_{i(V_r,T)}^o - \int_{V_r^o}^{\infty} \left(P - \frac{n_T RT}{V} \right) dV + \int_{V^o}^{\infty} \left(P - \frac{n_T RT}{V} \right) dV - n_T RT \ln \frac{V_i^o}{V_{i,r}^o} \qquad (A36)$$

With V_r for an *hypothetical ideal gas* at reference conditions

$$V_{i,r}^o = \frac{n_T RT}{P_r} \qquad (A37)$$

we get

$$A_{i(V,T)}^o = A_{i(V_r,T)}^o - \int_{V_r^o}^{\infty} \left(P - \frac{n_T RT}{V} \right) dV + \int_{V^o}^{\infty} \left(P - \frac{n_T RT}{V} \right) dV - n_T RT \ln \frac{P_r V_i^o}{n_T RT} \qquad (A38)$$

---

[9]   The lower and the higher integration boundaries are switched; therefore, the sign changes.

or with $v_i^o = V_i^o/n_T$ and going to the standard state of a *hypothetical ideal gas*

$$A_{i(V,T)}^o = A_{i(V_r,T)}^{o*} + \int_{V_i^o}^{\infty} \left(P - \frac{n_T RT}{V} \right) dV - n_T RT \ln \frac{P_r v_i^o}{RT} \tag{A39}$$

Using

$$G_{i(P,T)}^{o*} = A_{i(V,T)}^{o*} + PV_i^{o*} \tag{A40}$$

results in:

$$G_{i(P,T)}^o = A_{i(V_r,T)}^{o*} + \int_{V_i^o}^{\infty} \left(P - \frac{n_T RT}{V} \right) dV - n_T RT \ln \frac{P_r v_i^o}{RT} + PV_i^o \tag{A41}$$

Now for a pure fluid μ_i we have

$$\mu_i^0 = \frac{G_i^0}{n_T} = g_i^0 \tag{A42}$$

and Equation (A41) can be expressed by

$$g_{i(P,T)}^o = u_{i(V_r,T)}^{o*} - T s_{i(V_r,T)}^{o*} + \int_{V_i^o}^{\infty} \left(\frac{P}{n_T} - \frac{RT}{V} \right) dV - RT \ln \frac{P_r v_i^o}{RT} + P v_i^o \tag{A43}$$

To extract $\varphi_{i(V,T)}^o$, we have to transform Equation (A43) a bit further. For an ideal gas we can write $(Pv=RT)$

$$h_i^{o*} = u_i^{o*} + RT \tag{A44}$$

and therefore we have:

$$g_{i(P,T)}^o = h_{i(P_r,T)}^{o*} - T s_{i(P_r,T)}^{o*} + \int_{V_i^o}^{\infty} \left(\frac{P}{n_T} - \frac{RT}{V} \right) dV - RT \ln \frac{P_r v_i^o}{RT} + P v_i^o - RT \tag{A45}$$

Expanding with $RT \ln (P/P)$ we get:

$$g_{i(P,T)}^o = h_{i(P_r,T)}^{o*} - T s_{i(P_r,T)}^{o*} + \int_{V_i^o}^{\infty} \left(\frac{P}{n_T} - \frac{RT}{V} \right) dV - RT \ln \frac{P v_i^o}{RT} + P v_i^o - RT + RT \ln \frac{P}{P_r} \tag{A46}$$

With the general definition of z

$$z = \frac{Pv}{RT} \tag{A47}$$

we have:

$$g_{i(P,T)}^o = h_{i(P_r,T)}^{o*} - T s_{i(P_r,T)}^{o*} + \int_{V_i^o}^{\infty} \left(\frac{P}{n_T} - \frac{RT}{V} \right) dV - RT \ln z + RT (z-1) + RT \ln \frac{P}{P_r} \tag{A48}$$

From Equation (A48) we get the fugacity coefficient of a pure phase component as a function of V and T

$$\varphi_{i(V,T)}^o = \int_{V_i^o}^{\infty} \left(\frac{P}{n_T} - \frac{RT}{V} \right) dV - RT \ln z + RT (z-1) \tag{A49}$$

Fluid mixtures. As in (A23) for mixtures we get:

$$A_{(V,T)} = \sum_i n_i \left(a^{o*}_{i(V_r,T)} + RT \ln x_i \right) + \int_V^\infty \left(P - \frac{n_T RT}{V} \right) dV - n_T RT \ln \frac{V}{V_r} \quad \text{(A50)}$$

As for the pure fluid, in the following some transformations of Equation (A50) are performed, which have the purpose of obtaining an expression for μ_i as a function of P and T, so the derived values are compatible with thermodynamic properties with other phase components, e.g., solids. With V_r for an *hypothetical ideal gas* mixture at reference conditions

$$V_r = \frac{n_T RT}{P_r} \quad \text{(A51)}$$

we get:

$$A_{(V,T)} = \sum_i n_i \left(a^{o*}_{i(V_r,T)} + RT \ln x_i \right) + \int_V^\infty \left(P - \frac{n_T RT}{V} \right) dV - n_T RT \ln \frac{P_r V}{n_T RT} \quad \text{(A52)}$$

Incorporation of the ideal mixing entropy into the last term of Equation (A52)

$$A_{(V,T)} = \sum_i n_i a^{o*}_{i(V_r,T)} + \int_V^\infty \left(P - \frac{n_T RT}{V} \right) dV - RT \sum_i n_i \ln \frac{P_r V}{x_i n_T RT} \quad \text{(A53)}$$

and replacing the mole fraction x_i by its definition (n_i/n_T) we get:

$$A_{(V,T)} = \sum_i n_i a^{o*}_{i(V_r,T)} + \int_V^\infty \left(P - \frac{n_T RT}{V} \right) dV - RT \sum_i n_i \ln \frac{P_r V}{n_i RT} \quad \text{(A54)}$$

By applying Equation (A6) we have the expression for the chemical potential

$$\mu_{i(V,T)} = a^{o*}_{i(V_r,T)} + \int_V^\infty \left(\left(\frac{\partial P}{\partial n_i} \right)_{V,T,n_{i \neq j}} - \frac{RT}{V} \right) dV - RT \ln \frac{P_r V}{n_i RT} + RT \quad \text{(A55)}$$

or:

$$\mu_{i(V,T)} = \mu_{i(P,T)} = u^{o*}_{i(V_r,T)} + RT - Ts^{o*}_{i(V_r,T)} + \int_V^\infty \left(\left(\frac{\partial P}{\partial n_i} \right)_{V,T,n_{i \neq j}} - \frac{RT}{V} \right) dV - RT \ln \frac{P_r V}{n_i RT} \quad \text{(A56)}$$

Applying Equation (A44) we get:

$$\mu_{i(P,T)} = h^{o*}_{i(P_r,T)} - Ts^{o*}_{i(P_r,T)} + \int_V^\infty \left(\left(\frac{\partial P}{\partial n_i} \right)_{V,T,n_{i \neq j}} - \frac{RT}{V} \right) dV - RT \ln \frac{P_r V}{n_i RT} \quad \text{(A57)}$$

Expanding with $RT \ln (P/P)$ and n_T

$$\mu_{i(P,T)} = h^{o*}_{i(P_r,T)} - Ts^{o*}_{i(P_r,T)} + \int_V^\infty \left(\left(\frac{\partial P}{\partial n_i} \right)_{V,T,n_{i \neq j}} - \frac{RT}{V} \right) dV$$
$$- RT \ln \frac{n_T P_r V}{n_T n_i RT} + RT (\ln P - \ln P) \quad \text{(A58)}$$

and inserting the mole fraction x_i and the molar volume $v = V/n_T$ we get:

$$\mu_{i(P,T)} = h^{o*}_{i(P_r,T)} - Ts^{o*}_{i(P_r,T)} + \int\limits_{V}^{\infty}\left(\left(\frac{\partial P}{\partial n_i}\right)_{V,T,n_{i\neq j}} - \frac{RT}{V}\right)dV - RT\ln\frac{Pv}{RT} + RT\ln\frac{x_iP}{P_r} \qquad (A59)$$

Using Equation (A47) results in:

$$\mu_{i(P,T)} = h^{o*}_{i(P_r,T)} - Ts^{o*}_{i(P_r,T)} + \int\limits_{V}^{\infty}\left(\left(\frac{\partial P}{\partial n_i}\right)_{V,T,n_{i\neq j}} - \frac{RT}{V}\right)dV - RT\ln z + RT\ln\frac{x_iP}{P_r} \qquad (A60)$$

From (A60) we have the definition for the fugacity coefficient for the case that the EOS needs V and T as state variables

$$RT\ln\varphi_{i(V,T)} = \int\limits_{V}^{\infty}\left(\left(\frac{\partial P}{\partial n_i}\right)_{V,T,n_{i\neq j}} - \frac{RT}{V}\right)dV - RT\ln z \qquad (A61)$$

As in Equation (A30), we can also write

$$\mu_{i(P,T)} = h^{o*}_{i(P_r,T)} - Ts^{o*}_{i(P_r,T)} + RT\ln\frac{x_i\varphi_{i(V,T)}P}{P_r} \qquad (A62)$$

or with Equation (A29)

$$\mu_{i(P,T)} = h^{o*}_{i(P_r,T)} - Ts^{o*}_{i(P_r,T)} + RT\ln\frac{f_{i(V,T)}}{P_r} \qquad (A63)$$

Again, the only prerequisite for the definition of the fugacity coefficient is the ideal gas law. Identical relationships as in Equations (A31)-(A33) can be formulated by just replacing P by V in the subscript.

v and *T* as state variables

In some cases it is convenient to use molar quantities and mole fractions in the formulation of an EOS. In that case, a slightly different method can be used to calculate $\varphi_{i(V,T)}$. Let us start with a slightly different form of Equation (A53)

$$A_{(V,T,n)} = \sum_i n_i a^{o*}_{i(v_r,T)} + n_T\int\limits_{v}^{\infty}\left(P - \frac{RT}{v}\right)dv - RT\sum_i n_i\ln\frac{P_r v}{x_i RT} \qquad (A64)$$

and make it molar by dividing by the total number of moles n_T

$$\frac{A_{(V,T,n)}}{n_T} = a_{(v,T,x)} = \sum_i x_i a^{o*}_{i(v_r,T)} + \int\limits_{v}^{\infty}\left(P - \frac{RT}{v}\right)dv - RT\sum_i x_i\ln\frac{P_r v}{x_i RT} \qquad (A65)$$

It is obvious that:

$$A_{(V,T,n)} = n_T a_{(v,T,x)} \qquad (A66)$$

Here the molar Helmoltz free energy $a_{(v,T,x)}$ is a function of v, T, and x instead of V, T, and n. Now instead of differentiating the left side of Equation (A66), we use the right hand side and apply the product rule

$$\left(\frac{\partial A_{(V,T,n)}}{\partial n_i}\right)_{V,T,n_{j\neq i}} = \mu_{i(V,T,n)} = \left(\frac{\partial n_T a_{(v,T,x)}}{\partial n_i}\right)_{V,T,n_{j\neq i}} = a_{(v,T,x)} + n_T\left(\frac{\partial a_{(v,T,x)}}{\partial n_i}\right)_{V,T,n_{j\neq i}} \qquad (A67)$$

Because x as well as v^{10} are functions of n we have to apply the general chain rule for the differentiation:

$$\mu_{i(V,T,n)} = a_{(v,T,x)} + n_T \sum_j \left(\frac{\partial a_{(v,T,x)}}{\partial x_j}\right)_{v,T,x_{k \neq j}} \left(\frac{\partial x_j}{\partial n_i}\right)_{n_{k \neq i}} + n_T \left(\frac{\partial a_{(v,T,x)}}{\partial v}\right)_{T,x} \left(\frac{\partial v}{\partial n_i}\right)_{n_{k \neq i}} \quad (A68)$$

Inserting the appropriate definitions for x and v we have:

$$\mu_{i(V,T,n)} = a_{(v,T,x)} + \left(\frac{\partial a_{(v,T,x)}}{\partial x_i}\right)_{v,T,x_{k \neq i}} - \sum_j x_j \left(\frac{\partial a_{(v,T,x)}}{\partial x_j}\right)_{n_{k \neq j}} - n_T \left(\frac{\partial a_{(v,T,x)}}{\partial v}\right) \frac{v}{n_T} \quad (A69)$$

Equation (A65) can be split into an *ideal* and an *excess* term:

$$\mu_{i(V,T,n)} = \left(\frac{\partial A^{ideal}_{(V,T,n)}}{\partial n_i}\right)_{V,T,n_{j \neq i}} + \left(\frac{\partial A^{excess}_{(V,T,n)}}{\partial n_i}\right)_{V,T,n_{j \neq i}} \quad (A70)$$

or:

$$\mu_{i(V,T,n)} = a^{ideal}_{i(v,T,x)} + a^{excess}_{i(v,T,x)} + n_T \left(\frac{\partial a^{ideal}_{i(v,T,x)}}{\partial n_i}\right)_{V,T,n_{j \neq i}} + \left(\frac{\partial a^{excess}_{i(v,T,x)}}{\partial n_i}\right)_{V,T,n_{j \neq i}} \quad (A71)$$

The *ideal* term handles the terms which use only ideal gas properties

$$\left(\frac{\partial A^{ideal}_{(V,T,n)}}{\partial n_i}\right)_{V,T,n_{j \neq i}} = a^{o*}_{i(v,T,x)} + RT \ln x_i - RT \ln \frac{P_r v}{RT} + RT \quad (A72)$$

and the excess term the rest dealing with the real properties, the integral in Equation (A65)

$$\left(\frac{\partial A^{excess}_{(V,T,n)}}{\partial n_i}\right)_{V,T,n_{j \neq i}} = a^{excess} + \left(\frac{\partial a^{excess}}{\partial x_i}\right) - \sum_j x_j \left(\frac{\partial a^{excess}}{\partial x_j}\right) + Pv - RT \quad (A73)$$

with:

$$a^{excess} = \int_v^\infty \left(P - \frac{RT}{v}\right) dv \quad (A74)$$

Equations (A72) and (A73) both use Equation (A69). So we have:

$$\mu_{i(V,T,n)} = a^{o*}_{i(v,T,x)} + RT \ln x_i - RT \ln \frac{P_r v}{RT} + RT$$

$$+ a^{excess} + \left(\frac{\partial a^{excess}}{\partial x_i}\right) - \sum_j x_j \left(\frac{\partial a^{excess}}{\partial x_j}\right) + Pv - RT \quad (A75)$$

If we expand again with $RT \ln (P/P)$ and use Equation (A44) we get:

---

10   V is the relevant thermodynamic variable and $v = V/n_T$.

$$\mu_{i(V,T,n)} = h^{o*}_{i(v,T,x)} - T s^{o*}_{i(v,T,x)} + RT \ln x_i + RT \ln \frac{P}{P_r} \tag{A76}$$

$$+ a^{excess} + \left(\frac{\partial a^{excess}}{\partial x_i}\right) - \sum_j x_j \left(\frac{\partial a^{excess}}{\partial x_j}\right) - RT \ln z + Pv - RT$$

With the definition of z the fugacity coefficient $\varphi^m_{i(v,T,x)}$ is defined as:

$$RT \ln \varphi^m_{i(v,T,x)} = a^{excess} + \left(\frac{\partial a^{excess}}{\partial x_i}\right) - \sum_j x_j \left(\frac{\partial a^{excess}}{\partial x_j}\right) - RT \ln z + RT(z-1) \tag{A77}$$

or just as:

$$RT \ln \varphi^m_{i(v,T,x)} = \left(\frac{\partial n_T a^{excess}}{\partial n_i}\right)_{v,T,n_{j\neq i}} - RT \ln z + RT(z-1) \tag{A78}$$

ρ and T as state variables

In some cases the molar density ρ is used instead of v. In this case V is readily replaced by ρ:

$$V = \frac{n_T}{\rho} \tag{A79}$$

and:

$$\frac{\partial V}{\partial \rho} = -\frac{n_T}{\rho^2} \tag{A80}$$

As in Equation (A64)

$$A_{(V,T,n)} = \sum_i n_i a^{o*}_{i(\rho_r,T)} + n_T \int_0^\rho \frac{P - \rho RT}{\rho^2} d\rho - RT \sum_i n_i \ln \frac{P_r}{x_i \rho RT} \tag{A81}$$

and make it molar by dividing by the total number of moles n_T

$$\frac{A_{(V,T,n)}}{n_T} = a_{(v,T,x)} = \sum_i x_i a^{o*}_{i(\rho_r,T)} + \int_0^\rho \frac{P - RT\rho}{\rho^2} d\rho - RT \sum_i x_i \ln \frac{P_r}{x_i \rho RT} \tag{A82}$$

Because x as well as ρ are again functions of n:

$$\mu_{i(V,T,n)} = a_{(\rho,T,x)} + n_T \sum_j \left(\frac{\partial a_{(\rho,T,x)}}{\partial x_j}\right)_{v,T,x_{k\neq j}} \left(\frac{\partial x_j}{\partial n_i}\right)_{n_{k\neq i}} + n_T \left(\frac{\partial a_{(\rho,T,x)}}{\partial \rho}\right)_{T,x} \left(\frac{\partial \rho}{\partial n_i}\right)_{n_{k\neq i}} \tag{A83}$$

Inserting the appropriate definitions for x and v, we have:

$$\mu_{i(V,T,n)} = a_{(\rho,T,x)} + \left(\frac{\partial a_{(\rho,T,x)}}{\partial x_i}\right)_{v,T,x_{k\neq i}} - \sum_j x_j \left(\frac{\partial a_{(\rho,T,x)}}{\partial x_j}\right)_{n_{k\neq j}} - n_T \left(\frac{\partial a_{(\rho,T,x)}}{\partial v}\right)_{n_T} \frac{\rho}{n_T} \tag{A84}$$

The *ideal* term handles the terms which use only ideal gas properties

$$\left(\frac{\partial A^{ideal}_{(V,T,n)}}{\partial n_i}\right)_{V,T,n_{j\neq i}} = a^{o*}_{i(\rho,T,x)} + RT \ln x_i - RT \ln \frac{P_r}{\rho RT} + RT \tag{A85}$$

and the excess term handles the rest dealing with the real properties

$$\left(\frac{\partial A_{(V,T,n)}^{excess}}{\partial n_i}\right)_{V,T,n_{j\neq i}} = \int_0^\rho \frac{P-RT\rho}{\rho^2}d\rho + \left(\frac{\partial a^{excess}}{\partial x_i}\right) - \sum_j x_j \left(\frac{\partial a^{excess}}{\partial x_j}\right) + \frac{P}{\rho} - RT \qquad (A86)$$

with:

$$a^{excess} = \int_0^\rho \frac{P-RT\rho}{\rho^2}d\rho \qquad (A87)$$

Equations (A85) and (A86) both use Equation (A84). So we have:

$$\mu_{i(V,T,n)} = a_{i(\rho,T,x)}^{o*} + RT\ln x_i - RT\ln\frac{P_r}{\rho RT} + RT \qquad (A88)$$

$$+ a^{excess} + \left(\frac{\partial a^{excess}}{\partial x_i}\right) - \sum_j x_j \left(\frac{\partial a^{excess}}{\partial x_j}\right) + \frac{P}{\rho} - RT$$

If we expand again with $RT \ln (P/P)$ and use (A44) we get:

$$\mu_{i(V,T,n)} = h_{i(\rho,T,x)}^{o*} - T s_{i(\rho,T,x)}^{o*} + RT\ln x_i + RT\ln\frac{P}{P_r} \qquad (A89)$$

$$+ a^{excess} + \left(\frac{\partial a^{excess}}{\partial x_i}\right) - \sum_j x_j \left(\frac{\partial a^{excess}}{\partial x_j}\right) - RT\ln z + \frac{P}{\rho} - RT$$

With the definition of z the fugacity coefficient $\varphi_{i(\rho,T,x)}^m$ is defined as:

$$RT\ln\varphi_{i(\rho,T,x)}^m = a^{excess} + \left(\frac{\partial a^{excess}}{\partial x_i}\right) - \sum_j x_j \left(\frac{\partial a^{excess}}{\partial x_j}\right) - RT\ln z + RT(z-1) \qquad (A90)$$

U, H, and S as a function of the state variables V, P and T

In some cases we need the properties U, H, and S as a function of the state variables for a fluid. The following partial derivatives are required

$$\left(\frac{\partial S}{\partial V}\right)_{T,n_i} = \left(\frac{\partial P}{\partial T}\right)_{V,n_i} = -\left(\frac{\partial^2 G}{\partial T\partial V}\right)_{n_i} \qquad (A91)$$

$$\left(\frac{\partial S}{\partial P}\right)_{T,n_i} = -\left(\frac{\partial V}{\partial T}\right)_{P,n_i} = -\left(\frac{\partial^2 G}{\partial T\partial P}\right)_{n_i} \qquad (A92)$$

$$\left(\frac{\partial U}{\partial V}\right)_{T,n_i} = -P + T\left(\frac{\partial P}{\partial T}\right)_{V,n_i} \qquad (A93)$$

$$\left(\frac{\partial H}{\partial P}\right)_{T,n_i} = V - T\left(\frac{\partial V}{\partial T}\right)_{P,n_i} \qquad (A94)$$

U as function of V and T. Integration of (A93) gives:

$$U_{(V,T)} = U_{(V_r,T)} + \int_{V_r}^V \left(-P + T\left(\frac{\partial P}{\partial T}\right)_{V,n_i}\right) dV \qquad (A95)$$

For an ideal gas:

$$\left(\frac{\partial P}{\partial T}\right)_{V,n_i} = \frac{n_T R}{V} \tag{A96}$$

As a result, the integral in (A95) is zero for an ideal gas and therefore U is in this case not a function of V. This behavior is understandable, because there is no intermolecular interaction between particles. Expansion of Equation (A95), and splitting the term into an ideal gas and a deviation part like in Equation (A10) is, therefore, not necessary

$$\int_{V_r}^{V}\left(-P+T\left(\frac{\partial P}{\partial T}\right)_{V,n_i}\right)_{\text{ideal gas}} dV = \int_{V_r}^{V}\left(-\frac{n_T RT}{V}+\frac{n_T RT}{V}\right)dV = 0 \tag{A97}$$

However, expansion of the integration boundaries is again useful

$$U_{(V_r,T)} = U_{(V,T)} + \int_{V_r}^{\infty}\left(-P+T\left(\frac{\partial P}{\partial T}\right)_{V,n_i}\right)dV - \int_{V}^{\infty}\left(-P+T\left(\frac{\partial P}{\partial T}\right)_{V,n_i}\right)dV \tag{A98}$$

Combination of the first two terms of the right hand side of Equation (A98) yields U for a *hypothetical ideal gas*. In addition, for an ideal gas mixture, there is no excess U

$$U_{(V_r,T)} = U^{o*}_{(V_r,T)} + \int_{V}^{\infty}\left(P-T\left(\frac{\partial P}{\partial T}\right)_{V,n_i}\right)dV \tag{A99}$$

$$U_{(V_r,T)} = \sum_i n_i u^{o*}_{i(V_r,T)} + \int_{V}^{\infty}\left(P-T\left(\frac{\partial P}{\partial T}\right)_{V,n_i}\right)dV \tag{A100}$$

H as function of P and T. Integration of (A94) gives:

$$H_{(P,T)} = H_{(P_r,T)} + \int_{P_r}^{P}\left(V-T\left(\frac{\partial V}{\partial T}\right)_{P,n_i}\right)dP \tag{A101}$$

For an ideal gas:

$$\left(\frac{\partial V}{\partial T}\right)_{P,n_i} = \frac{n_T R}{P} \tag{A102}$$

As shown in (A103) the integral in (A101) is zero and H for an ideal gas is not a function of P

$$\int_{P_r}^{P}\left(V-T\left(\frac{\partial V}{\partial T}\right)_{P,n_i}\right)_{\text{ideal gas}} dP = \int_{P_r}^{P}\left(\frac{n_T RT}{P}-\frac{n_T RT}{P}\right)dP = 0 \tag{A103}$$

Expanding the lower integration boundary we get:

$$H_{(P,T)} = H_{(P_r,T)} - \int_{0}^{P_r}\left(V-T\left(\frac{\partial V}{\partial T}\right)_{P,n_i}\right)dP + \int_{0}^{P}\left(V-T\left(\frac{\partial V}{\partial T}\right)_{P,n_i}\right)dP \tag{A104}$$

Combination of the first two terms of the right hand side of (A104) yields H for a *hypothetical ideal gas*. In addition, for an ideal gas mixture there is no excess H

$$H_{(P_r,T)} = H^{o*}_{(P_r,T)} + \int_0^P \left(V - T\left(\frac{\partial V}{\partial T}\right)_{P,n_i} \right) dP \tag{A105}$$

$$H_{(P_r,T)} = \sum_i n_i h^{o*}_{i(P_r,T)} + \int_0^P \left(V - T\left(\frac{\partial V}{\partial T}\right)_{P,n_i} \right) dP \tag{A106}$$

S as function of V and T. For simplicity, we consider first a pure phase component i. Integration of (A91) yields

$$S^o_{i(V,T)} = S^o_{i(V_r,T)} + \int_{V_r}^V \left(\frac{\partial P}{\partial T}\right)_{V,n_i} dV \tag{A107}$$

Expansion in an ideal gas and an deviation term:

$$S^o_{i(V,T)} = S^o_{i(V_r,T)} + \int_{V_r}^V \left(\left(\frac{\partial P}{\partial T}\right)_{V,n_i} - \frac{n_T R}{V} \right) dV + \int_{V_r}^V \frac{n_T R}{V} dV \tag{A108}$$

Solving the last integral in (A108):

$$S^o_{i(V,T)} = S^o_{i(V_r,T)} + \int_{V_r}^V \left(\left(\frac{\partial P}{\partial T}\right)_{V,n_i} - \frac{n_T R}{V} \right) dV + n_T R \ln\frac{V}{V_r} \tag{A109}$$

Extending the lower integration limit:

$$S^o_{i(V,T)} = S^o_{i(V_r,T)} + \int_{V_r}^\infty \left(\left(\frac{\partial P}{\partial T}\right)_{V,n_i} - \frac{n_T R}{V} \right) dV - \int_V^\infty \left(\left(\frac{\partial P}{\partial T}\right)_{V,n_i} - \frac{n_T R}{V} \right) dV + n_T R \ln\frac{V}{V_r} \tag{A110}$$

The first two terms of (A110) are the S of an pure *hypothetical ideal gas*. If the system is a mixture ideal mixing entropy has to be added

$$S_{(V,T)} = \sum_i n_i s^{o*}_{i(V_r,T)} - \int_V^\infty \left(\left(\frac{\partial P}{\partial T}\right)_{V,n_i} - \frac{n_T R}{V} \right) dV + n_T R \ln\frac{V}{V_r} - R\sum_i n_i \ln x_i \tag{A111}$$

S as function of P and T. Integration of (A92) again, for pure phase component yields:

$$S^o_{i(P,T)} = S^o_{i(P_r,T)} - \int_{P_r}^P \left(\frac{\partial V}{\partial T}\right)_{P,n_i} dP \tag{A112}$$

Expansion in an ideal gas and a deviation term:

$$S^o_{i(P,T)} = S^o_{i(P_r,T)} - \int_{P_r}^P \left(\left(\frac{\partial V}{\partial T}\right)_{P,n_i} - \frac{n_T R}{P} \right) dP - \int_{P_r}^P \frac{n_T R}{P} dP \tag{A113}$$

Solving the last integral in (A113):

$$S^o_{i(P,T)} = S^o_{i(P_r,T)} - \int_{P_r}^P \left(\left(\frac{\partial V}{\partial T}\right)_{P,n_i} - \frac{n_T R}{P} \right) dP - n_T R \ln\frac{P}{P_r} \tag{A114}$$

Extending the lower integration limit:

$$S_{i(P,T)}^o = S_{i(P_r,T)}^o + \int_0^{P_r}\left(\left(\frac{\partial V}{\partial T}\right)_{P,n_i} - \frac{n_T R}{P}\right)dP - \int_0^P\left(\left(\frac{\partial V}{\partial T}\right)_{P,n_i} - \frac{n_T R}{P}\right)dP - n_T R\ln\frac{P}{P_r} \quad (A115)$$

The first two terms of (A110) are the S of a pure *hypothetical ideal gas*. If the system is a mixture ideal mixing entropy has to be added

$$S_{i(P_r,T)} = S_{i(P_r,T)}^{o*} - \int_0^P\left(\left(\frac{\partial V}{\partial T}\right)_{P,n_i} - \frac{n_T R}{P}\right)dP - n_T R\ln\frac{P}{P_r} - R\sum_j n_j \ln x_j \quad (A116)$$

$$S_{i(P_r,T)} = \sum_i n_i s_{i(P_r,T)}^{o*} - \int_0^P\left(\left(\frac{\partial V}{\partial T}\right)_{P,n_i} - \frac{n_T R}{P}\right)dP - n_T R\ln\frac{P}{P_r} - R\sum_j n_j \ln x_j \quad (A117)$$

APPENDIX B

The EOS by Anderko and Pitzer (1993) for H_2O-NaCl and the augmentation of CO_2 by Duan et al. (1995) are valuable tools for calculation of phase relationships in the system for H_2O-CO_2-NaCl and related phase equilibria involving solids. In both publications and in some further work using this EOS, the calculation of the fugacity coefficients is laid out in the accompanying appendixes. Unfortunately, there are some inaccuracies and misunderstandings in the equations, and few errors in the constants used that lead to some frustration if someone wants to program the EOS. With some effort it was, however, possible to extract the correct relations which I want to share in the following.[11]

Fugacity coefficients

According to Appendix A to calculate the fugacity coefficient we need to solve the following equation:

$$\ln \varphi_{i(v,T,x)}^m = \frac{a_{(v,T,x)}^{excess}}{RT} + \left(\frac{\partial\left(a_{(v,T,x)}^{excess}/RT\right)}{\partial x_i}\right) - \sum_j x_j\left(\frac{\partial\left(a_{(v,T,x)}^{excess}/RT\right)}{\partial x_j}\right) - \ln z + (z-1) \quad (A77)$$

To do this we need expressions for a^{excess}, z, and the derivatives of a^{excess}/RT with respect to x_i.

Excess Helmholtz free energy a^{excess}

In Anderko and Pitzer (1993) the molar Helmholtz free energy is split into an ideal, a hard-sphere, a dipole, and a perturbation contribution:

$$a = a^{ideal} + a^{rep} + a^{dip} + a^{per} \quad (B1)$$

So a^{excess} is:

$$a^{excess} = a^{rep} + a^{dip} + a^{per} \quad (B2)$$

In the Chapter above, the hard-sphere term is labeled a^{hs}. Anderko and Pitzer (1993) call it a^{rep}. To allow the comparison with Anderko and Pitzer (1993) and others, the original notation is

---

[11]   In the following equations in the Appendixes of Anderko and Pitzer (1993) and Duan et al. (1995) are labeled (Anderko xx) and (Duan xx), respectively. Corrected equations are marked with a double arrow.

used here. An exception is η, which is used with different definitions for the hard-sphere and for the dipole term. Therefore, y is used instead of η in the hard-sphere term. In the following the original equations from Anderko and Pitzer (1993) are listed. If needed, corrections are inserted and marked.

a^{ideal}. The ideal term a^{ideal} is:

$$\frac{a^{ideal}}{RT} = \frac{a_r^{o*}}{RT} - \ln\frac{v}{v_r} + \sum_{i=1} x_i \ln x_i \tag{B3}$$

a^{rep}. The repulsion term a^{rep} is:

$$\frac{a^{hs}}{RT} = \frac{a^{rep}}{RT} = \frac{\frac{3DE}{F}y - \frac{E^3}{F^2}}{1-y} + \frac{\frac{E^3}{F^2}}{(1-y)^2} + \left(\frac{E^3}{F^2} - 1\right)\ln(1-y) \tag{B4}$$

with:

$$D = \sum_{i=1} x_i\sigma_i, \qquad E = \sum_{i=1} x_i\sigma_i^2, \text{ and} \qquad F = \sum_{i=1} x_i\sigma_i^3 \tag{B5}$$

$$y = \frac{b}{4v} \tag{B6}$$

$$b = \frac{2}{3}\pi N_A F \tag{B7}$$

a^{dip}. The dipole term a^{dip} is:

$$\frac{a^{dip}}{RT} = \frac{A_2}{1 - A_3/A_2} \tag{B8}$$

with:

$$A_2 = -\frac{4}{3}\sum_{i=1}\sum_{j=1} x_i x_j \frac{b_i b_j}{b_{ij}^2}\eta_{ij}\tilde{\mu}_i^2\tilde{\mu}_j^2 I_{2(\eta)} \tag{B9}$$

$$A_3 = \frac{10}{9}\sum_{i=1}\sum_{j=1}\sum_k x_i x_j x_k \frac{b_i b_j b_k}{b_{ij}b_{ik}b_{jk}}\eta_{ijk}\tilde{\mu}_i^2\tilde{\mu}_j^2\tilde{\mu}_k^2 I_{3(\eta)} \tag{B10}$$

$$\eta_{ij} = \frac{b_{ij}}{4v} \tag{B11}$$

$$\eta_{ijk} = \frac{b_{ijk}}{4v} = \frac{(b_{ij}b_{ik}b_{jk})^{1/3}}{4v} \tag{B12}$$

$$b_{ij} = \left(\frac{b_i^{1/3} + b_j^{1/3}}{2}\right)^3 \tag{B13}$$

$$\eta = \frac{1}{4v}\sum_{i=1}\sum_{j=1} x_i x_j b_{ij} \tag{B14}$$

$$I_{2(\eta)} = 1 + c_1\eta + c_3\eta^2 + c_5\eta^3 \tag{B15}$$

$$I_{3(\eta)} = I_{2(\eta)}\left(1 + c_2\eta + c_4\eta^2 + c_6\eta^3\right) \tag{B16}$$

a^{per}. The dipole term a^{per} is:

$$\frac{a^{per}}{RT} = -\frac{1}{RT}\left(\frac{a}{v} + \frac{acb}{4v^2} + \frac{adb^2}{16v^3} + \frac{aeb^3}{64v^4}\right) \tag{B17}$$

with:

$$a = \sum_{i=1}\sum_{j=1} x_i x_j a_{ij} \tag{B18}$$

$$acb = \sum_{i=1}\sum_{j=1}\sum_{k=1} x_i x_j x_k (ac)_{ijk} b_{ijk} \tag{B19}$$

$$adb^2 = \sum_{i=1}\sum_{j=1}\sum_{k=1}\sum_{l=1} x_i x_j x_k x_l (ad)_{ijkl} b_{ijkl}^2 \tag{B20}$$

$$aeb^3 = \sum_{i=1}\sum_{j=1}\sum_{k=1}\sum_{l=1}\sum_{m=1} x_i x_j x_k x_l x_m (ae)_{ijklm} b_{ijklm}^3 \tag{B21}$$

$$b_{ijk} = \left(\frac{b_i^{1/3} + b_j^{1/3} + b_k^{1/3}}{3}\right)^3 \tag{B22}$$

$$b_{ijkl} = \left(\frac{b_i^{1/3} + b_j^{1/3} + b_k^{1/3} + b_l^{1/3}}{4}\right)^3 \tag{B23}$$

$$b_{ijklm} = \left(\frac{b_i^{1/3} + b_j^{1/3} + b_k^{1/3} + b_l^{1/3} + b_m^{1/3}}{5}\right)^3 \tag{B24}$$

$$a_{ij} = \left(a_i a_j\right)^{1/2} \alpha_{ij} \tag{B25}$$

$$(ac)_{ijk} = \left((ac)_i (ac)_j (ac)_k\right)^{1/3} \gamma_{ijk} \tag{B26}$$

Here is an inexactness. The parameter a_i is always positive, c_i is always negative, and the product $a_i c_i$ is, therefore, negative. Correct version of Equation (B26) is (B27):

$$\Rightarrow \qquad (ac)_{ijk} = -\left(\left|(ac)_i (ac)_j (ac)_k\right|\right)^{1/3} \gamma_{ijk} \tag{B27}$$

$$(ad)_{ijkl} = \left((ad)_i (ad)_j (ad)_k (ad)_l\right)^{1/4} \delta_{ijkl} \tag{B28}$$

$$(ae)_{ijklm} = \left((ae)_i (ae)_j (ae)_k (ae)_l (ae)_m\right)^{1/5} \varepsilon_{ijklm} \tag{B29}$$

The same situation as in equation (B26) arises here, e_i is always negative. Correct version of Equation (B29) is (B30):

$$\Rightarrow \qquad (ae)_{ijklm} = -\left(\left|(ae)_i (ae)_j (ae)_k (ae)_l (ae)_m\right|\right)^{1/5} \varepsilon_{ijklm} \tag{B30}$$

For H_2O and $NaCl$, explicit expression for a_i and c_i are given:

$$\left(ac\right)_i = a_i c_i, \quad \left(ad\right)_i = a_i d_i, \quad \left(ae\right)_i = a_i e_i \tag{B31}$$

In Duan et al. (1995) for CO_2, $(ac)_{CO_2}$, $(ad)_{CO_2}$, and $(ae)_{CO_2}$ are given implicit expressions.

For the subsystem H_2O-NaCl, α, γ, δ, and ϵ are highly temperature dependent, and they are handled by the function τ:

$$\alpha_{ij} = \alpha_{ij}^0 \tau, \quad \gamma_{ijk} = \gamma_{ijk}^0 \tau, \quad \delta_{ijkl} = \delta_{ijkl}^0 \tau, \quad \epsilon_{ijklm} = \epsilon_{ijklm}^0 \tau \tag{B32}$$

The parameters b_i, μ_i, a_i, c_i, d_i, and e_i, are specific for every phase component i and they are provided by the authors.

Compressibility factor z

Also required is the compressibility factor z, which can be calculated using the following equation:

$$z = -v\left(\frac{\partial\, a/RT}{\partial v}\right) \tag{B33}$$

As above, the molar Helmholtz free energy can be split into its contributions:

$$z = -v\left(\frac{\partial\left(a^{ideal} + a^{rep} + a^{dip} + a^{per}\right)/RT}{\partial v}\right) = 1 - v\left(\frac{\partial\left(a^{rep} + a^{dip} + a^{per}\right)/RT}{\partial v}\right) \tag{B34}$$

$$-v\frac{\partial a^{ideal}/RT}{\partial v} = \frac{Pv}{RT} = 1 \tag{B35}$$

The differential of the ideal term is 1 and therefore the Equation (Anderko A1) should read:

$$\Rightarrow \qquad\qquad z = 1 + z^{rep} + z^{dip} + z^{per} \tag{B36}$$

z^{rep}. Anderko and Pitzer (1993) included the ideal term implicitly into the repulsion term. This choice is unfortunate because it contradicts the later use of a^{rep}.

$$z^{rep} = \frac{\left(1 + \dfrac{3DE}{F}\right)y - \left(2 + \dfrac{3DE}{F} - \dfrac{3E^3}{F^2}\right)y^2 - \left(\dfrac{E^3}{F^2} - 1\right)y^3}{\left(1 - y\right)^3} \tag{B37}$$

z^{dip}. The contribution z^{dip} is:

$$z^{dip} = \eta\frac{\left(1 - \dfrac{2A_3}{A_2}\right)\left(\dfrac{\partial A_2}{\partial \eta}\right) + \left(\dfrac{\partial A_3}{\partial \eta}\right)}{\left(1 - \dfrac{A_3}{A_2}\right)^2} \tag{B38}$$

To perform the required differentiation in (B38) the Equations (B9) and (B10) have to be transformed.

$$A_2 = -\frac{4}{3}\sum_{i=1}\sum_{j=1} x_i x_j \frac{b_i b_j}{b_{ij} b^{dip}} \eta \tilde{\mu}_i^2 \tilde{\mu}_j^2 I_{2(\eta)} \tag{B39}$$

$$A_3 = \frac{10}{9}\sum_{i=1}\sum_{j=1}\sum_{k=1}x_i x_j x_k \frac{b_i b_j b_k}{b_{ij}b_{ik}b_{jk}}\frac{\left(b_{ij}b_{ik}b_{jk}\right)^{2/3}}{\left(b^{dip}\right)^2}\eta^2\tilde{\mu}_i^2\tilde{\mu}_j^2\tilde{\mu}_k^2 I_{3(\eta)} \tag{B40}$$

It is important that in these equations that the parameter b is not that of (B7) but is linked to Equation (B14):

$$b^{dip} = \sum_{i=1}\sum_{j=1}x_i x_j b_{ij} \tag{B41}$$

In the original equations this difference is not necessarily obvious. The derivatives are then:

$$\frac{\partial A_2}{\partial \eta} = -\frac{4}{3}\sum_{i=1}\sum_{j=1}x_i x_j \frac{b_i b_j}{b_{ij}b^{dip}}\tilde{\mu}_i^2\tilde{\mu}_j^2\left(I_{2(\eta)}+\eta\frac{\partial I_{2(\eta)}}{\partial \eta}\right) \tag{B42}$$

$$\frac{\partial A_3}{\partial \eta} = \frac{10}{9}\sum_{i=1}\sum_{j=1}\sum_{k=1}x_i x_j x_k \frac{b_i b_j b_k}{b_{ij}b_{ik}b_{jk}}\frac{\left(b_{ij}b_{ik}b_{jk}\right)^{2/3}}{\left(b^{dip}\right)^2}\tilde{\mu}_i^2\tilde{\mu}_j^2\tilde{\mu}_k^2\left(2\eta I_{3(\eta)}+\eta^2\frac{\partial I_{3(\eta)}}{\partial \eta}\right) \tag{B43}$$

with:

$$\frac{\partial I_{2(\eta)}}{\partial \eta} = c_1 + 2c_3\eta + 3c_5\eta^2 \tag{B44}$$

$$\frac{\partial I_{3(\eta)}}{\partial \eta} = I_{2(\eta)}\left(c_2 + 2c_4\eta + 3c_6\eta^2\right) + \frac{\partial I_{2(\eta)}}{\partial \eta}\left(1 + c_2\eta + c_4\eta^2 + c_6\eta^3\right) \tag{B45}$$

z^{per}. The contribution z^{per} is:

$$z^{per} = -\frac{1}{RT}\left(\frac{a}{v}+\frac{acb}{2v^2}+\frac{3adb^2}{16v^3}+\frac{aeb^3}{16v^4}\right) \tag{B46}$$

Derivatives of a^{excess} with respect to x_i

The remaining of the required terms are the partial derivatives with respect to the individual mole fractions.

Derivative of a^{rep}.

$$\frac{\partial\left(a^{rep}/RT\right)}{\partial x_i} = \frac{\partial\left(a^{rep}/RT\right)}{\partial D}\frac{\partial D}{\partial x_i}+\frac{\partial\left(a^{rep}/RT\right)}{\partial E}\frac{\partial E}{\partial x_i}+\frac{\partial\left(a^{rep}/RT\right)}{\partial F}\frac{\partial F}{\partial x_i} \tag{B47}$$

$$\frac{\partial\left(a^{rep}/RT\right)}{\partial D} = \frac{\frac{3E}{F}y}{1-y} \tag{B48}$$

$$\frac{\partial\left(a^{rep}/RT\right)}{\partial E} = \frac{\frac{3D}{F}y-\frac{3E^2}{F^2}}{1-y}+\frac{\frac{3E^2}{F^2}}{\left(1-y\right)^2}+\left(\frac{3E^2}{F^2}-1\right)\ln\left(1-y\right) \tag{B49}$$

Reminding us that y is also a function of F:

$$\frac{\partial\left(a^{rep}/RT\right)}{\partial F}=\frac{\dfrac{y}{F}+\dfrac{E^3}{F^3}(2-y)}{(1-y)}+\frac{\dfrac{3DE}{F^2}y^2-\dfrac{E^3}{F^3}(2+y)}{(1-y)^2}+\frac{\dfrac{2E^3}{F^3}y}{(1-y)^3}-\frac{2E^3}{F^3}\ln(1-y) \quad (B50)$$

with:

$$\frac{\partial D}{\partial x_i}=\sigma_i, \quad \frac{\partial E}{\partial x_i}=\sigma_i^2, \quad \frac{\partial F}{\partial x_i}=\sigma_i^3 \quad (B51)$$

Derivative of a^{dip}. Equation (Anderko A20) and (Duan A13) is:

$$\frac{\partial\left(n_T a^{dip}/RT\right)}{\partial n_i}=\frac{\left(1-\dfrac{2A_3}{A_2}\right)\dfrac{\partial\left(n_T A_2\right)}{\partial n_i}+\dfrac{\partial\left(n_T A_3\right)}{\partial n_i}}{\left(1-\dfrac{A_3}{A_2}\right)^2} \quad (B52)$$

Whereas this Equation (B52) is correct, the following two equations in the respective publications are not (Anderko A21, A22 and Duan A14, A15). In Anderko and Pitzer (1993), the equations are in error with respect to the running indexes and missing factors, both of which were corrected by Duan et al. (1995). But even the corrected equations are wrong in the sense that they are not the derivatives claimed to be in respect to n_i but in respect to x_i. So it is better to use:

$$\Rightarrow \qquad \frac{\partial\left(a^{dip}/RT\right)}{\partial x_i}=\frac{\left(1-\dfrac{2A_3}{A_2}\right)\dfrac{\partial A_2}{\partial x_i}+\dfrac{\partial A_3}{\partial x_i}}{\left(1-\dfrac{A_3}{A_2}\right)^2} \quad (B53)$$

with:

$$\Rightarrow \qquad \frac{\partial A_2}{\partial x_i}=-\frac{8}{3}I_2(\eta)\sum_{j=1}x_j\frac{b_i b_j}{b_{ij}^2}\eta_{ij}\tilde{\mu}_i^2\tilde{\mu}_j^2-\frac{4}{3}\frac{\partial I_2(\eta)}{\partial\eta}\sum_{j=1}\sum_{k=1}x_j x_k\frac{b_j b_k}{b_{jk}^2}\eta_{jk}\tilde{\mu}_j^2\tilde{\mu}_k^2 \quad (B54)$$

$$\Rightarrow \qquad \frac{\partial A_2}{\partial x_i}=\frac{10}{3}I_3(\eta)\sum_{j=1}\sum_{k=1}x_j x_k\frac{b_i b_j b_k}{b_{ij}b_{ik}b_{jk}}\eta_{ijk}^2\tilde{\mu}_i^2\tilde{\mu}_j^2\tilde{\mu}_k^2 \quad (B55)$$

$$+\frac{10}{9}\frac{\partial I_3(\eta)}{\partial\eta}\frac{\partial\eta}{\partial x_i}\sum_{j=1}\sum_{k=1}\sum_{l=1}x_j x_k x_l\frac{b_j b_k b_l}{b_{jk}b_{jl}b_{lk}}\eta_{jkl}^2\tilde{\mu}_j^2\tilde{\mu}_k^2\tilde{\mu}_l^2$$

$$\frac{\partial\eta}{\partial x_i}=\frac{1}{2v}\sum_{j=1}x_j b_{ij} \quad (B56)$$

It is important to note that, in Equations (B9) and (B10), the functions $I_2(\eta)$ and $I_3(\eta)$ are not part of the sum but factors, the respective terms in the derivatives (B54) and (B55) are written in front of the sums. The derivatives of $I_2(\eta)$ and $I_3(\eta)$ in respect to η are given in the Equations (B44) and (B45).

Derivative of a^{per}: The Equations (Anderko A26)-(Anderko A30) in Anderko and Pitzer (1993) and (Duan A19)-(Duan A23) in Duan et al. (1995) are in error in the given context. As above for the dipole terms, the derivatives (Anderko A27)-(Anderko A30) and (Duan A20)-(Duan A23) are not the ones claimed, but again are the derivatives with respect to x_i. The correct equations are:

$$\Rightarrow \qquad \frac{\partial\left(a^{per}/RT\right)}{\partial x_i} = -\frac{1}{RT}\left(\frac{1}{v}\frac{\partial a}{\partial x_i} + \frac{1}{4v^2}\frac{\partial acb}{\partial x_i} + \frac{1}{16v^3}\frac{\partial adb^2}{\partial x_i} + \frac{1}{64v^4}\frac{\partial aeb^3}{\partial x_i}\right) \qquad \text{(B57)}$$

with:

$$\Rightarrow \qquad \frac{\partial a}{\partial x_i} = 2\sum_j x_j a_{ij} \qquad \text{(B58)}$$

$$\Rightarrow \qquad \frac{\partial acb}{\partial x_i} = 3\sum_j \sum_k x_j x_k \left(ac\right)_{ijk} b_{ijk} \qquad \text{(B59)}$$

$$\Rightarrow \qquad \frac{\partial adb^2}{\partial x_i} = 4\sum_j \sum_k \sum_l x_j x_k x_l \left(ad\right)_{ijkl} b_{ijkl}^2 \qquad \text{(B60)}$$

$$\Rightarrow \qquad \frac{\partial aeb^3}{\partial x_i} = 5\sum_j \sum_k \sum_l \sum_m x_j x_k x_l x_m \left(ae\right)_{ijklm} b_{ijklm}^3 \qquad \text{(B61)}$$

Fugacity coefficients

To calculate φ_i^m the Equations (B2), (B36), and the appropriate partial derivatives of a^{excess} have to be inserted into Equation (A77).

Other versions and misprints of constants in parameter functions

Two versions for the parameter a_{NaCl} seem to exist. One is that of Anderko and Pitzer (1993), and the other is an extension used in a program which A. Anderko provided to me in 1995. The extended version is:

$$\Rightarrow \qquad a_{NaCl} = \left(14.412 + 5.644\exp\left(-0.4817(\theta - 8.959)^2\right)\right.$$

$$\left. -10\exp\left(-0.6154\,(\theta - 5.403)^{3.26}\right)\right)\times 10^6 \qquad \text{(B62)}$$

The following parameters by Anderko and Pitzer (1993) are in error. The correct parameters are:

$$\Rightarrow \qquad \gamma_{112}^o = 0.7865 + 0.1487\theta + 0.2537\exp\left(-0.1162\times 10^{-5}\,(\theta - 0.57315)^8\right) \qquad \text{(B63)}$$

$$\Rightarrow \qquad \gamma_{122}^o = 1.4194 + 0.06552\theta + 0.2058\exp\left(-1.604\,(\theta - 5.667)^2\right) \qquad \text{(B64)}$$

Reviews in Mineralogy & Geochemistry
Vol. 65, pp. 99-127, 2007
Copyright © Mineralogical Society of America

4

Liquid Immiscibility in Silicate Melts and Related Systems

Alan B. Thompson*, Maarten Aerts, Alistair C. Hack

Dept. Erdwissenschaften, ETH Zürich
Institute for Mineralogy & Petrology
Clausiusstrasse 25
Zürich, CH-8092, Switzerland
*also at Faculty of Mathematics and natural Sciences, University of Zürich, Switzerland
alan.thompson@erdw.ethz.ch   maarten.aerts@erw.ethz.ch   alistair.hack@erdw.ethz.ch

INTRODUCTION TO NATURAL IMMISCIBLE SYSTEMS

High temperature melts, fluids and gases progressively organize themselves structurally during cooling, usually causing separation of solids, liquids or gases. In many different chemical systems this phase separation results in distinct chemical separation (*immiscibility*), with associated contrasting physical properties in the separating phases. Because of the variety in chemistries, and relative changes in entropy and volumes of the natural mixtures compared to the separated phases, immiscibility can occur in different chemistries during heating and compression, as well as during cooling and decompression.

Although the main emphasis in this volume is on fluid–fluid equilibria, there are good examples and much literature on liquid–liquid equilibria in synthetic silicate melts and natural magmas. In fact much of our understanding of phase separation (for geochemists generally taken simply as liquid–vapor equilibria) actually originates from liquid–liquid immiscibility studies of silicate melts. For a one-component system there are three distinct regions in *PT* space where either solid or liquid or vapor exist as the stable phase. These are separated by three univariant curves (i) the solid–liquid curve, the solidus, (ii) the liquid–vapor curve, which can lead to the most commonest form of critical point (see Fig. 1a), and (iii) the solid–vapor curve, which reflects direct vaporization (sublimation) or condensation of solids from gas. The latter is relevant in some industrial processes and to condensation of stars at vacuum pressures. For each added component with solid, liquid and vapor phases, there is the possibility of mixing of each of the phases, or not. In silicates we are used to recognize immiscibility gaps (solvus) among chemically related minerals (e.g., alkali feldspars), which are miscible at high temperatures in the subsolidus region but are immiscible with cooling and undergo phase separation. We are less used to consider immiscibility between two anhydrous melts (silicate–silicate, silicate–carbonate, silicate–oxide, silicate–sulfide, etc.).

The occurrence of liquid immiscibility in temperature–composition (*T–X*) diagrams is represented by the crest of a two liquid (L_1–L_2) solvus lying at higher temperatures than the melting curve, which it intersects at a eutectic or peritectic (Figs. 1 and 2). By changing chemistry and pressure, it can be observed that the L–L solvus can occur at lower temperature. It appears to pass through the liquidus and in some cases can occur only metastably below the liquidus (Figs. 1e, 2a,b,c). The liquidus region here consists of *supercritical melt*. So the definitions of whether systems are immiscible (Fig. 2), miscible, supersolvus or supercritical depend upon the relative intersections of two curves, the solidus and the *P–T* trajectory of the solvus (critical curve). There are continuous changes in physical properties with chemical composition through critical points. In some liquid–gas systems (also in nature) there are

1529-6466/07/0065-0004$05.00   DOI: 10.2138/rmg.2007.65.4

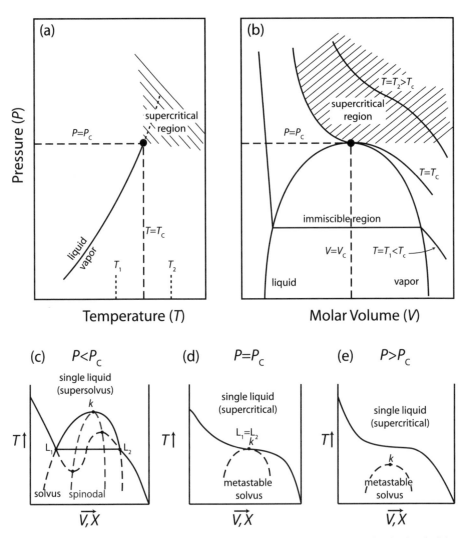

Figure 1. Pressure (*P*) – Temperature (*T*) – Volume (*V*) relations for simple systems showing immiscible, supersolvus and supercritical melt or fluid. (a) *P–T* diagram for a one component system showing the liquid-vapor equilibrium curve extending to the critical point (where $P=P_c$, $T=T_c$). At higher *P* and *T* than this, the single phase is the supercritical fluid. (b) *P–V* section for a one component (or *P–X* in a two-component) system with three isotherms, showing how immiscible liquid and vapor are distinct (where $(T=T_1<T_c)$. These phases are identified by quite different compositions in a binary join, with quite different physical properties (e.g., volume here, but could be density, enthalpy, etc.). *P–V* and *T–V* sections for one component fluid have similar form, but can even have opposite slope depending upon the d*P*/d*T* of the critical curve, (e.g., Levelt Sengers 2000). (c-e) Three *T–X* sections for a two component binary system at different pressures showing relations between solvus and solidus, (c) below P_c, (d) at P_c, (e) above P_c. The components could be oxide–SiO_2 in anhydrous systems, or rock–H_2O as discussed by Hack et al. (2007) . Note the metastable extensions of the solidus curve in (c) inflect at the spinodal (see Prigogine and Defay 1954, p.231).

some important changes expected when circulating in *P–T* in specific directions near to critical compositions (e.g., see Stanley 1971; Hack et al. 2007). But there are probably no rapid changes (spectacular phenomena) involving criticality alone, to be expected in rock+melt systems along common *PT* paths inside the Earth (at least not so far). In recent years there have been considerable advances on studying immiscibility in anhydrous molten silicate systems, but not all in the geological literature. We have not attempted to review the complete developments contained in other recent reviews mentioned here, but rather to integrate these new results into the two silicate melt situation.

Immiscible anhydrous silicate melts and magmas

Immiscible silicate melts have been of interest in geological sciences at least since the late 19th century (see Rosenbusch 1872), when nearby rocks of quite different chemical composition (notably rhyolite and basalt) were suggested to have formed by splitting of an intermediate composition into two-immiscible magmas. In the 1920's, experiments by Greig (1927a,b) and by Bowen (1928) suggested that the liquid immiscibility of binary silicate systems did not extend far into the natural range of rock compositions.

However, natural unmixing of anhydrous silicate melts is still considered to be responsible for a variety of natural rock associations (some mid-oceanic ridge basaltic (MORB) magmas, anorthosites, lunar and terrestrial tholeiitic to alkaline basaltic volcanics, lamprophyre dykes in granitoids, and some layered intrusions; see summary by Roedder 1978). Observations on natural silicate glasses and quenched melts are now complemented by a vast experimental and theoretical literature on liquid immiscibility in technological silicate melts and glasses. Systematic experimental studies of molten oxide-SiO_2 binary systems showed that two immiscible melts can occur in some cases. One very silica-rich melt (often labeled L_1 as in Fig. 1c, SiO_2 on the left; e.g. Barth 1962, p. 144) coexists with a second melt of intermediate composition, usually L_2 (as in Fig. 1c). In other cases, the two-immiscible melts occur only metastably below (at lower temperature than) the liquidus or even solidus. This reflects a transition from immiscible to supercritical melt behavior with increasing pressure or as a function of composition, as discussed further below. The meaning of the term liquidus is quite clear and separates the all melt region from that with solids. The meaning of the term solidus is less clear and is used to refer to the first signs of melt along a divariant melting loop or to a eutectic in binary or higher-degree systems, as well as to the univariant curve for pure substances in the *PT*-diagram (where solidus and liquidus are identical).

LIQUID IMMISCIBILITY IN SILICATE MELTS

Glassy globules of distinct mafic and felsic compositions in tholeiitic to alkaline basaltic rocks (Freestone 1978; Biggar 1979; Dixon and Rutherford 1979; Philpotts 1982; Phillpotts and Doyle 1983; Jakobsen et al. 2005) and ocelli in lamprophyre dykes (felsic blobs in a mafic matrix) are examples (Philpotts 1976; Foley 1984; Bedard 1994) of suggested immiscibility in low pressure volcanic rocks. Most evidence for liquid immiscibility in higher pressure plutonic rocks is much less convincing (e.g., see summary by Bogaerts and Schmidt 2006).

Silicate-oxide anhydrous molten binary systems and the role of network-forming and network-modifying cations

SiO_2 (silica)-rich melts are very viscous and easily suited to glass-working because during cooling they do not quench readily to crystals. It has been shown (e.g., papers in Stebbins et al. 1995) that such siliceous melts consist of a polymeric network of SiO_2, barely modified by additional oxides. Further addition of such oxides eventually (in some cases after passing through a two-liquid region) results in melts of much lower viscosity, where the additional oxides have severely modified the SiO_2-network. Kracek (1930, 1932) summarized the melting

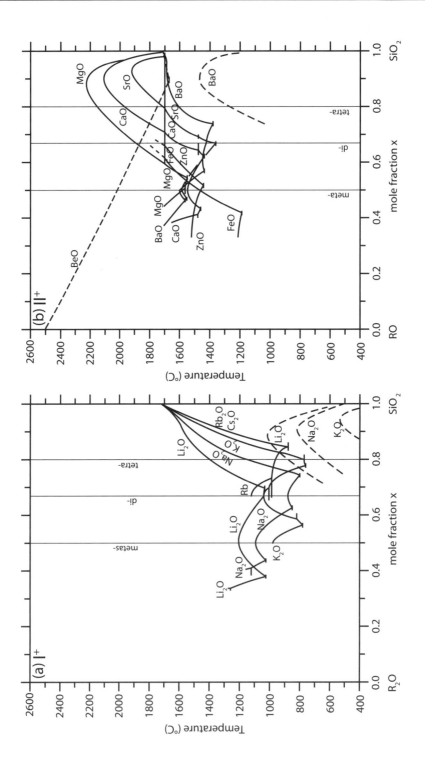

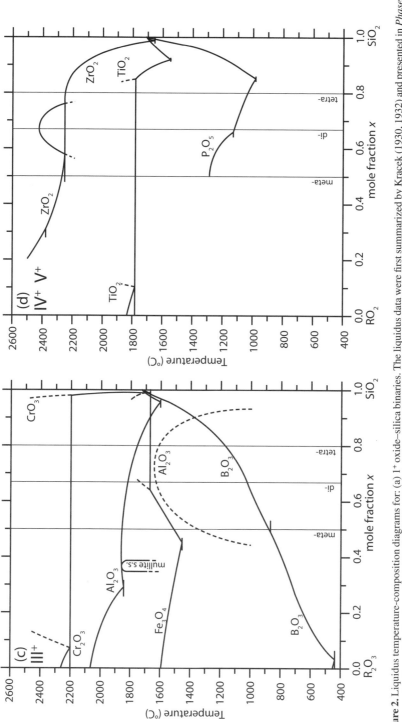

Figure 2. Liquidus temperature-composition diagrams for: (a) 1+ oxide–silica binaries. The liquidus data were first summarized by Kracek (1930, 1932) and presented in *Phase Diagrams for Ceramists* (Levin et al. 1964, Fig. 128; see also Veksler 2004 , p. 12). We have added recent subsolidus data on submerged solvi (see Table 1 for references), (b) 2+ oxide–silica binaries (data in Table 1). (c) 3+ oxide–silica binaries (data in Table 1). Note that Fe_3O_4–SiO_2 is not a true binary but an isobaric section through the system FeO–Fe_2O_3–SiO_2 in air at 1 atm (see Phillips and Muan 1959) (d) 4+ and 5+ oxide–silica binaries (data in Table 1). The stoichiometries for silicate compounds on a molar basis with examples for Na_2O–SiO_2, are according to Kracek (1930, 1932): ortho- (2:1, Na_4SiO_4); meta- (1:1, Na_2SiO_3); di- (1:2, $Na_2Si_2O_5$); tetra- (1:4, $Na_2Si_4O_9$).

diagrams of several oxide–silica binary systems in temperature–composition (T–x, SiO_2 on the right) diagrams. His original figures have been used, and added to, by several authors (see Table 1 and Figs. 2a to 2d)

The 1^+-oxides with SiO_2 show steepening of the slope of the liquidus and the movement of the composition of the first eutectic towards higher-SiO_2 in the order Li > Na > K > Rb > Cs (towards heavier/larger atoms). Recent studies (Table 1) have located the metastable two-liquid regions (identified by onset of opalescence for different prepared glass compositions, Fig 2a), with T_c (critical temperature, L_1=L_2) decreasing and X_c (critical composition) moving towards SiO_2 in the order Li > Na > K (Fig. 2a).

The T–x diagrams for 2^+-oxides with SiO_2 (Fig. 2b) show clearly two-immiscible melts (liquids), with one composition being very close to SiO_2 (L_1) and the other at intermediate compositions towards the oxide (L_2). The phase diagram includes a peritectic reaction with cristobalite (the high-temperature polymorph of SiO_2), and upon cooling the reaction of siliceous liquid (L_1) to cristobalite and the intermediate composition liquid (L_2) occurs (the peritectic reaction L_1 = SiO_2 + L_2). T_p is used to denote the temperature of such a peritectic reaction, where upon cooling one phase reacts to two phases. There is a distinct migration of the composition of L_1 towards SiO_2, and the decreasing temperature of the top of the two-liquid region (the critical temperature of the solvus, T_c), from Fe > Mg > Ca > Sr > Ba (i.e., towards heavier/larger atoms).

Table 1. References for melting equilibria in binary systems with SiO_2

I$^+$		II$^+$		II$^+$, VI$^+$, V$^+$	
Li$_2$O	Kim and Sanders 1991 Kracek 1930 Samsonov 1982 Moriya et al. 1967	BeO	Morgan and Hummel 1949 Budnikov and Cherepanov 1953	Cr$_2$O$_3$	Bunting 1930 (b)
Na$_2$O	Kim and Sanders 1991 Kracek 1930 Samsonov 1982 Haller et al. 1974	MgO	Wu et al. 1993 Michels and Wesker 1988 Ol'shanskii 1951 Kracek 1930	Fe$_2$O$_3$	Phillips and Muan 1959
K$_2$O	Kim and Sanders 1991 Kracek 1930 Moriya et al. 1967 Kawamoto and Tomozawa 1981 Samsonov 1982	CaO	Hageman et al. 1986 Hageman and Oonk 1986 Ol'shanskii 1951 Kracek 1930 Tewhey and Hess 1979 Taylor and Dinsdale 1990	Al$_2$O$_3$	MacDowell and Beall 1969 Staronka et al. 1968 Davis and Pask 1972 Aramaki and Roy 1962 Ball et al. 1993
Rb$_2$O	Kim and Sanders 1991 Kracek 1930 Alekseeva 1963	SrO	Huntelaar et al. 1993 Fields et al. 1972 Ghanbari and Brett 1988 (a, b) Ol'shanskii 1951 Kracek 1930	TiO$_2$	Kaufman 1988 DeVries et al. 1954
Cs$_2$O	Kim and Sanders 1991 Kracek 1930 Samsonov 1982 Charles 1966	BaO	Seward et al. 1968 (a, b) Kracek 1930 Ol'shanskii 1951	ZrO$_2$	Ball et al. 1993 Butterman and Foster 1967
		ZnO	Bunting 1930 (a, c)	MnO$_2$	Singleton et al. 1962 Glasser 1958 White et al. 1934
		FeO	Wu et al. 1993 Bowen and Schairer 1932	P$_2$O$_5$	Tien and Hummel 1962 Baret et al. 1991 Ryerson and Hess 1980

Figure 2c shows SiO_2 binaries with 3^+-oxides (Al_2O_3, Fe_2O_3, B_2O_3, Cr_2O_3), and Figure 2d the 4^+ oxides (TiO_2, ZrO_2) and a 5^+ oxide (P_2O_5) with SiO_2. In summary, relative to the 2^+-oxides–SiO_2 systems which show extensive two-liquid immiscibility, the 1^+-oxide–SiO_2 systems show increasing supercriticality with atomic number; the 3^+ oxide Al_2O_3–SiO_2 and the 5^+ oxide P_2O_5–SiO_2 do not show immiscibility because of compound formation (mullite at $3Al_2O_3 \cdot 2SiO_2$; and $P_2O_5 \cdot SiO_2$, $2P_2O_5 \cdot 3SiO_2$, $P_2O_5 \cdot 4SiO_2$) whereas Fe_2O_3–SiO_2, then Cr_2O_3–SiO_2, as well as TiO_2–SiO_2, ZrO_2–SiO_2, show extensive and progressively increasing L–L immiscibility towards higher temperatures.

High melting point compounds break up the composition join in some cases. Immiscibility is evidence that intermediate compounds of any state (S, L or V) are not stable over the immiscible range. Intermediate solid compounds limit the range over which L–L immiscibility can occur and establish eutectic, peritectic or minimum melting relations. Where intermediate compounds are stable, immiscibility is suppressed. These relations can be seen comparing the relative prevalence of intermediate compounds and immiscibility, compare 1^+ and 5^+ (P_2O_5) oxide–silica versus 2^+, 3^+, 4^+ oxide–silica (Fig. 2).

Factors controlling immiscibility or supercriticality in anhydrous silicate-oxide binary molten systems

The critical (or consolute) temperatures (T_c) of the two-melt fields observed in T–X melting diagrams (as noted with regard to Figs. 1 and 2), appear to correlate with Coulombic properties of the metal cations, such as ionic potential—the ratio of Z/r, where Z is the nominal charge and r is the atomic radius (Fig. 3a,b; see reviews by Hess 1980, 1995; Hudon and Baker 2002a; Veksler 2004; and references therein).

Veksler (2004) made a summary of Hess's (1995) conclusions, here paraphrased: the correlation between T_c and Z/r (Hess 1996, p. 2373, Fig. 9) can be viewed as an indication that the main forces at the atomic scale responsible for the phase separation are Coulombic in character and repulsive. In this view, bridging oxygen ions of highly polymerized silica networks provide poor shielding of network-modifying cations, which results in substantial Coulombic repulsions between the cations, and may eventually lead to phase separation. The higher the ionic potential of modifier cations, the greater are the Coulombic repulsions between them, and the larger are the two-liquid fields in oxide–silica binaries.

However, the detailed examination of liquid immiscibility phenomena in the mineral-silica binaries by Hudon and Baker (2002a,b), revealed more complex relationships between ionic potentials and T–X dimensions of the liquid–liquid miscibility gaps, and were summarized by Veksler (2004, p. 13), here paraphrased: Hudon and Baker (2002a,b) proposed that other factors, mostly related to configurations of electron shells of ions, affected liquid immiscibility. A plot of the critical (consolute) temperatures versus Z/r (Z, r are defined two paragraphs above) constructed by Hudon and Baker (2002a,b; here Fig. 3a) demonstrates that the relationships are parabolic, not linear. Homovalent cations (mono-, di-, trivalent, etc.) can be fitted each by a separate parabola. In relation to two-liquid immiscibility, Hudon and Baker (2002a,b) subdivided cations into four groups. The first group includes large cations ($r > 87.2$ pm in octahedral coordination), which have coordination numbers 5 and higher, cannot enter tetrahedral sites and can act in melt structures only as network modifiers. The typical representatives of the group are alkalis from Na to Cs, alkaline earths from Ca to Ba, light rare-earth elements, U and Th. The second group consists of amphoteric cations (e.g., Li^+, Mg^{2+}, Ga^{3+}, Al^{3+}, Ti^{4+} and Nb^{5+}), which have ionic radii larger than that of Si^{4+} (26 pm, Shannon 1976) but smaller than about 87.2 pm in six-fold coordination or 78.6 pm in four-fold coordination. Bonds formed by those cations with O^{2-} are characterized by substantial degree of covalency. This results in a better shielding of the cations and lower critical (consolute) temperatures than expected from the Z/r values. The third group includes cations with variable crystal field stabilization energies (commonly abbreviated VCFSE). They are formed by

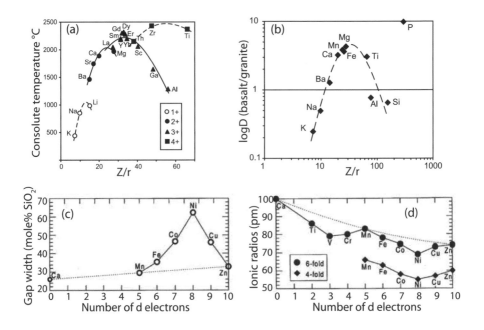

Figure 3. Effects of network-modifying cations on liquid immiscibility in anhydrous silicate systems (a) Consolute (critical) temperatures of the miscibility fields in the metal oxide – silica binaries as a function of the Coulombic properties of the metal cations (after Hudon and Baker 2002a,b); (b) Partitioning coefficients (D) of elements between anhydrous immiscible ferrobasaltic and granitic melts at atmospheric pressure, from experimental data by Longhi (1990) and Ryerson and Hess (1978). Dashed curve is a reference curve emphasizing the general trend of the data points, plotted as Z/r; Z—nominal electric charge, r—ionic radii in nanometers from Shannon (1976), see Veksler (2004) for discussion [Used by permission from Elsevier, from Veksler (2004), *Chemical Geology*, Vol. 210, Fig. 3, p. 14]; (c) miscibility gap widths of divalent cations with variable crystal field stabilization energies (VCFSE), and (d) Ionic radii of divalent cations with VCFSE. Cations adopting a low-spin state are expected to have very large immiscibility fields. The miscibility gap widths of Mn^{2+} and Ni^{2+} are assessed from the main trend defined by Hudon and Baker (2002a) [Used by permission from Elsevier, from Hudon and Baker (2002), *J Non-Cryst Solids*, Vol. 303, Fig. 5, p. 323].

elements that occupy the first row of transition elements in the periodic table (Fe^{2+}, Co^{2+}, Ni^{2+}, Cu^{2+}, V^{2+} and Cr^{2+}) and characterized by five d-electron orbitals. The orbitals are known to poorly shield the atomic nucleus (Hudon and Baker 2002a,b; and references therein), and the miscibility gaps associated with the cations in the binaries are larger than expected from simple Z/r relationships. Finally, highly polarizable cations with a lone pair of electrons (Pb^{2+}, Sn^{2+}, Bi^{3+}, Tl^{2+} and Te^{4+}) show better compatibility with polymerized silica-, borate- and germanate-networks and reduced miscibility gaps (Hudon and Baker 2002b).

Effect of higher pressure on liquid immiscibility in anhydrous molten silicate binaries

Although high pressure–temperature melting experiments have been conducted for some time (Bridgman 1927; Boyd and England 1960; see papers in Hemley 1999; and Hazen and Downs 2001) only recently has attention been focused on the anhydrous silicate binaries. From their experiments on MgO–SiO_2, CaO–SiO_2 and $CaMgSi_2O_6$–SiO_2, Hudon et al. (2004) have shown (Fig. 4a,c) that the critical curve (the P,T locus of the critical temperatures, T_c, where $L_1=L_2$) has a positive dP/dT slope (i.e., moves to higher T with increasing P). Moreover, because the peritectic reaction SiO_2-mineral+L_2=L_1 (at T_p) has a positive dP/dT slope that is less steep than the critical curve, the two curves intersect at higher pressure. This is shown in Figures 4a and 4c. This point shows the pressure where both T_c and T_p occur at the same temperature ($T_c = T_p$), it is a critical point. At lower pressures $T_c > T_p$, and at higher pressures $T_c < T_p$. The progress towards supercriticality, involves moving through the intersection (where $T_c = T_P$) of the critical curve and liquidus for the binary which is represented by the peritectic reaction, SiO_2-mineral+L_2=L_1. The SiO_2 liquidus becomes supercritical at $P \geq P$ ($T_c = T_P$), because the L_1+L_2 solvus submerges below the liquidus and becomes metastable. Hudon et al. (2004, page 12) suggest that the two liquid miscibility gaps close at 1.81 (MgO–SiO_2) and 1.33 GPa (CaMgSi$_2$O$_6$–SiO_2). Dalton and Presnall (1997) also suggested that MgO–SiO_2 becomes supercritical below 5.0 GPa. Figures 4a and 4b suggest that the MgO–SiO_2 L_1+L_2 miscibility gap is metastable (i.e., liquids are supercritical) above ca. 2 ($\pm$ 0.2) GPa and that CaMgSi$_2$O$_6$–SiO_2 is supercritical at pressures above ca. 1.5 ($\pm$ 0.2) GPa.

Hudon et al. (2004, p.14) and Hudon and Baker (2002a,b) noted that for silica-1$^+$oxide binaries, increasing pressure beyond P_c further decreases the size of the metastable miscibility gaps (example Li$_2$O–SiO_2), i.e. promotes supercriticality. For Ti^{4+} in TiO$_2$–SiO_2, pressure enlarges the size of immiscibility fields (experiments at 3 GPa by Circone and Agee 1995). Hudon et al. (2004, p.14, § 42) explain this as follows "Pressure appears to have a more pronounced effect on Ti^{4+} than on Mg^{2+} or Li$^+$, because there is probably more Ti^{4+} than Mg^{2+} or Li$^+$ in four-fold coordination in binary melts. Consequently, increasing pressure converts more Ti^{4+} than Mg^{2+} or Li$^+$ to six-fold coordination, which enlarges the TiO$_2$–SiO_2 miscibility gap more than the immiscibility fields associated with Mg^{2+} or Li$^+$. This explanation is supported by the ionic potentials (defined as Z/r; where Z is the valence and r is the ionic radius) of the cations: IV-fold Ti^{4+} has the largest with 9.52, IV-fold Mg^{2+} has 3.51, and IV-fold Li$^+$ has the smallest with 1.69; consequently Ti^{4+} is the most capable of polarizing the oxygen anions toward it to make covalent bonds and adopt a coordination of 4 while Li$^+$ is the least." Their conclusions are illustrated in Figures 3c and 3d.

Simplified representations of immiscibility, miscibility and supercriticality

Taken all together, these binary (oxide–SiO_2) melting diagrams show a quite remarkable progression of decreasing immiscibility with higher charge and atomic numbers, through BaO–SiO_2 (Greig 1927a, his Fig. 13) where the liquidus has a small flat (S-shaped portion, Fig. 2b), through to the 1$^+$-oxides–silica where the liquidus is inflected (S-shaped) and hides the metastable 2-liquid region (Figs. 1e and 4b). *The passing of the critical point (T_c) of the 2-liquid region through the liquidus locates the supercritical point.* Thus, the T–x phase diagrams for 1$^+$-oxide–silica binaries all show supercritical melt (liquid) behavior. The *subcritical region* is where immiscibility occurs and two liquids are related through a peritectic reaction with the end-member solid, and the *supercritical melt region* occurs where there is a continuity of melt compositions from SiO_2- end-member to eutectic with a second solid. Such nomenclature is discussed further by Hack et al. (2007) in the context of liquid-gas-fluid systems. In all such systems the difference between the two phases can be seen by plotting against temperature (T), a chemical compositional variable (T–X), or a physical property (T–V, volume, T–ρ, density, T–η, viscosity, as in Fig. 1b).

TERNARY AND HIGHER ANHYDROUS MOLTEN SILICATE SYSTEMS

Consideration of the binary T–x diagrams in Figures 2 and 4 indicates which combination in ternary or higher systems will lead to extended immiscibility or not. These anticipations can be compared with the experimentally produced phase diagrams (e.g., the compilations called *"Phase Diagrams for Ceramists"* from the American Ceramic Society, since 1964, *http://www. acers.org/publications/phasecdform.asp*).

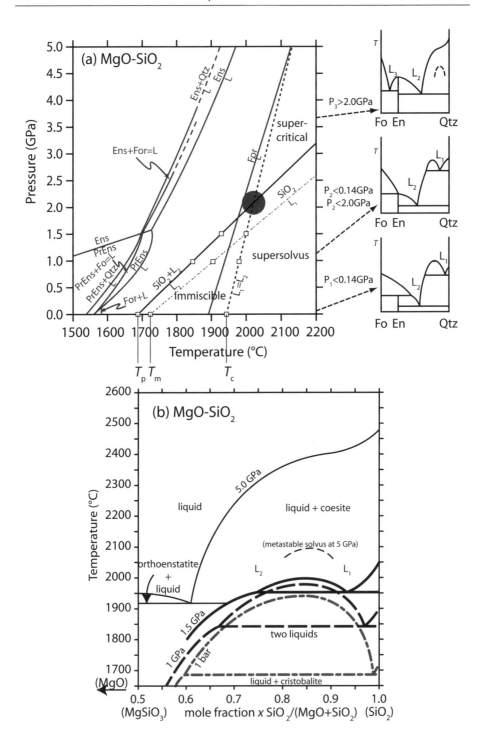

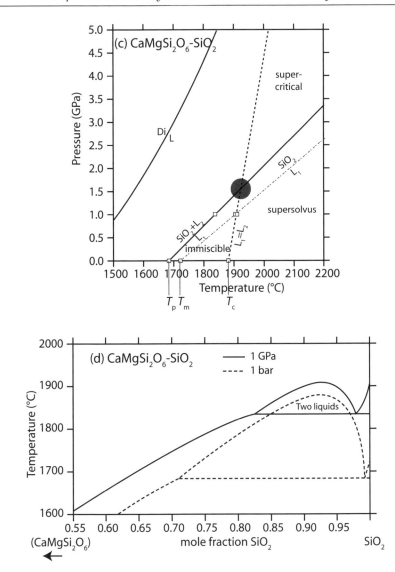

Figure 4 (*on facing page and above*). *P-T* and *T-x* diagrams for MgO–SiO$_2$ and CaMgSi$_2$O$_6$–SiO$_2$: (a) *P-T* diagram for Mg$_2$SiO$_4$–SiO$_2$ showing the +d*P*/d*T* for the melting curve of pure silica (*T$_m$*SiO$_2$), the critical curve for the two liquid region (*T$_c$*, where L$_1$ = L$_2$), and the labeled peritectics (*T$_P$*; on cooling, L$_1$ = L$_2$ + SiO$_2$-mineral). The intersections near 2000 °C, 1.5 GPa and 2020 °C, 2.1 GPa show respectively, the temperatures where *T$_c$* = *T$_m$*SiO$_2$, and the anticipated anhydrous critical end points (grey shaded region where *T$_c$* = *T$_P$*). Immiscible regions are distinguished from miscible regions (and distinguishing supersolvus from supercritical), as are supersolvus from supercritical. Reactions involving proto-enstatite (±quartz) and forsterite melting (lighter, grey curves) are adapted from Morse (1980, p. 363) using data of Boyd et al. (1964). Reactions Ens+For=L and PrEns+For=L from Chen and Presnall (1975), (b) *T-x* diagram for MgO–SiO$_2$ at several pressures superimposed, constructed from Morse[*] (1980, fig. 18.15 p. 362), from Hudon et al. (2004), with data at 5 GPa from Dalton and Presnall (1997), (c) *P-T* diagram for CaMgSi$_2$O$_6$–SiO$_2$, (d) *T-x* diagram for CaMgSi$_2$O$_6$–SiO$_2$, modified from Hudon et al. (2004), diopside liquidus from Boyd and England (1963). Open symbols at 1 atm, 1.0, and 1.5 GPa are data points from Wu et al. (1993), Jung (2003) and Hudon et al. (2004, Figs. 3 and 4). [*]*Note*: Morse (1980) in his Figure 18.15 (p. 362) shows Ens+For=L occurring at a lower *T* than Ens+Qtz=L. (a) is constructed according to the relative *P-T* positions for the reactions Ens+For=L and Ens+Qtz=L, as reported in Chen and Presnall (1975) and Boyd et al. (1964), respectively.

Two binaries both with immiscibility will produce an immiscibility tunnel across the ternary system (e.g., MgO–FeO–SiO$_2$, Bowen and Schairer 1935). Addition of a higher charge cation (e.g., Ti^{4+}) will extend the immiscibility of a 2$^+$-oxide–SiO$_2$ binary (e.g., TiO$_2$–MgO–SiO$_2$, Greig 1927a). Addition of a lower-charged cation (e.g., Li$^+$, Na$^+$) will decrease the immiscibility region of a 2$^+$-oxide–SiO$_2$ or higher system. These observations formed one of the original objections by Greig (1927b) to widespread silicate melt immiscibility in more complex systems compared to basic ternaries. Even though 2-liquid immiscibility is observed in many anhydrous silicate melting systems, they remain in compositions never, or only very rarely, found in nature. In Irvine's (1975b) synthesis of immiscibility effect in magmas, he noted that mafic binaries with silica maintained immiscibility, whereas felsic binaries with silica had eutectic and probably "submerged" critical phenomena (Fig. 5). Addition of anorthite (Andersen 1915;

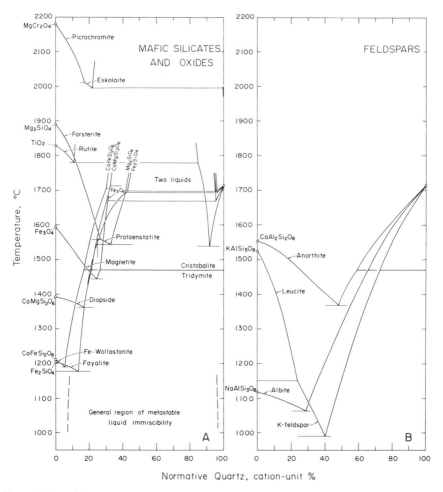

Figure 5. Plot of the 1 atmosphere experimental liquid relations between silica and end-members of the common rock-forming mafic silicates, oxide and feldspar minerals. Note that silica-rich melt shows immiscibility with all mafic minerals and oxide melts but complete miscibility with the feldspar melts, as plotted by Irvine (1975b), using data from Levin et al. (1964; Figs. 80, 82, 113, 266, 412, 508, 586, 599, 715 and 957). [Used by permission of Carnegie Institution of Washington, from Irvine (1975b), *Carnegie Institution of Washington – Yearbook*, Vol. 74, Fig. 57, p. 486.]

Irvine 1975a; see Morse 1980, p. 162), or diopside (Bowen 1914; Kushiro 1972; see Morse 1980, p. 185) to the join Mg_2SiO_4–$MgSiO_3$–SiO_2 slightly extends the MgO–SiO_2 anhydrous binary immiscibility into the ternary systems. The immiscible region then submerges beneath the solidus lowered by the additional component that builds ternary cotectics and eutectics.

The double role of Al_2O_3 in silicate melts

Of greater interest in silicate systems to be applied to natural examples are pseudo-ternary systems where mineral species consisting of two or more oxide components form at least one apex. Very important here are aluminosilicates normally with an alkali element (commonly feldspars and feldspathoids). The role of Al in silicate melts is two-fold, mainly because of its diverse nature (behaving differently in different chemistries and at different pressures). At low pressures, Al mostly forms a four (IV)-coordinated species and joins Si in corner-shared aluminosilicate tetrahedral networks, but it may also act as a network modifier in six (VI)-fold octahedral coordination (at higher pressures).

As is seen in Figure 2c, the Al_2O_3–SiO_2 binary does not show melt immiscibility at lower pressure (1 atm). The solid phases (cristobalite, mullite and corundum), coexisting with silicate liquids, are all very refractory. A metastable solvus has been found at 1 atmosphere beneath the cristobalite liquidus (see MacDowell and Beall 1969; Hudon and Baker 2002a,b). Addition of alkali -or alkaline earth- oxides, causes enormous lowering of liquidus temperatures, eliminates the metastable immiscibility in alumina–silica glasses (MacDowell and Beall 1969) and causes precipitation of feldspars and/or feldspathoids. Hess (e.g., 1995) notes that aluminosilicate units contain tetrahedrally (IV)-coordinated Al^{3+}, which appears to charge-balance mono-or divalent network modifiers. Thus, the molar ratio of Na, K, Ca, (or the typical network modifiers) to Al is an important chemical parameter controlling immiscibility in silicate systems. New spectroscopic work needs to be done to determine the nature and lifetime of certain structural units in the melt (see Farnan and Stebbins 1990; Hans Keppler, personal communication).

Silicate melt immiscibility should be enhanced in "alkaline" compositions (where the sum of the mole-fractions of Na + K + 0.5 Ca + 'network-formers' > the mole fraction of Al) because the excess of network modifiers will tend to form a separate liquid phase (as in the oxide binaries with silica). Conversely, "aluminous" compositions (Na + K + 0.5 Ca + 'network-formers' < Al) will precipitate refractory aluminous crystalline phases from a single homogenous melt, and not show stable immiscibility. The condition where molar Na + K + 0.5 Ca = Al, is special because it defines the compositions of crystalline feldspars and feldspathoids.

Immiscibility in mineral ternary alkali-aluminosilicate melts

Mineral ternary systems thus represent sections through polycomponent composition space relevant to the compositions of magmatic (and metamorphic) rock evolution. Several of such *mineral ternary systems* in *T–X* diagrams show portions with a flat liquidus (like BaO–SiO_2) and thus, presumably hide metastable ternary solvi (e.g., leucite–SiO_2–diopside, nepheline–SiO_2–diopside; as well as albite–fayalite; leucite–forsterite–silica, orthoclase–diopside; Holgate 1954; Barth 1962; Roedder 1978, p. 1611). One particular mineral ternary system (leucite–fayalite–silica) has received a lot of experimental attention because it does indeed show ternary silicate liquid immiscibility (Roedder 1951, 1979). The ternary immiscibility occurs because the lowering of the solidus by the third component is larger than the ternary suppression of the solvus (Barron 1991). The immiscible compositions affect granitic and ferrobasaltic magmatic rocks (enriched in FeO, K_2O and SiO_2) discovered in lunar basalts and in a number of terrestrial igneous rocks (Fig. 6).

Important here are two coincident effects, one is the high FeO-content and the second is the incongruent melting at low pressure of $KAlSi_3O_8$ to leucite ($KAlSi_2O_6$) + SiO_2-rich melt. This permits a re-emergence of the two-liquid immiscibility of the "FeO"–SiO_2 binary into

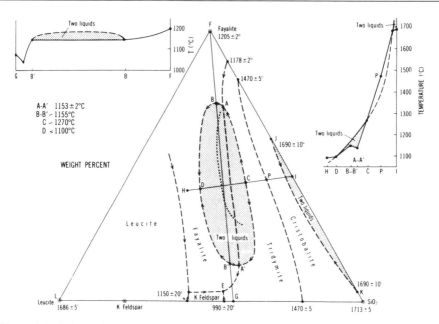

Figure 6. Preliminary diagram of the system leucite–fayalite–SiO₂ (Roedder 1951), showing fields of immiscibility (shaded) at high temperature (along J–K), and at low temperature (A–B–DB′–A′–CA). Dotted line is the 1180 °C isotherm on the upper surface of two-liquid solvus (Watson 1976). The inset figures are *T–X* sections along the lines G-F and H-I. [Used by permission of Elsevier, from Roedder (1978), *Geochimica et Cosmochimica Acta*, Vol. 42, Fig. 2, p. 1601.]

the ternary systems with leucite, along the ternary tridymite + fayalite cotectic. High FeO-contents are significant here because Fe end-members of Fe–Mg silicate solid solutions have low crystallization temperatures (at 1 atmosphere, fayalite (Fe_2SiO_4) melts at 1205 °C, whereas forsterite (Mg_2SiO_4) melts at 1890 °C). Furthermore, pure ferrosilite pyroxene ($FeSiO_3$) is not stable at low pressures, and thus crystallization is suppressed such that polymerized FeO-silica-rich melt develops instead of crystallizing Fe-rich pyroxene. This example demonstrates well the key role of network modifiers (here Fe^{2+}) and the competitive relationships between immiscibility and crystallization (Fig. 6). The analogous Na-system with nepheline ($NaAlSiO_4$)–fayalite–silica (Bowen 1937) does not develop stable ternary immiscibility, partly reflecting compound formation (albite, $NaAlSi_3O_8$) and eutectic behavior between nepheline–albite and albite–SiO₂. The Na- chemical analogue to the feldspathoid leucite ($KAlSi_2O_6$) is the pyroxene mineral jadeite ($NaAlSi_2O_6$), which is not stable at 1 atmosphere pressure with silicate melt.

Tie-lines connecting coexisting immiscible melts run lengthwise along the two-liquid dome in the middle of the fayalite–tridymite field (profile GF in Fig. 6). The SiO₂-rich and the lower-SiO₂ melts could be called felsic and mafic melts respectively. They represent granite and ferrobasalt (fayalite, plagioclase, ilmenite, etc.) immiscibility. Melts arriving by crystallizing tridymite (SiO₂) meet the solvus dome (see inset on top right in Fig. 6, along profile IH) and move either towards fayalite or towards $KAlSi_2O_6$ + SiO₂ depending where they meet relative to point C. They could end their crystallization history at A or A′ respectively, if these points are truly ternary eutectics. Melts beginning on the leucite–fayalite side will meet the dome between B and B′, depending upon initial composition, and end crystallization at D. There is some discussion about the relative temperatures of the possible eutectics A and A′ and the temperature maxima B and B′ along the join GF (e.g., Biggar 1983; Roedder 1983).

There have been several studies of how the other common rock-forming oxide components, and also pressure, affect the two liquid immiscibility in leucite–fayalite–SiO$_2$. Replacing some FeO by MgO increases the liquidus temperature and forces the immiscible region below the liquidus (Bogaerts and Schmidt 2006). Nakamura (1974) suggests that the immiscibility gap in leucite–fayalite–SiO$_2$ is closed at 1.5 GPa. The *supercritical point* describing the transition should be found at some intermediate pressure. Adding small amounts of CaO depresses both the liquidus and the two-melt field, but at higher CaO Hoover and Irvine (1978) noted a re-emergence of immiscibility with ferrobustamite (to about 10 °C above the liquidus). Adding TiO$_2$ causes a small expansion of the two-melt field (Visser and Koster van Groos 1979b). P$_2$O$_5$ expands the two-melt field and depresses the liquidus (Watson 1976; Visser and Koster van Groos 1979a,c; Bogaerts and Schmidt 2006). This occurs despite the fact that no stable immiscibility was reported in the P$_2$O$_5$–SiO$_2$ binary (Levin et al. 1964; Ryerson and Hess 1980). Increased pressure suppresses two-melt immiscibility in the systems just referred to, as in the oxide–SiO$_2$ binary systems discussed above.

It is likely that terrestrial magmas usually contain H$_2$O whose effect on immiscibility we have not considered. However, lunar lavas are likely to have been anhydrous. Roedder (1978, p. 1613, his Fig. 8) has plotted the analyses of lunar lavas (obtained with Paul Weiblen) on a more general diagram than leucite–fayalite–SiO$_2$, i.e. using molar units as (Na$_2$O+K$_2$O+Al$_2$O$_3$)–(CaO+MgO+FeO+TiO$_2$+P$_2$O$_5$)–SiO$_2$ in Figure 7.

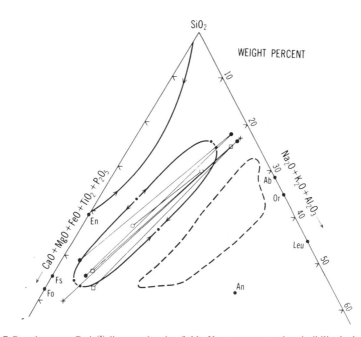

Figure 7. Pseudoternary Greig[*] diagram showing field of low-temperature immiscibility in the system leucite–fayalite–SiO$_2$. and tie lines for various conjugate melt pairs, adapted from Weiblen and Roedder (1973, 1977). All compositions recalculated on the basis of plotted oxides only: symbols • = coexisting glasses in lunar basalts from Apollo 11, 12 and 15; + = coexisting glasses in lunar basalt 14310; o = synthetic Apollo 11 sample after equilibration at 1045 °C; = Apollo 15 grain. The last item is from Switzer (1975); all other data from Roedder and Weiblen (1970, 1971 and 1972) and Weiblen and Roedder (1973). Other lunar samples are similar but have been omitted for clarity. Most analyzed volcanic rock suites fall within the dashed field (Brooks and Gelinas 1975). (*) _Note_ that (Roedder (1979, p.1613) attributes the diagram to Greig (1927b), but Roedder (1978) himself appears to have added TiO$_2$ and P$_2$O$_5$ to the CFM corner. [Used by permission of Elsevier, from Roedder (1978), *Geochimica et Cosmochimica Acta*, Vol. 42, Fig. 8, p. 1613.]

It is quite remarkable that the lunar lavas straddle the two-melt immiscibility dome in the Greig diagram (Fig. 7), a "serendipitous confirmation" of the occurrence of some two-silicate-melt immiscibility in lunar nature. The apparent compositional shift of natural rocks away from silica relative to the field for leucite–fayalite–SiO_2 mainly reflects the unequal effects of the additional components upon immiscibility (e.g., see Holgate 1954).

Summary for anhydrous silicate melt systems

Both increasing pressure and temperature generally decrease silicate immiscibility in anhydrous magmas, so that anhydrous magmas at high temperatures and depth can be of continuous composition from intermediate to siliceous. Distinct chemical effects are also obvious, because increasing CaO and MgO decreases immiscibility, whereas increasing TiO_2 or P_2O_5 promotes immiscibility (Ryerson and Hess 1980; Ryerson 1985; Mysen 1990; Bogaerts and Schmidt 2006). At microscopic scale and with regard to melt structure, the main factors controlling immiscibility phenomena in silicate melts appear to be the Coulombic properties of network-modifying cations (in natural compositions represented mainly by alkalis and alkaline earths) and their relationships to Al (e.g., see Hess 1995: Mysen 2004).

These immiscibility/supercriticality discussions so far have not considered volatile species. The most important magmatic volatile in the Earth is H_2O (e.g., see Hack et al. 2007), but there are important examples where the magmatic volatile is CO_2. Our next section considers anhydrous melts with the addition of CO_2, and with it carbonate melts that sometimes coexist with anhydrous silicate melts.

MOLTEN SILICATE-CARBONATE SYSTEMS

The occurrence of magmatic carbonatites with specific silicate magmatic rocks (alkali basalts, nephelinites, melilitites), and the occurrence of carbonate ocelli in some mantle xenoliths have been suggested as evidence for magmatic silicate–carbonate liquid immiscibility (e.g., Ferguson and Currie 1971; Rankin and Le Bas 1974; Romanchev and Sokolov 1979; Gittins 1989; Bailey 1989, 1993; Barker 1989; Kjarsgaard and Hamilton 1989; Kjarsgaard and Peterson 1991; Church and Jones 1995; Bell 1998). Other occurrences of magmatic carbonate are ascribed to primary carbonate magma from some parts of the mantle, or originate during fractional crystallization of silicate magma (see Bell et al. 1998).

Experimental studies at high pressures, firstly in the system $CaO–MgO–SiO_2–CO_2$ (= CMS+CO_2, e.g., Wyllie and Huang 1976, Wyllie 1977, and Eggler 1978a,b) addressed the origin of calcic and dolomitic carbonatite magmas and coexisting silicate magmas. Later studies in more complex synthetic systems and natural compositions (e.g., Falloon and Green 1989; Thibault et al. 1992; Dalton and Wood 1993) revealed the extent of the silicate–carbonate miscibility also with sodic carbonatite magmas.

The systematic effects of composition (including Al/Si) and pressure on the silicate carbonate miscibility gaps have been summarized by Lee and Wyllie (1998a,b). Figure 8 shows their reconstruction for phase relations in quite complex silicate–carbonate systems near 1.0 GPa. The 2-liquid field separates silicate liquids from Na-Ca carbonate liquids.

The subsolidus reaction olivine + clinopyroxene + CO_2 = enstatite + dolomite (with forsterite and enstatite in the CMS+CO_2 model system) has a positive dP/dT slope, and produces dolomite when CO_2 is added to mantle rock with increasing depth, and decarbonation (and melting) occurs with increasing temperature. CO_2 replenishment of the mantle wedge is possible by subduction of carbonates. Therefore, beginning at depths > 70 km at a T of ~ 1000 °C, a dolomitic liquid can be produced as a primary melt from carbonated mantle. Figure 9 shows that the solidus curve for a carbonate-bearing peridotite has a sharp kink or ledge at a depth of ~ 70 km (ca. 2 GPa; Wyllie 1978, 1987 1989; Eggler 1978a,b, 1989; Brey et al. 1983; Green and

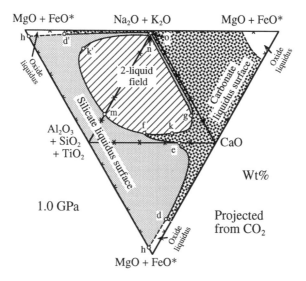

Figure 8 Field boundaries intersected by the end-member triangles of the tetrahedron CaO–(MgO+FeO*)–(Na$_2$O+K$_2$O)–(SiO$_2$+Al$_2$O$_3$+TiO$_2$), projected from CO$_2$, at 1.0 GPa. Three major fields are defined for the silicate–carbonate liquid miscibility gap (two liquid field), silicate liquidus surface, and carbonate liquidus surface. A small oxide field near the (MgO+FeO*) apex is also sketched. k′ is the critical point for the CaO-free miscibility gap m–k′–n. Point d′ between the carbonate and oxide fields in the CaO-free triangle corresponds to d. [Used by permission of Oxford University Press, from Lee and Wyllie (1998a), *Journal of Petrology*, Vol. 39, Fig. 4, p. 502.]

Wallace 1988; Falloon and Green 1989, 1990) corresponding to the above peridotite + dolomite decarbonation reaction. This ledge is much shallower than that observed experimentally for hydrous mantle, where the ledge for hydrated mantle is at about 95 km (3 GPa, Fig. 9b). Thus the lithosphere-asthenosphere boundary would be interpreted to be able to occur at different depths if the mantle is carbonated (70 km) compared to if the mantle is hydrated (95 km). The "oceanic intraplate" geotherm reintersects the hydrated solidus at a minimum depth of ca. 150 km (base of asthenosphere), hence permitting production of small amounts of a carbonate (still dolomitic and then magnesitic) liquid between 95 and 150 km. A reaction between the dolomitic carbonate liquid and mantle lherzolites produces a calcite carbonate liquid, leaving behind wehrlites (olivine + clinopyroxene) through a reaction simplified as dolomite + orthopyroxene = clinopyroxene + olivine + 2CO$_2$ in the complex mantle chemical system. These wehrlites are presumed to act as pipes (conduits) for the calcic-carbonate liquids to pass through on their ascent.

All hypothetical models for the origin of carbonatites invoke the production of a carbonate melt by very small degrees of partial melting reactions involving carbonate minerals in the mantle at rather great depths (200–300 km). Some hypotheses consider primary carbonatite magma, others consider carbonate exsolving as an immiscible liquid phase from or residual to silicate magma during fractionation. At higher pressures (ca. 2.5 GPa) cotectics between carbonate and silicate liquidus surfaces are stable over a considerable range of temperatures. With decreasing pressure, the carbonate and silicate liquidus surfaces retract from one another and become separated due to developing L–L immiscibility (e.g., Fig. 10). So a transition from supercriticality to immiscibility between carbonate and silicate melts occurs during decompression.

Several quite diverse origins have been proposed for carbonatites (e.g., see Bell et al. 1998), some involving immiscibility between carbonate and silicate liquids increasing with decompression. Some carbonatites are proposed as products of crystallization from a primary

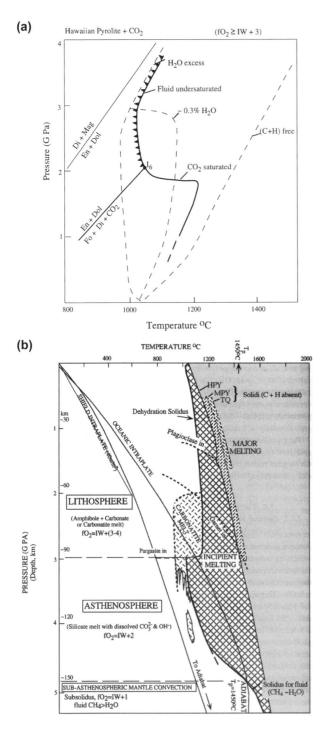

Figure 9. (a) *PT* diagram showing the solidi for Hawaiian pyrolite (a synthetic mixture of 0.25 MORB and 0.75 peridotite) with (C+H+O) compared with the (C+H)-free solidus and the H_2O-saturated and H_2O-undersaturated solidi. These three solidi are shown as short dashed lines. The figure shows the experimentally determined solidus for Hawaiian pyrolite + CO_2 (heavy solid line). The solidus is CO_2-saturated at pressures lower than the intersection of the carbonation reaction with the solidus at I_6 (see Wyllie 1978, 1987). At higher pressures, dolomite is present as a subsolidus phase, the solidus is fluid-undersaturated, and the melt phase is sodic dolomitic carbonatite. (b) The *PT*- experimental data plotted as a function of depth (downwards into the Earth). The oceanic geotherm meets the ledge of the mantle solidus near 95 km, thus defining the base of the lithosphere. The intersection at 150 km defines the base of the asthenosphere in the oceanic mantle. [Used by permission of Cambridge University Press, from Green and Falloon (1998), *The Earth's Mantle – Composition, Structure and Evolution*, Jackson (ed), Chapter 7, Fig. 7.5a, p. 336.]

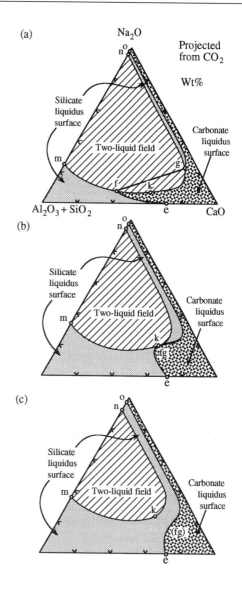

Figure 10. Variation of the geometrical arrangement of silicate–carbonate phase field boundaries as a function of composition. The decrease in size of the miscibility gap (two liquid field) and the nature of the intersection of the two surfaces (silicate, carbonate) occurs from a to b to c, with increasing Al/Si ratio or decreasing pressure; and also decreases with increasing Mg/Ca at constant pressure, (a) The silicate–carbonate liquidus field boundary (e–f, g–n) intersects the miscibility gap field boundary (m–f–k–g–n), at *P* near 1.0 GPa (at 2.5 GPa, point f lies towards point m). (b) The liquidus field boundary is tangential to the miscibility gap field boundary due to increased increasing Al/Si. (c) the two field boundaries are further separated from each other. [Used by permission of Oxford University Press, from Lee and Wyllie (1998a), *Journal of Petrology*, Vol. 39, Fig. 2, p. 449.]

carbonate melt, but experiments have shown that such melts are dolomitic whereas carbonatites are predominantly calcitic in nature). Carbonatite magma produced by these melting reactions and rising through the mantle will solidify at depth prior to reaching the surface, and through cooling release the available volatiles. The association with alkalic rocks reflects widespread fenitization—release of dissolved alkalis and silica from the cooling magma into the surrounding rock, which when of suitable composition reacts with the metasomatising fluids. The shallower... the depth of CO_2 release, the greater the explosion of decarbonation.

Fractional crystallization of an alkalic magma (produced by slightly larger degrees of partial melting in the mantle) does not simply explain why carbonatites are enriched in REE, Zr, Nb, Ta, etc, (although these trace elements could remain incompatible to solids by staying in the liquid during fractionation). If natrocarbonatites are products of liquid immiscibility from high-Na regions of the mantle, this requires the presence of a significant amount of Na_2CO_3 (at least 5% and up to 20%) in the source region.

Ca-carbonatities and Na-carbonatites show quite different immiscibility behavior (Koster van Groos and Wyllie 1963; Freestone and Hamilton 1980; Kjarsgaard 1998; Lee and Wyllie 1998a; Petibon et al. 1998). The formation of latter being probably more important at low pressures and perhaps related to the fractionation of calcic plagioclase in the silicate magma permitting enrichment of Na in the later carbonatite magma.

Because carbonate liquid is characterized by a very-low viscosity and very-low surface tension (e.g., Rabinowicz et al. 2002), it could be mobilized even at low degrees of partial melting (0.01 vol%).

In sodium-bearing systems (the system $Na_2O–CaO–SiO_2–Al_2O_3–CO_2$), a silicate melt is immiscible with a Na-carbonate melt, with the result that three phases may coexist: a silicate melt, a Na > Ca carbonate melt and an alkali–rich volatile fluid.

ANHYDROUS MOLTEN SILICATE SYSTEMS WITH PHOSPHOROUS, FLUORINE, CHLORINE, BORON, SULFUR

Additions of even minor amounts of non-silicate anions (sometimes collectively called salts; of phosphorous, fluorine, chlorine, boron, sulfur) appear to have significant effects on liquid immiscibility, also in anhydrous systems. For example, enhancement of liquid immiscibility between silicate melts and liquids of non-silicate anions have been shown in experimental studies on fluorides (Kogarko and Krigman 1970; Gramenitskiy and Shekina 1994; Gramenitskiy et al. 1993; Veksler et al. 1998a,b); chlorides (Webster and DeVivo 2002) and borates (Veksler et al. 2002a,b, 2004, p.17). Experiments indicate that two sulfide liquids, one metal-poor (M/S ~1.00) and one metal-rich (M/S ~1.14) may exist in parts of the Cu–Ni–Fe–S system (Craig and Kullerud 1969; Kullerud et al. 1969; Cabri 1973). These melts may exist within a crystallized silicate mineral matrix, e.g. sulfides, prominently (Fe, Ni)S solutions, are likely to be molten where present in the upper mantle even at sub-silicate solidus conditions.

Visser and Koster van Groos (1979b) proposed that P_2O_5 strongly increases the miscibility gap between ferrobasaltic and granitic compositions (Fig. 6; Krigman and Krot 1991; Suk 1998). P_2O_5 has a quite different role to the other oxides concerning the modification of silicate melt structure. Ryerson and Hess (1980) noted that P_2O_5 in silicate melts not only produces a marked decrease in silica activity, but co-polymerizes the SiO_2-network rather than modifying the net-work. In more complex melts with other additional cations (as oxides M_xO_y), $M–O–P$ bonds are produced which are stronger than $M–O–Si$ bonds. In immiscible granitic+ferrobasaltic melts, P_2O_5 would bond with high charge density cations Fe^{3+}, Mg^{2+}, Mn^{2+}, Mn^{3+}, Ca^{2+}, Cr^{3+} Ti^{4+} in (depolymerized) lower silica melts (Ryerson and Hess 1980). The binaries of these with silica ($TiO_2–SiO_2$, $Fe_2O_3–SiO_2$, $Cr_2O_3–SiO_2$, $MnO–SiO_2$, $Mn_3O_4–SiO_2$, etc) have fields of liquid immiscibility at high temperatures that cover almost the whole range of composition (Fig. 2). The El Laco magnetite lava flow in Northern Chile (Park 1961), may owe its formation to phosphorous-enhanced immiscibility and the associated fractionation effects (Sillitoe and Burrows 2002).

Veksler (2004) considers that the general rules and regularities of salt–silicate unmixing and the decisive role of network-modifying cations are the same as in the pure silicate systems, paraphrased: in comparison to the network-modifying cations, the effects of the non-silicate anions on liquid immiscibility are subordinate, indirect and secondary. Their role may be better envisaged as fluxing components which suppress crystallization. They may thus reveal the tendencies for unmixing in silicate melts that are otherwise hidden below liquidus surfaces. Non-silicate anions (P, S, F, Cl, and B) provide a means for network modifiers to separate in the form of salt melts, often with very low-SiO_2 concentrations.

SUPERCRITICAL OR SUPERSOLVUS MELTS IN ANHYDROUS SILICATE ROCK SYSTEMS AT HIGHER PRESSURE?

The available experimental data from binary and ternary systems with silica may be used to explore supercritical phenomena in anhydrous silicate rock systems, relevant to planetary interiors. We have seen that distinct classes of miscibility are found at distinct pressures depending upon the charge and size of the cation in the oxide component with SiO_2. We will examine the available anhydrous melting data to see how miscibility might change with pressure in the Earth, in simplified anhydrous peridotite (model upper mantle), basalt (model oceanic crust) and by comparison for granite (model continental crust).

Simplified peridotite mantle

Anhydrous mantle peridotite can be modeled by olivine + pyroxene. In $MgO–FeO–SiO_2$ (MFS) that is the olivine+orthopyroxene region. Both the systems $MgO–SiO_2$ (MS) and $FeO–SiO_2$ (FS) show regions of two melts at 1 atmosphere (Bowen and Anderson 1914; Bowen and Schairer 1932: summarized in Fig. 2b). The ternary phase diagram MFS contains a two-melt tunnel (Bowen and Schairer 1935), between very SiO_2-rich melt (L_1 in Fig. 4b) and intermediate "basaltic" melt (L_2 in Figure 4b). Two reactions are important; one (at ca. 1695 °C at 1 atm) involving these two melts with one solid, at composition between $MgSiO_3$–SiO_2, which with cooling can be written as the eutectic:

L_1 (SiO_2-rich melt) → L_2 (intermediate melt) + SiO_2-mineral (cristobalite/quartz/coesite)

The other is peritectic and involves two solids, enstatite + forsterite, and one melt (basaltic). The peritectic occurs at lower temperatures (1557 °C) than the eutectic (at ca. 1695 °C, at 1 atm), and can be written with cooling as :

forsterite + L_2 (intermediate "basaltic" melt) → enstatite

It is this second peritectic reaction which generates intermediate "basaltic" melt from lower-SiO_2 compositions by precipitation of forsterite. Melt L_2 finally crystallizes at the eutectic (at 1543 °C):

L_2 ("basaltic" melt) → enstatite + SiO_2-mineral (cristobalite, then quartz)

With increasing pressure in the anhydrous $MgO–SiO_2$ system, enstatite melting becomes congruent near 0.14 GPa (e.g., Boyd et al. 1964: see Morse 1980, Fig. 18.10, p. 358) so that now forsterite and enstatite are involved in eutectic melting rather than the above-mentioned peritectic. This eutectic (enstatite + forsterite → L_3, Fig. 11d) generates ultramafic liquids upon mantle melting. These ultramafic liquids (L_3, Fig. 11d) have nothing to do with the two-melt liquid immiscibility which involves L_1 and L_2. The two-melt field is confined to the region between the enstatite + cristobalite eutectic and the silica-rich side. Dalton and Presnall (1997) have confirmed that the two-melt field has disappeared below the $MgSiO_3$ + SiO_2 solidus at 5 GPa. They note that the flat liquidus most likely reflects the submerged two-melt solvus (their Fig. 5, p. 2372). According to the data from Hudon et al. (2004), the system $MgO–SiO_2$ (MS, Fig. 4a) still shows two-melt immiscibility at 1.5 GPa. Our extrapolation (Fig. 4a) for MS shows supercritical intersection between solvus and solidus near 2 GPa. The displacements due to FeO in upper mantle concentrations may lower the critical solvus line and the solidus by similar amounts (lighter lines in Fig. 11a). Thus there is no reason to assume that the equivalent intersection in $FeO–SiO_2$ (FS) should occur at pressure higher than 2 GPa. This means that melt behavior at high-SiO_2 in anhydrous MFS should also go from immiscible (below ca. 2 GPa) to supercritical at quite shallow mantle depths in the Earth (ca. 75 km). The (T–x) diagram in Figure 11d show that at supercritical pressures, a single-melt field is present between $MgSiO_3$ (at B) –SiO_2, and that such anhydrous melts will enter into the two-melt region of coexistence of silicic melt (L_1) with "basaltic" melt (L_2) upon decompression.

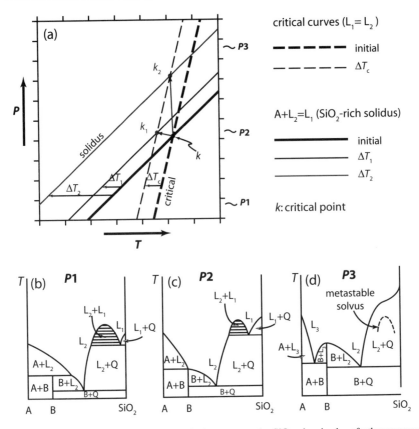

Figure 11. (a) Schematic *PT*-diagram for the anhydrous system A – SiO₂ , showing how further components will influence the location of melting curves (solidi) and solvi, depending on how they partition between solid and melt, relative to the end-member systems illustrated here. Greater solubility will result in a larger ΔT shift ($\Delta T_2 > \Delta T_1$) and thus displacement of the impure critical point k to higher pressures at lower temperatures. FeO for example will partition into melt more than mineral (x_{Fe} liquid > x_{Fe} olivine, which is equivalent to x_{Mg} olivine > x_{Mg} liquid; also for pyroxene). The net effect is to displace the solidus to lower T more than the solvus (ΔT_1 and ΔT_2 are > than ΔT_C). The net effect is to move the critical point for the end-member system to lower T but higher P in the impure system. Schematic *T–x* sections (b–d) for the anhydrous system A – SiO₂ with intermediate compound B, at $P_3 > P_2 > P_1$. For example in Mg₂SiO₄ (composition A) –SiO₂ for which P_2 = ca. 0.14 GPa, there is a switch from incongruent anhydrous melting of enstatite (composition B, P_1 < 0.14 GPa) to congruent (P_3 > 0.14 GPa). Q refers to cristobalite as the low-pressure liquidus phase, and quartz as the high-pressure liquidus phase.

While at $P > 2$ GPa all initially siliceous liquids would end up at the "basaltic" eutectic orthopyroxene (enstatite) + SiO₂, this would only be relevant for subducted intermediate and felsic compositions at mantle depths but not for ultramafic mantle rocks.

Simplified basaltic crust

At lower pressure, basalts can be modeled by plagioclase + clinopyroxene + olivine for which there is some approximate data on immiscibility from the appropriate mineral–silica anhydrous binaries (Fig. 5, from Irvine 1975b). The mafic+SiO₂ versus felsic+SiO₂ diagrams of Figure 4 show that basaltic compositions can be modeled by the *P–T* relations for supercritical phenomena in olivine with the solidus lowered by plagioclase. We discussed earlier how addition to the join Mg₂SiO₄–MgSiO₃–SiO₂ of anorthite (Andersen 1915; Irvine 1975a, see

Morse 1980, p. 162), or diopside (Bowen 1914; Kushiro 1972, see Morse 1980, p. 185) extends the MgO–SiO$_2$ anhydrous binary immiscibility slightly into the ternary systems.

We may consider how additional basaltic components to MgO–SiO$_2$ (Figs. 2b, 4a,b and 11a) might displace the ultramafic (simplified as forsterite + enstatite in MSH) critical phenomena, where with increasing pressure the two melts merge at around 2 GPa, when the solvus and solidus intersect (Fig. 4a). It is the SiO$_2$-rich melt which becomes miscible with feldspar components to large extent, while the intermediate melt takes in some alkali to become "basaltic".

It is likely that plagioclase and clinopyroxene components added to MgO–SiO$_2$ to make "basalt", may lower the two-melt solvus slightly in temperature, but with much larger T lowering of the anhydrous basalt solidus (simplified as cristobalite + enstatite = L$_2$ in MSH) compared to peridotite (forsterite + enstatite = L$_3$; Figs. 4, 11).

As shown by the various arrows in Figure 11a, additional components which partition into the melts will displace the MgO–SiO$_2$ (L$_1$ = L$_2$) critical point to higher pressures but lower temperatures in a basaltic system. This suggested behavior of a higher P critical intersection between solvus and solidus, can be seen in the anhydrous CaMgSi$_2$O$_6$–SiO$_2$ system in Figure 4c. Basalts are eclogitic at higher pressure. At pressures higher than plagioclase stability, garnet + clinopyroxene show a eutectic (Yoder and Tilley 1962; Pertermann and Hirschmann 2003) so that any solvus lies submerged at lower T than the liquidus, i.e. anhydrous eclogite liquids are supersolvus melts.

Simplified felsic crust

Simplified anhydrous granitic compositions can be modeled by anhydrous melting of albite + orthoclase + quartz and considered with reference to the alkali feldspar solvus in "Petrogeny's residua system" (Schairer and Bowen 1955, 1956; Tuttle and Bowen 1958). This name was actually applied by these workers to the water-bearing system with the allusion that all common silicate magmas migrate towards these components following fractional crystallization of higher temperature (and more mafic) minerals.

In the anhydrous system, the alkali feldspar minimum lies at higher temperature than the critical temperature at the top of the solvus, at all pressures. This temperature gap between solvus and solidus introduces supersolvus conditions where the solids are no longer immiscible as they are at lower temperatures.

Thus for the anhydrous system, the alkali feldspars are completely miscible at the solidus. The "minimum" melting for an azeotropic system is supersolvus (called hypersolvus by Tuttle and Bowen 1950) as distinct from supercritical, which occurs following intersection of solvus and solidus (defined in Figs. 4c and 11a). Around 0.4 GPa, the intersection of solidus and solvus (see Tuttle and Bowen 1950; Yoder et al. 1957; Morse 1980 p. 396) allows two feldspars to coexist. The granites are no longer supersolvus, but now eutectic, and we can refer to subcritical melting as that where albite + orthoclase are not completely miscible in the melting assemblage. Hence we are discussing here the relation of critical phenomena (sub-solvus compared to supersolvus) in solidus mineral melting, not melt L–L immiscibility, as in previous examples.

The P–T slope of the critical line for the alkali feldspar solvus has been estimated (Waldbaum and JB Thompson 1969). It does intersect the alkali feldspar to jadeite + quartz subsolidus reaction near 2.5 GPa and 1000 °C, but continues to lie at much lower temperatures than the anhydrous granite solidus (e.g., Thompson and Thompson 1976, p. 260). So that at higher pressure anhydrous granite melting is supersolvus with respect to the alkali feldspar minerals but not supercritical, because of lack of intersection of reactions. Felsic compositions buried or subducted to great depths (at least 2 GPa ca. 75 km) at anhydrous melting, will show continuous melt compositions from SiO$_2$ towards eutectics with feldspathic or intermediate compositions, but no alkali feldspar unmixing (or L–L phenomena).

At low pressures, the alkali feldspar solvus lies some 200 °C lower than the anhydrous solidus. Anhydrous granites are "hypersolvus" (supersolvus). Complete liquid crystallization at low pressures produces a single feldspar at the solidus (albite or orthoclase rich, because of the minimum/eutectic), which upon cooling can exsolve to make (anti-)perthite. Increasing H_2O-pressure lowers the temperature of the solidus enormously with very little increase of the solidus pressure (Tuttle and Bowen 1958).

CONCLUDING REMARKS

Whether oxide–SiO_2 binary systems are immiscible (two coexisting melts) or miscible depends upon size and charge of the other oxide cation. Systems are supercritical when the solidus is at higher T than the critical L=L solvus T. Supercritical melts have the property that they may coexist with the melting solids up to the liquidus of the system end-members without unmixing. Most 1^+ oxide–SiO_2 systems show submerged solvi, whereas 2^+ oxide–SiO_2 systems show clearly that L–L solvi are stable at higher T than solidi. 3^+ oxide–silica and higher (4^+, 5^+) oxide–silica systems (Fig. 2) show often the formation of intermediate compounds with high melting points which divide up the oxide–silica join, but still submerged L–L solvi are sometimes found at the high-SiO_2 end.

Binary immiscibility disappears in most ternary, and more component systems, with the exception of Fe_2SiO_4–$KAlSi_2O_6$–SiO_2. Here the lowering of the solidus by the FeO-component relative to the K_2O-component is greater than the T-lowering of the L–L solvus. This causes a reemergence of the binary two-melt field in the middle of the ternary system. Further oxide components added to fayalite–leucite–silica diminish the two-melt immiscibility.

Increased pressure in anhydrous silicate systems has the general tendency to cause intersection of solidi and L–L solvi, so that immiscible systems at low pressure become supercritical at higher pressures.

The systematic P–T displacement effects of additional components can be predicted for anhydrous natural compositions. Anhydrous magma systems during cooling and fractionation override the intermediate immiscibility of low-pressure binary systems, with the effect that only in the late stages of anhydrous magma fractionation liquid–liquid immiscibility involving silicate-only components is likely to be common. At higher pressures supercritical melt phenomena may also be found in multi-component magma systems at mantle depths, but the very-high temperatures required are probably not found in the current Earth.

Other volatile components (S, C, F, P, Cl, H_2O) can induce and extend immiscibility from anhydrous silicate systems. These components may well be responsible for some documented examples of immiscibility in terrestrial as well as lunar lavas.

The terminology and examples of supercritical behavior has become much more developed in volatile systems than in anhydrous silicate systems. The essential terrestrial ingredient – water, is involved in many inorganic natural systems including silicate melts in magmas, and in the fluids/vapors they evolve upon cooling. This is witnessed by most of the papers in this volume.

ACKNOWLEDGMENTS

ABT wishes to acknowledge Ed Roedder and JB Thompson for discussions over several years. We thank Michael Bogaerts, Peter Ulmer and Max Schmidt for their discussions on immiscibility and supercriticality, Ursula Stidwill for help with the text, and Hans Keppler and Axel Liebscher for their reviews and suggestions.

This work was supported by the Swiss National Science Foundation.

REFERENCES

Alekseeva ZD (1963) Phase diagram of the system Rb_2SiO_3-SiO_2. Russ J Inorg Chem 8:1426-1430 (in Russian)

Andersen O (1915) The system anorthite-forsterite-silica. Am J Sci 39:407-454

Aramaki S, Roy R (1962) Revised phase diagram for the system Al_2O_3-SiO_2. J Am Ceram Soc 45:229-242

Bailey DK (1989) Carbonate melt from the mantle in the volcanos of southeast Zambia. Nature 338:415-418

Bailey DK (1993) Carbonate magmas. J Geol Soc London 150:637-651

Ball RGJ, Mignanelli MA, Barry TI, Gisby JA (1993) The calculation of phase-equilibria of oxide core concrete systems. J Nucl Mater 201:238-249

Baret G, Madar R, Bernard C (1991) Silica-based oxide systems. 1. Experimental and calculated phase-equilibria in silicon, boron, phosphorus, germanium, and arsenic oxide mixtures. J Electrochem Soc 138:2830-2835

Barker DS (1989) Field relations of carbonatites. *In*: Carbonatites: Genesis and Evolution. Bell K (ed) Unwin Hyman, p 38-69

Barron LM (1991) A possible solvus geometry for liquation in quartz-fayalite-leucite. Geochim Cosmochim Acta 55: 761-767

Barth TFW (1962) Theoretical Petrology, 2nd edition. John Wiley

Bedard JH (1994) Mesozoic east North American alkaline magmatism.1. Evolution of Monteregian lamprophyres, Quebec, Canada. Geochim Cosmochim Acta 58:95-112

Bell K (1998) Radiogenic isotope constraints on relationships between carbonatites and associated silicate rocks - a brief review. J Petrol 39:1987-1996

Bell K, Kjarsgaard BA, Simonetti A (1998) Carbonatites - into the twenty-first century. J Petrol 39:1839-1845

Biggar GM (1979) Immiscibility in tholeiites. Mineral Mag 43:543-544

Biggar GM (1983) A reassessment of phase-equilibria involving 2 liquids in the system K_2O-Al_2O_3-FeO-SiO_2. Contrib Mineral Petrol 82:274-283

Bogaerts M, Schmidt MW (2006) Experiments on silicate melt immiscibility in the system Fe_2SiO_4-$KAlSi_3O_8$-SiO_2-CaO-MgO-TiO_2-P_2O_5 and implications for natural magmas. Contrib Mineral Petrol 152:257-274

Bowen NL (1914) The ternary system: diopside – forsterite – silica. Am J Sci 38:207-264

Bowen NL (1928) The Evolution of the Igneous Rocks. Dover Publications Inc.

Bowen NL (1937) Recent high-temperature research on silicates and its significance in igneous geology. Am J Sci 33:1-21

Bowen NL, Andersen O (1914) The binary system MgO-SiO_2. Am J Sci 37:487-500

Bowen NL, Schairer JF (1932) The system FeO-SiO_2. Am J Sci 24:177-213

Bowen NL, Schairer JF (1935) The system MgO-FeO-SiO_2. Am J Sci 29:151-217

Boyd FR, England JL (1960) Apparatus for phase-equilibrium measurements at pressures up to 50 kilobars and temperatures up to 1750 °C. J Geophys Res 65:741-748

Boyd FR, England JL (1963) Effect of pressure on the melting of diopside, $CaMgSi_2O_6$, and albite, $NaAlSi_3O_8$, in the range up to 50 kilobars. J Geophys Res 68:311-323

Boyd FR, England JL, Davis BTC (1964) Effects of pressure on the melting and polymorphism of enstatite, $MgSiO_3$. J Geophys Res 69:2101-2109

Brey G, Brice WR, Ellis DJ, Green DH, Harris KL, Ryabchikov ID (1983) Pyroxene-carbonate reactions in the upper mantle. Earth Planet Sci Lett 62:63-74

Bridgman PW (1927) The breakdown of atoms at high pressure. Phys Rev 29:0188-0191

Brooks C, Gelinas L (1975) Immiscibility and ancient and modern volcanism. Carnegie Institution of Washington Yearbook 74:240-247

Budnikov PP, Cherepanov AM (1953) Lithium-silicate compounds. Russ Chem Rev 22:821-837 (in Russian)

Bunting EN (1930a) Phase equilibria in the system SiO_2-ZnO. J Am Ceram Soc 13:5-10

Bunting EN (1930b) Phase equilibria in the system Cr_2O_3-SiO_2. J Res Nat Bur Stand 5:325-327

Bunting EN (1930c) Phase-equilibria in the system SiO_2-ZnO. J Res Nat Bur Stand 4:131-136

Butterman WC, Foster WR (1967) Zircon stability and ZrO_2-SiO_2 phase diagram. Am Mineral 52:880-885

Cabri LJ (1973) New data on phase relations in the Cu-Fe-S system. Econ Geol 68:443-454

Charles RJ (1966) Metastable liquid immiscibility in alkali metal oxide-silica systems. J Am Ceram Soc 49:55-62

Chen CH, Presnall DC (1975) The system Mg_2SiO_4-SiO_2 at pressures up to 25 kbar. Am Mineral 60:398-406

Church AA, Jones AP (1995) Silicate-carbonate immiscibility at Oldoinyo-Lengai. J Petrol 36:869-889

Circone S, Agee CB (1995) Effect of pressure on cation partitioning between immiscible liquids in the system TiO_2-SiO_2. Geochim Cosmochim Acta 59:895-907

Craig JR, Kullerud G (1969) Phase relations in the Cu-Fe-Ni-S system and their application to magmatic ore deposits. *In*: Magmatic Ore Deposits. Vol 4. Wilson HDB (ed) Economic Geology Publ., p 344-358

Dalton JA, Presnall DC (1997) No liquid immiscibility in the system $MgSiO_3$-SiO_2 at 5.0 GPa. Geochim Cosmochim Acta 61:2367-2373

Dalton JA, Wood BJ (1993) The compositions of primary carbonate melts and their evolution through wallrock reaction in the mantle. Earth Planet Sci Lett 119:511-525

Davis RF, Pask JA (1972) Diffusion and reaction studies in system Al_2O_3-SiO_2. J Am Ceram Soc 55:525-531

DeVries RC, Roy R, Osborn EF (1954) The system TiO_2-SiO_2. Brit Ceram Trans J 53:525-540

Dixon S, Rutherford MJ (1979) Plagiogranites as late-stage immiscible liquids in ophiolite and mid-ocean ridge suites : an experimental study. Earth Planet Sci Lett 45:45-60

Eggler DH (1978a) Effect of CO_2 upon partial melting of peridotite in system Na_2O-CaO-Al_2O_3-MgO-SiO_2-CO_2 to 35 Kb, with an analysis of melting in a peridotite-H_2O-CO_2 system. Am J Sci 278:305-343

Eggler DH (1978b) Stability of dolomite in a hydrous mantle, with implications for mantle solidus. Geology 6:397-400

Eggler DH (1989) Carbonatites, primary melts, and mantle dynamics. *In*: Carbonatites: Genesis and Evolution. Bell K (ed) Unwin Hyman, p 561-579

Falloon TJ, Green DH (1989) The solidus of carbonated, fertile peridotite. Earth Planet Sci Lett 94:364-370

Falloon TJ, Green DH (1990) Solidus of carbonated fertile peridotite under fluid-saturated conditions. Geology 18:195-199

Farnan I, Stebbins JF (1990) High-temperature ^{29}Si NMR investigation of solid and molten silicates. J Am Chem Soc 112:32-39

Ferguson J, Currie KL (1971) Evidence of liquid immiscibility in alkaline ultrabasic dikes at Callander Bay, Ontario. J Petrol 12:561-586

Fields JM, Dear PS, Brown JJ (1972) Phase equilibria in the system BaO-SrO-SiO_2. J Am Ceram Soc 55:585-588

Foley SF (1984) Liquid immiscibility and melt segregation in alkaline lamprophyres from Labrador. Lithos 17:127-137

Freestone IC (1978) Liquid immiscibility in alkali-rich magma. Chem Geol 23:115-123

Freestone IC, Hamilton DL (1980) The role of liquid immiscibility in the genesis of carbonatites: an experimental study. Contrib Mineral Petrol 73:105-117

Ghanbari-Ahari K, Brett NH (1988a) Phase-equilibria and microstructure in the system ZrO_2-MgO-SiO_2-SrO: 1. The ternary Systems ZrO_2-SiO_2-SrO and ZrO_2-MgO-SrO. Calphad 87:27-32

Ghanbari-Ahari K, Brett NH (1988b) Phase-equilibria and microstructure in the system ZrO_2-MgO-SiO_2-SrO: 2. The ternary system MgO-SiO_2-SrO. Calphad 87:103-106

Gittins J (1989) The origin and evolution of carbonatite magmas. *In*: Carbonatites: Genesis and Evolution. Bell K (ed) Unwin Hyman, , p 580-599

Glasser FP (1958) The system MnO-SiO_2. Am J Sci 256:398-412

Gramenitskiy YN, Shekina TI (1994) Phase relationships in the liquidus part of a granitic system containing fluorine. Geochem Int 31:52–70

Gramenitskiy YN, Shekina TI, Berman DP, Popenko DP (1993) Lithium concentration by aluminofluoride melt in a granitic system containing fluorine. Trans Russian Acad Sci 331A:139-144

Green DH, Falloon TJ (1998) Pyrolite: a ringwood concept and its current expression. *In*: The Earth's Mantle: Structure, Composition and Evolution. Jackson I (ed) Cambridge University Press, p 311-378

Green DH, Wallace ME (1988) Mantle metasomatism by ephemeral carbonatite melts. Nature 336:459-462

Greig JW (1927a) Immiscibility in silicate melts, part 1. Am J Sci 13:133-154

Greig JW (1927b) Immiscibility in silicate melts, part 2. Am J Sci 13:1-44

Hack AC, Thompson AB, Aerts M (2007) Phase relations involving hydrous silicate melts, aqueous fluids, and minerals. Rev Mineral Geochem 65:129-185

Hageman VBM, Oonk HAJ (1986) Liquid immiscibility in the SiO_2 + MgO, SiO_2 + SrO, SiO_2 + La_2O_3, and SiO_2 + Y_2O_3 systems. Phys Chem Glasses 27:194-198

Hageman VBM, Van den Berg GJK, Janssen HJ, Oonk HAJ (1986) A reinvestigation of liquid immiscibility in the SiO_2-CaO system. Phys Chem Glasses 27:100-106

Haller W, Blackburn DH, Simmons J (1974) Miscibility gaps in alkali-silicate binaries – data and thermodynamic interpretation. National Bureau of Standards, Washington DC 20234:120-126

Hazen RM, Downs RT (eds) (2001) High-Temperature and High-Pressure Crystal Chemistry. Reviews in Mineralogy and Geochemistry. Vol 41. Mineralogical Society of America

Hemley RJ (ed) (1999) Ultrahigh-Pressure Mineralogy: Physics and Chemistry of the Earth's Deep Interior. Reviews in Mineralogy. Vol 37. Mineralogical Society of America

Hess PC (1980) Polymerization model for silicate melts. *In*: Physics of Magmatic Processes. Hargraves RB (ed) Princeton Press, p 3-48

Hess PC (1995) Thermodynamic mixing properties and the structure of silicate melts. Rev Mineral 32:145-190

Hess PC (1996) Upper and lower critical points: Thermodynamic constraints on the solution properties of silicate melts. Geochim Cosmochim Acta 60:2365-2377

Holgate NJ (1954) The role of liquid immiscibility in igneous petrogenesis. J Geology 62:439-480

Hoover JD, Irvine TN (1978) System Mg_2SiO_4-Fe_2SiO_4-$CaMgSi_2O_6$-$CaFeSi_2O_6$-$KAlSi_3O_8$-SiO_2 and its bearing on silica enrichment trends. Trans-Am Geophys Union 59:1217-1217

Hudon P, Baker DR (2002a) The nature of phase separation in binary oxide melts and glasses. I. Silicate systems. J Non-Cryst Solids 303:299-345

Hudon P, Baker DR (2002b) The nature of phase separation in binary oxide melts and glasses. II. Selective solution mechanism. J Non-Cryst Solids 303:346-353

Hudon P, Jung IH, Baker DR (2004) Effect of pressure on liquid-liquid miscibility gaps: A case study of the systems CaO-SiO$_2$, MgO-SiO$_2$, and CaMgSi$_2$O$_6$-SiO$_2$. J Geophys Res-Sol Ea 109:B03207

Huntelaar ME, Cordfunke EHP, Scheele A (1993) Phase relations in the SrO-SiO$_2$-ZrO$_2$ System: 1. The System SrO-SiO$_2$. J Alloys Compd 191:87-90

Irvine TN (1975a) Olivine-pyroxene-plagioclase relations in the system Mg$_2$SiO$_4$-CaAl$_2$Si$_2$O$_8$-KAlSi$_3$O$_8$-SiO$_2$ and their bearing on the differentiation of stratiform intrusions. Carnegie I Wash 74:492-500

Irvine TN (1975b) The silica immiscibility effect in magmas. Carnegie I Wash 74:484-492

Jakobsen JK, Veksler IV, Tegner C, Brooks CK (2005) Immiscible iron- and silica-rich melts in basalt petrogenesis documented in the Skaergaard intrusion. Geology 33:885-888

Jung I (2003) Critical evaluation and thermodynamic modeling of phase equilibria in multicomponent oxide systems. Ph.D. thesis, Ecole Polytech. de Montreal, Canada

Kaufman L (1988) System TiO$_2$-SiO$_2$. Physica B & C 150:99-114

Kawamoto Y, Tomozawa M (1981) Prediction of immiscibility boundaries of the systems K$_2$O-SiO$_2$, K$_2$O-Li$_2$O-SiO$_2$, K$_2$O-Na$_2$O-SiO$_2$, and K$_2$O-BaO-SiO$_2$. J Am Ceram Soc 64:289-292

Kim SS, Sanders TH (1991) Thermodynamic modeling of phase diagrams in binary alkali silicate systems. J Am Ceram Soc 74:1833-1840

Kjarsgaard BA (1998) Phase relations of a carbonated high-CaO nephelinite at 0.2 and 0.5 GPa. J Petrol 39:2061-2075

Kjarsgaard BA, Hamilton DL (1989) The genesis of carbonatites by liquid immiscibility. *In*: Carbonatites: Genesis and Evolution. Bell K (ed) Unwin Hyman, p 388-404

Kjarsgaard BA, Peterson T (1991) Nephelinite-carbonatite liquid immiscibility at Shombole Volcano, East-Africa - petrographic and experimental evidence. Mineral Petrol 43:293-314

Kogarko LN, Krigman LD (1970) Phase equilibria in system nepheline-NaF. Geochem Int 7:103-107

Koster van Groos AF, Wyllie PJ (1963) Experimental data bearing on role of liquid immiscibility in genesis of carbonatites. Nature 199:801-802

Kracek FC (1930) The cristobalite liquidus in the alkali oxide-silica systems and the heat of fusion of cristobalite. J Am Chem Soc 52:1436-1442

Kracek FC (1932) The ternary system, K$_2$SiO$_3$-Na$_2$SiO$_3$-SiO$_2$. J Phys Chem 36:2529-2542

Krigman LD, Krot TV (1991) Stable phosphate aluminosilicate liquation in magmatic melts. Geokhimiya 11:1548-1560 (in Russian)

Kullerud G, Yund RA, Moh GH (1969) Phase relations in the Cu-Fe-S, Cu-Ni-S, and Fe-Ni-S systems. *In*: Magmatic Ore Deposits. Vol 4. Wilson HDB (ed) Economic Geology Publ., p 323-343

Kushiro I (1972) Determination of liquidus relations in synthetic silicate systems with electron microprobe analysis: the system forsterite-diopside-silica at 1 atmosphere. Am Mineral 57:1260-1271

Lee WJ, Wyllie PJ (1998a) Petrogenesis of carbonatite magmas from mantle to crust, constrained by the system CaO-(MgO+FeO*)-(Na$_2$O+K$_2$O)-(SiO$_2$+Al$_2$O$_3$+TiO$_2$)-CO$_2$. J Petrol 39:495-517

Lee WJ, Wyllie PJ (1998b) Processes of crustal carbonatite formation by liquid immiscibility and differentiation, elucidated by model systems. J Petrol 39:2005-2013

Levelt Sengers JMH (2000) Supercritical Fluids: Their properties and applications. *In*: Supercritical Fluids: Vol 366. Kiran E, Debenedetti PG, Peters CJ (eds) NATO Science Series E Applied Sciences, p 1-30

Levin EM, Robbins CR, McMurdie HF (1964) Phase Equilibria Diagrams. American Ceramic Society

Longhi J (1990) Silicate liquid immiscibility in isothermal crystallization experiments. Proc Lunar Planet Sci Conf 20:13-24

MacDowell JF, Beall GH (1969) Immiscibility and crystallization in Al$_2$O$_3$-SiO$_2$ glasses. J Am Ceram Soc 52:17-25

Michels MAJ, Wesker E (1988) A network model for the thermodynamics of multicomponent silicate melts: 1. Binary mixtures MO-SiO$_2$. Calphad 12:111-126

Morgan RA, Hummel FA (1949) Reactions of BeO and SiO$_2$ - synthesis and decomposition of phenacite. J Am Ceram Soc 32:250-255

Moriya Y, Warrington DH, Douglas RW (1967) A study of metastable liquid-liquid immiscibility in some binary and ternary alkali silicate glasses. Phys Chem Glasses 8:19-25

Morse SA (1980) Basalts and Phase Diagrams. Springer-Verlag

Mysen BO (1990) Relationships between silicate melt structure and petrologic processes. Earth-Sci Rev 27:281-365

Mysen BO (2004) Element partitioning between minerals and melt, melt composition and melt structure. Chem Geol 213:1-16

Nakamura Y (1974) The system SiO$_2$-H$_2$O-H$_2$ at 15 kbar. Carnegie I Wash 73:259-263

Ol'shanskii YI (1951) Equilibrium of two immiscible liquids in silicate systems of alkali-earth metals. Dokl Akad Nauk SSSR 76:93-96 (in Russian)

Park CF (1961) A magnetite "flow" in northern Chile. Econ Geol 56:431-436

Pertermann M, Hirschmann MM (2003) Anhydrous partial melting experiments on MORB-like eclogite: Phase relations, phase compositions and mineral-melt partitioning of major elements at 2-3 GPa. J Petrol 44:2173-2201

Petibon CM, Kjarsgaard BA, Jenner GA, Jackson SE (1998) Phase relationships of a silicate-bearing natrocarbonatite from Oldoinyo Lengai at 20 and 100 MPa. J Petrol 39:2137-2151

Phillips B, Muan A (1959) Phase equilibria in the system CaO-iron oxide-SiO_2 in air. J Am Ceram Soc 42:413-423

Philpotts AR (1976) Silicate liquid immiscibility: its probable extent and petrogenetic significance. Am J Sci 276:1147-1177

Philpotts AR (1982) Compositions of immiscible liquids in volcanic rocks. Contrib Mineral Petrol 80:201-218

Philpotts AR, Doyle CD (1983) Effect of magma oxidation-state on the extent of silicate liquid immiscibility in a tholeiitic basalt. Am J Sci 283:967-986

Prigogine I, Defay R (1954) Chemical Thermodynamics. Longmans

Rabinowicz M, Ricard Y, Gregoire M (2002) Compaction in a mantle with a very small melt concentration: Implications for the generation of carbonatitic and carbonate-bearing high alkaline mafic melt impregnations. Earth Planet Sci Lett 203:205-220

Rankin AH, Le Bas MJ (1974) Liquid immiscibility between silicate and carbonate melts in naturally occurring ijolite magma. Nature 250:206-209

Roedder E (1951) Low temperature liquid immiscibility in the system K_2O-FeO-Al_2O_3-SiO_2. Am Mineral 36:282-286

Roedder E (1978) Silicate liquid immiscibility in magmas and in system K_2O-FeO-Al_2O_3-SiO_2 - example of serendipity. Geochim Cosmochim Acta 42:1597-1617

Roedder E (1979) Silicate Liquid Immiscibility in Magmas. *In*: The Evolution of the Igneous Rocks, Fiftieth Anniversary Perspectives. Yoder HSJ (ed) Princeton University Press, p 15-57

Roedder E (1983) Discussion of "A re-assessment of phase equilibria involving two liquids in the system K_2O-Al_2O_3-FeO-SiO_2", by G.M. Biggar. Contrib Mineral Petrol 82:284-290

Roedder E, Weiblen PW (1970) Lunar petrology of silicate melt inclusions. Apollo 11 rocks. Proc Apollo 11 Lunar Sci Conf, Geochim Cosmochim Acta Suppl 1 1:801-837

Roedder E, Weiblen PW (1971). Petrology of silicate melt inclusions, Apollo 11 and Apollo 12 and terrestrial equivalents. Proc Second Lunar Sci Conf, Geochim Cosmochim Acta Suppl. 2 1:507-528

Roedder E, Weiblen PW (1972). Petrographic features and petrologic significance of melt inclusions in Apollo 14 and I5 rocks. Proc Third Lunar Sci Conf, Geochim Cosmochim Acta Suppl 3 1:251-279

Roedder E, Weiblen PW (1977). Compositional variation in late-stage differentiates in mare lavas, as indicated by silicate melt inclusions. Proc Eighth Lunar Sci Conf, Geochim Cosmochim Acta Suppl. 8 2:1767-1783

Romanchev BP, Sokolov SV (1979) Liquation in the production and geochemistry of the rocks in carbonatite complexes. Geochem Int 16:125-135

Rosenbusch H (1872) Mikroskopische Physiographie der Mineralien und Gesteine. Schweizerbartsche Verlagsbuchhandlung

Ryerson FJ (1985) Oxide solution mechanisms in silicate melts: systematic variations in the activity coefficient of SiO_2. Geochim Cosmochim Acta 49:637-649

Ryerson FJ, Hess PC (1978) Implications of liquid-liquid distribution coefficients to mineral-liquid partitioning. Geochim Cosmochim Acta 42:921-932

Ryerson FJ, Hess PC (1980) The role of P_2O_5 in silicate melts. Geochim Cosmochim Acta 44:611-624

Samsonov GV (1982) The Oxide Handbook, 2nd ed. IFI/Plenum Publishing, New York

Schairer JF, Bowen NL (1955) The system K_2O-Al_2O_3-SiO_2. Am J Sci 253:681-746

Schairer JF, Bowen NL (1956) The system Na_2O-Al_2O_3-SiO_2. Am J Sci 254:129-195

Seward TP, Uhlmann DR, Turnbull D (1968a) Phase separation in system BaO-SiO_2. J Am Ceram Soc 51:278-285

Seward TP, Uhlmann DR, Turnbull D (1968b) Development of 2-phase structure in glasses with special reference to system BaO-SiO_2. J Am Ceram Soc 51:634-643

Shannon RD (1976) Revised effective ionic radii and systematic studies of interatomic distances in halides and chalcogenides. Acta Crystallogr A32:751–767

Sillitoe RH, Burrows DR (2002) New field evidence bearing on the origin of the El Laco magnetic deposit, Northern Chile. Econ Geol 97:1101-1109

Singleton EL, Carpenter L, Lundquist RV (1962) Studies of the MnO-SiO_2 binary system. U.S. Bureau of Mines - Report Investigation 5938:1-31

Stanley E (1971) Introduction to phase transitions and critical phenomena. Oxford University Press

Staronka A, Pham H, Rolin M (1968) Étude du système silice-alumine par la méthode des courbes de refroidissement. Rev Int Haut Temp Refract 5:111-115

Stebbins JF, McMillan PR, Dingwell DB (eds) (1995) Structure, Dynamics and Properties of Silicate Melts. Reviews in Mineralogy. Vol 32. Mineralogical Society of America

Suk NI (1998) Distribution of ore elements between immiscible liquids in silicate – phosphate systems (experimental investigation). Acta Univ Carol Geol 42:138–140

Switzer GS (1975) Composition of three glass phases present in an Apollo 15 basalt fragment. *In*: Mineral Sciences Investigations 1972-1973. Smithsonian Contribution to Earth Sciences Vol 14. Switzer GS (ed), Smithsonian Institution Press, p 25-30

Taylor JR, Dinsdale AT (1990) Thermodynamic and phase diagram data for the CaO-SiO$_2$ system. Calphad 14:71-88

Tewhey JD, Hess PC (1979) 2-Phase region in the CaO-SiO$_2$ system: experimental data and thermodynamic analysis. Phys Chem Glasses 20:41-53

Thibault Y, Edgar AD, Lloyd FE (1992) Experimental investigation of melts from a carbonated phlogopite lherzolite: implications for metasomatism in the continental lithospheric mantle. Am Mineral 77:784-794

Thompson JB Jr, Thompson AB (1976) A model system for mineral facies in pelitic schists. Contrib Mineral Petrol 58:243-277

Tien TY, Hummel FA (1962) The system SiO$_2$-P$_2$O$_5$. J Am Ceram Soc 45:422-424

Tuttle OF, Bowen NL (1950) High temperature albite and contiguous feldspars. J Geology 58:572-583

Tuttle OF, Bowen NL (1958) Origin of granite in the light of experimental studies in the system NaAlSi$_3$O$_8$-KAlSi$_3$O$_8$-SiO$_2$-H$_2$O. Geol Soc Am Mem 74:1-153

Veksler IV (2004) Liquid immiscibility and its role at the magmatic-hydrothermal transition: a summary of experimental studies. Chem Geol 210:7-31

Veksler IV, Dorfman AM, Dingwell DB, Zotov N (2002a) Element partitioning between immiscible borosilicate liquids: a high-temperature centrifuge study. Geochim Cosmochim Acta 66:2603–2614

Veksler IV, Fedorchuk YM, Nielsen TFD (1998a) Phase equilibria in the silica -undersaturated part of theKAlSiO$_4$–Mg$_2$SiO$_4$–Ca$_2$SiO$_4$–SiO$_2$–F system at 1 atm and the larnite-normative trend of melt evolution. Contrib Mineral Petr 131:347-363

Veksler IV, Petibon C, Jenner G, Dorfman AM, Dingwell DB (1998b) Trace element partitioning in immiscible silicate and carbonate liquid systems: an initial experimental study using a centrifuge autoclave. Journal of Petrology 39:2095-2104

Veksler IV, Thomas R, Schmidt C (2002b) Experimental evidence of three coexisting immiscible fluids in synthetic granite pegmatite. Am Mineral 87:775–779

Visser W, Koster van Groos AF (1979a) Effect of pressure on liquid immiscibility in the system K$_2$O-FeO-Al$_2$O$_3$-SiO$_2$-P$_2$O$_5$. Am J Sci 279:1160-1175

Visser W, Koster van Groos AF (1979b) Effects of P$_2$O$_5$ and TiO$_2$ on liquid-liquid equilibria in the system K$_2$O-FeO-Al$_2$O$_3$-SiO$_2$. Am J Sci 279:970-988

Visser W, Koster van Groos AF (1979c) Phase relations in the system K$_2$O-FeO-Al$_2$O$_3$-SiO$_2$ at 1 atmosphere with special emphasis on low temperature liquid immiscibility. Am J Sci 279:70-91

Waldbaum DR, Thompson JB Jr (1969) Mixing properties of sanidine crystalline solution. IV. Phase diagrams from equations of state. Am Mineral 54:1274-1298

Watson EB (1976) Two-liquid partition coefficients: Experimental data and geochemical implications. Contrib Mineral Petrol 56:119-134

Webster JD, DeVivo B (2002) Experimental and modeled solubilities of chlorine in aluminosilicate melts, consequences of magma evolution, and implications for exsolution of hydrous chloride melt at Mt. Somma-Vesuvius. Am Mineral 87:1046-1061

Weiblen PW, Roedder E (1973) Petrology of melt inclusions in Apollo samples 15598 and 62295, and of clasts in 67915 and several lunar soils. Proc Fourth Lunar Sci Conf, Geochim Cosmochim Acta Suppl 4:681-703

White J, Howatt DD, Hay R (1934) The binary system MnO-SiO$_2$. J Royal Tech Coll (Glasgow) 3:231-240

Wu P, Eriksson G, Pelton AD, Blander M (1993) Prediction of the thermodynamic properties and phase diagrams of silicate systems: evaluation of the FeO-MgO-SiO$_2$ System. Isij Int 33:26-35

Wyllie PJ (1977) Mantle fluid compositions buffered by carbonates in peridotite-CO$_2$-H$_2$O. J Geol 85:187-207

Wyllie PJ (1978) Mantle fluid compositions buffered in peridotite-CO$_2$-H$_2$O by carbonates, amphibole, and phlogopite. J Geol 86:687-713

Wyllie PJ (1987) Volcanic rocks: boundaries from experimental petrology. Fortschr Mineral 65:249-284

Wyllie PJ (1989) Origin of carbonatites: evidence from phase equilibrium studies. *In*: Carbonatites. Genesis and Evolution. Bell K (ed) Unwin Hyman, p 500-545

Wyllie PJ, Huang WL (1976) Carbonation and melting reactions in system CaO-MgO-SiO$_2$-CO$_2$ at mantle pressures with geophysical and petrological applications. Contrib Mineral Petrol 54:79-107

Yoder HSJ, Stewart DB, Smith JV (1957) Ternary feldspars. Carnegie I Wash 56:206-214

Yoder HSJ, Tilley CE (1962) Origin of basalt magmas: an experimental study of natural and synthetic rock systems. J Petrol 3:342-53

Reviews in Mineralogy & Geochemistry
Vol. 65, pp. 129-185, 2007
Copyright © Mineralogical Society of America

5

Phase Relations Involving Hydrous Silicate Melts, Aqueous Fluids, and Minerals

Alistair Campbell Hack, Alan Bruce Thompson[(*)], Maarten Aerts

Dept. Erdwissenschaften, Institute for Mineralogy & Petrology
ETH Zurich
Zürich, CH-8092, Switzerland

also at Faculty of Mathematics and Natural Sciences, University of Zürich, Switzerland

alistair.hack@erdw.ethz.ch   alan.thompson@erdw.ethz.ch   maarten.aerts@erdw.ethz.ch

IMMISCIBILITY IN NATURAL SYSTEMS INVOLVING HYDROUS SILICATE MELTS

The importance of magma degassing and separation of magmatic fluids (aqueous as well as other components) has been recognized in many geologic observations at least since Niggli (1912; see also Morey and Niggli 1913; Bowen 1928; Morey 1957). Improved analytical techniques in recent years have permitted micron scale examination of melt and fluid inclusions inside magmatic minerals. In these inclusions, concentrated aqueous solutions and extreme enrichment of melts and fluids in volatile components (S, CO_2, B_2O_3, Cl, F, and metals) have been observed (Roedder 1992; Webster 1997; Thomas et al. 2000; Kamenetsky et al. 2004). Pegmatite formation reflects the role of exsolution of aqueous fluid from silicate melt (Jahns and Burnham 1969) at the magmatic–hydrothermal transition (e.g., Veksler 2004). Fluid–gas immiscibility is clearly involved in formation of pegmatites, and for some other fluids originating in magmatic or metamorphic environments.

Certain natural fluid systems rich in volatile components are immiscible and others show completely mixed compositions at geological temperatures and pressures (e.g., Roedder 1984; hydrocarbons, Pedersen and Christensen 2007; geothermal and submarine hydrothermal systems, Arnórsson et al. 2007, Foustoukos and Seyfried 2007; ore-forming fluids, Hedenquist and Lowenstern 1994; Heinrich 2007a; mixed metamorphic fluids, Trommsdorff and Skippen 1986; Heinrich 2007b; and magmatic fluids, Webster and Mandeville 2007). This has encouraged the idea that certain common natural melt–fluid–gas systems pass from immiscible (with multiple coexisting fluids) to miscible behavior (a single homogeneous fluid) with changing pressure and/or temperature (increasing or decreasing) or changing composition (see recent summary by Veksler 2004). The term supercritical is applied to any solution (liquid or solid) occurring at a temperature or pressure above its critical point. In the case of supercritical fluid this corresponds to conditions where liquid and vapor are completely miscible. Such completely miscible behavior has recently been discussed in detail for high-pressure fluids and melts such as might be found arising from subducted oceanic lithosphere and percolating through overlying mantle (Kerrick and Connolly 2001; Poli and Schmidt 2002; Manning 2004; Schmidt et al. 2004; Hermann et al. 2006; Hirschmann 2006).

We review and consider some aspects of immiscible and miscible fluid behavior that are not thoroughly emphasized in the recent papers by Veskler (2004), who in particular reviewed un-translated Russian literature up to that date (e.g., Ravich 1974; Valyashko 1990a). We will also examine cases of unmixing at low pressure in geological systems with examples from melt–melt (= liquid–liquid), as well as liquid–gas, and gas–gas systems. We will try to summarize known systematics behind these phase separations from a chemical and geochemical viewpoint. We

1529-6466/07/0065-0005$05.00   DOI: 10.2138/rmg.2007.65.5

will also question what happens at higher pressure and consider whether supercritical (here, synonymous with completely miscible) fluids versus liquid-immiscibility are prominent features inside the Earth. Thompson et al. (2007) addressed molten anhydrous silicate systems, here we consider hydrous melt–fluid systems, then addition of other volatile components, then phase separation versus continuous behavior in liquid–vapor systems versus supercritical fluids.

Supercritical H₂O

Most of the physical properties of natural fluids reflect the critical behavior of H_2O modified by dissolved salts, minerals and gases. Most significant is that water undergoes liquid–vapor phase separation quite close to the Earths surface. This behavior is governed by the location of the critical point for H_2O ($T_{crit} = 647$ K $= 374$ °C; $P_{crit} = 22.1$ MPa, see Haar et al. 1984).

As water is heated towards its critical point, it undergoes a transformation considerably more dramatic than that of most other substances, changing from the familiar polar liquid to an almost *non-polar* fluid. The change, also seen in density, occurs over a relatively wide temperature range; even at 200 °C, the density has dropped to 0.8 g cm⁻³. Massive changes are seen in the relative static dielectric constant (ε) and the self-ionization (pK_W) values as the critical temperature and pressure are approached (e.g., Franck 1987). These physical properties have a determining effect on polarity and acid/base-catalytic properties of solvent water. Chemical diffusivity increases and the acidity is enhanced more than would be expected purely on the basis of higher temperatures, favoring ionic processes over radical or purely thermal pathways (e.g., Weingärtner and Franck 2005). At supercritical conditions water becomes miscible in all proportions with organic compounds and with oxygen. Thus, these compounds can be oxidized in a homogeneous supercritical phase. Thus, industrial processes which carefully control how pressure and temperature are changed around the critical point can carefully modify solubility in and transport behavior of aqueous fluids. Then by either dropping the pressure, or by cooling, the products precipitate from solution in a controlled fashion (Schneider 1978; Rizvi et al. 1986; Levelt Sengers 1993; Eckert et al. 1996). The switchover from polar to non-polar results in increased stability of neutral, polymerized species over solvated ions, and correspondingly, silicate mineral solubilities are enhanced.

Critical behavior systematics for H₂O with added components

The effect of adding another component to pure H_2O is to displace the *critical point* for the aqueous solution along the *critical line* (a critical curve where the L (liquid) + V (vapor) solvus closes) in a *P-T* diagram. The amount of displacement is proportional to the concentration of the added component and the distribution coefficient between liquid and vapor. The critical curve separates an immiscible two-phase region from a supercritical- or miscible single-phase region. The new dP/dT slope of the critical curve extending from the critical point of pure H_2O depends on the L to V partitioning behavior of the added component. The critical curve for a certain binary, A–H_2O, can lie on either side of the critical density line (isochore) for H_2O (depending whether the added solute prefers the liquid or the vapor, Levelt Sengers 1991; Anisimov et al. 2004). It is useful to consider the kind of chemical interaction involved because the observations can be used predicatively.

Non-volatile solutes that strongly interact with and dissolve into liquid H_2O have critical curves falling near but on the higher-temperature side of the critical isochore for H_2O (dashed curve in Fig. 1). Examples include alkali-salts and silicate components. Such critical curves run to higher temperature and pressure (generally less than a few hundred MPa). In this case the immiscible L–V region lies on the lower-pressure (and higher-temperature) side of the critical curve, indicated on Figure 1. The addition of volatile solutes (e.g., CO_2, noble gases) to H_2O has the opposite effect on the dP/dT slope of the critical curve. These components interact weakly with the liquid and strongly favor the vapor. This brings the critical curves towards the low-temperature side of the H_2O critical isochore. In these systems L–V immiscibility is restricted to the lower-temperature side of the binary critical curve.

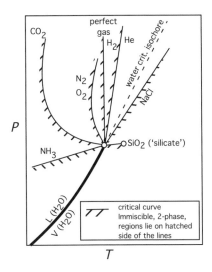

Figure 1. Summary diagram of *P, T* systematics for critical curves (L=V; thin solid curves) for some binary systems A–H$_2$O where A = NH$_3$, CO$_2$, N$_2$, O$_2$, H$_2$, He, NaCl or SiO$_2$). Solid black curve is the boiling (or L+V coexistence) curve of pure H$_2$O, critical point of pure H$_2$O (open circle), and dashed curve is the critical H$_2$O isochore (modified from Levelt Sengers 1991).

Phase separation in subsolidus and super-solidus systems

In fluid systems, understanding the range of physical (i.e., pressure - *P*, temperature - *T*) and chemical conditions (fluid compositions - *X*) where L (liquid) – V (vapor) immiscibility may occur, is of geochemical interest. This is because coexisting L and V have different chemical compositions and physical properties, and because of contrasting densities thus have the potential to separate spatially and thereby physically fractionate the dissolved mass. The occurrence of L+V immiscibility is well-appreciated in low-pressure (< ca. 150 MPa) hydrothermal environments where H$_2$O–CO$_2$–NaCl–KCl largely dominate fluid phase relations and are reasonably well-known (Barnes 1979; Bischoff and Pitzer 1985; Arnórsson et al. 2007; Heinrich 2007a; Liebscher 2007). At higher temperature (> 500 °C), magma degassing is an important example of L+V immiscibility and plays a central role fractionating volatile components (e.g., S, N, Cl, C, noble gases, H$_2$O; see Carroll and Holloway 1994) and metals (e.g., Re, Au, Cu; see Hedenquist and Lowenstern 1994; Sun et al. 2003) from silicate liquids.

The general role of H$_2$O and other volatiles in silicate melt immiscibility

Immiscibility and supercritical phenomena are known or suspected in several water-saturated silicate melt systems. These two immiscible phases are also usually abbreviated L and V, meaning hydrous silicate melt (L) and aqueous fluid (V), although the latter has more liquid-like than vapor-like densities at high pressures. With increasing pressure, mineral solubilities in aqueous fluids (V) and volatile solubilities in silicate melts (L) increase significantly, such that L and V may become completely miscible at high pressure (e.g., Kennedy et al. 1962; Boettcher and Wyllie 1969a; Shen and Keppler 1997; Bureau and Keppler 1999). The occurrence of L–V supercriticality has a number of important consequences: (1) L–V immiscibility becomes impossible and mineral-saturated fluids experiencing changes in *P, T* conditions vary composition continuously (via dissolution and precipitation) between dilute, solute-poor fluid and solute-rich, hydrous melt; and, (2) a wet solidus (fixing the initial location of V-saturated melting or minimum magma degassing) is no longer defined at conditions where supercritical fluid exists (abbreviated as L=V to emphasize a single phase).

The possibility of supercritical phenomena and termination of conventional melting on wet solidi has been recognized for some time (e.g., Wyllie and Tuttle 1960; Kennedy et al. 1962; Boettcher and Wyllie 1969a; Paillat et al. 1992). However, it was for a long time only recognized in a few restricted petrological systems, owing largely to difficulties in characterizing supercriticality from quenched experimental samples. Thus, the extent and importance of supercriticality in the Earth was unclear to these early workers. Since then, experimental advances have made it possible to study supercritical behavior in more detail for a wider range of systems than previously possible (e.g., Shen and Keppler 1997; Kessel et al. 2004; Spandler et al. 2006; Mibe et al. 2007). As a result there has been renewed interest in the nature and significance of supercritical fluids with hydrous melt-like properties in geological processes (Shen and Keppler 1997; Bureau and Keppler 1999; Stalder et al. 2000, 2001; Manning 2004), because P, T conditions in subduction zones might be suitable to form such fluids (Massonne 1992; Sharp et al. 1993; Scambelluri and Philippot 2001; Hermann 2003; Audétat and Keppler 2004; Kessel et al. 2004; Schmidt et al. 2004; Ferrando et al. 2005; Kessel et al. 2005a,b; Spandler et al. 2006; Mibe et al. 2007). Here we attempt to supplement these recent discussions by focusing attention on L–V phase relations in simple hydrous silicate systems, by discussing the effect of various additional components, reviewing what we know of the physical properties of fluids, and then by identifying rocks and tectonic environments where supercritical and immiscible fluids may be important.

Brief history of supercritical fluid research

Long before critical phenomena were considered in the Earth sciences, liquid–vapor immiscibility behavior was known in many chemical systems. The first experiments carried out on L–V (im)miscibility were by a French engineer, Charles Cagniard de la Tour, in 1822, who heated up a sealed cannon barrel which inside contained some liquid and a silica sphere (Cagniard de la Tour 1822). By rocking the cannon barrel back and forth, he listened for the sphere splashing into the liquid as it passed the liquid–vapor interface. At a certain temperature, he no longer heard the splash when the cannon was inverted, and concluded that L and V must have become completely miscible. The earliest mention of a critical point on a boiling curve appears when Andrews (1869) describes the continuity of states around the critical point of carbon dioxide. Ten years after that, Hannay and Hogarth (1879) were probably the first to observe radical changes in solubility close the critical point of the solvent (potassium iodide in alcohol). Smits (1903) provides what appears to be the first quantitative report of a critical curve intersecting and terminating a three-phase (S+V=L) curve—the first solid-saturated critical endpoint was found in the binary system diethylether–anthraquinone.

Much early work was inspired by Van der Waals' equation of state (EOS) for fluid mixtures (1890; English translation by Rowlinson 1988). Korteweg (1891a,b) first used the EOS in its original form to calculate fluid phase equilibria and many studies followed. However, it was not until the late nineteen seventies that van Konynenburg and Scott (1980), thanks to increasing computational capabilities, demonstrated that almost all experimentally determined types of phase relations could be reproduced by calculations based on Van de Waals' equation. In the 25 years after, further topology types have been identified (see below), and alternate criteria for type subdivision have been proposed (e.g., Bolz et al. 1998). Computer simulations have also lead to the discovery of additional unanticipated phase relation types (e.g., Boshkov and Mazur 1986). A thorough treatment is given by Levelt Sengers (2002).

Niggli (1912) and Morey and Niggli (1913) were two of the earliest Earth scientists to adopt discoveries by the chemists. In particular, Niggli already suggested that in late-stage magmatic crystallization, the transition between the magmatic and hydrothermal stage may be continuous. Van Nieuwenburg and Van Zon (1935) appear to be the first to study the solubility of quartz in supercritical water. Their work was inspired by the experimental findings of Greig et al. (1933) who suggested silica could be transported under conditions of increased pressure and temperature in the presence of a vapor. Since then, many more solubility studies in geological systems followed. It suffices to say, there is a vast current and historical literature on this subject and it is virtually impossible to summarize but a tiny fraction here.

INTRODUCTION TO FLUID PHASE RELATIONS:
LIMITATIONS OF SOLID–WATER SYSTEMS AS IDEAL BINARIES

Solid–water binaries fall in two major groups of general and geological interest—one where highly soluble solids tend to dissolve increasingly in water with increasing temperature (e.g., $NaCl-H_2O$), and another where sparingly soluble solids show decreasing or only a weakly increasing solubility with increasing temperature (fluoride–H_2O, sulfate–H_2O and most silicate–H_2O). Only recently has it become apparent that the effects of pressure on solubility may be enormous and might cause reversal in solubility behavior (see below, and Caciagli and Manning 2003). For increasingly soluble solid-bearing systems with increasing temperature there can be completely continuous solubility, up to 100% at solid (or "salt") melting. For weakly soluble systems with increasing temperature there can even be a progress towards practically zero solubility close to the critical point of water. A solid solubility threshold somewhere between 5 and 10 mol% of dissolved solid in vapor-saturated liquid at temperatures above 200 °C appears to determine whether the system progresses to "low-" or "high-solubility" behavior at higher temperatures (Valyashko 1990a; Veksler 2004). Additional L–L and L–V complexity is known to occur in fluids but which have not yet been recognized in the geological literature, we also draw attention to these. This is done for both completeness and because it is likely that phase relations in some natural fluids are more complex than currently appreciated.

Two basic phase relation topologies of A–H_2O binary systems

Here we review simple ideal binary phase relations to illustrate fundamental aspects of relations between solids, critical and immiscible L–V solutions. The underlying concepts are built on and used in later sections to interpret observations in mineral- and rock–H_2O systems. In such simple A–H_2O binary systems, A is non-volatile, no immiscibility occurs in liquid-A, and no intermediate hydrous compounds form. Although some of this discussion is also treated in Liebscher (2007) and Liebscher and Heinrich (2007), we emphasize here certain aspects relevant to high-pressure (P) conditions, as well as high-T conditions relevant to hydrous (wet) melting. For completeness we also discuss some other system topologies that are important for understanding the theoretical and practical basis for industrial extraction, and note that the relevance of these phase relations to natural fluids has not yet been fully assessed.

One-component systems, like H_2O or A, each have three coexistence curves in P-T diagrams, for pairs of solid–liquid–vapor. In particular they have a liquid + vapor coexistence curve that at high-temperature terminates at a critical point (Fig. 2, see Table 1 for symbol explanations). With increasing temperature along this curve, the properties of the liquid and vapor converge to the extent that liquid and vapor become indistinguishable at the critical point (analogous to the closing of a miscibility gap). At T and P above the critical point, a single phase, *supercritical fluid* is stable. In the two-component system A–H_2O, a critical curve connects the critical points in the end-member systems. Along this critical curve only a single-phase supercritical fluid exists. By analogy with behavior near a critical point, in higher-order systems coexisting liquid and vapor become completely miscible as the critical curve is approached. The critical curve represents the maximal P, T and X stability of the liquid–vapor miscibility gap (in the following sometimes abbreviated as *MG*); it defines the field of L+V coexistence versus that of a single-phase supercritical fluid. At a critical point in a one-component system and on critical curves in higher-order systems, all properties (P, T, composition, density, etc.) of the supercritical fluid are fixed.

Niggli (1912), among others, showed that complete phase relations for ideal A–H_2O binary systems can be divided into two main types depending on the stability of the critical curve and the three-phase solubility curve ($S_A + V = L$). Numerous discussions can be found in the literature on these topologies (Ricci 1951; Morey 1957; Wyllie and Tuttle 1960; Fyfe et al. 1978; Paillat et al. 1992; Manning 2004), so only a brief discussion is justified here.

Type-1 and -2 system phase relations are shown in Figures 2 and 3 (constructed after Morey 1957). When both the critical curve and the solubility curve are continuous (or non-intersecting) in P-T-X space, the system is said to be of type-1 (Fig. 2). If the two curves intersect, we speak of a type-2 system (Fig. 3). The first type of behavior is characteristic of systems in which solid-A is highly soluble in H_2O; an important geological example is NaCl–

Table 1. Symbols used and their meaning.

Symbols and usage	
P	Pressure 1 bar = 10^5 Pa, 1 GPa = 10^4 bar = 10 kbar
T	Temperature (usually as celsius for experiments, otherwise kelvin)
X	Composition (usually by weight in experiments, otherwise molar)
x	Composition (mole fraction)
A	silicate, the non-volatile component, in the solid state at room conditions
S_A	pure A in the solid state*
L_A	pure A in the liquid state
G_A	pure A in the vapor state
H_2O	water, in the liquid state at ambient T,P, the volatile component
ice	pure H_2O in the solid state
steam	pure H_2O in the vapor state
A–H_2O	simple binary system of interest
V	aqueous Vapor (=H_2O with dissolved S_A)
L	hydrous Liquid/melt (=L_A with dissolved H_2O)
$E(H_2O$–$S_A)$	H_2O–S_A eutectic point, where ice-S_A+L+V are in equilibrium
E_L	eutectic composition of L
E_V	eutectic composition of V
trH_2O	triple point of pure H_2O
trA	triple point of pure A
kH_2O	L=V critical point on the boiling curve of H_2O
kA	L=V critical point on boiling curve of pure A
$k_1(L=V)A$	lower L=V critical endpoint in the binary A–H_2O
$k_2(L=V)A$	upper L=V critical endpoint in the binary A–H_2O
SCF	supercritical fluid, L and V completely miscible

*A is refractory (or has low-volatility) as it is solid (S_A) under most of the P–T conditions considered (unless specified).

H_2O (e.g., Sourirajan and Kennedy 1962). The second type, where the critical curves and solubility curves intersect, is characteristic of systems in which solid-A is sparingly soluble in H_2O (for example SiO_2–H_2O, see Kennedy et al. 1962). In this review we are mostly concerned with the second type as it is appropriate to most silicate mineral–H_2O systems.

In type-1 systems, the critical composition along the critical curve varies continuously from H_2O to A between the respective critical points. A second curve (termed the solubility curve) is defined by the coexistence of three phases, solid-A + liquid + vapor, and extends from a low-temperature A–H_2O eutectic point (E) to the invariant triple point for A (solid + liquid + vapor) where it terminates.

Type-2 phase relations are characterized by an intersection of the solubility and critical curves. This produces two separate P-T regions where a liquid–vapor miscibility gap is stable. As a result of their intersection, the solubility and critical curves terminate (or vanish) at the so-called first and second critical endpoints. In the geological literature, the high-temperature and low-temperature segments of the solubility curve are commonly called the wet solidus and the vapor-saturation curve, respectively.

T-X and *P-X* projections (Fig. 2b,c and 3b,c) show L and V compositions along the three phase curves. It can also be seen that L and V converge in composition at the triple points trH_2O and trA and kA, along the critical curve, and in type-2 systems also at the two critical endpoints $k_1(L=V)A$ and $k_2(L=V)A$.

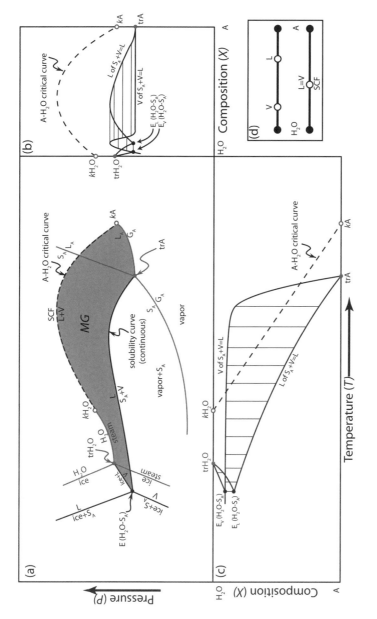

Figure 2. Phase relations for a type-1 binary system A–H$_2$O, where: i) component A is highly soluble and less volatile than H$_2$O; ii) no intermediate solids occur; iii) critical curve and solubility curve are continuous and do not intersect in the *P-T-X* space, and, iv) no immiscibility occurs in pure liquid-A. (a) *P-T* projection: grey curves refer to the unary or end-member systems A and H$_2$O, black curves refer to phase boundaries in A–H$_2$O. *k*A and *k*H$_2$O are the critical points of pure A and H$_2$O, respectively, and are connected by a continuous critical curve in the binary system A–H$_2$O. trA and trH$_2$O are the triple (S+L+V) points of pure A and H$_2$O, respectively. E(H$_2$O–S$_A$) is the eutectic of the A–H$_2$O, where solid-A (S$_A$) and ice coexist with liquid (L) and vapor (V). The solubility curve (S$_A$+V=L) is continuous from E(H$_2$O–S$_A$)

to trA. Grey shading represents the *P-T* field where the L + V miscibility gap (*MG*) is stable. (b) *P-X* projection of liquid–vapor relations along the three-phase coexistence (S$_A$+V=L), where the L + V miscibility gap (*MG*) is stable. (b) *P-X* projection of liquid–vapor relations along the three-phase coexistence (S$_A$+V=L) occurs as two separate curves, denoting coexisting V and L. L and V compositions converge at the triple points for the pure A and H$_2$O. For clarity, only tie-lines between coexisting L and V are shown. (c) *T-X* projection of liquid–vapor relations along the three-phase coexistence (S$_A$+V=L, solid lines) and critical curves (dashed). (d) Schematic relations between A, L, V and H$_2$O where a liquid–vapor miscibility gap is stable (upper) and where L and V have converged in composition and are completely miscible (lower) (after Morey 1957).

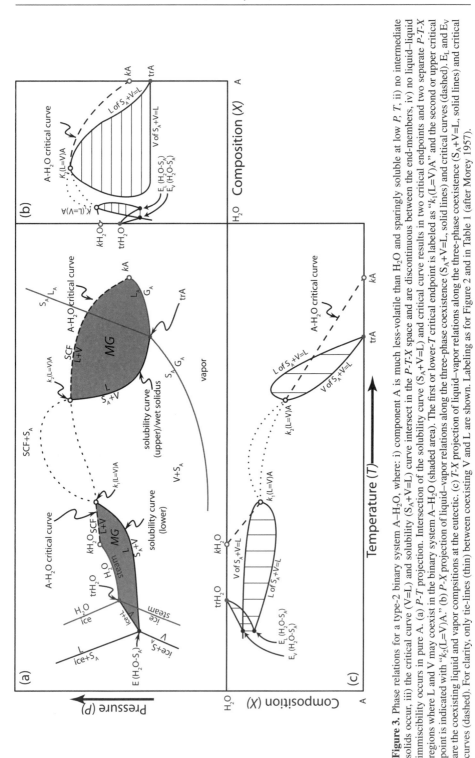

Figure 3. Phase relations for a type-2 binary system A–H₂O, where: i) component A is much less-volatile than H₂O and sparingly soluble at low P, T, ii) no intermediate solids occur, iii) the critical curve (V=L) and solubility (S_A+V=L) curve intersect in the P-T-X space and are discontinuous between the end-members, iv) no liquid–liquid immiscibility occurs in pure A. (a) P-T projection. Intersection of the solubility curve (S_A+V=L) and critical curve results in two critical endpoints and two separate P-T-X regions where L and V may coexist in the binary system A–H₂O (shaded area). The first or lower-T critical endpoint is labeled as "k_1(L=V)A" and the second or upper critical point is indicated with "k_2(L=V)A." (b) P-X projection of liquid–vapor relations along the three-phase coexistence (S_A+V=L, solid lines) and critical curves (dashed). E_L and E_V are the coexisting liquid and vapor compositions at the eutectic. (c) T-X projection of liquid–vapor relations along the three-phase coexistence (S_A+V=L, solid lines) and critical curves (dashed). For clarity, only tie-lines (thin) between coexisting V and L are shown. Labeling as for Figure 2 and in Table 1 (after Morey 1957).

Critical points (L=V) are presented as open circles in Figures 2 and 3. kH_2O occurs at $T_{crit} = 647$ K $= 374$ °C; $P_{crit} = 22.1$ MPa (Haar et al. 1984). For our refractory component A, the critical point kA occurs at much higher T than kH_2O. L=V critical conditions and L+V and S+V coexistence curves for almost all silicates are poorly constrained, though the latter are of interest to those concerned with condensation of planetary materials from accretionary solar disks.

A–H_2O has a low-temperature eutectic, $E(H_2O–S_A)$. At the eutectic, ice+S_A+L+V coexist, with L and V in their respective eutectic compositions E_L and E_V. At the eutectic, four equilibria originate, derivable from the binary relations shown in Figure 2d following Schreinemakers' notation (1915–1925) of the absent phase noted in brackets (see also Zen 1966):

$$S_A + ice\ (H_2O) = L\quad [V]$$

$$S_A + ice = V\qquad [L]$$

$$ice + L = V\qquad [S_A]$$

$$S_A + V = L\qquad [ice]$$

Figures 4b to d show conventional isobaric *T-X* sections intersecting the wet solidus and upper or second critical region of a type-2 A–H_2O system (Figs 4a and 3a). At $P_1(<P_C)$, the L+V miscibility gap is stable, and closes at higher temperature (T_2). This closure is given by the critical curve in the *P-T* plane (Fig. 4a). Where fluids are supercritical, L–V immiscibility is a metastable phenomenon. L and V compositions coexisting on the wet solidus (T_1) at this P_1 can be read from the isobaric section in Figure 4b. T_3 is the temperature of the dry liquidus for A on the isobar. At P_C, T_C, the critical endpoint k_2(L=V)A occurs, it is where the critical L=V condition (or L–V solvus crest) intersects the wet solidus. The wet solidus is only defined where L and V are immiscible. The composition of this critical point can be read from the isobaric section in Figure 4c. At higher pressure (e.g., P_2), the wet solidus is absent as the L+V gap is shown metastable below the solubility curve. At P_2, there is continuous solubility of mineral-A from dry liquid-A (at T_5) down temperature towards pure H_2O (Fig. 4d).

Further types of L–V phase relations

We are probably restricting our views of natural processes when we think that there is only a single natural salty fluid (a brine) and a single natural gas (e.g., H_2O or CO_2-rich). Other volatile components do occur in natural systems and these can in certain circumstances lead to gas–gas and fluid–fluid immiscibility, and hence critical curves related to supercritical behavior. Such components lead to quite new classes of critical phenomena *additional* to the two types discussed above, and might easily have been overlooked so far.

Other types of volatile phase relations are certainly known from the field of physical chemistry at least since the early work of Van Laar (1905). Six main types of binary fluid phase diagrams have been found experimentally, reflecting the occurrence of different forms of critical curve complexity and the occurrence of three-phase equilibria (e.g., Zernike 1955; Van Konynenburg and Scott 1980; McGlashan 1985; Young 1986; Calado and Lopes 1999; Schneider 2002).

The six types of fluid systems in Figure 5 are all derived from liquid + vapor equilibria and thus do not consider solid–volatile equilibria, the general concern of geologists. The introduction of solids introduces the possibility for considerably more complexity, a new component adds additional degrees of freedom, manifest as further three-phase curves, azeotropy and invariant quadruple points in *P-T* projections of binary phase relations. Details of various possible system types involving solid–volatile equilibria are outlined by several authors (Wenzel et al. 1980; Koningsveld and Diepen 1983; Lu and Zhang 1989; Valyashko 1990b, 1995, 1997; Garcia and Luks 1999; Urusova and Valyashko 2001; Valyashko 2002a,b; Valyashko and Churagulov 2003; Valyashko and Urusova 2003).

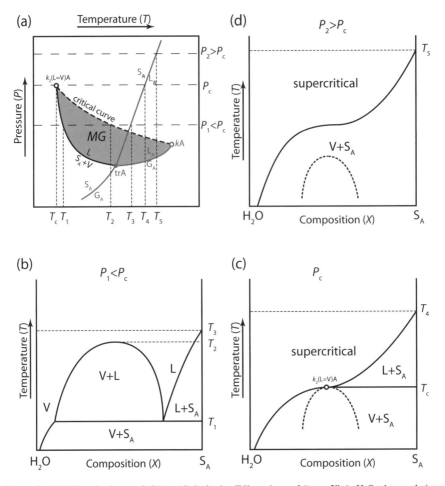

Figure 4. (a) *P-T* projection, and (b) to (d) isobaric, *T-X* sections of "type-2" A–H$_2$O phase relations (see Figure 3) involving wet melting and critical behavior. At *P* below the second critical endpoint a wet solidus is defined. (b) L–V are immiscible between T_1–T_2. (c, d) Above the second critical pressure there is no stable L–V immiscibility, fluid may transform continuously between L and V depending on *T*. With increasing *P* the L–V solvus vanishes (here, below the solidus) reflecting increasing mutual solubility of V and L with increasing *P* on the wet solidus.

Veksler (2004) in a recent review of liquid immiscibility in magmatic to hydrothermal environments, stresses the possible importance of low-temperature three-phase volatile equilibria becoming stable under special compositional circumstances, which subsequently lead to otherwise unusual immiscibility phenomena at low *P, T* conditions. Such phenomena described by Veksler (2004) may explain certain complex immiscible and miscible fluid phase relations observed in some systems (Tuttle and Friedman 1948; Friedman 1950; Morey and Hesselgesser 1952). In the system Na$_2$SiO$_3$–SiO$_2$–H$_2$O, for example, it has been reported by several workers that both solid quartz and sodium disilicate can each stably coexist with a vapor phase and two liquids of different compositions (a dilute one and a more concentrated one) down to very low-temperatures and pressures (200 °C and 80 MPa, Valyashko and Kravchuk 1978; Kravchuk and Valyashko 1979). It remains unclear if such observations reflect equilibrium or slow(?) reaction kinetics (< 300 °C) leading to metastable immiscibility.

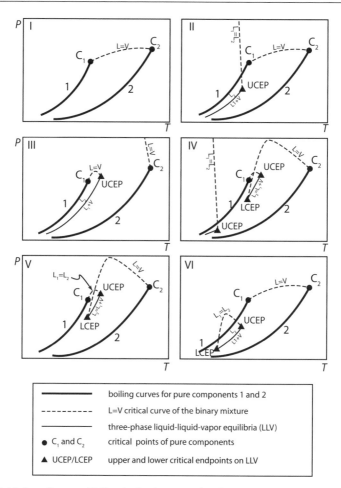

Figure 5. Fluid phase diagrams (*P-T* projections) representing six types (I to VI) of binary mixtures. The thick lines represent either the saturation curves of the two pure components or three-phase lines of their mixtures. The latter end at the critical points of each pure substance (circles) or at the upper/lower critical endpoints (triangles) of the mixtures. The thin lines represent L=L or V=L critical curves. Nomenclature used here differs slightly from ours (Table 1): C_1 and C_2 refer to the critical points for pure components 1 and 2, respectively; UCEP and LCEP refer to upper and lower critical points on L+V coexistence curves occurring in the binary 1–2 systems, respectively (modified from Calado and Lopes 1999, adapted originally from Van Konynenburg and Scott 1980).

For those concerned, the wealth of fluid phase equilibria information in the physical chemistry literature is remarkable. Further studies of silicate solubility in natural fluids may reveal the importance of other types of multiple fluid–fluid phase equilibria governing certain geological processes where additional volatiles or solids are present.

MINERAL–H₂O SYSTEMS

Mineral–H₂O systems provide a basis for understanding phase relations in more complicated rock–H₂O systems. Here we review the available experimental data on phase relations, liquid–vapor (im)miscibility and wet melting at high-pressure in the systems SiO_2 (quartz)–H_2O and

NaAlSi$_3$O$_8$ (albite)–H$_2$O, as well as the effect of additional volatile components. These two binaries are extensively studied and appear to possess all the defining characteristics of "type-2" systems (see above, Fig. 3). We also highlight where uncertainties remain in both systems, and what the consequences arise from certain assumptions used to resolve these gaps. Table 2 contains a representative selection of source data used in this section.

SiO$_2$(quartz)–H$_2$O

Mineral solubility and wet melting relations in the SiO$_2$–H$_2$O system have been the subject of numerous investigations (Table 2), which are summarised in Figure 6a. The two critical endpoints k_1(L=V)Qtz and k_2(L=V)Qtz are located near kH$_2$O (374 °C, 22.1 MPa; Haar et al. 1984) and at 1100 °C, 0.97 GPa (Kennedy et al. 1962), respectively. The critical point for pure silica, kSiO$_2$, is at 11,700 °C, ca. 0.2 GPa and a critical density of 0.58 g cm^{-3} (Guissani and Guillot 1996).

Liquid and vapor compositions coexisting with the stable SiO$_2$-polymorph are given by isoplethal contours in mole fraction H$_2$O in Figure 6b. The experimental data of Kennedy et al. (1962) for L and V compositions on the quartz wet solidus and the quartz solubility data of Anderson and Burnham (1965), Nakamura (1974) and Manning (1994) at 1.5 GPa provide the constraints on the isopleth positions at P and T above the termination of the wet solidus at a critical endpoint (k_2(L=V)Qtz). Quartz solubility isopleths for x(H$_2$O) $\geq$ 0.97, < 900 °C are consistent with the solubility equation of Manning (1994). Elsewhere, the isopleths are fit by hand to the available experimental data mentioned on Figure 6a. Quartz solubility near the lower S+L+V curve segment (up to 500 °C, 0.1 GPa) is shown as isopleths in Figure 6c and are fit by hand to the data of Kennedy (1950). No quartz solubility data appear to be available for V compositions on the "lower" S+L+V curve or in the "gaseous" supercritical region.

The upper P-T region in which L–V immiscibility occurs in SiO$_2$–H$_2$O is in some respects not well constrained. The location of the critical curve is not presently determined (dashed curve, Fig. 6d). However, it is known that the quartz wet solidus (Qtz+V=L) terminates at 1100 °C, 0.97 GPa and x(SiO$_2$) = 0.5 by intersection with the critical curve (Kennedy et al. 1962). At higher temperature, the critical curve is expected to intersect the critical point of pure silica (see above, Figs. 3 and 4). The location of the critical endpoint of the quartz wet solidus and critical point for pure silica indicates that the critical curve has an overall slightly negative dP/dT slope (ca. −0.07 MPa K^{-1}, for linear dP/dT).

Figure 7 shows a set of T-x sections in the SiO$_2$–H$_2$O system that illustrate the manner in which solubility behavior and L–V phase relations vary with pressure. Figure 7a shows the composition of V-saturated silica liquid (L) at different temperatures for a series of pressures, using the data of Kennedy et al. (1962) and Holtz et al. (2000). Below 0.4 GPa there is little or no temperature dependence of H$_2$O solubility in the V-saturated hydrous liquid. Above 0.5 GPa, as k_2(L=V)Qtz is approached, H$_2$O solubility becomes strongly prograde (increasing with temperature). On a contoured P-T projection (e.g., Fig. 6b), this corresponds to a change in the slope of the respective isopleths (long dashes in Fig. 6b) with increasing pressure from dP/dT near zero to negative. L and V compositions converge to one critical composition (or supercritical fluid, SCF) at the upper or second critical endpoint on the wet solidus k_2(L=V)Qtz. Figures 7b and c clearly illustrates the continuity between L and V compositions as k_2(L=V)Qtz is passed. Figure 7b shows SiO$_2$-polymorph solubility in H$_2$O at 1.5 GPa as a function of temperature, using data from Nakamura (1974) and Manning (1994). The L+V miscibility gap is gone at this pressure (above P of k_2(L=V)Qtz). There is a continuous transition from dry SiO$_2$-liquid (L$_{SiO2}$) to pure H$_2$O-liquid, without passing through a wet melting reaction. L and V are completely miscible and a single fluid coexists with the stable SiO$_2$-polymorph at these conditions; the fluid is supercritical. At lower pressure L and V are immiscible as shown on Figure 7c, an isobaric T-X section through Figure 6b at 0.4 GPa using the data of Anderson and Burnham (1965), Kennedy et al. (1962) and Holtz et al. (2000).

Table 2. Experimental studies in SiO_2-H_2O and $NaAlSi_3O_8$-H_2O

SiO_2–H_2O		$NaAlSi_3O_8$–H_2O	
Dry melting			
Jackson (1976)		Schairer and Bowen (1956)	
Kanzaki (1990)		Birch and le Comte (1960)	
Hudon et al. (2002)		Boyd and England (1963)	
		Boettcher et al. (1982)	
		Nekvasil and Carroll (1996)	
Wet melting			
Tuttle and England (1955)		Yoder (1958)	
Kennedy et al. (1962)		Bohlen et al. (1982)	
Stewart (1967)		Goldsmith and Jenkins (1985)	
Boettcher (1984)		Luth and Boettcher (1986)	
Luth and Boettcher (1986)		Stalder et al. (2000)	
Solubility studies (numbers explained in notes)			
Kennedy (1944)	2	Goranson (1938)	4
Kennedy (1950)	1,2	Burnham and Jahns (1962)	4
Morey (1951)	2	Orlova (1964)	4
Wasserburg (1958)	2	Khitarov et al. (1963)	4
Wood (1958)	2	Luth et al. (1964)	4
Fournier (1960)	1	Currie (1968)	2
Kitahara (1960)	1	Kadik and Lebedev (1968)	4
van Lier et al. (1960)	1	Boettcher and Wyllie (1969)	3,4
Morey et al. (1962)	1	Morse (1970)	4
Siever (1962)	1	Burnham and Davis (1971)	4
Kennedy et al. (1962)	3,4	Davis (1972)	2
Weill and Fyfe (1964)	2	Burnham and Davis (1974)	4
Anderson and Burnham (1965)	2	Eggler and Kadik (1979)	4
Sommerfeld (1967)	2	Voight et al. (1981)	4
Stewart (1967)	3	Day and Fenn (1982)	4
Semenova and Tsiklis (1970)	2	Vidale (1983)	2
Crerar and Anderson (1971)	1,2	Hamilton and Oxtoby (1986)	4
Mackenzie and Gees (1971)	1	Richet et al. (1986)	4
Volosov et al. (1972)	1	Woodland and Walther (1987)	2
Nakamura (1974)	2,3	Paillat et al. (1992)	3,4
Hemley et al. (1980)	1	Behrens (1995)	4
Fleming and Crerar (1982)	1	Shen and Keppler (1997)	3
Fournier and Potter (1982)	1,2	Stalder et al. (2000)	2
Ragnarsdóttir and Walther (1983)	2	Hauzenberger et al. (2001)	2
Walther and Orville (1983)	2	Pak et al. (2003)	2
Manning (1994)	2		
Manning and Boettcher (1994)	2		
Guissani and Guillot (1996)	5		
Rimstidt (1997)	1		
Holtz et al. (2001)	4		
Wang et al. (2004)	2		

(1) mineral solubility along the water liquid + vapor coexistence curve; (2) mineral solubility in supercritical H_2O; (3) L+V critical phenomena; (4) solubility of H_2O in vapor-saturated melt; (5) pure anhydrous (dry) boiling curve and critical point

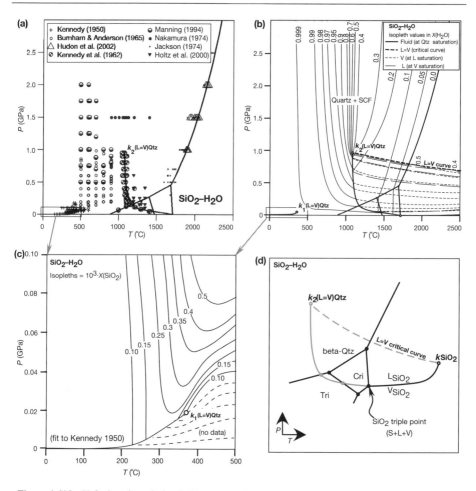

Figure 6. SiO₂–H₂O phase boundaries (bold curves) and solubility surfaces (in mole fraction H₂O, x(H₂O)). (a) *P, T* conditions experimentally investigated in SiO₂–H₂O (silica polymorph stability studies not included). (b) *P-T* projection of L and V compositions (shown as constant compositions isopleths) in equilibrium with the stable SiO₂-polymorph (thin solid lines), also shown are isopleths representing coexisting L (thin long dash lines) and V (thin short dash lines) compositions. Note although it is hardly visible at this scale the position of the critical curve occurs at the intersection of coexisting L and V isopleths with identical composition. The L+V isopleth geometry is shown more clearly elsewhere (Hack et al. 2007). Isopleth geometry derived from data sources in (a). (c) Quartz solubility isopleths near the lower three-phase (Qtz+V=L) curve, fit to data of Kennedy (1950, solid lines), and dashed curves where inferred. k_1(L=V)Qtz is the first critical endpoint for SiO₂–H₂O and at this scale appears coincide with kH₂O. (d) Schematic SiO₂–H₂O phase relations near the SiO₂ wet solidus (grey curve) and relation to SiO₂-polymorph stability (black curves). kSiO₂ is the pure SiO₂ critical point. Grey dashed curve is the SiO₂–H₂O critical curve. Abbreviations: Tri = tridymite, Cri = cristobalite, L$_{SiO2}$ = pure SiO₂-liquid; V$_{SiO2}$ = pure SiO₂-vapor.

Some controversy surrounds the location of a critical endpoint on the quartz wet solidus. Stewart (1967) suggested that the endpoint of the quartz wet solidus occurs at $P > 1$ GPa. He reported that at the wet solidus at 1.0 GPa the vapor contains 22±2 wt% SiO₂ and the coexisting liquid contains 38±2 wt% H₂O, thus he questioned the work of Kennedy et al. (1962). Mysen (1998) also questioned the occurrence of a critical endpoint for SiO₂–H₂O near 1 GPa. He investigated a portion of K₂O–SiO₂–H₂O from 0.8–2.0 GPa, 700–1000 °C and

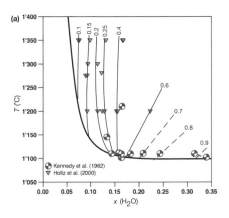

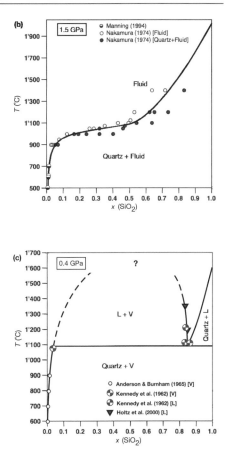

Figure 7. *T-x* sections through SiO_2–H_2O phase relations and solubility surfaces. (a) H_2O solubility (mole fraction, $x(H_2O)$) in liquid silica as a function of *T* for various isobars (thin solid and dashed curves, numbers at curve ends indicates isobar in GPa). Thick curve is the quartz wet solidus (Qtz+V=L). H_2O solubility in silica liquid increases strongly with increasing *T* and *P* as k_2(L=V)Qtz is approached. (b) Quartz solubility in H_2O as mole fraction SiO_2 at 1.5 GPa as observed experimentally by Nakamura (1974) and Manning (1994) [where, $x(SiO_2) = 1 - x(H_2O)$, calculated for mw H_2O=18, mw SiO_2=60]. At this pressure SiO_2-rich and H_2O-rich liquids are completely miscible up to quartz-saturation, no wet solidus occurs. (c) Isobaric *T-x* phase relations at 0.4 GPa. There is a miscibility gap separating L and V at the wet solidus. The solvus is experimentally poorly-defined (dashed curve), except at the wet solidus. The consolute temperature of the solvus is experimentally unknown.

found that a L–V miscibility gap was present up to 2.0 GPa in this system. In his experiments increasing amounts of K_2O lead to a shrinkage of the L–V miscibility gap. Extrapolating these results to pure SiO_2–H_2O he concluded immiscibility would persist to 2 GPa, *unless* solubility mechanisms were different in the absence of K_2O. However, observations by Nakamura (1974) showing the continuity of quartz solubility with increasing temperature (to the melting point) leave little doubt that in SiO_2–H_2O the quartz wet solidus is no longer defined at 1.5 GPa. Moreover, it has probably not been emphasized sufficiently by previous studies that the different quartz solubility data sets are in good general agreement and support the melting relations and critical endpoint on the wet solidus observed by Kennedy et al. (1962). Figure 7b shows the agreement on the *T*-dependence of supercritical quartz solubility in H_2O at 1.5 GPa as measured by Nakamura (1974) and Manning (1994). Similar agreement is observed between Anderson and Burnham (1965), Kennedy et al. (1962) and Holtz et al. (2000) at pressures where the wet solidus is still present (e.g., Figs. 7a,c).

$NaAlSi_3O_8$(albite)–H_2O

Experimental data for $NaAlSi_3O_8$–H_2O, together with our knowledge of SiO_2–H_2O phase topology, underpins much of our understanding of supercritical and wet melting relations as might apply in hydrous silicate systems. These undoubtedly provided much of the recent motivation to search for the occurrence of similar behavior in more complex systems and in natural rocks. Figure 8 summarizes the experimental data and inferred phase relations for the system NaAlSi$_3$O$_8$–H$_2$O. Like SiO$_2$–H$_2$O, NaAlSi$_3$O$_8$–H$_2$O shows a type-2 system topology, with a critical endpoint terminating the wet solidus of albite, k_2(L=V)Alb.

Figure 8a indicates experimentally investigated P, T conditions in the albite–water (NaAlSi$_3$O$_8$–H$_2$O) system (a representative selection from Table 2). Figure 8b shows a schematic P-T projection of the albite–H$_2$O phase relations. The location of the critical curve is determined between 1.0 to 1.5 GPa by Shen and Keppler (1997, see Fig. 8a,b), and has an

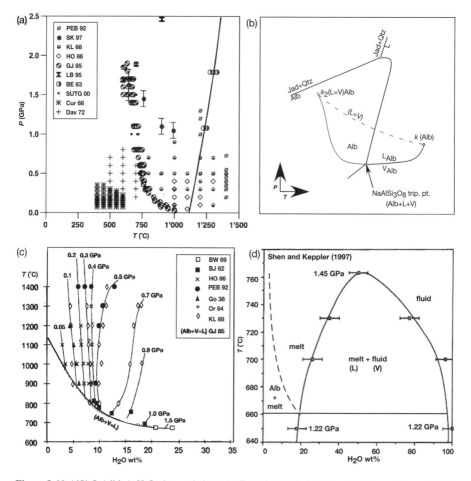

Figure 8. NaAlSi$_3$O$_8$(albite)–H$_2$O phase relations. (a) Experimentally investigated P,T conditions. (b) P-T projection of inferred phase relations where the NaAlSi$_3$O$_8$(albite)–H$_2$O critical curve intersects the wet solidus Alb+V=L (Paillat et al. 1992; Shen and Keppler 1997; and Stalder et al. 2000; L–V relations are more complex if anhydrous L$_{Alb}$ vaporization is incongruent). (c) H$_2$O solubility (as wt% H$_2$O) in liquid albite (L) as a function of T for a series of values of P (numbers at ends of *steep curves*) from 0.05 to 1.5 GPa. (d) L+V solvus data from Shen and Keppler (1997). Note the data are polybaric, L+V data at 1.22 GPa occur at a T below the wet solidus (660 °C) and are metastable. Data sources: PEB 92: Paillat et al. (1992); SK 97: Shen and Keppler (1997); KL 68: Kadik and Lebedev (1968); HO 86: Hamilton and Oxtoby (1986); GJ 85: Goldsmith and Jenkins (1985); LB 95: Liu and Bohlen (1995); BE 63: Boyd and England (1963); SUTG 00: Stalder et al. (2000); Cur 68: Currie 1968; Dav 72: Davis 1972; BW 69: Boettcher and Wyllie (1969b); BJ 62: Burnham and Jahns (1962); Go 38: Goranson (1938); Or 64: Orlova (1964) (c - modified from Paillat et al. 1992 ; d - modified from Shen and Keppler 1997).

overall negative dP/dT slope (ca. −0.55 MPa K^{-1}, assuming linear dP/dT). The intersection of this critical curve with the three-phase curve (Alb+V=L) occurs at 700 °C, 1.6 GPa (Stalder et al. 2000).

Figure 8c shows H_2O solubility in albite liquid (L) as a function of temperature for a series of pressures. Up to 0.4 GPa H_2O solubility in the liquid decreases with increasing T or is neutral, and all data agree well. Above 0.5 GPa H_2O solubility becomes increasingly prograde with increasing pressure; at 0.7 and 0.9 GPa only the highest values are plotted and these represent a minimum H_2O concentration in melt (L), since exsolution of the dissolved vapor upon quenching made accurate measurements on the glasses difficult (see Paillat et al. 1992). The values reported by Paillat et al. (1992) at and above 1.0 GPa where inferred by chemographic methods.

Figure 8d shows the *T-x* structure of L+V solvus in $NaAlSi_3O_8$–H_2O (polybaric) mapped out by Shen and Keppler (1997).

Paillat et al. (1992) constructed a schematic three-dimensional illustration of the *P-T-x* relations of the albite–water system, based on experimental observations (Table 2; Fig. 8) and thermodynamic considerations (Fig. 9). Figure 9 has been modified from the original illustration to include the occurrence of a critical endpoint on the wet solidus at 1.6 GPa. At 1.7 GPa, the highest pressure shown in Figure 9, the L–V miscibility gap has become metastable and albite can coexist with continuous range of fluid compositions, from pure dry liquid albite to pure water, depending on temperature.

Several aspects of the $NaAlSi_3O_8$–H_2O system are currently poorly-constrained. They include a paucity of information on subsolidus albite solubility in supercritical water.

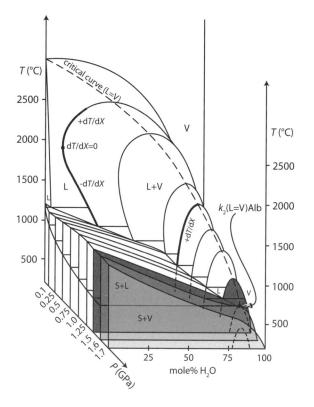

Figure 9. Schematic illustration of *P-T-x* relations for the system $NaAlSi_3O_8$(albite)–H_2O, highlighting critical behavior and termination of the wet solidus. Note changes in solubility behavior in different *P-T* regions can be related to the structure of the solvus (modified from Paillat et al. 1992).

Experimentally the system is complicated by incongruent dissolution of albite. Solubility studies include, Currie (1968, as reviewed by Fyfe et al. 1978), Anderson and Burnham (1983, an important source of unpublished data from Davis 1972), Shmulovich et al. (2001), and Antignano and Manning (2003). Vapor compositions (i.e., the dissolved silicate components) coexisting with hydrous albite liquids, and generally, are poorly-known (a valuable compilation of various solubility data is Clark 1966, p. 415–436).

Effect of added volatiles on critical behavior in SiO_2–H_2O and $NaAlSi_3O_8$–H_2O

General systematics. The addition of H_2O to the various anhydrous (dry) systems addressed by Thompson et al. (2007) leads to several different types of phase behavior, depending whether the solid components are quite soluble or relatively insoluble in water and to what extent. Each of the systems above with H_2O are often referred to as salt–water systems in the literature, even if silicates are dealt with (e.g., Veksler 2004). Earlier work on binary systems considered the case involving one volatile substance (that could exist either as liquid, solid or gas), and one solid substance which might have more than one solid polymorph and which might melt (but not vaporize). More recently, systems have been investigated where there might be more than one liquid or vapor in binary or multi-component systems (e.g., Veksler et al. 2002). With the extension into supercritical regions, it can be envisaged that there might be more than one type of supercritical fluid in some systems.

Added volatile components usually partition into vapor relative to the coexisting liquid. The effect is to expand the region of L–V immiscibility to higher pressure and lower temperature, whereas components partitioning more into the liquid, stabilize the liquid and expand the region of L–V immiscibility to higher temperature and lower pressure. These effects are discussed with reference to a L+V coexistence curve that has *positive* dP/dT, with liquid on the higher-pressure side, and V on the higher-temperature side.

Only few studies have experimentally investigated the addition of a second volatile component to a hydrous silicate system and its effect on melting relations and critical phenomena. As mentioned earlier (see Fig. 1) an added component will have a pronounced effect on the critical curve, depending on its degree of interaction with the solvent and its preference for either the vapor or the liquid.

Unequal partitioning of a component between phases will tend to maintain a miscibility gap, while components that dissolve equally into both phases may enhance mutual solubilities of the other components, as it implies formation of intermediate complexes common to both phases, and thereby driving the closure of chemical miscibility gaps.

SiO_2–H_2O–CO_2. Boettcher (1984) studied the effect of CO_2 on quartz–water melting relations up to 2.8 GPa from 1000 to 1650 °C (Fig. 10a). Beginning of melting was observed to increase with increasing CO_2 concentration. The dP/dT slope of the solidus also changes from negative to positive with increasing amounts of CO_2 added to the system. Addition of 0.05 $X(CO_2)$ to SiO_2–H_2O displaces the critical endpoint k_2(L=V)Qtz by about 1 GPa towards higher in pressure (to about 1.9 GPa). These results clearly show that CO_2 displaces critical phenomena towards higher pressures compared to SiO_2–H_2O. This is because CO_2 partitions into the vapor rather than liquid, expanding the two-phase field toward higher pressure and temperature, relative to an H_2O-saturated solidus with negative dP/dT.

SiO_2–H_2O–H_2. The investigation of Luth and Boettcher (1986) in the system SiO_2–H_2O–H_2 serves as a further example of how partitioning of added components may displace L–V immiscibility and melting curve critical points (Fig. 10b). The results show that higher fH_2/fH_2O displaces the critical point from ~1 to ~2.9 GPa at ~30 mol% H_2 (buffered at iron-wustite (IW)) and the L+V immiscible region towards higher pressures and higher temperatures relative to an H_2O-saturated solidus with negative dP/dT. This is because H_2 partitions into the vapor relative to the liquid. The location of the quartz wet solidus at 1.5 GPa in the presence of H_2O–H_2 (IW)

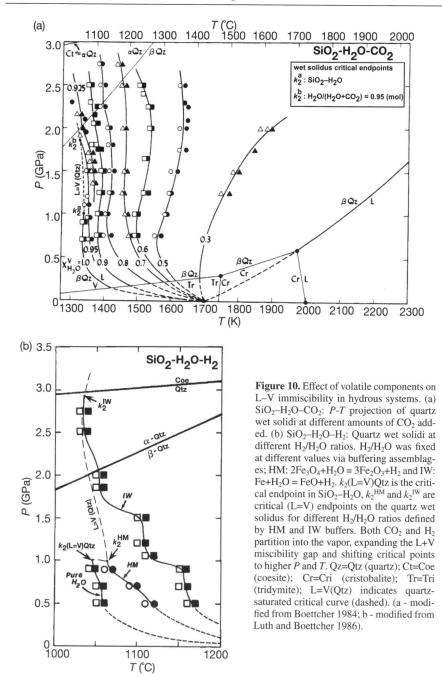

Figure 10. Effect of volatile components on L–V immiscibility in hydrous systems. (a) SiO_2–H_2O–CO_2: P-T projection of quartz wet solidi at different amounts of CO_2 added. (b) SiO_2–H_2O–H_2: Quartz wet solidi at different H_2/H_2O ratios. H_2/H_2O was fixed at different values via buffering assemblages; HM: $2Fe_3O_4+H_2O = 3Fe_2O_3+H_2$ and IW: $Fe+H_2O = FeO+H_2$. k_2(L=V)Qtz is the critical endpoint in SiO_2–H_2O, k_2^{HM} and k_2^{IW} are critical (L=V) endpoints on the quartz wet solidus for different H_2/H_2O ratios defined by HM and IW buffers. Both CO_2 and H_2 partition into the vapor, expanding the L+V miscibility gap and shifting critical points to higher P and T. Qz=Qtz (quartz); Ct=Coe (coesite); Cr=Cri (cristobalite); Tr=Tri (tridymite); L=V(Qtz) indicates quartz-saturated critical curve (dashed). (a - modified from Boettcher 1984; b - modified from Luth and Boettcher 1986).

found by Luth and Boettcher (1986) is in agreement with the observations of Nakamura (1974). The abrupt changes in slope of the vapor-saturated solidus of SiO_2–H_2O–H_2 (IW) are intriguing and may be explained by pressure-induced structural transformations in either the liquid and/or the vapor (Luth and Boettcher 1986).

NaAlSi₃O₈–H₂O–CO₂. In the case of NaAlSi₃O₈–H₂O–CO₂ (Eggler and Kadik 1979; Bohlen et al. 1982; Sykes and Holloway 1987) the addition of CO₂ not only displaces the wet solidus to higher temperature with increasing CO₂, it expands the immiscible L+V coexistence field to higher pressure than in the CO₂-free system (the upper limit of which was reported by Shen and Keppler 1997). As with SiO₂–H₂O–CO₂, the dramatic enhancement of L–V immiscibility and suppression of melting to higher temperatures, reflects the operation of different chemical complexing mechanisms by which CO₂ dissolves in silicate liquid and aqueous solvents, and also points to a general inability for CO₂ to easily build chemical bridging structures which would lead to miscibility between silicate melts and aqueous fluids. CO₂ generally promotes immiscibility in silicate systems.

Effects of non-volatile components on critical behavior in NaAlSi₃O₈–H₂O and SiO₂–H₂O

Sowerby and Keppler (2002), expanding on the work of Shen and Keppler (1997), found that addition of Na₂O, B₂O₃ and F₂O₋₁ to NaAlSi₃O₈–H₂O displaces the critical curve to lower pressure and lower temperature (relative to an H₂O-saturated solidus with negative d*P*/d*T*, Fig. 11). Sowerby and Keppler (2002) report that additions of 5 and 10 wt% Na₂O to hydrous NaAlSi₃O₈ glass + H₂O lowers the upper critical curve by 143 and 247 °C at 1 GPa, respectively. The direction of the displacement (down temperature) indicates these added components fractionate strongly into the liquid (melt) relative to the coexisting vapor.

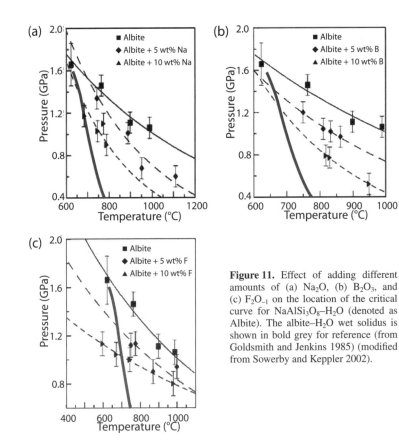

Figure 11. Effect of adding different amounts of (a) Na₂O, (b) B₂O₃, and (c) F₂O₋₁ on the location of the critical curve for NaAlSi₃O₈–H₂O (denoted as Albite). The albite–H₂O wet solidus is shown in bold grey for reference (from Goldsmith and Jenkins 1985) (modified from Sowerby and Keppler 2002).

Urusova and Valyashko (2001) summarized the experimental data and discussed L–V immiscibility and critical phenomena in the SiO_2–Na_2SiO_3–H_2O system. They showed that the addition of Na_2SiO_3 to SiO_2–H_2O displaces the quartz-saturated upper critical point to significantly lower P, T conditions. This effect is the opposite to that of CO_2 and other volatile components described above. Displacement of the quartz wet solidus towards lower T occurs because Na_2SiO_3 dissolves readily into hydrous silica liquid relative to the vapor at SiO_2-rich compositions. Increasing Na_2SiO_3 content is required to move along the quartz-saturated ternary critical curve to lower T. Because Na_2SiO_3 is completely miscible with H_2O above 0.18 GPa, liquids in this system remain supercritical to much lower pressure (Valyashko and Kravchuk 1978; Kravchuk and Valyashko 1979). Alternatively, L–V immiscibility in Na_2SiO_3–H_2O is only possible at pressure less than 0.18 GPa. The hydrous persodic liquids studied may be considered to be broadly pegmatitic in character. If so, such phase relations allow for the possibility of a supercritical fluid origin for some relatively low-pressure, low-temperature pegmatites.

ROCK–H₂O SYSTEMS

Here we consider theoretical ternary phase relations and critical curve and wet melting data in the model systems, $NaAlSiO_4$–SiO_2–H_2O, haplogranite–H_2O, MgO–SiO_2–H_2O. We also consider the role of fluxing and (im)miscibility in the formation of granitic pegmatites, as well as trace-element partitioning and major element compositions of fluids and melts in a more complex eclogite–H_2O system.

General A–B–H₂O ternary phase relations involving L–V supercriticality

Up to this point we have dealt with relatively uncomplicated melting and L–V phase relations in hydrous binaries that are broadly recognized as models for more complex silicate systems, and introduced the concept of displacing critical phenomena with additional components. Our purpose here is to outline a more complete theoretical basis for understanding phase relations involving supercriticality in ternary and higher-order systems.

Figure 12 shows phase relations in a hypothetical A–B–H_2O system. An important feature shown on Figure 12 is that critical endpoints on different wet solidi must be systematically connected by critical curves along which composition varies. To understand what gives rise to these, possibly unfamiliar phase relations, consider the position of the A wet solidus reaction (A + V = L). As component B is added, the wet solidus is displaced to lower T, until the system saturates in solid-B at the A+B wet solidus. Now consider the critical endpoint k_2(L=V)A, as B is added the critical L=V or consolute point is displaced together with the A-saturated wet solidus to lower T until saturation in solid-B occurs. In this way a critical line, along which solid-A is stable and the critical fluid composition varies, joins k_2(L=V)A and k_2(L=V)A+B (Fig. 12). These concepts were discussed in detail by Boettcher and Wyllie (1969a) in the context of a hypothetical ternary MO–SiO_2–H_2O system. In our view, Boettcher and Wyllie (1969a) outlined a simple and robust framework with which to unravel critical phenomena systematics, where additional components are added, and an appropriate projection of the system can be made. We will return to these concepts of critical phenomena displacement and continuity in P-T-X when we analyze melting relations for more complex "rock"–water systems (below).

At P_1 (Fig. 12c) the L–V miscibility gap extends across all A/B ratios, and wet melting occurs at all cotectics. The dashed line indicates coexisting L and V compositions at this P on the A+B wet solidus. With increasing pressure the L–V miscibility gap closes quickly on the A–H_2O join, but retreats more slowly elsewhere in the system. P_2 corresponds to the critical pressure on the A–H_2O wet solidus. At and above this pressure the A–H_2O join is supercritical, solid-A may dissolve continuously into H_2O with increasing temperature up to the melting point of solid-A.

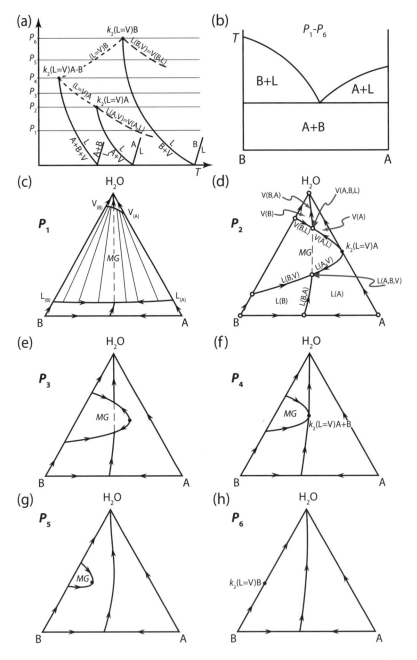

Figure 12. A–B–H$_2$O ternary phase relations, including critical behavior, and shrinkage of the L–V miscibility gap with increasing P. (a) P-T projection. (b) schematic T-X projection of the join A–B, indicating dry melting is eutectic at all pressures (A+B=L). (c) to (h) are isobaric (P_1–P_6), polythermal ternary projections of liquidus–fluidus relations. Arrows indicate direction of falling T. Coexisting L + V tie lines are shown on (c) but are omitted on subsequent sections for clarity. MG=Miscibility Gap. (c) to (e) a dashed tie line indicates the coexisting L+V at the A+B wet solidus (symbols and figure follow Boettcher and Wyllie 1969a).

With increasing pressure, the miscibility gap continues to shrink as vapor and liquid compositions approach each other and the critical point moves away from A–H_2O towards B–H_2O. At P_4
the critical point touches the A+B cotectic, producing a second critical endpoint k_2(L=V)A+B.
For $P > P_4$ the A+B cotectic does not intersect the immiscible region so fluids saturated in A+B
are incapable of unmixing at any T. P_6 corresponds to the critical endpoint on the join B–H_2O
where the wet solidus B+V=L intersects the B–H_2O critical curve, and is the vanishing point of
L–V immiscibility in the ternary system. The movement of the critical point on the miscibility
gap across the polythermal ternary projection appears on the *P-T* projection as a continuous
critical curve joining wet solidi critical endpoints in the ternary system. Ternary critical curves,
unlike for the simple binary case, are also saturated with a primary liquidus/fluidus phase.

The same rules apply for even more complex systems like natural rocks and their melts.
For each added component it needs to be determined whether this partitions more into liquid
than vapor. For many silicate melts, where the added component goes into melt rather than
solids or vapor, the relationship remains constant.

$NaAlSiO_4$–SiO_2–H_2O (nepheline–quartz–H_2O)

Establishing the nature of the phase relations in simple end-member systems such as NaAl-
SiO_4–SiO_2–H_2O is important. This is because they have been and probably will be further used
to calibrate new thermodynamic datasets incorporating melts and fluids (e.g., Holland Powell
2001; Evans and Powell 2006). Here, we consider the wet melting relations in the system NaAl-
SiO_4–SiO_2–H_2O in light of the recent observations of complete liquid–vapor miscibility in the
various constitutive binary sub-systems (i.e., SiO_2–H_2O, Kennedy et al. 1962; $NaAlSi_3O_8$–H_2O,
Shen and Keppler 1997; Paillat et al. 1992; Stalder et al. 2000; $NaAlSi_2O_6$–H_2O and NaAl-
SiO_4–H_2O, Bureau and Keppler 1999). Wet melting is only defined where L (melt) and V (fluid)
can be distinguished, hence where L and V are not completely miscible. Our analysis reveals
how more recent studies identifying supercriticality call into question interpretations of widely
accepted earlier observations regarding the occurrence of classical wet melting reactions at high
pressure in this system (e.g., Boettcher and Wyllie 1969b).

Boettcher and Wyllie (1969b) investigated subsolidus and wet melting phase relations in
$NaAlSiO_4$–SiO_2–H_2O up to 3.5 GPa (Fig. 13). In that study no occurrence of complete liquid–
vapor miscibility, or termination of any wet melting reaction was reported, except in SiO_2–H_2O
(Kennedy et al. 1962). In conflict with the Boettcher and Wyllie (1969b) topology are, however,
more recent investigations of Shen and Keppler (1997), Bureau and Keppler (1999) and Stalder
et al. (2000) in the same system which all report evidence of complete L–V miscibility on all
binary mineral–H_2O joins in the $NaAlSiO_4$–SiO_2–H_2O plane at conditions within the albite
stability field (< 2 GPa) (Fig. 14).

As wet melting in the binaries Alb–H_2O and Jad–H_2O is reported as congruent, extrapolation of the respective binary critical curves (Shen and Keppler 1997; Bureau and Keppler 1999)
should result in critical endpoints at their intersection with the corresponding wet solidi (determined by extrapolation where necessary) (Fig. 14a). In contrast, wet melting is *incongruent*
in Nepheline–H_2O (Boettcher and Wyllie 1969b), thus there is no reason for the critical curve
in the $NaAlSiO_4$–H_2O join to intersect and terminate the Nep+V=L (wet solidus) at a critical
endpoint. Accordingly, observations of Boettcher and Wyllie (1969b) and Bureau and Keppler
(1999) can be in agreement on wet nepheline melting relations, but are in clearly disagreement
on the other joins.

Figure 14b shows $NaAlSi_3O_8$–SiO_2–H_2O phase relations that are consistent with critical
phenomena observations by Shen and Keppler (1997), Bureau and Keppler (1999) and Stalder
et al. (2000). This figure was constructed by us assuming the simplest *P-T-X* melt–vapor
miscibility gap behavior necessary, i.e., a single immiscible L+V field in the system which
shrinks and eventually vanishes with increasing pressure (Fig. 15).

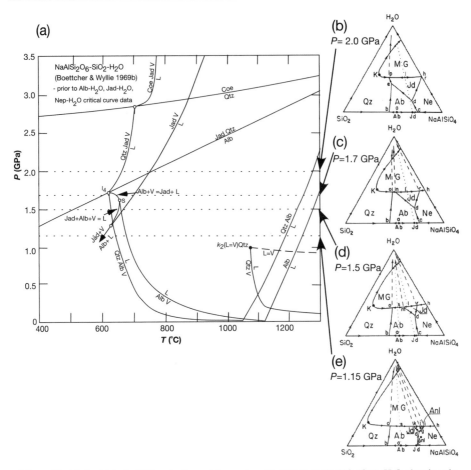

Figure 13. (a) An historic *P-T* projection of the system NaAlSi₂O₆(Jad)–SiO₂(Qtz)–H₂O showing the interpretation of wet melting and L–V (im)miscibility relations. No critical (L = V) behavior terminates any of the wet solidi to *P* ≤ 4 GPa (except in SiO₂–H₂O, following Kennedy et al. 1962, the critical curve location (L=V, dashed) is speculative). (b) to (e) Ternary (NaAlSiO₄–SiO₂–H₂O) polythermal projection of mineral liquidus–fluidus relations occurring at various isobars (dotted lines). *MG*=L–V Miscibility Gap; K=ternary isobaric invariant critical point (L=V); Jad=Jd=NaAlSi₂O₆; Nep=Ne=NaAlSiO₄; Alb=Ab=NaAlSi₃O₈; Anl=NaAlSi₂O₆.H₂O; Qtz=Qz=SiO₂ (modified after Boettcher and Wyllie 1969b).

Figure 15 shows simplified polythermal, isobaric ternary sections for these two different interpretations of liquid–vapor immiscibility relations in the ternary system NaAlSiO₄–SiO₂–H₂O. Figure 15a is based on the experimental observations of Boettcher and Wyllie (1969b; their Fig. 11), whereas Figure 15b is constructed by us and represents a synthesis of the recent experimental findings of Paillat et al. (1992), Shen and Keppler (1997), Bureau and Keppler (1999), and Stalder et al. (2000). Figure 15a is consistent with Figure 13 and Figure 15b is consistent with Figure 14. The main differences between Boettcher and Wyllie's (1969b) interpretation of phase relations and that required by the observations of more recent workers are: 1) that L–V *MG* now shrinks with increasing pressure more rapidly, thereby terminating most wet solidi at pressures less than ~2 GPa; 2) the system becomes supercritical in both silica-saturated and silica-undersaturated compositions (rather than in silica-rich compositions only); and 3) the range of immiscible compositions at higher *P* (e.g., > 2 GPa) is vastly

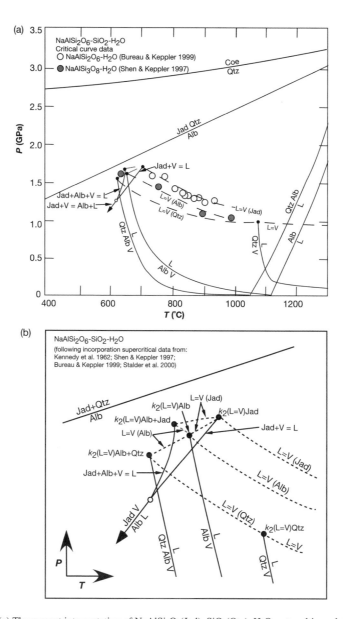

Figure 14. (a) The present interpretation of NaAlSi$_2$O$_6$(Jad)–SiO$_2$(Qtz)–H$_2$O wet melting relations where critical behavior terminates wet solidi (producing continuous or supercritical mineral-saturated cotectics at high pressure) for Qtz–H$_2$O, Alb–H$_2$O, Alb–Qtz–H$_2$O, Jad–Qtz–H$_2$O, Jad–H$_2$O. Critical relations were constructed from the observations of Kennedy et al. (1962), Paillat et al. (1992), Shen and Keppler (1997), Bureau and Keppler (1999) and Stalder et al. (2000). (b) Detailed relations in NaAlSi$_2$O$_6$–SiO$_2$–H$_2$O near 1.5 GPa between wet melting equilibria, critical endpoints and critical curves. The mineral(s) parenthesized on the various critical curves are stable in the presence of the relevant supercritical fluid, e.g., quartz is stable at the critical endpoint on the quartz wet solidus and also along the critical curve extending from critical endpoints k_2(L=V)Qtz to k_2(L=V)Qtz+Alb.

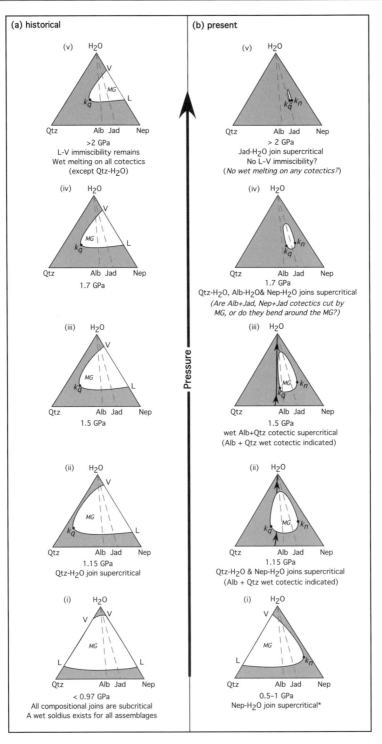

Figure 15. *(caption on facing page)*

reduced. Hence, concepts of high-pressure melting and fluids in the $NaAlSiO_4–SiO_2–H_2O$ ternary require significant revision from those proposed by Boettcher and Wyllie (1969b), which do not incorporate extensive supercriticality.

An important conclusion coming from Figures 14 and 15b is that the non-binary wet solidi (Qtz + Alb + V = L, Jad + Alb + V = L) are likely to terminate at critical endpoints at an intermediate *P* between the critical endpoints of adjacent binary wet solidi, i.e., *P* of k_2(L=V)Qtz < *P* of k_2(L=V)Qtz+Alb < *P* of k_2(L=V)Alb. Other melting relations are permissible in the $NaAlSiO_4–SiO_2–H_2O$ ternary but would require more complex L+V behavior, such as the occurrence of multiple L+V miscibility gaps. Presently there is little evidence to support this.

The phase relations shown in Figure 15a,b assume that all liquid and vapor compositions lie in the $NaAlSiO_4–SiO_2–H_2O$ plane. It remains possible (but uncertain) that dissolution and melting relations (e.g., Jad + V = L) are incongruent so liquid and/or vapor compositions fall outside the $NaAlSiO_4–SiO_2–H_2O$ plane (i.e., Na/Al ≠1). Example, there is some evidence that melting on $Nep–H_2O$ is incongruent (Boettcher and Wyllie 1969b). If this is also more generally the case, critical curves located by Bureau and Keppler (1999) would not intersect and terminate the respective wet solidi (as shown on Fig. 14). This could explain why Boettcher and Wyllie (1969b) observed the persistence of wet melting in $NaAlSiO_4–SiO_2–H_2O$ to at least 3.5 GPa. If relations are strongly incongruent, what are the L and or V compositions (in particular their Na/Al ratio) that coexist with and without the various solids? Available mineral solubility data in this system suggests that fluids coexisting with albite can be distinctly peralkaline (Na/Al > 1, e.g., Currie 1968; Antignano and Manning 2003) and thus lie off the $NaAlSiO_4–SiO_2–H_2O$ compositional plane (in which case relations on Fig. 15 would be projected). Our analysis of the available data indicates much remains to be clarified regarding liquid and especially vapor compositions in this system.

Haplogranite–water (quartz+albite+K-feldspar+H_2O±anorthite)

Wet melting relations and critical phenomena in the synthetic system "haplogranite"–H_2O have been investigated by many workers (Tuttle and Bowen 1958; Luth et al. 1964; Merrill et al. 1970; Huang and Wyllie 1975; Johannes 1984; Ebadi and Johannes 1991; Holtz et al. 1992a,b, 1995). Figure 16 shows a *P-T* projection of selected phase relations in this system.

A simple linear extrapolation of the "haplogranite"–H_2O critical curves as experimentally observed by Bureau and Keppler (1999) suggests their intersection with the wet "granite" solidus (Huang and Wyllie 1975) around 2.2 to 3 GPa, depending on the anorthite-component in the plagioclase. Note that the anorthite-component has virtually no effect on the position of the "haplogranite" wet solidus but significantly enlarges the region of L+V immiscibility to higher pressures.

Figure 15 (*on facing page*). Simplified polythermal, isobaric ternary sections depicting contrasting historical (a) and present (b) interpretations of L+V immiscibility relations for the ternary system $NaAlSiO_4$(Nep)–SiO_2(Qtz)–H_2O. (a) the interpretation of Boettcher and Wyllie (1969b), i.e., prior supercriticality studies; and (b) the present interpretation that incorporates the supercritical phenomena reported recently by Paillat et al. (1992), Shen and Keppler (1997), Bureau and Keppler (1999) and Stalder et al. (2000). Shaded areas a(i–v) and b(i–v) are polythermal regions in which a single solution is stable (with or without coexisting minerals) at the *P* of interest. Binary composition joins, Alb–H_2O and Jad–H_2O, are shown as dashed lines. Abbreviations: *MG* = Miscibility Gap (L+V); V = vapor; L = liquid; k_q and k_n = different critical points in the ternary system, denotes the maximum compositional extent of the *MG* at the *P* of interest; Qtz = quartz; Alb = Albite; Jad = Jadeite; Nep = Nepheline. Critical endpoints on wet solidi occur where *k* intersects the wet binary or ternary cotectic. *Note, although Bureau and Keppler (1999) showed that the join Nep–H2O is supercritical to low pressures, it is uncertain whether the Nep–H2O critical curve intersects the Nepheline wet solidus as implied, because there is evidence for incongruent behavior (see text for further discussion).

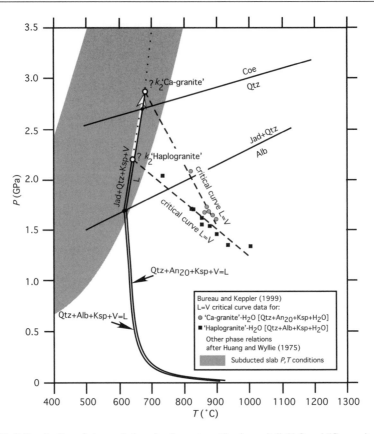

Figure 16. *P-T* projection of phase relations for the systems "haplogranite"–H₂O and "Ca-granite"–H₂O. Solid lines are experimentally determined wet solidi (Huang and Wyllie 1975). Dashed lines are critical curves (Bureau and Keppler 1999). k_2"haplogranite" and k_2"Ca-granite" are inferred second critical endpoints for the respective wet solidi. Dotted black curve represents the extension of the wet solidi to higher pressures. Grey shaded area shows a range of typical subducted slab P, T conditions. The arrow indicates the large $P-T$ shift of the wet haplogranite critical endpoint caused by addition of an anorthite (An_{20}) component. Alb=Ab=$NaAlSi_3O_8$; Jad=Jd=$NaAlSi_2O_6$; Qtz=Coe=SiO_2; Ksp=K-feldspar= $KAlSi_3O_8$; An=anorthite=$CaAl_2Si_2O_8$; V=vapor (fluid); L=liquid (melt); k_2"x"=critical endpoint on wet solidus for "x"-H₂O.

Figure 16 could be read as suggesting anorthite-poor felsic rocks in some subducted slabs may experience P, T conditions where critical phenomena/supercritical hydrous fluids may play a role. It is worth cautioning however that it is uncertain whether the critical curves observed intersect and terminate the wet solidi as illustrated. This uncertainty arises because L and V compositions must be identical on both the critical curve and wet solidus at the P, T condition of their intersection for termination of the wet solidus at a critical endpoint to occur. This was discussed above in the context of NaAlSiO₄–SiO₂–H₂O. We do not know whether L and V compositions on the haplogranite wet solidus are lying in the same compositional joins studied by Bureau and Keppler (1999). In this case L–V immiscibility and classical wet melting would persist in hydrous haplogranitic mineral-saturated systems to much higher P than implied by Figure 16.

Within continental crust most P-T paths observed occur at pressures less than 2 GPa (50–60 km), and so are restricted to shallower depths (lower pressure) than current estimates of the onset on supercriticality. Consequently even in double-thickened crust, L–V immiscibility and melting or degassing at a wet solidus can be expected in haplogranitic compositions. Below we develop on this discussion of wet granite phase relations in the context of pegmatite formation.

Pegmatites

Classical models of late-stage magma evolution, e.g., pegmatite formation (Jahns and Burnham 1969), emphasized the role of exsolution of aqueous fluid from silicate melt, and the evidence from fluid/melt inclusions inside minerals (Roedder 1992) implied that other types of fluid immiscibility may also be important at the magmatic–hydrothermal transition.

The effect of a second volatile on immiscibility in wet mineral and granite systems. The effect of a second volatile component following H_2O in mineral systems can be well illustrated by the system $NaAlSi_3O_8$–H_2O–X (Fig. 17a, Wyllie and Tuttle 1964). Greater temperature lowering of the melting point of $NaAlSi_3O_8$–H_2O (from 800 °C in experiments at 0.275 GPa) occurs in the order HCl (down to 710 °C), SO_3 (680 °C), P_2O_5 (650 °C), HF (600 °C), whereas NH_3 causes slight increase in the melting point. These temperature effects reflect the relative partitioning into melt (HCl, SO_3, P_2O_5, HF) versus vapor (NH_3). Figure 17a shows flattening of the saturation surface for HCl and P_2O_5 (Wyllie and Tuttle 1964). This could reflect fluid/melt phase immiscibility as melt saturation with a new fluid phase).

Whether a second volatile component after H_2O in rock systems causes increase or decrease of melting temperature reflects relative solubility in melt or aqueous vapor. For example in the system granite–H_2O–X (compared to albite–H_2O; from 665 °C in experiments at 0.275 GPa, Fig. 17b), SO_3 apparently has no effect, both NH_3 and now HCl cause temperature increase, whereas P_2O_5, HF and Li_2O cause a decrease in vapor-saturation temperature. The switch in behavior for HCl must reflect interaction with the components SiO_2 and $KAlSi_3O_8$ in granite compared to $NaAlSi_3O_8$–H_2O (Fig. 17, Wyllie and Tuttle 1964).

Note that the wet solidus displacements in *P*-*T* by added components just discussed also influences the *P, T* location of the critical endpoint and thus the occurrence of supercriticality and the extent of immiscibility. If the additional component shifts the wet solidus (with negative d*P*/d*T*) to higher temperature and the critical curve (with negative d*P*/d*T*) is unaffected by the

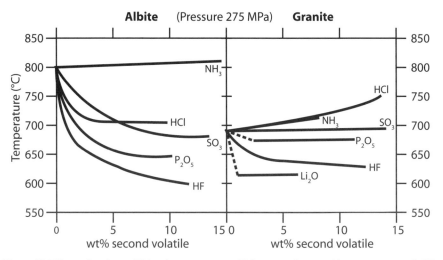

Figure 17. Effects of various additional components on H_2O-saturated wet melting temperature of albite and granite reflects L–V partitioning of the added component (modified after Wyllie and Tuttle 1964).

added component, the critical endpoint will shift to lower pressure and higher temperature. Alternatively, if this critical curve had positive dP/dT slope, the critical endpoint would shift to higher pressure and higher temperature. These simple geometric relations reflect the underlying chemical interactions and can be used to grossly predict critical endpoint shifts in *P-T*.

Late stage (im)miscible fractions of crystallizing granite magma: origins of granitic pegmatites. Granitic pegmatites are generally accepted as late-residual fractions derived from larger masses of granite melt (e.g., London 2005). Aqueous vapor plays an essential role in the formation of pegmatite (e.g., the experimental studies presented by Jahns and Tuttle 1963, p. 90). Segregation of major alkalis can occur during crystallization of granitic pegmatitic magma when it becomes saturated with volatile constituents, so that both silicate melt and vapor are present in the system, according to the model of Jahns and Burnham (1969). Their experimental results indicate that potassium is extracted from the liquid into the vapor in preference to sodium, and that potassium and other feldspathic constituents can diffuse rapidly through the vapor in response to temperature gradients or compositional gradients.

It is very significant that studies of pegmatite mineral assemblages and fluid inclusions often indicate temperatures of crystallization well below the accepted solidus temperatures of hydrous granitic melts (near 650 to 700 °C), for the inferred depths (related to pressure).

A favored mechanism relating the igneous composition of pegmatites with their apparent low temperatures of formation, is the role of specific natural chemical fluxes which lower the liquidus temperatures of pegmatite-forming melts. The commonly cited fluxing components in pegmatite magmas (in addition to H_2O) are, B, F, and P (Jahns 1953; London 1997). These lower the melting and crystallization temperatures compared to water-saturated granitic liquids (e.g., London et al. 1989, 1993; London 1997). Furthermore, the fluxes importantly enhance miscibility among otherwise less soluble constituents (e.g., London 1986a; London et al. 1987; Keppler 1993; Thomas et al. 2000; Sowerby and Keppler 2002).

Because of their strong chemical complexing with H_2O, accumulation of B, F, and P to high concentrations in residual silicate melts can promote H_2O solubility towards complete miscibility and inhibit aqueous vapor saturation of melts (London 1986a,b; Thomas et al. 2000).

The sources of the volatile elements that can act as fluxes are not clearly formulated. Their end locations in pegmatite minerals are well described (Li in mica; Be in beryl; B in tourmaline; F in fluorite, topaz; Na in feldspar, feldspathoid, zeolite, mica; P in xenotime, apatite; S in sulfate, sulfide, scapolite; K in mica, feldspar; Cl in halite, mica, feldspathoid). The intermediate stage in the magmatic phase may well involve gaseous species with or without H_2O. The source regions must also ultimately involve minerals, which melt or dissolve in other volatile species, in processes which rely upon greatly enhanced mineral/rock solubility at lower crustal and upper mantle pressures (Ague 1997; Philippot and Rumble 2000; Kessel et al. 2005a).

Peridotite–water: MgO–SiO$_2$–H$_2$O (forsterite+enstatite+quartz+H$_2$O)

Parts of the MgO–SiO$_2$ (MS) system have been consistently used as models for subsolidus metasomatic processes, both anhydrous (dry by J. B. Thompson, Jr. 1959) and with H_2O (wet by Greenwood 1975; Brady 1977; Fyfe et al. 1978, p. 168; Manning 1995).

The system MgO–SiO$_2$–H$_2$O (MSH) has been investigated experimentally by numerous workers as a simplified model for melting and solubility in natural hydrous mantle (e.g., Kushiro et al. 1968; Ryabchikov et al. 1982; Inoue 1994; Luth 1995; Irifune et al. 1998; Zhang and Frantz 2000; Ohtani et al. 2000; Stalder et al. 2001; Mibe et al. 2002; Yamada et al. 2004; Kawamoto et al. 2004; Melekhova et al. 2007). A selection of phase relations is shown in the *PT*-projection in Figure 18. In this hydrous system forsterite, enstatite, and quartz (amongst others) all occur as stable phases and supercriticality has been described on the quartz- and also forsterite+enstatite-wet solidi (Kennedy et al. 1962; Stalder et al. 2001; respectively). The

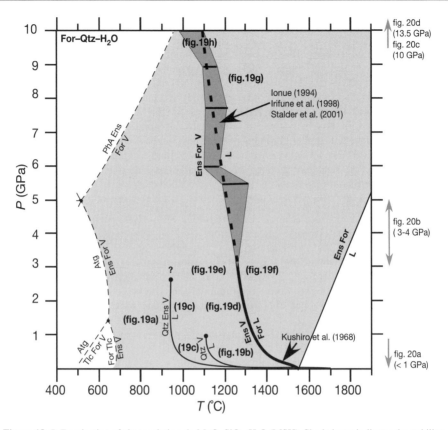

Figure 18. *P-T* projection of phase relations in MgO–SiO$_2$–H$_2$O (MSH). Shaded area indicates the stability For + Ens assemblages in the hydrous system. Abbreviations: For=Forsterite; Ens=Enstatite (includes polymorphs); Qtz=Quartz (includes polymorphs); L=Liquid; V=Vapor; Atg=Antigorite; Tlc=Talc; PhA=Phase-A. Sources not indicated on plot: Atg=Tlc+For+V from Evans et al. (1976); For+Tlc=Ens+V (Pawley 1998); Atg=Ens+For+V from Bromiley and Pawley (2003); PhA+Ens=For+V from Luth (1995); Qtz+V=L from Kennedy et al. (1962); Qtz+Ens+V=L from Kushiro (1969). For clarity, silica and enstatite polymorph stabilities are not shown.

conditions where L–V immiscibility and complete miscibility occur remain in debate (see Stalder 2004 and Mibe et al. 2004a), not least because the problems bear directly on the nature of wet magma generation and slab fluxes at convergent margins, and ultimately the primary mechanism through which continental crust and to some lesser(?) extent the atmosphere/hydrosphere is extracted from the Earth's mantle. We will outline the concepts of solubility and melting in MSH and the evidence available to constrain the phase relations. Central to this issue, is the pressure above which the peridotite (in MSH, simply forsterite + enstatite) wet solidus terminates due to complete L–V miscibility.

As already discussed, any fluid in the SiO$_2$–H$_2$O join is supercritical at pressures above 1 GPa (Kennedy et al. 1962, see Fig. 6). Thus, at upper mantle pressures the liquid–vapor miscibility gap has already started closing in MgO–SiO$_2$–H$_2$O and some silica-rich portions of the system are already completely miscible. Consideration of the available experimental data for fluids and melts in MgO–SiO$_2$–H$_2$O leads to conclusion the compositional range of the liquid–vapor miscibility gap continues shrinking with increasing pressure, such that for model mantle compositions wet melting (or liquid–vapor immiscibility) ceases to be relevant at depths approaching 300 km in the Earth (e.g., Stalder et al. 2001; Mibe et al. 2002; Melekhova et al. 2007). We outline the evidence further below.

Vapor-saturated forsterite (For) + enstatite (Ens) melting is well known at low pressure (< 3 GPa, Kushiro et al. 1968). Inoue (1994) found that melting in $MgSiO_3$–H_2O switches from incongruent (Ens + V = For + L, Kushiro et al. 1968) to congruent (Ens + V = L) somewhere between 4 and 6 GPa. Stalder et al. (2001) measured fluid and melt compositions along the forsterite + enstatite wet solidus. With increasing pressure the Mg/Si ratio increases in both L and V (consistent with the observations of Inoue 1994 and Yamada et al. 2004). Likewise, the solute contents in the fluid and H_2O solubility in the melt increase with increasing pressure and temperature. Above 9 GPa forsterite + enstatite -saturated fluid compositions vary continuously from dilute to solute-rich over a range of T-X conditions consistent with complete L–V miscibility on the forsterite + enstatite cotectic. Accordingly, Stalder et al. (2001) concluded that the forsterite + enstatite wet solidus terminated at a critical endpoint near 10 GPa.

Mineral–fluid–melt phase relations in the Mg_2SiO_4–SiO_2–H_2O system are summarized by us in series of isothermal, isobaric ternary sections (Fig. 19). This figure, although schematic, is consistent with available experimental data (Kennedy et al. 1962; Kushiro et al. 1968; Ryabchikov et al. 1982; Inoue 1994; Luth 1995; Irifune et al. 1998; Zhang and Frantz 2000; Ohtani et al. 2000; Stalder et al. 2001; Mibe et al. 2002; Yamada et al. 2004; Kawamoto et al. 2004). The figure highlights wet melting relations and the potential for the simultaneous occurrence of L+V coexistence and supercritical fluid–mineral relations in a ternary (or higher) system at a given P, T condition. Recently, Melekhova et al. (2007) presented an analysis of fluid–melt relations extending into the MgO-saturated part of MSH. They also report experimental data on fluid compositions coexisting with dense hydrous magnesium silicates stable in the MgO-rich (Mg/Si > 2) part of the system at pressures 11 to 13.5 GPa and temperatures to 1350 °C. The balance of data available in the MgO–SiO_2–H_2O system indicates that the ternary L–V MG closes with increasing pressure, first in SiO_2–H_2O at ~1 GPa (Kennedy et al. 1962), then supercriticality migrates to more MgO-rich compositions with increased pressure (Stalder et al. 2001; Mibe et al. 2002; Melekhova et al. 2007) and eventually the L–V miscibility gap is no longer stable and all liquids in the system are supercritical. Melekhova et al. (2007) report that complete supercriticality in MSH (i.e., the L–V miscibility gap has vanished entirely from the system) is attained at a pressure of 13.5 GPa.

Available constraints on MSH phase relations can be explained relatively simply by a destabilizing of a single ternary L–V solvus with increasing pressure. Figure 20, constructed by us from a synthesis of the available data, summarizes (very) simplistically liquidus and fluidus cotectics and the retreat of the MSH L–V MG with increasing pressure and is consistent with the phase relations given in Figures 18 and 19. Note the emphasis in this Figure 19 rests again on the concept that a L–V miscibility gap can migrate within a system with changing pressure, and that mineral cotectics and liquidus and fluidus surfaces may or may not intersect the miscibility gap at a given pressure. Where cotectics intersect the MG a wet solidus is defined. The sweeping of cotectics from silica-rich to MgO-rich compositions with increasing pressure is consistent with the available experimental data (Kushiro et al. 1968; Inoue 1994; Irifune et al. 1998; Ohtani et al. 2000; Stalder et al. 2001; Mibe et al. 2002; Yamada et al. 2004; Melekhova et al. 2007).

In summary, although a reasonably complete P-T-X model of MSH phase relations incorporating supercritical behavior can be constructed, some aspects of MSH melting and solubility require clarification. For example there are too few data and insufficient agreement between forsterite+enstatite-saturated fluid compositions reported by different workers at the same or nearby P, T conditions to be confident that critical phenomena have been correctly located in P-T-X (e.g., at 3 and 8–10 GPa, Stalder et al. 2001; Mibe et al. 2002; also see discussion: Stalder 2004; Mibe 2004a). The difficulty comes when geological processes are considered.

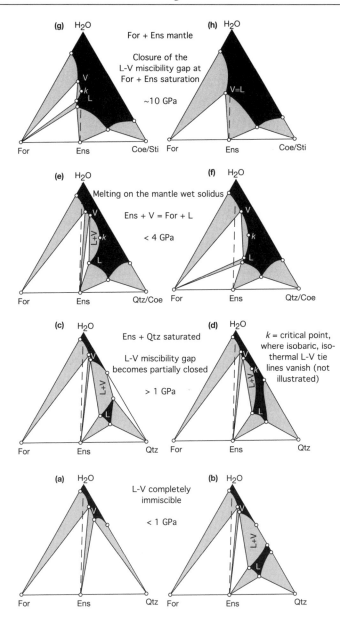

Figure 19. Schematic series of isothermal, isobaric phase relations in MSH corresponding to selected *P, T* conditions (shown on Fig. 17). The sections (a–h) highlight melting relations and transition to supercritical behavior as L–V miscibility gap shrinks with increasing pressure. Abbreviations as used in Figure 18.

Observations of Stalder et al. (2001) suggest that "melt production" in the upper mantle above subduction zones occurs via a discrete melting reaction at a "classical fluid-saturated solidus" and persists in the upper mantle to depths around 300 km. In contrast, higher solubilities for coexisting forsterite + enstatite observed by Mibe et al. (2002) suggest solute-rich supercritical fluids (not L–V immiscibility) should predominate to much shallower depths in the mantle. Such disagreement probably relates to the different experimental methods used (see below).

In this case, which study is flawed remains unresolved. Nonetheless, there is also some broad agreement in the available data such as forsterite + enstatite solubility in H_2O increases rapidly above 4 GPa and fluids become more Mg-rich (molar $MgO/SiO_2 = M/S > 1$), whereas at lower P they are strongly enriched in SiO_2 ($M/S < 1$). For subduction zones this suggests that the degree of SiO_2-metasomatism of the mantle wedge is strongly coupled to depth and the extent of the metasomatic imprint will be limited by and proportional to the integrated fluid flux at a given depth.

Mineral solubility and melting relations must be investigated further, especially in light of the newly determined peridotite wet solidus (by Grove et al. 2006), wet peridotite critical curve (Mibe et al. 2007) and eclogite wet solidus (Kessel et al. 2005a,b). It does seem that MSH is not a sufficient analogue for the mantle despite its common use in the literature.

A note on high-pressure fluids and experimental methods: High-pressure fluids are difficult to study because they are highly reactive and commonly unquenchable, such that a fluid which is homogeneous will on quenching separate into glass, quench crystals, and a water-rich liquid. Different approaches have been used in MSH (and elsewhere) to determine fluid compositions and avoid quenching problems: conventional phase equilibrium methods (Nakamura and Kushiro 1974; Ryabchikov et al. 1982; Zhang and Frantz 2000; Mibe et al. 2002); diamond-trap experiments where bulk analysis of fluid quench material is obtained by LA-ICPMS (Stalder et al. 2001; Kessel et al. 2004); *in situ* X-ray diffraction to determine the mineral coexisting with fluid at known *P-T-X* conditions (Kawamoto et al. 2004). Although

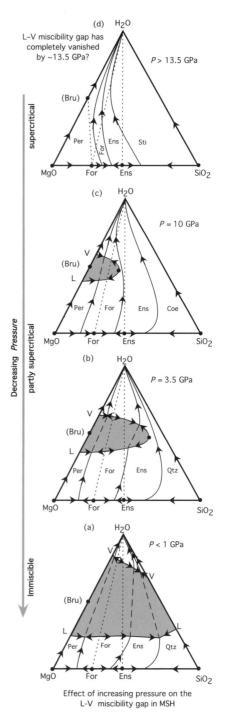

Figure 20. Schematic polythermal, isobaric projections summarizing MSH solubility and melting relations with increasing *P* to ca. 13.5 GPa. Note the relation between mineral-saturated cotectics and the shrinkage of the L–V miscibility gap with increasing *P*. Arrows indicate the direction of decreasing temperature on the cotectics. Dashed lines connect coexisting V and L compositions at the wet solidus. Construction notes: 1) MgO–H_2O melting relations are assumed (no data available); 2) fluid cotectics involving hydrous MSH phases, e.g., talc, antigorite, "alphabet phases", are not shown; and, 3) MSH hydrous minerals are not assumed to be stable at melting *P, T* conditions.

not yet applied in MSH, *in situ* observations of high pressure melt–fluid critical phenomena have been made using hydrothermal diamond anvil cells (Shen and Keppler 1997; Bureau and Keppler 1999; Sowerby and Keppler 2002) and also in multi-anvil high-pressure apparatus using X-ray radiography (Mibe et al. 2004b, 2007). Weight-loss is another well-established technique for determining mineral solubilities (Kennedy 1950; Anderson and Burnham 1965; Manning 1994; Newton and Manning 2002). For this technique an individual crystal is weighed before the experiment, then sealed in a capsule containing a fluid and subjected to a given pressure and temperature. After the run the crystal is extracted and re-weighed, the reduction in weight corresponds to its solubility. Mineral-hosted synthetic fluid inclusions have also been used to study solubility (Loucks and Mavrogenes 1999; Hack and Mavrogenes 2006a,b). Here, initially fractured minerals are allowed to anneal hydrothermally, during which ambient fluids can be trapped *in situ* as tiny inclusions which can be later analyzed by LA-ICPMS and other methods to determine their chemical composition and volumetric properties. It should be emphasized each technique has its strengths and disadvantages. Perhaps not surprisingly, results obtained from different techniques do not always agree and the reasons for the discrepancies are not always clear (see review by Keppler and Audétat 2005).

Eclogite–water (garnet+omphacite+coesite+kyanite+rutile+H_2O)

Kessel et al. (2005a,b) used the diamond-trap technique (Kessel et al. 2004) to determine the major element composition of fluids and melts (and mineral-fluid trace element partitioning) along an eclogite wet solidus up to 6 GPa. The composition studied was always saturated in garnet (Gar), omphacite (Omp), coesite (Coe), kyanite (Kya) and rutile (Rut) and was based on a K_2O-free MORB composition. They showed that vapor and liquid converge in composition with increasing P along the wet solidus. Kessel et al. (2005a,b) found that at a pressure between 5 and 6 GPa and $T \sim 1050$ °C vapor and liquid could no longer be distinguished on the basis of composition. This was interpreted as evidence of complete mutual L–V miscibility and correspondingly, the termination of the eclogite wet solidus at an upper critical endpoint (Figs. 21 and 22). Above 5 GPa fluids in K-free eclogite are supercritical. Such supercritical relations terminating classical concepts of wet melting in eclogite–water were suspected and discussed earlier by Schmidt et al. (2004). The significance of the Kessel et al. (2005a,b) studies is that it demonstrated experimentally the termination of a wet solidus in a multiply saturated, multi-component system, by measuring fluid and melt compositions and observing their convergence. The findings constitute important evidence that supercritical phenomena and termination of conventional wet melting may be of significance in "real" rock–H_2O systems and is not just a phase relation curiosity restricted to hypothetical and simple systems. The consequences for processes in the Earth's mantle are introduced below.

Trace element partitioning and liquid–vapor phase relations. Trace-element partitioning between a K-free eclogite mineral assemblage and coexisting melt, vapor and supercritical fluid was measured by Kessel et al. (2005b) using the diamond-trap technique. Mineral and fluid (the bulk quench in trap) trace element compositions were measured using LA-ICPMS. The starting material was doped with a suite of petrogenetically significant trace elements including Cs. The fluid element concentration ratios measured by LA-ICPMS were quantified based on the initial Cs concentration because it is highly incompatible with the solid residue and thus retained almost exclusively in the liquid phase.

The data of Kessel et al. (2005b) show that most trace element partition coefficients increase significantly with increasing temperature, e.g., high-field strength elements, Ti, Nb, Ta, Zr, and Hf rise approximately 2 orders of magnitude from 700 to 900 °C and switch from compatible ($D_{fluid/residue} < 1$) to incompatible ($D_{fluid/residue} > 1$). Another interesting aspect of the partitioning data is that at 4 GPa the significant T-dependence (considering measurement uncertainties) obscures any obvious partitioning discontinuity between subsolidus fluids and higher temperature hydrous melts for certain elements. Figure 23 is an Arrhenius plot showing liquid/

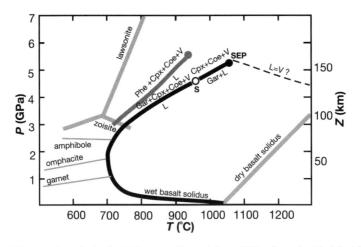

Figure 21. *P-T* projection of eclogite–H₂O wet melting relations. SEP = Second critical End Point. S = singularity on wet solidus (garnet switches side). The critical curve (L=V; dashed) position is speculative but consistent with the projected isobaric *T-X*(H₂O) relations at 4, 5 and 6 GPa in Figure 22 (modified after Kessel et al. 2005a, their Fig. 8).

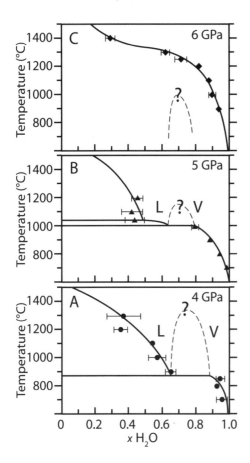

Figure 22. *T–x*(H₂O) diagram (in mol fraction) for the liquid phase at 4, 5, and 6 GPa in equilibrium with K–free MOR eclogite. Superimposed on the experimentally derived data are schematic locations of the solidi and liquidi at each pressure. Experimental data require that the system changes from eutectic to peritectic between 4 and 5 GPa and suggest supercritical behavior above 5 GPa. The interpretation is approximated in that all solid phases are anhydrous and project onto the left axis of this diagram (the peritectic phase therefore does not appear on the diagram). Errors are not shown where smaller than the symbol size. L–V solvus is dashed, it is shown as metastable at 6 GPa (modified after Kessel et al. 2005a).

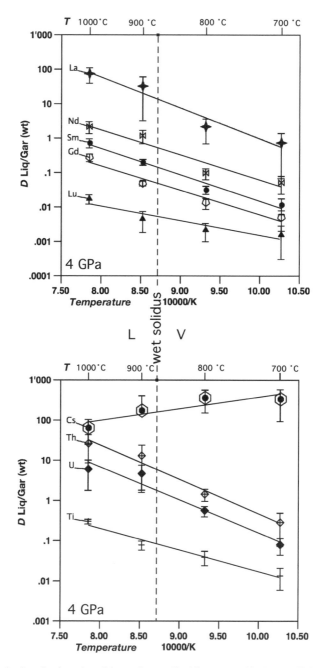

Figure 23. Arrhenius plot for selected trace element liquid/garnet partitioning coefficients (by weight) at 4 GPa for 700–1000 °C for (a) La, Nd, Sm, Gd, Lu; and (b) Cs, Th, U and Ti (HFSE). Note that no obvious discontinuity in partitioning between L and V occurs at the 4 GPa wet solidus for the elements shown (Fig. 22). Linear solid curves represent unweighted least square fits to the partitioning data, dashed curve indicates wet solidus (ca. 875 °C) at 4 GPa reported by Kessel et al. (2005a,b) for the eclogite–H_2O composition studied (data only from Kessel et al. 2005b).

garnet partitioning as a function of temperature at 4 GPa for La, Nd, Sm, Gd, Lu (Fig. 23a), and Cs, Th, U and Ti (Fig. 23b) as examples. On the basis of liquid/garnet (*D*) trace element partitioning, V and L cannot be distinguished at the wet solidus, within the uncertainty of the data. Whereas, a discontinuity in major element compositions coherently identifies the wet solidus at 4 GPa. A simple linear relation, log*D* = a/*T* + b, describes the isobaric temperature-dependence for trace element liquid/garnet (*D*) partitioning. Mineral solubilities in both the sub-solidus and supercritical regions continually increase with increasing *T* and increasing *P*.

Immiscible liquid and vapor that approach conditions of complete miscibility must develop common chemical structures due to the increasingly similar character of the dissolved silicate components in both phases. Where trace elements are able to form (near)identical complexes with the major solute species in both L and V, partitioning coefficients may become insensitive to differences, in say H_2O-content, between the coexisting phases. This type of behavior in immiscible phases, reflecting proximity to supercriticality, could explain the trace element partitioning data of Kessel et al. (2005b). L+V activity–concentration relations need to be formulated to express this behavior. Alternatively, do the Kessel et al. (2005b) trace-element data indicate that continuous solubility occurs at 4 and 5 GPa (implying no L–V solvus and supercritical behavior)? More precise data would help clarify the situation. Manning (2004) suggested that polymerized, silicate melt-like complexes predominate solute speciation in aqueous fluids at higher pressure (*P* > ca. 1 GPa). Although the details remain uncertain Kessel et al.'s trace element data and the reported closure of the L–V *MG* tend to support this concept. The results of Kessel et al. (2005b) are especially intriguing and deserve further attention, as they speak to the importance of thermal-control on trace element patterns and element retention and release in subduction zones, and assign lesser roles to wet melting and strictly supercritical fluid phenomena.

FLUID PHYSICAL PROPERTIES, COMPOSITION AND *P-T* PATHS

Rates of fluid-mediated mass transport are dependant on the physical properties of the fluid. Lower density fluids have greater buoyancy than the surrounding rock, thus permitting ascent. Lower viscosity fluids can flow faster. In this section we review some of the theoretical aspects of how fluid physical properties and phase relations are coupled, and the information we have available for silicate-bearing fluids. An attempt (in part necessarily speculative) is also made to combine the somewhat fragmentary observations from the different systems in a rigorous manner. Our approach involves considerable extrapolation outside controlling experimental data, but nonetheless provides a basis for discussing fluid flow in geological environments, and gives limits on the possible range of *P-T-X*-viscosity conditions which need to be considered. We draw attention to the differences between the contrasting physical properties of immiscible fluids compared to single-phase (supercritical) fluid.

Clapeyron slope of critical curves and fluid density

Unfortunately very few data for physical properties (for example, density) are available for high-pressure (> 0.5 GPa) silicate-bearing aqueous fluids. Nonetheless, the fundamental relations for volumetric and other properties of supercritical and immiscible fluid mixtures are known from simple thermodynamic relations (for example, Rowlinson 1983; Levelt Sengers 1990, 2000; Fig. 24).

The geometry of critical phase boundaries contains information about the physical properties of the phases involved. For example, at constant temperature, if a mixture of L+V becomes completely miscible with increasing pressure, it is because the homogeneous phase has a higher density than the immiscible L+V mixture. The slope of an equilibrium boundary, such as the critical curve, is defined by the Clapeyron relation: d*P*/d*T* = Δ*S*/Δ*V*. If the boundary is a straight line, then the values of Δ*S* (entropy of reaction) and Δ*V* (volume of reaction) are

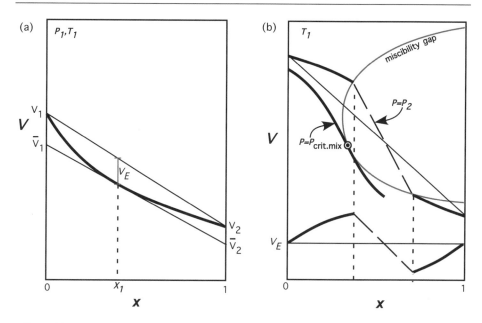

Figure 24. Isothermal, isobaric, solution volume–composition (*V–x*) relations traversing continuous to immiscible conditions. (a) Molar volume *V* of a solution as a function of mole fraction *x* composition, plotted at constant pressure $P = P_1$ and temperature $T = T_1$. For the composition x_1, the tangent construction defining the partial molar volumes $\bar{V}_1$ and $\bar{V}_2$, and the chord construction leading to the excess volume V_E are shown. (b) *V–x* relations at $T=T_1$ for a pressure ($P = P_2$) where a miscibility gap is stable. As a result of the immiscible region (long dash lines), excess volume as a function of composition displays two discontinuities in slope (shown in lower part of Figure). Also shown are relations at $P = P_{\text{crit.mix}}$ which correspond to the position of the L = V critical curve at $T = T_1$. The stability region of the miscibility gap at T_1 is shown (modified after Levelt Sengers 1990).

not dependent upon *T* and *P*. A negative d*P*/d*T* slope implies that Δ*S* and Δ*V* have opposite signs, while for positive d*P*/d*T*, Δ*S* and Δ*V* have the same sign. Thus, the slope (and curvature) of critical curves has a simple thermodynamic significance and can provide information on the nature of the relative compressibilities of L+V and supercritical fluids.

So far, critical curves reported in hydrous silicate systems show negative d*P*/d*T* slopes and are linear, within the experimental uncertainty (NaAlSi$_3$O$_8$(albite)–H$_2$O, Shen and Keppler 1997; NaAlSiO$_4$(nepheline)–H$_2$O, NaAlSi$_2$O$_6$(jadeite)–H$_2$O, "haplogranite"–H$_2$O, Bureau and Keppler 1999; NaAlSi$_3$O$_8$–H$_2$O±Na$_2$O±B$_2$O$_3$±F$_2$O$_{-1}$, Sowerby and Keppler 2002). In these cases the supercritical region lies on the high-pressure side of the critical curve. Accordingly, we expect high-pressure critical fluids to be more compressible than equivalent immiscible L+V mixtures. A negative d*P*/d*T* for these curves also implies that cooling of a supercritical solution results in L+V unmixing.

A critical curve possessing a temperature minimum or maximum requires a density inversion amongst the phases occurs at this point as Δ*V* = 0. A density inversion need not accompany a significant change in the coexisting fluid compositions. Thus, with increasing pressure the hydrous silicate liquid would become relatively less dense and eventually more buoyant than the H$_2$O-rich vapor at high *P*. Example, L–V density inversion occurs in the binary CO$_2$–H$_2$O system close to 0.23 GPa pressure (Tödheide and Franck 1963; Takenouchi and Kennedy 1964; see also Fig. 1e in Liebscher 2007, this volume). Boettcher and Wyllie (1969a) present a theoretical analysis of critical curves indicating that *P, T* minima (and maxima) are practically unavoidable in non-binary, higher-order systems (e.g., ternary $MO-SiO_2-H_2O$, Fig. 12). Such critical curve geometries, however, have not yet been documented experimentally in silicate–H_2O systems.

Many wet solidi appear to display a T minimum with increasing P (e.g., Lambert and Wyllie 1972; Schmidt et al. 2004). In many cases the change in wet solidus dP/dT slope can be shown to be due to phase transitions specific to the minerals, and along the solidus there may be a change in the melt reaction stoichiometry whereby some minerals switch from reactant to product. However, given the involvement L and V in wet melting reactions, a dramatic increase in vapor-density relative to melt-density could be an alternate explanation for curvature in P-T, and even reversal of Clapeyron slope of some wet solidi.

Cooling or decompression paths crossing critical curves

The connection between dP/dT slopes of critical curves and density was just discussed. By linking critical curve P-T-X locations to P-T flow paths it may be possible to identify where such fluids form immiscible mixtures, and have greater opportunities to separate. It is of interest to ask where in P-T space the fluid is more likely to be a single phase and whether it is relatively more or less dense/buoyant than the immiscible mixture. Further knowledge of critical behavior in silicate–H_2O systems would facilitate more precise statements on processes occurring in natural rocks.

Viscosity of silicate-bearing aqueous fluids

In comparison to hydrous silicate melts (Shaw 1972; Dingwell 1998; Giordano et al. 2006) our current knowledge of the viscosity of silicate-bearing aqueous fluids is far more limited. The majority of information available comes from a pioneering study by Audétat and Keppler (2004), who measured viscosities of aqueous fluids containing 10 to 80 wt% dissolved silicate at various conditions between 1–2 GPa and 600–950 °C (Fig. 27). The results were obtained from *in situ* observation of falling spheres in a hydrothermal diamond anvil cell (HDAC). Most of their measurements were of $NaAlSi_3O_8$ (albite)–H_2O solutions, but they also studied fluid compositions occurring in the joins $KAlSi_2O_6$ (leucite)–H_2O and $Ca_2NaSi_3O_8OH$ (pectolite)–H_2O. Figure 25a is an Arrhenius plot of Audétat and Keppler's fluid viscosity measurements and additional viscosity data compiled from the literature by them for anhydrous $NaAlSi_3O_8$(albite) and hydrous $NaAlSi_3O_8$–H_2O melts (obtained at various pressures). The plot shows that for all compositions investigated, viscosity decreases with temperature, and that for $NaAlSi_3O_8$–H_2O solutions, viscosities become increasingly temperature-dependent at lower H_2O contents, the effects of pressure were considered negligible. Figure 25b shows the P, T coordinates of the Audétat and Keppler (2004) fluid viscosity measurements.

Audétat and Keppler (2004) fit their fluid viscosity measurements together with the literature data with a continuous viscosity-composition-temperature function (Fig. 26). However, it may be noted from Figure 25b that many of Audétat and Keppler's (2004) higher T measurements (> 640 °C) occur within the immiscible L+V region. Miscibility gaps should be apparent as discontinuities in both chemical and physical properties. Indeed, reexamination of the polybaric $NaAlSi_3O_8$–H_2O viscosity–composition data, referred to a fixed T of 800 °C (replotted by us in our Fig. 26a) shows that discontinuity in slope could be present, but was not discussed by Audétat and Keppler (2004).

The polybaric nature of the $NaAlSi_3O_8$–H_2O viscosity data set of Audétat and Keppler (2004) is significant and that data speaks to a distinct separation of V and L viscosity regimes. H_2O solubility in silicate liquids is strongly P-dependent (e.g., Paillat et al. 1992; McMillan 1994) as is the upper extent of L–V immiscibility (Shen and Keppler 1997; Bureau and Keppler 1999; Sowerby and Keppler 2002). The degree of separation of L and V regimes depends on the proximity of supercriticality (e.g., to a wet solid critical endpoint, or critical curve).

The effect of composition on fluid viscosity is poorly understood. A conference abstract, Panero et al. (2003), reports the viscosity of aqueous fluid in equilibrium with quartz at 950

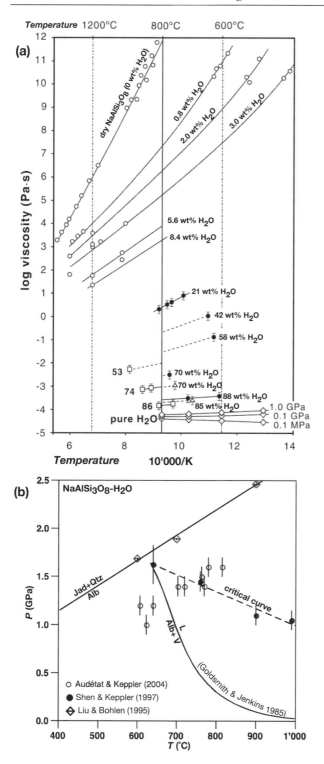

Figure 25. (a) Arrhenius plot showing the viscosity of NaAlSi$_3$O$_8$–H$_2$O solutions in the entire range from pure silicate melt to pure H$_2$O. Diamonds and open circles correspond to literature values for pure H$_2$O and NaAlSi$_3$O$_8$(albite) melts with H$_2$O (as compiled by Audétat and Keppler 2004). Measurements by Audétat and Keppler (2004) are represented by filled circles (NaAlSi$_3$O$_8$–H$_2$O), open squares (KAlSi$_2$O$_6$–H$_2$O), and open triangles (Ca$_2$NaSi$_3$O$_8$OH–H$_2$O). Water contents in weight percent H$_2$O are also indicated. Also shown are curves (solid and dashed) fitted to the measured viscosities for constant composition. The position of these curves at fixed temperature of 800 °C (vertical solid line), allows the construction of Figure 26. (b) *P-T* projection of NaAlSi$_3$O$_8$–H$_2$O phase relations and conditions at which NaAlSi$_3$O$_8$–H$_2$O solution viscosity measurements by Audétat and Keppler (2004) were made. A pressure uncertainty of ± 0.1 GPa was assumed based on uncertainties reported in similar studies by Shen and Keppler (1997), Bureau and Keppler (1999), and Sowerby and Keppler (2002) (a - modified after Audétat and Keppler 2004).

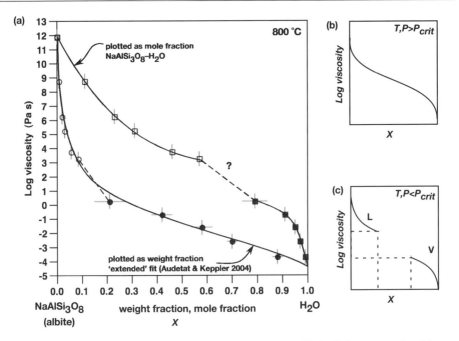

Figure 26. Extrapolated fluid and melt viscosities at 800 °C in the NaAlSi$_3$O$_8$–H$_2$O system, replotted from Figure 25. Square symbols show composition in mole fraction and circles are same data replotted in weight fraction; solid symbols are H$_2$O–rich fluid viscosities from Audétat and Keppler (2004); open symbols are hydrous albite melt data from the literature (and are those compiled by Audétat and Keppler 2004). Uncertainties on extrapolated points are from Audétat and Keppler (2004). The solid curve through the data plotted in weight fraction units represents the continuous "extended" fit equation given by Audétat and Keppler (2004). The discontinuous solid curve through the same data plotted as mole fraction were fit by eye, a linear short dashed line indicate the position of a possible gap between melt and fluid measurements. The miscibility gap is also apparent when plotted in weight concentration units (linear short dashed line). (b) Where a solution is supercritical a continuous viscosity–composition relation is expected, whereas, (c) a discontinuous relation occurs where a miscibility gap is stable at the P, T conditions.

K, 1 GPa (4 wt% dissolved SiO$_2$) is three orders of magnitude higher than pure H$_2$O at the same conditions. These results when considered with those of Audétat and Keppler (2004) indicate fluid viscosities, in addition to P, T and solute concentration are also likely to depend significantly on the chemical composition.

Precipitation and dissolution on flow paths, L–V immiscibility, single-phase fluids and metasomatism

Fluids may evolve compositionally via unmixing (L–V immiscibility), solute precipitation or wall-rock dissolution. The general metasomatic characteristics in rocks that experience the last two fluid processes are grossly predictable. The general effect of mineral precipitation from a fluid is to increase the solvent concentration, whereas dissolution drives the fluid to higher total solute and lower solvent concentration. In a rock these processes equate to enrichment or depletion of dissolved constituents.

Figure 27a shows the general effects of mineral precipitation and wall-rock dissolution by a supercritical fluid along a hypothetical fluid ascent path (b). Figure 27 highlights the location of the L–V miscibility gap (projected from saturation with some mineral assemblage) and that extensive solute loss, via precipitation or dissolution in the supercritical region can prevent fluids from under going L–V immiscibility and phase separation, where fluids remain in single-

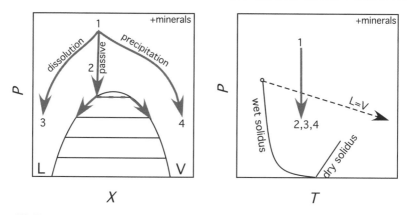

Figure 27. Illustration of some mineral-saturated fluid (composition) evolution paths originating in the single-phase region. Fluid ascending on the *P-T-X* path (1→2) intersects the L + V miscibility gap (at a *P* below the critical curve). Phase separation lowers the systems degrees of freedom by one so composition is fixed at *P* and *T*. Whereas composition is not fixed where fluids stay single phase (1→3, and 1→4). It is possible that single-phase fluids may evolve continuously over a wide range of composition via precipitation of solutes (fluid becomes more H_2O-rich, 1→4) or via dissolution of enclosing rocks (fluid becomes solute-rich, H_2O-poor, 1→3).

phase regions of composition. The compositional evolution of fluids may be complex, and erratic, cryptic metasomatic evidence may indicate the passage of single-phase fluids.

Supercritical fluids could show evolution across the whole compositional range in a direction in composition space determined by which minerals are dissolving or precipitating. Immiscible fluids would show more restricted evolution of fluid composition from regions close in composition to the solvus limbs.

A more restricted range of metasomatic chemical enrichment and/or depletion signatures should characterize rocks where L+V coexist, as the L will dominate the metasomatic signature owing to a higher and fairly compositionally restricted dissolved silicate component (as L can evolve on a well-defined V-saturated cotectic). Because vapors are characterized by very H_2O-rich compositions, they are of minor importance in terms of major element transport (except where solubilities are very high) so that modal metasomatism will be dominated by interaction with L and hence relatively independent of the L/V mass ratio. The composition of vapor-saturated liquids (melts) derived from wet melting of various rock types (e.g., pelite, eclogite, peridotite) is far better constrained than the composition of the coexisting vapors. In contrast, supercritical metasomatism, where only a single fluid is stable on the *P-T* path, is potentially more difficult to identify because for a given rock the fluid composition can be sensitive to H_2O/rock ratio and may vary continuously along the *P-T* path followed.

Because mineral solubility varies systematically with *P* and *T*, fluid evolution along specific paths may be well-defined in those cases where the coexisting minerals are known and the assemblage completely buffers or fixes the fluid composition along the path (e.g., simple low-variance assemblages like quartz + orthoclase + muscovite + aqueous fluid are buffered at fixed *P, T* conditions). A complexity to this, however, arises where considering high-variance (components >> phases) rocks, like typical eclogites where assemblages consist of a few minerals possessing extensive solid solutions like garnet + omphacite, as here the coexisting fluid composition also depends on the rock/fluid ratio. This effect can be imagined for high rock/fluid ratios using Figures 19 (MSH) and 15 (Nep–Qtz–H_2O) where the two-phase assemblages in the ternary systems can be seen very clearly to depend upon the position of the bulk composition.

FLUID EVOLUTION IN LARGE SCALE TECTONIC PROCESSES

At crustal pressures, phase relations in natural rock–H_2O systems involve low-density aqueous fluids (supercritical with respect to the endpoint of the H_2O liquid–vapor curve) and/ or high-density hydrous melts. The wide miscibility gap between these two liquids leads to distinctly different mobile phases with quite distinct element solubilities and geochemical signatures. As pressure increases, the fluid–melt miscibility gap closes at ever lower temperatures, until the crest of the miscibility gap intersects the fluid-saturated solidus at its endpoint, leaving a single liquid that has chemical and physical properties continuously evolving with temperature, and which is supercritical with respect to the endpoint of the fluid-saturated solidus.

Experiments by Kessel et al. (2005a) on a K-free MOR Basalt system constrain "classical melting" at 4 and 5 GPa but indicate a continuously evolving liquid composition at 6 GPa (also summarized in Fig. 28 here). The subsolidus assemblage consisted always of omphacitic clinopyroxene + grossular-rich garnet + coesite + rutile ± kyanite. Major element compositions of the fluid/melt phase evolve at all pressures from peralkaline, H_2O-rich, "granitic" compositions to metaluminous, "andesitic" to basaltic compositions with increasing temperature. Locating the endpoint of the fluid-saturated solidus between 5 and 6 GPa, and just above 1000 °C indicates that at higher pressures (see Fig. 28), the distinction of fluid versus melt ceases to exist in the subducted oceanic crust.

Experiments by Kessel et al. (2005b) on trace element partitioning between minerals, fluid and melt for a K-free MOR basalt system were investigated by measuring trace element partitioning between omphacite, garnet and liquid, the latter either being an aqueous fluid, a hydrous melt, or a supercritical liquid. Hydrous melts and supercritical liquids (the latter down to at least 200 °C below the hypothetical extension of the solidus) are almost indistinguishable with their trace element pattern.

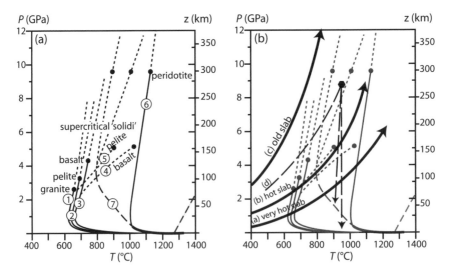

Figure 28. (a) *P,T* locations for wet solidi (and critical endpoints, solid circles) for various rock types. Extrapolations are critical isopleths (light dash). Numbers 1 to 7 refer to wet solidi for: 1 = granite (Huang and Wyllie 1981; Holtz et al. 2001); 2 = pelite (Nichols et al. 1994); 3 = basalt (Lambert and Wyllie 1972); 4 = basalt (Kessel et al. 2005a); 5 = pelite (Schmidt et al. 2004); 6 = peridotite (Green 1973); 7 = peridotite (Grove et al. 2006). (b) *P-T* trajectories for different slab subduction paths (heavy solid curves) superimposed on phase equilibria (curves 1–7, in grey).

A critical solidus endpoint has consequences for the trace element characteristics of the metasomatizing agents emanating from the subducted oceanic crust (those produced by the breakdown of hydrous MORB phases or by fluxing with H_2O-rich fluids originating from dehydration of the underlying serpentinites). Thus, recycling rates of relevant trace elements are not indicative of melting, and in the fast and steep circum-pacific subduction zones, they most likely testify for production of a mobile phase from the subducting crust in the supercritical liquid regime (beyond the endpoint of the solidus), i.e., at pressure in excess of 5 GPa corresponding to depth of 160 km.

Abrupt crossing of a solidus, by decompression or heating, produces effects that can be noticeable, whereas similar crossing of the critical line will generate rather continuous change in physical properties of a supercritical fluid/melt. Gradual changes will be manifest in composition (simply as H_2O to oxide components) and the resulting continuous changes in density and viscosity. Lower density fluids have greater buoyancy against enclosing rock, thus permitting ascent. Lower viscosity fluids can flow faster.

There are various large-scale tectonic processes involving natural fluids or melts, which sometimes encounter immiscible fluids and other times supercritical fluids/melts. Small-scale processes with fluids/melts are discussed elsewhere in this Volume so we will concentrate here on larger-scale processes.

Firstly we will consider subduction of oceanic lithosphere.

Oceanic lithosphere subduction environments

An average subducting oceanic lithosphere of ca. 100 km thickness, has about 8 km thickness of basalt (hydrated to greenschist or amphibolite facies) with patches of oceanic sediments (modeled here as pelite) above hydrated peridotite (perhaps as much as 12 km), atop 80 km of non-hydrated mantle (e.g., Helffrich et al. 1989; Peacock et al. 1994; Poli and Schmidt 2002). Dehydration reactions in the subducting slab release aqueous fluid appropriate to composition and depth for the superimposed (model) thermal structure (Helffrich et al. 1989; Peacock et al. 1994; Furukawa 1993). The thermal structure is modeled after Stein and Stein (1992) for hot mantle geotherms (Peacock et al. 1994, Fig. 5) and by Iwamori (1997) for today's average mantle geotherms, including convective flow. Figure 28 shows three typical subduction paths shown for the top of the slab in each case, defined for three slab thermal regimes (a) very hot, young, closely after mid-oceanic-ridge subduction; (b) hot, relatively recent lithosphere; and (c) old, cold lithosphere beneath a large ocean.

The contact between subducting oceanic crust and overlying mantle is the coldest location in the thermal profile, because the top of the subducting oceanic crust (initially near 0 °C) is progressively heated by downward conduction of heat from progressively warmer overlying mantle as subduction continues, also with possible frictional contributions (shear heating). Thus slab temperatures rise both due to downwards conduction from overlying mantle as well as upwards from underlying mantle, and gradually heats the sinking slabs up to ambient mantle temperatures at depths of 660 to 1200 km, depending upon the subduction rate.

A selection of H_2O-saturated solidi are shown for four rock compositions. These are taken from the compilation by Hermann et al. (2006, their Fig. 2) and augmented with Grove et al. (2006) observations on the peridotite wet solidus (Fig. 28a, curve 7). The water-saturated solidi shown terminate at supposed critical endpoints (CEPs). These CEPs (solid points drawn along water-saturated solidi) are not particularly well-located, and could lie at even higher pressures than indicated by the extrapolated wet solidi along the dashed lines. As discussed above the locations of these endpoints may be debated. Here we will simply consider them first to be correctly located, then consider how the interpretation might change if the actual locations were at higher or at lower pressures.

Burial along subduction paths between (a) and (b) in Figure 28b meets the water-saturated solidi for each of pelite, basalt, and granite, at depths of 30 to 90 km. The former two representing sediment and oceanic crust, and the granite representing subducting continental crust (attached to slab which thus can subduct despite density difference). Here felsic melts (almost "granitic") would be produced in amount proportional to the original amount of free water present (Burnham 1979, 1997; Clemens and Vielzeuf 1987; Thompson 1988). The density and viscosity of such melts would decrease strongly with increased H_2O-content, as discussed above. Significant temperature overstepping of the illustrated wet solidi and/or additional sources of water (e.g., dehydration mineral melting) would be needed to produce enough melt to segregate. Only young and hot slabs can produce enough "granitic melt," by melting of hydrated basaltic (tonalite) or pelitic (S-type granite) oceanic crust, that might segregate (> 30 vol%; e.g., Peacock et al. 1994). How far the melts might ascend depends upon the rate of migration compared to rate of cooling (e.g., Weinberg 1996; Thompson 1999; Petford et al. 2000). Subduction paths from (a) to (b) would indeed meet these three solidi, but progressively less melt would be generated for the cooler subduction paths. Small melt fractions generally remain close to their sites of origin and do not travel with their fluids into the overlying mantle wedge. Differences in behavior from the above would occur for the recently suggested higher P,T location of CEPs (Hermann et al. 2006, their Fig. 2), for pelite (based upon experiments of Schmidt et al. 2004) and for eclogite (based upon experiments of Kessel al. 2005a). The dotted lines between 3 and 5 GPa and 700 to ca. 1000 °C in the lower part of Figure 28a, show that the solidus curves would extend up to these pressures compared to the previous interpretation. Interestingly, the subduction P-T path (b) is sub-parallel to these up-P extended solidi. For this exact path (b) basalt remains entirely in the subsolidus (always dehydrating to produce supercritical fluid and never water-saturated melt), whereas pelite could cross the water-saturated solidus near 2.5 GPa.

Supercritical fluids in the earth's mantle?

Melting of water-bearing mantle peridotite can only occur for "very hot" paths (e.g., Thompson 1992; equivalent to "young" of Iwamori 1997), which cross the wet solidus for mantle peridotite from 4 to 6 GPa (120 to 180 km depth). Again the degree of wet solidus mantle melting would be proportional to the amount of free water. The location of the crossing of the wet mantle solidus depends upon which experimental data is chosen. Hermann et al. (2006, their Fig. 2) used a wet peridotite solidus based upon the data of Green (1973, his Fig. 1 p. 40) which steepens near 1000 °C (called here "the 1000 °C mantle solidus"). Recently, Grove et al. (2006, their Fig. 6, p. 84) have proposed that a much lower temperature wet solidus is preferred (close to that originally determined by Mysen and Boettcher 1975). Importantly, this solidus at a depth of 100 to 150 km is at about 200 °C lower temperature than the other solidus steepening at about 800 °C. For the "800 °C mantle solidus", all paths from hotter than (a) to colder than (b) would cut the wet mantle solidus. The implied widespread melting with corresponding lack of evidence for massive mantle melting above slabs of a whole range of thermal ages, would thus probably speak to limited water availability.

Subduction paths from "old" to "hot" pass at higher pressures than the Hermann et al. (2006, their Fig. 2) inferred-CEPs for subducted oceanic and continental crust components. In the subsolidus regions for these P-T paths, minerals continue to dissolve in available water because silicate mineral solubility continues to increase with increased P and T (e.g., Manning 2004). By crossing the critical curves from the subsolidus into supercritical region, would simply continue mineral dissolution. The progressively more concentrated solutions may be considered more melt-like than the more dilute fluid-like solutions from the lower P-T subsolidus region. These supercritical fluids would be less-viscous and less-dense than water-saturated solidus melts and have the potential for upward migration more than wet melts. There are quite a variety of hydrous minerals that continue to be stable in the cold parts of subducted slabs deeper than 100 km (e.g., Frost 2006, references therein). Their dehydration would occur at deeper levels (and higher temperatures) than the H_2O-saturated solidi shown in Figure 28 (Poli and Schmidt 2002). Also a time progression is suggested where early in the history of a young slab, wet solidus melting is possible (depending upon water availability). Later in the subduction history as the subducting slab becomes colder, no melting takes place and slab fluids progressively dissolve more minerals into a supercritical fluid which is probably less-dense and less-viscous than an equivalent water-saturated melt. Such concentrated aqueous fluids ("mantle soup" of various flavors) might well have more potential for upward migration than wet melts.

We can simplify our views of fluid migration in the mantle either as a vertical path from the top of the slab into the wedge (where the rising fluid or melt would be rapidly heated to the local mantle temperature) or back up along the mantle slab interface (where the rising fluid would not heat). Both such paths are decompression paths and are shown only simply in Figure 28. For fluids passing up the slab-wedge interface in response to buoyancy forces, they would encounter progressively lower temperatures at depth. Depending upon their rate of ascent they could advect heat and mass if they rise fast, whereas gradual rise would tend towards maintaining chemical equilibrium.

Water-bearing melts rising upwards from subduction paths from (a) to (b) in Figure 28b, would increase their water contents as the wet solidi are approached, encouraging a play off between decreased temperature but increased water content on decreased density and decreased viscosity. The solidi locations (800 vs. 1000 °C) would mark the shallowest depths for the hypothetical rising magmas. Further rise of fluids separated at the crossing of the water-saturated solidi would result in continued mineral precipitation. Returning back up each of the subduction paths between (b) and (c), would result in continuously decreased solubility of silicate in supercritical fluid through to "subsolidus" fluid.

As one model for a subduction path which then subsequently involved fluid/melt ascent through the wedge, can be illustrated with the hypothetical cool subduction path (d), located between (b) and (c). This hypothetical path generates successive supercritical fluids for granite, pelite then basalt dehydration between ca. 140 to 170 km. For the Green (1973, his Fig. 1) 1000 °C wet peridotite solidus this path remains subsolidus until deeper than 450 km. Whereas for the Grove et al. (2006; cf. Mysen and Boettcher 1975) 800 °C wet peridotite solidus, path (d) could undergo water-saturated mantle melting near 200 km producing "pseudo-basaltic" melts. Such water-saturated pseudo-basaltic melts could theoretically rise through a mantle crack – one such hypothetical path is shown rising from the solid hexagon near 280 km, 900 °C. An isothermal and an adiabatic path are shown (the latter with a slope of +0.3 °C/km). Such hypothetical water-saturated mantle melts rising up cracks would re-cross the 800 °C Grove et al. (2006) wet peridotite solidus between 70 km (adiabat) and 30 km (isotherm). Such crystallised former wet mantle melts would release their dissolved fluids at the crossing of the illustrated solidus and this could be available for melting of pelitic granitic and basaltic compositions in the arc zone.

NATURAL SYSTEMS: WHERE ARE IMMISCIBILITY, SUPERCRITICALITY LIKELY TO OCCUR?

Natural systems

At crustal pressures, higher-temperature phase relations in natural rock–H_2O systems involve low-density aqueous fluids, which can be supercritical with respect to the endpoint of the H_2O liquid+vapor curve, and/or high-density hydrous melts. The wide miscibility gap between these two liquid phases leads to a distinctly different mobile phases with quite distinct element solubilities and geochemical signatures. As pressure increases, the fluid–melt miscibility gap closes at ever lower temperatures, until the crest of the miscibility gap intersects the fluid-saturated solidus at its endpoint, leaving a single liquid that has chemical and physical properties continuously evolving with temperature, and which is supercritical with respect to the endpoint of the fluid-saturated solidus.

Extent of immiscibility and supercriticality in natural processes

Supercriticality seems to be a high-pressure phenomenon. Transition from supersolvus to supercritical occurs when a critical curve for L–V (or L–L) solvus cuts the relevant solidus. Phase separation at depth is proportional to H_2O-system plus the effects of dissolved components (i.e., deeper, and hotter than 212 bars, e.g., > 600 m in rock). Immiscibility on cooling is a common feature of fluids, wet magmas, and also in some anhydrous magmatic systems. Processes involving melts to concentrated solutions may be enhanced when additional components strongly partitioning into melt (liquid) rather than vapor (e.g., sodium disilicate (peralkaline components), lithium, fluorine, beryllium, boron).

What to do next?

Fluid inclusions provide the most direct evidence for addressing the occurrence of immiscibility processes in natural systems. The study of inclusions occurring in pegmatites (e.g., review by Roedder 1992; Thomas et al. 2000) and UHP rocks (e.g., Frezzotti et al. 2004; Touret and Frezzotti 2003; Ferrando et al. 2005) is encouraged, as it provides much needed information on types and ranges of natural fluid (and liquid) compositions, and where immiscibility and supercriticality occur at low and higher pressure in natural environments, and guides theory and the experimenters.

Mineral solubility in supercritical water has received much experimental attention in recent years (Manning 1994; Stalder et al. 2000). Such studies suggest significant changes in solute speciation occur above ca. 0.5 GPa, leading to non-linearly enhanced solubilities (Zotov and Keppler 2000; Newton and Manning 2003). The experimental study of solubility of more complex mineral assemblages is needed, as dominant solute complexes that may form need not be those identified in the chemically simpler fluids.

Melting experiments determining phase equilibria, melting and critical phenomena suffer from ambiguous interpretation of the appearance of textural evidence of quenched hydrous liquids as opposed to quenched solute from aqueous fluids (e.g., Mysen and Boettcher 1975; Niida and Green 1999; Grove et al. 2006). Experimental work focused on chemical and physical property measurements, that can distinguish more clearly continuous versus discontinuous behavior, may provide results less prone to divergent opinion (Stalder et al. 2001; Kessel et al. 2005a,b; Audétat and Keppler 2004). *In situ* studies have obvious advantages, but at the moment prove difficult for *in situ* chemical analysis in complex chemical systems on the micron scale.

Integration of solubility and physical property measurements and extraction of thermodynamic properties would allow extension of thermodynamic databases (e.g., Berman 1988; Holland and Powell 1998, 2001) and calculation of phase relations that include melt–fluid miscibility and solubility.

Fractionation of trace-elements between liquid and vapor at high pressures appears, at this stage, to be at least partly decoupled from H_2O partitioning (at least for eclogite–water, Kessel et al. 2005a,b), but reasons for this are not fully understood. What role is required of accessory minerals for trace-element fluxes in natural fluid systems?

Physical property measurements over compositionally relevant ranges have application to flow transport, thermochemistry and phase equilibria. Densities of aqueous fluids are (unlike hydrous melts) poorly known, but would reflect the nature of solute speciation. Density variation with X as well as P-T needs to be determined to physically model fluid flow at mantle and lower crustal depths. Viscosity measurements of silicate-bearing fluids under high P-T conditions remain scarce, but it is clearly a fruitful area for future work (Audétat and Keppler 2004).

ACKNOWLEDGMENTS

We wish to thank Chris Heinrich and Axel Liebscher for inviting us to contribute this review chapter, Mark Caddick, Christian Liebske, Peter Ulmer and Max Schmidt for constructive discussions on natural immiscible and supercritical fluids, Ursula Stidwill for help in preparing the manuscript, Hans Keppler, Daniel Vielzeuf and Axel Liebscher for their constructive reviews and comments.

This work was supported by the Swiss National Science Foundation.

REFERENCES

Ague JJ (1997) Crustal mass transfer and index mineral growth in Barrow's garnet zone, northeast Scotland. Geology 25:73-76

Anderson GM, Burnham CW (1965) The solubility of quartz in supercritical water. Am J Sci 263:494-511

Anderson GM, Burnham CW (1983) Feldspar solubility and the transport of aluminum under metamorphic conditions. Am J Sci 283A:283-297

Andrews T (1869) On the continuity of the gaseous and liquid states of matter. Philos Trans R Soc London 159A:575-590

Anisimov MA, Sengers JV, Levelt Sengers JMH (2004) Near-critical behavior of aqueous systems. *In*: Aqueous Systems at Elevated Temperatures and Pressures: Water, Steam, and Hydrothermal Solutions. Palmer DA, Fernández-Prini R, Harvey A (eds) Elsevier, p 29-71

Antignano A, Manning CE (2003) Solubility of albite + paragonite ± quartz in H_2O at 1 GPa, 580°C: Implications for metamorphic fluids. AGU Fall Meeting V22D-0613

Arnórsson S, Stefánsson A, Bjarnason JÖ (2007) Fluid-fluid interactions in geothermal systems. Rev Mineral Geochem 65:259-312

Audétat A, Keppler H (2004) Viscosity of fluids in subduction zones. Science 303:513-516

Barnes HL (ed) (1979) Geochemistry of hydrothermal ore deposits (2nd edition). John Wiley & Sons

Behrens H (1995) Determination of water solubilities in high-viscosity melts: An experimental study on $NaAlSi_3O_8$ and $KAlSi_3O_8$ melts. Eur J Mineral 7:905-920

Berman RG (1988) Internally consistent thermodynamic data for minerals in the system Na_2O-K_2O-CaO-MgO-FeO-Fe_2O_3-Al_2O_3-SiO_2-TiO_2-H_2O-CO_2. J Petrol 29:445-522

Birch F, LeComte P (1960) Temperature-pressure plane for albite composition. Am J Sci 258:209-217

Bischoff JL, Pitzer KS (1985) Phase relations and adiabats in boiling seafloor geothermal systems. Earth Planet Sc Lett 75:327-338

Boettcher AL (1984) The system SiO_2-H_2O-CO_2: Melting, solubility mechanisms of carbon, and liquid structure to high pressures. Am Mineral 69:823-833

Boettcher AL, Wyllie PJ (1969a) The system CaO-SiO_2-CO_2-H_2O III. Second critical end-point on the melting curve. Geochim Cosmochim Acta 33:611-632

Boettcher AL, Wyllie PJ (1969b) Phase relationships in the system $NaAlSiO_4$-SiO_2-H_2O to 35 kilobars pressure. Am J Sci 267:875-909

Boettcher AL, Burnham CW, Windom KE, Bohlen SR (1982) Liquids, glasses, and the melting of silicates to high-pressures. J Geol 90:127-138

Bohlen SR, Boettcher AL, Wall VJ (1982) The system albite-H_2O-CO_2 - a model for melting and activities of water at high-pressures. Am Mineral 67:451-462

Bolz A, Deiters UK, Peters CJ, De Loos TW (1998) Nomenclature for phase diagrams with particular reference to vapour-liquid and liquid-liquid equilibria. Pure Appl Chem 70:2233-2257

Boshkov LZ, Mazur VA (1986) Phase-equilibria and critical lines of Lennard-Jones molecules in binary-mixtures. Russ J Phys Chem 60:29-33

Bowen NL (1928) The Evolution of the Igneous Rocks. Dover Publications Inc.

Boyd FR, England JL (1963) Effect of pressure on melting of diopside, $CaMgSi_2O_6$, and albite, $NaAlSi_3O_8$, in the range up to 50 kilobars. J Geophys Res 68:311-323

Brady JB (1977) Metasomatic zones in metamorphic rocks. Geochim Cosmochim Acta 41:113-125

Bromiley GD, Pawley AR (2003) The stability of antigorite in the systems MgO-SiO_2-H_2O (MSH) and MgO-Al_2O_3-SiO_2-H_2O (MASH): The effects of Al^{3+} substitution on high-pressure stability. Am Mineral 88:99-108

Bureau H, Keppler H (1999) Complete miscibility between silicate melts and hydrous fluids in the upper mantle: experimental evidence and geochemical implications. Earth Planet Sc Lett 165:187-196

Burnham CW (1979) Magmas and hydrothermal fluids. *In*: Geochemistry of Hydrothermal Ore Deposits. Barnes HL (ed) John Wiley and Sons, p 71-136

Burnham CW (1997) Magmas and hydrothermal fluids. *In*: Geochemistry of Hydrothermal Ore Deposits. Barnes HL (ed) John Wiley and Sons, New York, p 63-124

Burnham CW, Davis NF (1971) Role of H_2O in silicate melts: 1. P-V-T relations in the system $NaAlSi_3O_8$-H_2O to 10 kilobars and 1000 °C. Am J Sci 270:54-79

Burnham CW, Davis NF (1974) Role of H_2O in silicate melts: 2. Thermodynamic and phase relations in the system $NaAlSi_3O_8$-H_2O to 10 kilobars, 700 ° to 1100 °C. Am J Sci 274:902-940

Burnham CW, Jahns RH (1962) A method for determining the solubility of water in silicate melts. Am J Sci 260:721-745

Caciagli NC, Manning CE (2003) The solubility of calcite in water at 6-16 kbar and 500-800 °C. Contrib Mineral Petrol 146:275-285

Cagniard de la Tour C (1822) Exposé de quelques resultats obtenus par l'action combinée de la chaleur et de la compression sur certaines liquides, tels que l'eau, l'éther sulfurique et l'essence de pétrole rectifié. Ann Chim Phys 21:127-132 and 178-182

Calado JCG, Lopes JNC (1999) The building-up of phase diagrams. Pure Appl Chem 71:1183-1196

Carroll MR, Holloway JR (eds) (1994) Volatiles in Magmas. Rev Mineral Geochem Vol 30. Mineral Society of America

Clark SPJ (ed) (1966) Handbook of Physical Constants. Mineralogical Society of America

Clemens JD, Vielzeuf D (1987) Constraints on melting and magma production in the crust. Earth Planet Sc Lett 86:287-306

Crerar DA, Anderson GM (1971) Solubility and solvation reactions of quartz in dilute hydrothermal solutions. Chem Geol 8:107-122

Currie KL (1968) On solubility of albite in supercritical water in range 400 to 600 °C and 750 to 3500 bars. Am J Sci 266:321-341

Davis NF (1972). Experimental studies in the system $NaAlSi_3O_8$-H_2O: the apparent solubility of albite in supercritical water. PhD Dissertation. Pennsylvania State University, Philadelphia, Pennsylvania

Day HW, Fenn PM (1982) Estimating the P-T-X_{H2O} conditions during crystallization of low-calcium granites. J Geol 90:485-507

Dingwell DB (1998) Melt viscosity and diffusion under elevated pressure. Rev Mineral Geochem 37:397-424

Ebadi A, Johannes W (1991) Beginning of melting and composition of first melts in the system Qz-Ab-Or-H_2O-CO_2. Contrib Mineral Petrol 106:286-295

Eckert CA, Knutson BL, Debenedetti PG (1996) Supercritical fluids as solvent for chemical and materials processing. Nature 383:313-318

Eggler DH, Kadik AA (1979) System $NaAlSi_3O_8$-H_2O-CO_2 to 20 kbar pressure. 1. Compositional and thermodynamic relations of liquids and vapors coexisting with albite. Am Mineral 64:1036-1048

Evans K, Powell R (2006) A method for activity calculations in saline and mixed solvent solutions at elevated temperature and pressure: A framework for geological phase equilibria calculations Geochim Cosmochim Acta 70:5488-5506

Evans BW, Johannes W, Oterdoom H, Trommsdorff V (1976) Stability of chrysotile and antigorite in the serpentine multisystem. Schweiz Miner Petrog 56:79-93

Ferrando S, Frezzotti ML, Dallai L, Compagnoni R (2005) Multiphase solid inclusions in UHP rocks (Su-Lu, China): Remnants of supercritical silicate-rich aqueous fluids released during continental subduction. Chem Geol 223:68-81

Fleming BA, Crerar DA (1982) Silicic-acid ionization and calculation of silica solubility at elevated-temperature and pH. Application to geothermal fluid processing and reinjection. Geothermics 11:15-29

Fournier RO (1960) The solubility of quartz in water in the temperature interval from 25 °C to 300 °C. Geol Soc Am Bull 71:1867-1868

Fournier RO, Potter RW (1982) An equation correlating the solubility of quartz in water from 25 ° to 900 °C at pressures up to 10,000 bars. Geochim Cosmochim Acta 46:1969-1973

Foustoukos DI, Seyfried Jr. WE (2007) Fluid phase separation processes in submarine hydrothermal systems. Rev Mineral Geochem 65:213-239

Franck EU (1987) Fluids at high pressures and temperatures. Pure Appl Chem 59:25-34

Frezzotti ML, Cesare B. Scambulluri M (2004) Fluids at extreme P-T metamorphic conditions: the message from high-grade rocks. Periodico Mineral 73:209

Friedman II (1950) Some aspects of the system H_2O-Na_2O-SiO_2-Al_2O_3. J Geol 59:19-31

Frost DJ (2006) The stability of hydrous mantle phases. Rev Mineral Geochem 62:243-271

Furukawa Y (1993) Magmatic processes under arcs and formation of the volcanic front. J Geophys Res-Sol Ea 98:8309-8319

Fyfe WS, Price NJ, Thompson AB (1978) Fluids in the Earth's crust. Elsevier Scientific Publishing Company

Garcia DC, Luks KD (1999) Patterns of solid-fluid phase equilibria: new possibilities? Fluid Phase Equilibr 161:91-106

Giordano D, Mangiacapra A, Potuzak M, Russell JK, Romano C, Dingwell DB, Di Muro A (2006) An expanded non-Arrhenian model for silicate melt viscosity: A treatment for metaluminous, peraluminous and peralkaline liquids. Chem Geol 229:42-56

Goldsmith JR, Jenkins DM (1985) The hydrothermal melting of low and high albite. Am Mineral 70:924-933

Goranson RW (1938) Silicate-water systems: phase equilibria in the $NaAlSi_3O_8-H_2O$ and $KAlSi_3O_8-H_2O$ systems at high temperatures and pressures. Am J Sci, Day Volume 35A:71-91

Green DH (1973) Experimental melting studies on a model upper mantle composition at high pressure under water-saturated and water-undersaturated conditions. Earth Planet Sc Lett 19:37-53

Greenwood HJ (1975) Buffering of pore fluids by metamorphic reactions. Am J Sci 275:573-593

Greig JW, Merwin HE, Shepherd ES (1933) Notes on the volatile transport of silica. Am J Sci 225:61-73

Grove TL, Chatterjee N, Parman SW, Medard E (2006) The influence of H_2O on mantle wedge melting. Earth Planet Sc Lett 249:74-89

Guissani Y, Guillot B (1996) A numerical investigation of the liquid-vapor coexistence curve of silica. J Chem Phys 104:7633-7644

Haar L, Gallagher JS, Kell GS (1984) Steam Tables. Thermodynamic and Transport Properties and Computer Programs for Vapour and Liquid States of Water in SI Units. Hemisphere Publishing

Hack AC, Mavrogenes JA (2006a) A synthetic fluid inclusion study of copper solubility in hydrothermal brines from 525 to 725 °C and 0.3 to 1.7 GPa. Geochim Cosmochim Acta 70:3970-3985

Hack AC, Mavrogenes JA (2006b) A cold-sealing capsule design for synthesis of fluid inclusions and other hydrothermal experiments in a piston-cylinder apparatus. Am Mineral 91:203-210

Hack AC, Hermann J, Mavrogenes JA (2007) Mineral solubility and hydrous melting relations in the deep Earth: Analysis of some binary $A-H_2O$ system pressure-temperature-composition topologies. Am J Sci, (in press)

Hamilton DL, Oxtoby S (1986) Solubility of water in albite-melt determined by the weight-loss method. J Geol 94:626-630

Hannay JB, Hogarth J (1879) On the solubility of solids in gases. Proc Royal Soc 29:324-326

Hauzenberger CA, Baumgartner LP, Pak TM (2001) Experimental study on the solubility of the "model"-pelite mineral assemblage albite + K-feldspar + andalusite + quartz in supercritical chloride-rich aqueous solutions at 0.2 GPa and 600 °C. Geochim Cosmochim Acta 65:4493-4507

Hedenquist JW, Lowenstern JB (1994) The role of magmas in the formation of hydrothermal ore deposits. Nature 370:519-527

Heinrich CA (2007a) Fluid – fluid interactions in magmatic-hydrothermal ore formation. Rev Mineral Geochem 65:363-387

Heinrich W (2007b) Fluid immiscibility in metamorphic rocks. Rev Mineral Geochem 65:389-430

Helffrich GR, Stein S, Wood BJ (1989) Subduction zone thermal structure and mineralogy and their relationship to seismic-wave reflections and conversions at the slab mantle interface. J Geophys Res 94:753-763

Hemley JJ, Montoya JW, Marinenko JW, Luce RW (1980) Equilibria in the system $Al_2O_3-SiO_2-H_2O$ and some general implications for alteration-mineralization processes. Econ Geol 75:210-228

Hermann J (2003) Experimental evidence for diamond-facies metamorphism in the Dora-Maira massif. Lithos 70:163-182

Hermann J, Spandler C, Hack A, Korsakov AV (2006) Aqueous fluids and hydrous melts in high-pressure and ultra-high pressure rocks: Implications for element transfer in subduction zones. Lithos 92:399-417

Hirschmann MM (2006) Water, melting, and the deep Earth H_2O cycle. Annu Rev Earth Planet Sci 34:629-653

Holland TJB, Powell R (1998) An internally consistent thermodynamic data set for phases of petrological interest. J Metamorph Geol 16:309-343

Holland T, Powell R (2001) Calculation of phase relations involving haplogranitic melts using an internally consistent thermodynamic dataset. J Petrol 42:673-683

Holtz F, Behrens H, Dingwell DB, Taylor RP (1992a) Water solubility in aluminosilicate melts of haplogranite composition at 2 kbar. Chem Geol 96:289-302

Holtz F, Pichavant M, Barbey P, Johannes W (1992b) Effects of H_2O on liquidus phase relations in the haplogranite system at 2 and 5 kbar. Am Mineral 77:1223-1241

Holtz F, Behrens H, Dingwell DB, Johannes W (1995) H_2O solubility in haplogranitic melts - compositional, pressure, and temperature dependence. Am Mineral 80:94-108

Holtz F, Roux J, Behrens H, Pichavant M (2000) Water solubility in silica and quartzofeldspathic melts. Am Mineral 85:682-686

Holtz F, Becker A, Freise M, Johannes W (2001) The water-undersaturated and dry Qz-Ab-Or system revisited. Experimental results at very low water activities and geological implications. Contrib Mineral Petrol 141:347-357

Huang WL, Wyllie PJ (1975) Melting reactions in the system $NaAlSi_3O_8-KAlSi_3O_8-SiO_2$ to 35 kilobars, dry and with excess water. J Geol 83:737-748

Huang WL, Wyllie PJ (1981) Phase-relationships of S-type granite with H_2O to 35 kbar - muscovite granite from Harney Peak, South-Dakota. J Geophys Res 86:515-529

Hudon P, Jung I, Baker DR (2002) Melting of ß-quartz up to 2.0 GPa and thermodynamic optimization of the silica liquidus up to 6.0 GPa. Phys Earth Planet Int 130:159-174

Inoue T (1994) Effect of water on melting phase relations and melt composition in the system Mg_2SiO_4-$MgSiO_3$-H_2O up to 15 GPa. Phys Earth Planet Int 85:237-263

Irifune T, Kubo N, Isshiki M, Yamasaki Y (1998) Phase transformations in serpentine and transportation of water into the lower mantle. Geophys Res Lett 25:203-206

Iwamori H (1997) Heat sources and melting in subduction zones. J Geophys Res 102:14803-14820

Jackson I (1976) Melting of silica isotypes SiO_2, BeF_2 and GeO_2 at elevated pressures. Phys Earth Planet Int 13:218-231

Jahns RH (1953) The genesis of pegmatites: I. Occurrence and origin of giant crystals. Am Mineral 38:563-598

Jahns RH, Burnham CW (1969) Experimental studies of pegmatite genesis: 1. A model for derivation and crystallization of granitic pegmatites. Econ Geol 64:843-864

Jahns RH, Tuttle OF (1963) Layered pegmatite-aplite intrusives. Mineral Soc Am - special paper 1:78-92

Johannes W (1984) Beginning of melting in the granite system Qz-Or-Ab-An-H_2O. Contrib Mineral Petrol 86:264-273

Kadik AA, Lebedev YB (1968) Temperature dependence of the solubility of water in an albite melt at high pressures. Geochem Int 5:1172-1181

Kamenetsky VS, Naumov VB, Davidson P, van Achterbergh E, Ryan CG (2004) Immiscibility between silicate magmas and aqueous fluids: a melt inclusion pursuit into the magmatic-hydrothermal transition in the Omsukchan Granite (NE Russia). Chem Geol 210:73-90

Kanzaki M (1990) Melting of silica up to 7.0 GPa. J Am Ceram Soc 73:3706-3707

Kawamoto T, Matsukage KN, Mibe K, Isshiki M, Nishimura K, Ishimatsu N, Ono S (2004) Mg/Si ratios of aqueous fluids coexisting with forsterite and enstatite based on the phase relations in the Mg_2SiO_4-SiO_2-H_2O system. Am Mineral 89:1433-1437

Kennedy GC (1944) The hydrothermal solubility of SiO_2. Econ Geol 39:25-36

Kennedy GC (1950) A portion of the system silica-water. Econ Geol 45:629-653

Kennedy GC, Heard HC, Wasserburg GJ, Newton RC (1962) The upper 3-phase region in system SiO_2-H_2O. Am J Sci 260:501-521

Keppler H (1993) Influence of fluorine on the enrichment of high field strength trace elements in granitic rocks. Contrib Mineral Petrol 114:479-488

Keppler H, Audétat A (2005) Fluid-mineral interaction at high pressure. In: Mineral behaviour at extreme conditions. EMU notes in Mineralogy. Vol 7. Miletich R (ed), Eötvös University Press, Budapest, p 225-252

Kerrick DM, Connolly JAD (2001) Metamorphic devolatilization of subducted marine sediments and the transport of volatiles into the Earth's mantle. Nature 411:293-296

Kessel R, Ulmer P, Pettke T, Schmidt MW, Thompson AB (2004) A novel approach to determine high-pressure high-temperature fluid and melt compositions using diamond-trap experiments. Am Mineral 89:1078-1086

Kessel R, Ulmer P, Pettke T, Schmidt MW, Thompson AB (2005a) The water-basalt system at 4 to 6 GPa: Phase relations and second critical endpoint in a K-free eclogite at 700 to 1400 °C. Earth Planet Sc Lett 237:873-892

Kessel R, Schmidt MW, Ulmer P, Pettke T (2005b) Trace element signature of subduction-zone fluids, melts and supercritical liquids at 120-180 km depth Nature 437:724-727

Khitarov NI, Kadik AA, Lebedev YB (1963) Estimate of the thermal effect of the separation of water from felsic melts based on data for the system albite-water. Geokhimiya 7:637-649

Kitahara S (1960) The solubility of quartz in water at high temperatures and pressures. Rev Phys Chem Jap 30:109-137

Koningsveld R, Diepen GAM (1983) Supercritical phase equilibria involving solids. Fluid Phase Equilibria 10:159-172

Korteweg DJ (1891a) Sur les points de plissement. Arch neerl 24:57-98

Korteweg DJ (1891b) La theorie generale des plis. Arch neerl 24:295-368

Kravchuk KG, Valyashko VM (1979) Phase diagram of the system SiO_2-$Na_2Si_2O_5$-H_2O. In: Methods of Experimental Investigations of Hydrothermal Equilibria. Godovikov AA (ed) Nauka, Novosibirsk, p 105-117

Kushiro I (1969) System forsterite-diopside-silica with and without water at high pressures. Am J Sci 267A:269-294

Kushiro I, Yoder HS, Nishikawa M (1968) Effect of water on melting of enstatite. Geol Soc Am Bull 79:1685-1692

Lambert IB, Wyllie PJ (1972) Melting of gabbro (quartz eclogite) with excess water to 35 kilobars, with geological applications. J Geol 80:693-708

Levelt Sengers JMH (1990) Thermodynamic properties of aqueous solutions at high temperatures: needs, method and challenges. Int J Thermophys 11:399-415

Levelt Sengers JMH (1991) Solubility near the solvent's critical point. J Supercrit Fluid 4:215-222

Levelt Sengers JMH (1993) Critical behavior of fluids: concepts and applications. *In:* Supercritical Fluids, Fundamentals for Application. NATO ASI Vol E 273. Kiran E, Levelt Sengers JMH (eds) Kluwer Acad Publ, p 1-38

Levelt Sengers JMH (2000) Supercritical fluids: Their properties and applications. *In:* Supercritical fluids. Vol 366. Kiran E, Debenedetti PG, Peters CJ (eds) NATO Science Series E Applied Sciences, p 1-30

Levelt Sengers JMH (2002) How fluids unmix - Discoveries by the School of Van der Waals and Kamerlingh Onnes. Koninklijke Nederlandse Akademie van Wetenschappen, Amsterdam

Liebscher A (2007) Experimental studies in model fluid systems. Rev Mineral Geochem 65:15-47

Liebscher A, Heinrich CA (2007) Fluid–fluid interactions in the Earth's lithosphere. Rev Mineral Geochem 65:1-13

Liu J, Bohlen SR (1995) Mixing properties and stability of jadeite-acmite pyroxene in the presence of albite and quartz. Contrib Mineral Petrol 119:433-440

London D (1986a) Formation of tourmaline-rich gem pockets in miarolitic pegmatites. Am Mineral 71:396-405

London D (1986b) The magmatic-hydrothermal transition in the Tanco rare-element pegmatite: evidence from fluid inclusions and phase equilibrium experiments. Am Mineral 71:376-395

London D (1987) Internal differentiation of rare-element pegmatites - effects of boron, phosphorus, and fluorine. Geochim Cosmochim Acta 51:403-420

London D (1997) Estimating the abundances of volatile and other mobile components in evolved silicic melts through mineral-melt equilibria. J Petrol 38:1691-1706

London D (2005) Granitic pegmatites: an assessment of current concepts and directions for the future. Lithos 80:281-303

London D, Morgan GB, Hervig RL (1989) Vapor-undersaturated experiments in the system macusanite-H_2O at 200 MPa, and the internal differentiation of granitic pegmatites. Contrib Mineral Petrol 102:1-17

London D, Morgan GB, Babb HA, Loomis JL (1993) Behavior and effects of phosphorus in the system Na_2O-K_2O-Al_2O_3-SiO_2-P_2O_5-H_2O at 200 MPa (H_2O). Contrib Mineral Petrol 113:450-465

Loucks RR, Mavrogenes JA (1999) Gold solubility in supercritical hydrothermal brines measured in synthetic fluid inclusions. Science 284:2159-2163

Lu BCY, Zhang D (1989) Solid-supercritical fluid phase equilibria. Pure Appl Chem 61:1065-1074

Luth RW (1995) Is phase A relevant to the Earth's mantle? Geochim Cosmochim Acta 59:679-682

Luth RW, Boettcher AL (1986) Hydrogen and the melting of silicates. Am Mineral 71:264-276

Luth WC, Jahns RH, Tuttle OF (1964) The granite system at pressures of 4 to 10 kilobars. J Geophys Res 69:759-773

Mackenzie FT, Gees R (1971) Quartz: synthesis at Earth surface conditions. Science 173:533-535

Manning CE (1994) The solubility of quartz in H_2O in the lower crust and upper mantle. Geochim Cosmochim Acta 58:4831-4839

Manning CE (1995) Phase-equilibrium controls on SiO_2 metasomatism by aqueous fluids in subduction zones: Reaction at constant pressure and temperature. Int Geol Rev 37:1074-1093

Manning CE (2004) The chemistry of subduction-zone fluids. Earth Planet Sc Lett 223:1-16

Manning CE, Boettcher SL (1994) Rapid-quench hydrothermal experiments at mantle pressures and temperatures. Am Mineral 79:1153-1158

Massonne HJ (1992) Evidence for low-temperature ultrapotassic siliceous fluids in subduction zone environments from experiments in the system K_2O-MgO-Al_2O_3-SiO_2-H_2O (KMASH). Lithos 28:421-434

McGlashan ML (1985) Phase equilibria in fluid mixtures. J Chem Thermodyn 17:301-319

McMillan PF (1994) Water solubility and speciation models. Rev Mineral Geochem 30:132-156

Melekhova E, Schmidt MW, Ulmer P, Pettke T (2007) The composition of liquids coexisting with dense hydrous magnesium silicates at 11 - 13.5 GPa and the endpoints of the solidi in the MgO-SiO_2-H_2O system. Geochim Cosmochim Acta, doi: 10.1016/j.gca.2007.03.034 (in press)

Merrill RB, Robertson JK, Wyllie PJ (1970) Melting reactions in system $NaAlSi_3O_8$-$KAlSi_3O_8$-SiO_2-H_2O to 20 kilobars compared with results for other Feldspar-Quartz-H_2O and Rock-H_2O Systems. J Geol 78:558-569

Mibe K, Fujii T, Yasuda A (2002) Composition of aqueous fluid coexisting with mantle minerals at high pressure and its bearing on the differentiation of the Earth's mantle. Geochim Cosmochim Acta 66:2273-2285

Mibe K, Fujii T, Yasuda A (2004a) Response to the comment by R. Stalder on "Composition of aqueous fluid coexisting with mantle minerals at high pressure and its bearing on the differentiation of the Earth's mantle". Geochim Cosmochim Acta 68:929-930

Mibe K, Kanzaki M, Kawamoto T, Matsukage KN, Fei Y, Ono S (2004b) Determination of the second critical end point in silicate-H_2O systems using high-pressure and high-temperature X-ray radiography. Geochim Cosmochim Acta 68:5189-5195

Mibe K, Kanzaki M, Kawamoto T, Matsukage KN, Fei YW, Ono S (2007) Second critical endpoint in the peridotite-H_2O system. J Geophys Res-Sol Ea 112:B03201, doi:10.1029/2005JB004125

Morey GW (1951) The solubility of quartz and some other substances in superheated steam at high pressures. Am Mineral 36:322-322

Morey GW (1957) The solubility of solids in gases. Econ Geol 51:225-251

Morey GW, Hesselgesser JM (1952) The System H_2O-Na_2O-SiO_2 at 400 °C. Am J Sci 250:343-371

Morey GW, Niggli P (1913) The hydrothermal formation of silicates, a review. J Am Chem Soc 35:1086-1130

Morey GW, Fournier RO, Rowe JJ (1962) The solubility of quartz in water in the temperature interval from 25° to 300 °C. Geochim Cosmochim Acta 26:1029-1043

Morse SA (1970) Alkali feldspars with water at 5 kb pressure. J Petrol 11:221-251

Mysen BO (1998) Interaction between aqueous fluid and silicate melt in the pressure and temperatures regime of the Earth's crust and upper mantle. Neues Jb Miner Abh 172:227-244

Mysen BO, Boettcher AL (1975) Melting of a hydrous mantle: 1. Phase relations of natural peridotite at high-pressures and temperatures with controlled activities of water, carbon-dioxide, and hydrogen. J Petrol 16:520-548

Nakamura Y (1974) The system SiO_2-H_2O-H_2 at 15 kbar. Carnegie I Wash 73:259-263

Nakamura Y, Kushiro I (1974) Composition of the gas phase in Mg_2SiO_4-SiO_2-H_2O at 15 kbar. Carnegie I Wash 73:255-258

Nekvasil H, Carroll W (1996) Experimental constraints on the compositional evolution of crustal magmas. Journal of the Royal Society of Edinburgh: Earth Sciences 87:139-146

Newton RC, Manning CE (2002) Solubility of enstatite plus forsterite in H_2O at deep crust/upper mantle conditions: 4 to 15 kbar and 700 to 900 °C. Geochim Cosmochim Acta 66:4165-4176

Newton RC, Manning CE (2003) Activity coefficient and polymerization of aqueous silica at 800 °C, 12 kbar, from solubility measurements on SiO_2-buffering mineral assemblages. Contrib Mineral Petrol 146:135-143

Nichols GT, Wyllie PJ, Stern CR (1994) Subduction zone-melting of pelagic sediments constrained by melting experiments. Nature 371:785-788

Niggli P (1912) Die gasmineralisatoren im magma. Z Anorg Chem 75:161-188

Niida K, Green DH (1999) Stability and chemical composition of pargasitic amphibole in MORB pyrolite under upper mantle conditions. Contrib Mineral Petrol 135:18-40

Ohtani E, Mizobata H, Yurimoto H (2000) Stability of dense hydrous magnesium silicate phases in the systems Mg_2SiO_4-H_2O and $MgSiO_3$-H_2O at pressures up to 27 GPa. Phys Chem Mineral 27:533-544

Orlova GP (1964) The solubility of water in albite melts. Int Geol Rev 6:254-258

Paillat O, Elphick SC, Brown WL (1992) The solubility of water in $NaAlSi_3O_8$ melts: a re-examination of Ab-H_2O phase relationships and critical behaviour at high pressures. Contrib Mineral Petrol 112:490-500

Pak TM, Hauzenberger CA, Baumgartner LP (2003) Solubility of the assemblage albite + K-feldspar + andalusite + quartz in supercritical aqueous chloride solutions at 650 °C and 2 kbar. Chem Geol 200:377-393

Panero WR, Davis MK, Boyd O, Stixrude LP (2003) Viscosity of silica-rich water at high pressure and temperature. AGU Fall Meeting abstract #S21E-0362

Pawley AR (1998) The reaction talc + forsterite = enstatite + H_2O: New experimental results and petrological implication. Am Mineral 83:51-57

Peacock SM, Rushmer T, Thompson AB (1994) Partial melting of subducting oceanic crust. Earth Planet Sc Lett 121:227-244

Pedersen KS, Christensen P (2007) Fluids in hydrocarbon basins. Rev Mineral Geochem 65:241-258

Petford N, Cruden AR, McCaffrey KJW, Vigneresse JL (2000) Granite magma formation, transport and emplacement in the Earth's crust. Nature 408:669-673

Philippot P, Rumble D (2000) Fluid-rock interactions during high-pressure and ultrahigh-pressure metamorphism. Int Geol Rev 42:312-327

Poli S, Schmidt MW (2002) Petrology of subducted slabs. Annu Rev Earth Pl Sc 30:207-235

Ragnarsdóttir KV, Walther JV (1983) Pressure sensitive silica geothermometer determined from quartz solubility experiments at 250 °C. Geochim Cosmochim Acta 47:941-946

Ravich MI (1974) Water-Salt systems at elevated temperatures and pressures. Nauka, Moscow. (In Russian)

Ricci JE (1951) The Phase Rule and Heterogeneous Equilibrium. Dover Publications Inc, New York

Richet P, Roux J, Pineau F (1986) Hydrogen isotope fractionation in the system H_2O liquid $NaAlSi_3O_8$: New data and comments on D/H fractionation in hydrothermal experiments. Earth Planet Sci Lett 78:115-120

Rimstidt JD (1997) Quartz solubility at low temperatures. Geochim Cosmochim Acta 61:2553-2558

Rizvi SSH, Benado AL, Zollweg JA, Daniels JA (1986) Supercritical fluid extraction: Fundamental principles and modeling methods. Food Technol-Chicago 40:55-65

Roedder E (1984) Fluid Inclusions. Rev Mineral Vol 12, Mineralogical Society of America

Roedder E (1992) Fluid inclusion evidence for immiscibility in magmatic differentiation. Geochim Cosmochim Acta 56:5-20

Rowlinson JS (1983) Critical and supercritical fluids. Fluid Phase Equilibr 10:135-139

Rowlinson JS (1988) J. D. Van der Waals, On the continuity of the gaseous and the liquid states. Studies in statistical mechanics. Vol XIV. Lebowitz JL (ed) North Holland, Amsterdam

Ryabchikov ID, Schreyer W, Abraham K (1982) Compositions of aqueous fluids in equilibrium with pyroxenes and olivines at mantle pressures and temperatures. Contrib Mineral Petrol 79:80-84

Scambelluri M, Philippot P (2001) Deep fluids in subduction zones. Lithos 55:213-227

Schairer JF, Bowen NL (1956) The system $Na_2O-Al_2O_3-SiO_2$. Am J Sci 254:129-195

Schmidt MW, Vielzeuf D, Auzanneau E (2004) Melting and dissolution of subducting crust at high pressures: the key role of white mica. Earth Planet Sc Lett 228:65-84

Schneider GM (1978) Physicochemical principles of extraction with supercritical gases. Angew Chem Int Edit 17:716-727

Schneider GM (2002) Aqueous solutions at pressures up to 2 GPa: gas-gas equilibria, closed loops, high-pressure immiscibility, salt effects and related phenomena. Phys Chem Chem Phys 4:845-852

Schreinemakers FAH (1915-1925) In-, mono-, and divariant equilibria. K Ned Akad Van Wet 18-28:(29 separate articles in the series)

Semenova AI, Tsiklis DS (1970) Solubility of silicon dioxide in steam at high pressure and temperatures. Russ J Phys Chem+ 44:1420-1422

Sharp ZD, Essene EJ, Hunziker JC (1993) Stable isotope geochemistry and phase equilibria of coesite-bearing whiteschists, Dora-Maira Massif, Western Alps. Contrib Mineral Petrol 114:1-12

Shaw HR (1972) Viscosities of magmatic silicate liquids; an empirical method of prediction. Am J Sci 272:870-893

Shen AH, Keppler H (1997) Direct observation of complete miscibility the albite-H_2O system. Nature 385:710-712

Shmulovich K, Graham C, Yardley B (2001) Quartz, albite and diopside solubilities in $H_2O-NaCl$ and H_2O-CO_2 fluids at 0.5-0.9 GPa. Contrib Mineral Petrol 141:95-108

Siever R (1962) Silica solubility, 0°–200 °C, and the diagenesis of siliceous sediments. J Geol 70:127-150

Smits A (1903) Die Löslichkeitskurve in der kritischen Gegend. Z Elektrochem 9:663-666

Sommerfeld RA (1967) Quartz solution reaction: 400–500 °C, 1000 bars. J Geophys Res 72:4253-4257

Sourirajan S, Kennedy GC (1962) The system $H_2O-NaCl$ at elevated temperatures and pressures. Am J Sci 260:115-141

Sowerby JR, Keppler H (2002) The effect of fluorine, boron and excess sodium on the critical curve in the albite-H_2O system. Contrib Mineral Petrol 143:32-37

Spandler C, Mavrogenes JA, Hermann J (2006) Experimental constraints on element mobility from subducted sediments using high-P synthetic fluid/melt inclusions Chem Geol doi:10.1016/j.chemgeo.2006.10.005:

Stalder R (2004) Comment on K. Mibe, T. Fujii, and A. Yasuda (2002) "Composition of aqueous fluid coexisting with mantle minerals at high pressure and its bearing on the differentiation of the Earth's mantle," Geochimica et Cosmochimica Acta 66:2273-2285. Geochim Cosmochim Acta 68:927-928

Stalder P, Ulmer P, Thompson AB, Gunther D (2000) Experimental approach to constrain second critical end points in fluid/silicate systems: Near-solidus fluids and melts in the system albite-H_2O. Am Mineral 85:68-77

Stalder R, Ulmer P, Thompson AB, Günther D (2001) High pressure fluids in the system $MgO-SiO_2-H_2O$ under upper mantle conditions. Contrib Mineral Petrol 140:607-618

Stein CA, Stein S (1992) A model for the global variation in oceanic depth and heat-flow with lithospheric age. Nature 359:123-129

Stewart DB (1967) Four-phase curve in the system $CaAl_2Si_2O_8-SiO_2-H_2O$ between 1 and 10 kilobars. Schweiz Miner Petrog 47:35-59

Sun WD, Bennett VC, Eggins SM, Kamenetsky VS, Arculus RJ (2003) Enhanced mantle-to-crust Rhenium transfer in undegassed arc magmas. Nature 422:294-297

Sykes ML, Holloway JR (1987) Evolution of granitic magmas during ascent: Constraints from thermodynamics and phase equilibria. Magmatic processes: Physicochemical principles, Special Pub. 1. B. O. Mysen. New York, The Geochemical Society: 447-461.

Takenouchi S, Kennedy GC (1964) Binary system H_2O-CO_2 at high temperatures + pressures. Am J Sci 262:1055-1074

Thomas R, Webster JD, Heinrich W (2000) Melt inclusions in pegmatite quartz: complete miscibility between silicate melts and hydrous fluids at low pressure. Contrib Mineral Petrol 139:394-401

Thompson AB (1988) Dehydration melting of crustal rocks. Rendiconti della societa italiana di mineralogia e petrologia 43:41-60

Thompson AB (1992) Water in the Earth's upper mantle. Nature 358:295-302

Thompson AB (1999) Some time-space relationships for crustal melting and granitic intrusion at various depths. *In*: Understanding Granites: Integrating New and Classical Techniques. Vol 168. Fernández CA, Vigneresse JL (eds) Geological Society, p 7-25

Thompson AB, Aerts M, Hack AC (2007) Liquid immiscibility in silicate melts and related systems. Rev Mineral Geochem 65:99-127

Thompson JBJ (1959) Local equilibrium in metasomatic processes. *In*: Researches in Geochemistry. Abelson PH (eds) John Wiley & Sons, p 427-457

Tödheide K, Frank EU (1963) Das Zweiphasengebiet und die kritische Kurve im System Kohlendioxid-Wasser bis zu Drucken von 3500 bar. Z Phys Chem Neue Fol 37:387-401

Touret JLR, Frezzotti ML (2003) Fluid inclusions in high pressure metamorphic rocks. *In*: Ultrahigh Pressure Metamorphism. EMU Notes in Mineralogy, 5. Carswell DA, Compagnoni R (eds) European Mineralogical Union, p 467-487

Trommsdorff V, Skippen G (1986) Vapor loss (boiling) as a mechanism for fluid evolution in metamorphic rocks. Contrib Mineral Petrol 94:317-322

Tuttle OF, Bowen NL (1958) The origin of granite in light of experimental studies in the system $NaAlSi_3O_8$-$KAlSi_3O_8$-SiO_2-H_2O. Geol Soc Am Bull, Memoir 74:153

Tuttle OF, England JL (1955) Preliminary report on the system SiO_2-H_2O. Geol Soc Am Bull 66:149-152

Tuttle OF, Friedman II (1948) Liquid immiscibility in the system H_2O-Na_2O-SiO_2. J Am Chem Soc 70:919-926

Urusova MA, Valyashko VM (2001) Solubility and immiscibility behavior in ternary hydrothermal systems with critical phenomena in saturated solutions. High Pressure Res 20:447-455

Valyashko VM (1990a) Phase equilibria and properties of hydrothermal systems. Nauka, Moscow. (In Russian)

Valyashko VM (1990b) Subcritical and supercritical equilibria in aqueous-electrolyte solutions. Pure Appl Chem 62:2129-2138

Valyashko VM (1995) Strategy, Methods and equipment for experimental studies of water-salt systems under superambient conditions. Pure Appl Chem 67:569-578

Valyashko VM (1997) Phase behavior in binary and ternary water-salt systems at high temperatures and pressures. Pure Appl Chem 69:2271-2280

Valyashko VM (2002a) Fluid phase diagrams of ternary systems with one volatile component and immiscibility in two of the constituent binary mixtures. Phys Chem Chem Phys 4:1178-1189

Valyashko VM (2002b) Derivation of complete phase diagrams for ternary systems with immiscibility phenomena and solid-fluid equilibria. Pure Appl Chem 74:1871-1884

Valyashko VM, Churagulov BR (2003) Solubilities of salt-water systems at high temperatures and pressures. *In*: Wiley Series in Solution Chemistry: The Experimental Determination of Solubilities. Vol 6. Hefter GT, Tomkins RPT (eds) John Wiley & Sons, p 359-435

Valyashko VM, Kravchuk KG (1978) P-T-X parameters of critical phenomena in solutions of the system SiO_2-Na_2O-H_2O. Dokl Akad Nauk SSSR 242:1104-1107 (In Russian)

Valyashko VM, Urusova MA (2003) Solubility behavior in ternary water-salt systems under sub- and supercritical conditions. Monatsh Chem 134:679-692

Van der Waals JD (1890) Molekulartheorie eines Körpers, der aus zwei verschiedenen Stoffen besteht. Z Phys Chem 5:133-173

Van Konynenburg PH, Scott RL (1980) Critical lines and phase equilibria in binary Van der Waals mixtures. Philos Trans Royal Soc London A 298:495-540

Van Laar JJ (1905) On the shape of the plait point curves for mixtures of normal substances. (second communication). Proc Kon Akad Ned 8:33-48

van Lier JA, Debruyn PL, Overbeek JTG (1960) The solubility of quartz. J Phys Chem 64:1675-1682

van Nieuwenburg CJ, Van Zon PM (1935) Semi-quantitative measurements of the solubility of quartz in super-critical steam. Recl Trav Chim Pay-B 54:129-132

Veksler IV (2004) Liquid immiscibility and its role at the magmatic-hydrothermal transition: a summary of experimental studies. Chem Geol 210:7-31

Veksler IV, Thomas R, Schmidt C (2002) Experimental evidence of three coexisting immiscible fluids in synthetic granite pegmatite. Am Mineral 87:775-779

Vidale R (1983) Pore solution compositions in a pelitic system at high pressures and salinities. Am J Sci 283A:298-313

Voight DE, Bodnar RJ, Blencoe JG (1981) Water solubility in melts of alkali feldspar composition at 5 kb, 950 °C. EOS 62:428

Volosov AG, Khodakov IL, Ryzhenko BN (1972) Equilibrium in system SiO_2-H_2O at elevated temperatures along the lower three-phase curve. Geochem Int 9:362-377

Walther JV, Orville PM (1983) The extraction quench technique for determination of the thermodynamic properties of solute complexes: Application to quartz solubility in fluid mixtures. Am Mineral 68:731-741

Wang HM, Henderson GS, Brenan JM (2004) Measuring quartz solubility by *in situ* weight-loss determination using a hydrothermal diamond cell. Geochim Cosmochim Acta 68:5197-5204

Wasserburg GJ (1958) The solubility of quartz in supercritical water as a function of pressure. J Geol 66:559-578

Webster JD (1997) Exsolution of magmatic volatile phases from Cl-enriched mineralizing granitic magmas and implications for ore metal transport. Geochim Cosmochim Acta 61:1017-1029

Webster J, Mandeville C (2007) Fluid immiscibility in volcanic environments. Rev Mineral Geochem 65:313-362

Weill DF, Fyfe WS (1964) The solubility of quartz in H_2O in the range 1000-4000 bars and 400-550 °C. Geochim Cosmochim Acta 28:1243-1255

Weinberg RF (1996) Ascent mechanisms of felsic magmas: news and views. Trans Royal Soc Edin-Earth 87:95-103

Weingärtner H, Franck EU (2005) Supercritical water as a solvent. Angew Chem Int Edit 44:2672-2692

Wenzel J, Limbach U, Bresonik G, Schneider GM (1980) Kinetics of phase-separation in binary-liquid mixtures. J Phys Chem 84:1991-1995

Wood JA, Jr. (1958) The solubility of quartz in water at high temperatures and pressures. Am J Sci 256:40-47

Woodland AB, Walther JV (1987) Experimental determination of the solubility of the assemblage paragonite, albite, and quartz in supercritical H_2O. Geochim Cosmochim Acta 51:365-372

Wyllie PJ, Tuttle OF (1960) The system CaO-CO_2-H_2O and the origin of carbonatites. J Petrol 1:1-46

Wyllie PJ, Tuttle OF (1964) Experimental investigation of silicate systems containing 2 volatile components - Part 3 - Effects of SO_3, P_2O_5, HCl and Li_2O in addition to H_2O on melting temperature of albite + granite. Am J Sci 262:930-935

Yamada A, Inoue T, Irifune T (2004) Melting of enstatite from 13 to 18 GPa under hydrous conditions. Phys Earth Planet In 147:45-56

Yoder HS, Jr. (1958) Effect of water on the melting of silicates. Carnegie I Wash 57:189-191

Young CL (1986) Phase equilibria in fluid mixtures at high pressures. Pure Appl Chem 58:1561-1572

Zen EA (1966) Construction of pressure-temperature diagrams for multi-component systems after the method of Schreinemakers - a geometrich approach. Geol Soc Am Bull 1225:p56

Zernike J (1955) Chemical Phase Theory – A Comprehensive Treatise on the Deduction, the Applications and the Limitations of the Phase Rule. Kluwer

Zhang YG, Frantz JD (2000) Enstatite-forsterite-water equilibria at elevated temperatures and pressures. Am Mineral 85:918-925

Zotov N, Keppler H (2000) *In situ* Raman spectra of dissolved silica species in aqueous fluids to 900 °C and 14 kbar. Am Mineral 85:600-603

Reviews in Mineralogy & Geochemistry
Vol. 65, pp. 187-212, 2007
Copyright © Mineralogical Society of America

6

Numerical Simulation of Multiphase Fluid Flow in Hydrothermal Systems

Thomas Driesner

Institute of Isotope Geochemistry and Mineral Resources
ETH Zurich
ETH Zentrum NW, 8092 Zurich, Switzerland
thomas.driesner@erdw.ethz.ch

Sebastian Geiger

Institute of Petroleum Engineering
and
Edinburgh Collaborative of Subsurface Science and Engineering (ECOSSE)
Heriot-Watt University
Edinburgh, EH14 4AS, United Kingdom
sebastian.geiger@pet.hw.ac.uk

INTRODUCTION

Ore-forming hydrothermal systems often show signs of boiling. In certain classes of deposits—such as low-sulfidation epithermal Au-Ag veins—the chemical consequences of boiling are often considered the cause of ore mineral precipitation, while in others the role of boiling may rather lie in enforcing a favorable pressure-temperature regime. In yet other cases it may, however, be just a passive record of the conditions during ore formation, with little or no direct causal relation to the actual metal enrichment.

A quantitative understanding of the occurrence, the spatial and temporal extent, and the dynamic evolution of boiling zones in hydrothermal systems on a generic level, as well as for specific systems, can help in determining the role of boiling and is thus a potentially valuable tool in economic geology. However, while numerical simulation of the simultaneous flow of coexisting fluid phases (water, oil and gas) is a routinely used key tool in hydrocarbon exploration and reservoir management (Gerritsen and Durlofsky 2005), it has not yet found a similar position in the study and exploration of mineral resources. This is at least partly due to the fact that the geological context of hydrothermal mineral deposits is more difficult to resolve with geophysical methods and, hence, the overall geometry for which simulation of fluid flow should be carried out, is often less clear.

Nevertheless, proper simulation methods have emerged rather recently and have found some application in the hydrothermal regime. Simulations are now a tool to understand and manage individual geothermal fields. O'Sullivan et al. (2001) provide a recent review and Pruess (1990) gives a summary of the basic simulation methods. So-called high-enthalpy geothermal systems often operate under conditions of boiling and are considered to be at least hydrologically equivalent to epithermal hydrothermal systems (Hedenquist et al. 1992). However, studies that have specifically investigated the systematics behind fluid-fluid coexistence (in particular vapor+liquid) in high-temperature hydrothermal systems are scarce. With the recent development of new algorithms for multiphase fluid flow and equation of state models covering wide ranges of temperature (T), pressure (P), and composition (X), new studies

1529-6466/07/0065-0006$05.00   DOI: 10.2138/rmg.2007.65.6

with unprecedented physical and geological realism are emerging, many of which have only been published as conference contributions. "Multiphase" here and in the following refers to the simultaneous flow of multiple fluid phases through the same volume of permeable rock and does NOT refer to several temporal phases of fluid flow.

This paper deliberately reviews only the small number of systematic numerical studies of system-scale multiphase aqueous fluid flow in mostly magmatic-hydrothermal settings. These papers will be discussed in some detail to outline the fundamental factors influencing the occurrence, extent and dynamics of regions of fluid-fluid immiscibility. In particular, we will rely on the paper by Hayba and Ingebritsen (1997) with additional insights from Hurwitz et al. (2003) to emphasize the outstanding role of large-scale host rock permeability on the emergence, structure and dynamics of boiling hydrothermal systems. How simulation results can be used in conjunction with fluid inclusion data to quantitatively reconstruct the evolution of fossil ore-forming hydrothermal systems will be shown using an example by Kostova et al. (2004). Finally, we will present results from some reconnaissance studies by Geiger et al. (2005b) on multiphase flow of saline fluids in magmatic-hydrothermal systems that illustrate a potential direction of research in the next few years.

The studies that we review are restricted to flow of fluids that are either pure H_2O or H_2O-NaCl. Multiphase flow of other fluid systems such as H_2O-CO_2, H_2O-NaCl-CO_2, is likely to be important in other types of ore-forming systems such as orogenic gold lodes, and even in deep-seated magmatic systems (e.g., forming iron-oxide-copper-gold deposits) with complex fluids there is evidence that abrupt physical changes may induce phase separation (Oliver et al. 2006). Due to the lack of adequate flow algorithms and equations of state, no systematic numerical studies of phase separation in these systems have been carried out so far.

METHODS

Multiphase fluid flow can be described by a set of coupled non-linear equations, the application of which typically acts on fluids in a complex geometry of porous rocks and fractures. Since analytical solutions can only be derived for idealized systems (e.g., Young 1993), numerical simulation is usually the tool of choice. In the following a brief review of the governing equations and a short introduction into how they are solved is given. For more detailed information, a number of papers are available (Faust and Mercer 1979a,b; Huyakorn and Pinder 1983; Pruess 1991; Hayba and Ingebritsen 1994; Geiger et al. 2006a).

Model discretization

In principle, the governing equations have to be solved at every point in the system of interest. Of course, this infinite number of points cannot be achieved with a computer's finite memory and computational speed. Therefore, one "discretizes" the geological model in both space and time by solving the equations only at points separated from each other by some distance and for successive time steps of finite duration.

Figure 1 shows a simple geological model with a fault in a matrix. This geometry has been discretized using a program called "Triangle" (Shewchuk 1996; the program is publicly available from various internet sources), which constructs a set of triangles that share corner points. The size of the triangles is automatically adjusted to capture the spatial extent of the two units "matrix" and "fault" in the model. If one were simulating fluid flow in this discretization, the equations would be solved at the nodes (i.e., the corner points) of the triangles and/or at integration points (i.e., points within the triangle). Irregular sets of triangles in principle offer greatest geometric flexibility. However, most computer codes used in hydrothermal simulations use rectangular grids because this allows a more straightforward implementation for the respective solution algorithms.

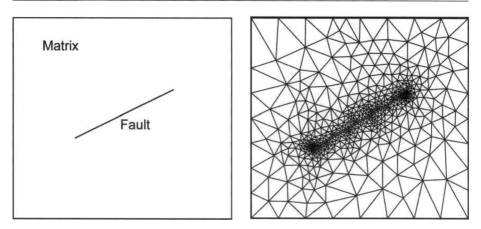

Figure 1. Discretization of a simple model with triangular elements. Modified from Geiger et al. (2004).

Mass conservation equations

Probably the most basic equation is that enforcing mass conservation. For a simple groundwater model with liquid as the only fluid phase, and assuming constant density of liquid water and constant porosity, the mass conservation for fluid in a given volume of porous rock can be expressed as

$$\phi \frac{\partial S_l}{\partial t} = -\nabla \cdot v_l + q_l \tag{1}$$

Here, ϕ is the porosity of the rock, S_l the saturation (volume fraction) of liquid (indicated by subscript l) in the pore space, t denotes time, v_l is the Darcy velocity with which the fluid flows and q_l is a source term that accounts for any fluid produced or consumed in the volume, e.g., by pumping or chemical reactions. The (constant) fluid density does not explicitly appear in this equation since it is present on both sides and simply cancels out.

If the density of fluid and the porosity are allowed to vary, Equation (1) becomes

$$\frac{\partial}{\partial t}(\phi \rho_l S_l) = -\nabla \cdot (\rho_l v_l) + q_l \tag{2}$$

with ρ_l denoting density of the liquid. If both liquid and vapor (subscripts l and v) are present (e.g., in boiling hydrothermal systems), the equation changes to:

$$\frac{\partial}{\partial t}\left[\phi(\rho_l S_l + \rho_v S_v)\right] = -\nabla \cdot (\rho_l v_l) - \nabla \cdot (\rho_v v_v) + q_l + q_v \tag{3}$$

Momentum conservation

Momentum conservation is typically expressed by an appropriate version of Darcy's law. In the simplest case of an incompressible, single phase fluid without gravity:

$$v_l = -\frac{k}{\mu} \nabla P \tag{4}$$

Here, k and μ are material properties, k being the rock's intrinsic permeability (with units of m^2; note that 1 Darcy, D, equals 10^{-12} m^2) and μ being the fluid's dynamic viscosity (with units of Pa s). P is pressure and has to be replaced by a property called hydraulic head if gravity is taken into account (for a more detailed discussion, see, e.g., Ingebritsen et al. 2006).

In the case of a flowing two-phase fluid mixture, and when density and gravity effects are included, one usually computes the velocities of the individual phases i (i.e., l or v) as

$$v_i = -k\frac{k_{ri}}{\mu_i}\left[\nabla P - \rho_i g\right] \tag{5}$$

Here, k_r is the so-called relative permeability that accounts for the fact that in a multiphase mixture the resistance to flow of the individual phases depends on their relative amounts in the pore space, which in turn depends on their wetting properties, temperature, pressure, etc., and g is the gravitational constant. These Darcy equations for the individual phases can then be inserted for the velocities v_i in the mass conservation Equation (3).

Energy conservation equation

Changes in temperature result from heat being conducted and advected into and out of the volume of interest and being absorbed/liberated by phase changes. This is accounted for by the energy conservation equation, which is typically written in terms of the specific enthalpy, h_l, of the fluid. For a single phase incompressible fluid, the conservation of energy equation looks very similar to the mass conservation equation except that heat is exchanged with the rock and so the specific enthalpy of the rock, h_r, has to be included. Assuming that the time for thermal equilibration between the rock matrix and fluid in the pores is short compared to a typical time step in a simulation, one arrives at:

$$\frac{\partial}{\partial t}\left[(1-\phi)h_r + \phi S_l h_l\right] = -\nabla \cdot v_l h_l + \nabla \cdot (K\nabla T) + Q \tag{6}$$

where K is the rock's thermal diffusivity and Q is a source term for thermal energy. If only a low-temperature liquid is present in the pores, this can be simplified to show the temperature evolution with time using the heat capacity. This is because the isobaric heat capacities, defined as $c_P = (\partial h/\partial T)_P$, of liquid and rock are essentially constant at low temperatures. Then, the thermal evolution is described by

$$\left[(1-\phi)c_{P,r} + \phi c_{P,l}\right]\frac{\partial T}{\partial t} = -\nabla \cdot v_l h_l + \nabla \cdot (K\nabla T) + Q \tag{7}$$

Solving this equation for the temperature change over a time step can in principle be done if the right hand side is computed and then divided by the term in square brackets from the left hand side. It would then be multiplied with the current simulation time step Δt as an approximation to the infinitesimally small ∂t in the equation, to arrive at the temperature change ΔT.

The situation becomes more complicated if boiling, condensation, and density variations are considered. Then, the expression becomes

$$\frac{\partial}{\partial t}\left[(1-\phi)\rho_r h_r + \phi\left(S_l \rho_l h_l + S_v \rho_v h_v\right)\right] = -\nabla \cdot (\rho_l v_l h_l) - \nabla \cdot (\rho_v v_v h_v) + \nabla \cdot (K\nabla T) + Q \tag{8}$$

A comprehensive discussion of this equation and some additional terms to be considered in certain cases is provided by Faust and Mercer (1979a) and Huyakorn and Pinder (1983). In this equation, the partial derivatives on the left hand side can no longer be resolved in a simple fashion like in Equation (7) since the saturations on the left hand side are coupled to the right hand side and cross-terms would arise. Also, the temperature derivatives become strongly temperature-dependent, can diverge at the boiling curve of water, and are coupled to each other (e.g., $\partial S/\partial T$ is strongly coupled to $\partial H/\partial T$). Solving for ΔT therefore requires iterative methods to find a solution.

Computation of pressure changes

The pressure change over a time step can be computed in various ways. Several codes, such as TOUGH2 (Pruess 1991) and HYDROTHERM (Hayba and Ingebritsen 1994) use an approach in which various derivatives of the above expressions are used in iterative algorithms to find a solution which satisfies all governing equations. Another approach is to decouple the computation of mass transport, temperature and pressure change (Geiger et al. 2006a). Then, the pressure computation can be derived from the definition of compressibility, β:

$$\beta = \frac{1}{\rho} \frac{\partial \rho}{\partial P} \tag{9}$$

which can be re-arranged to give

$$\partial P = \frac{1}{\rho \beta} \partial \rho \tag{10}$$

which for finite changes, Δ, rather than infinitesimal changes, ∂, can be expressed as

$$\Delta P = \frac{1}{\rho \beta} \Delta \rho \tag{11}$$

Hence, the pressure change is a response to the change in density. The latter is a response to a change in mass due to fluid being advected into and out of the volume, and to density changes due to changes in temperature and/or salinity. This leads to the general description (assuming that the rock matrix is incompressible and no changes due to mineral reactions occur)

$$\phi \rho_f \beta_f \frac{\partial P}{\partial t} = \nabla \cdot \left[k \left(\frac{k_{r,l}}{\mu_l} \rho_l + \frac{k_{r,v}}{\mu_v} \rho_v \right) \nabla P \right] + k \left(\frac{k_{r,l}}{\mu_l} \rho_l^2 + \frac{k_{r,v}}{\mu_v} \rho_v^2 \right) g \nabla z + \phi \left(\alpha_f \frac{\partial T}{\partial t} - \gamma_f \frac{\partial X}{\partial t} \right) \tag{12}$$

Notice that $\rho_f = S_l \rho_l + S_v \rho_v$ is the phase-averaged fluid density, and β_f and α_f are the respective compressibility and thermal expansivity for this mixture. γ_f is a term that Geiger et al. (2006a) named "chemical expansivity" and accounts for the density change with salinity X. Since vapor can condense into liquid upon pressurization, liquid evaporate upon depressurization and latent heat is released or consumed, these are NOT simply the compressibilities or expansivities of the individual phases in the pore space weighted by their relative proportions. A detailed derivation for the compressibility of a liquid-vapor water mixture was given by Grant and Sorey (1979). They showed that the proper compressibility is orders of magnitude higher that the weighted compressibilities of the individual phases, and even much higher than the pure vapor compressibility.

Solving the equations

The governing equation for heat and mass transport discussed above can only be solved linearised at the discrete points which represent the geological model in time and space. These discrete points are usually generated with finite difference (Faust and Mercer 1979b), finite element (Huyakorn and Pinder 1983), integrated finite difference (Narasimhan and Witherspoon 1976), or a combination of finite element and finite volume methods (Geiger et al. 2006a). Each method has distinct advantages and disadvantages regarding its computational efficiency and flexibility to represent geometrically complex geological structures. These are discussed in detail by Geiger et al. (2006a). The resulting discretized equations form a system of (possibly coupled) algebraic equations. In addition, boundary conditions (e.g., a fixed pressure and temperature at the land surface) and initial conditions (e.g., a hydrostatic pressure distribution and geothermal temperature gradient) are needed to solve the system of algebraic equations. Discretized equations and boundary conditions can then be written in matrix form as $\mathcal{A} \cdot \mathbf{x} = \mathbf{b}$.

$\mathcal{A}$ is a sparse matrix containing the system of discretized algebraic equations, $\mathbf{x}$ is the solution vector (e.g., for fluid pressure and/or temperature), and $\mathbf{b}$ a right-hand side vector containing the boundary and initial conditions. Matrix $\mathcal{A}$ must then be inverted at each time step to obtain the solution $\mathbf{x}$ for the given boundary and initial conditions $\mathbf{b}$. Note that once the simulation has evolved in time, the initial conditions are replaced by the respective results from the previous time step. It is obvious that a more efficient matrix solution method will allow us to invert a matrix $\mathcal{A}$ with more entries, meaning a finer spatial and/or temporal discretization of the geological model. The most efficient matrix solvers that are currently available are so-called algebraic multigrid solvers. They do not need any information on the geometry of the problem and the computation time scales linearly with the size of matrix $\mathcal{A}$ (Stueben 2001).

PERMEABILITY AND THERMAL EVOLUTION OF HYDROTHERMAL SYSTEMS

Important processes that drive fluid flow in a magmatic-hydrothermal system are buoyancy, thermal pressurization, liberation of magmatic fluids, and fluid release from contact metamorphism. The most important physical factors determining the evolution of a hydrothermal system are the magmatic heat supply and the permeability structure (Cathles 1977; Norton and Knight 1977; Hanson 1995, 1996; Hayba and Ingebritsen 1997; Hurwitz et al. 2003).

Host rock permeability determines whether heat transport around a cooling intrusion will be dominantly conductive (low permeability) or advective (moderate and high permeability). The boundary between the two regimes lies somewhere near a permeability of 10^{-16} m^2 (0.1 md) (Cathles 1977, 1990; Norton and Knight 1977; Hayba and Ingebritsen 1997; Hurwitz et al. 2003). As a rule of thumb, low permeability host rocks are found at depths >5 km. Such intrusions produce contact aureoles under pressures that may reach lithostatic, the overall amount of fluid flow is rather small and directed outward from the intrusion, and time scales are long (10^4 to 10^5 yr; Hanson 1995, 1996). Shallow intrusions typically invade rather permeable host rocks, drive convection of groundwater in the host rock with large amounts of fluid (rule of thumb: approximately 1:1 volume of water : volume of pluton; see Eqn. 1 of Cathles 1983) with a strong flow towards and above the intrusion. Hydrothermal alteration zones are well-developed and classical contact aureoles are absent. Pressures are near-hydrostatic and time scales are short (10^3 to 10^4 yr) (Cathles 1977; Norton and Knight 1977; Hayba and Ingebritsen 1997).

Epithermal mineral deposits in the more distal parts of magmatic-hydrothermal systems often show signs of fluid-fluid coexistence ("boiling," see Hedenquist et al. 1992 for a summary) that formed in zones of advective heat and mass transport. Hence, the overall permeability was likely at least moderate, which agrees with the empirical observation that many formed in extensional tectonic settings. As will be shown below, only a very narrow range of system-scale permeabilities allows the development of extensive boiling zones, and the vertical interval over which boiling occurs is a severe constraint on system-scale permeability at the time of formation.

In the proximal part of magmatic-hydrothermal systems near the cooling magma, permeability issues appear to be somewhat different. There is ample evidence, for example, that the mineralization of porphyry copper type deposits formed from boiling fluids released from the magma (see Heinrich 2007). While the magma itself can be considered impermeable for convecting external fluids, exsolved magmatic fluids will be under near-lithostatic pressure, and can rise as bubbles in the magma and probably collect at the highest point of the system, which is often a cupola or stock rising out of the magma chamber. The high fluid pressure can lead to rock failure in the crystallized carapace and create permeability by hydrofracturing. The magmatic fluids that are released move through a steep pressure-temperature gradient through this fracture network and deposit minerals. The transition from lithostatic to hydrostatic pressure and a change in temperature of several hundred degrees typically occurs over distances of tens to hundreds of meters. The influence of convecting meteoric waters appears to be restricted to later stages in the system development. The extent to which the actual convection of the meteoric water plays a role in the mineralization has yet to be solved in a quantitative manner, and numerical simulations are likely to play a key role in solving this problem.

Permeability, discharge, recharge and efficiency of heat transfer

Heat transfer. There is a general agreement that the permeability of a given rock will decrease substantially once temperatures between ca. 350 °C and 400 °C are reached (Cathles 1983; Fournier 1999). In this temperature interval, rocks start to be ductile and fluid pathways tend to close. One should, however, be aware that fluid overpressures, certain deformation styles, mineral reactions, etc., may still be able to create transient permeability. In a magmatic body at temperatures above ~400 °C, permeability can be considered low. Heat transport will be dominated by conduction in the already crystallized outer zones while magma convection is restricted to the inner, molten parts. This means that any fluid flow that cools the heat source will occur in the outermost regions where temperatures are below ~400 °C. Heat from the hotter inner part will be transported to the fluid flow zone by conduction in the hot solidified magma (Fig. 2).

How hot a fluid flowing can get is therefore a question of how fast the heat can be transported from within the heat source to the permeability boundary along which fluid flow takes place relative to the velocity with which the fluid is flowing. A number of studies have dealt with various aspects of this theme (e.g., Cathles 1993; Jupp and Schultz 2000, 2004). The basic principle is rather simple: if the fluid flows too fast, it will stay relatively cold since the heat supply cannot keep pace; if the fluid flows very slowly, heat transfer to the fluid will be optimized. The temperature at which the fluid leaves the source does not, however, tell how hot it will remain

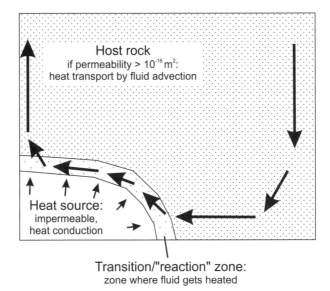

Figure 2. Heat transfer around an impermeable heat source in a porous, permeable rock. Within the heat source (e.g., a magma body), heat transfer is by conduction and slow. Convecting fluids pick up heat in the "reaction zone" where the already cooled material has become permeable at temperatures below ca. 400 °C.

in the upflow zone of a hydrothermal cell. Here just the opposite is true: if fluid flow is fast, it loses little heat by conduction. If fluid flow is slow, it will tend to lose heat and the temperature distribution will approach the local geotherm. Figure 3 gives an illustration of this principle.

Driving force and flow velocity. For a given host rock permeability structure and in the absence of significant topography, the flow velocity is a function of how hot the rising fluid column is, because the main driving force for hydrothermal circulation is the buoyancy force resulting from the expansion of fluid upon heating. More quantitatively, the driving force will be the difference between the cold hydrostatic head in the downflow/recharge zone and the hot hydrostatic head in the upflow zone divided by the distance between the two head/pressure regimes (which is of course not a strictly defined distance due to some horizontal spreading of both down- and upflow zones). The hotter the fluid in the upflow zone, the smaller its density

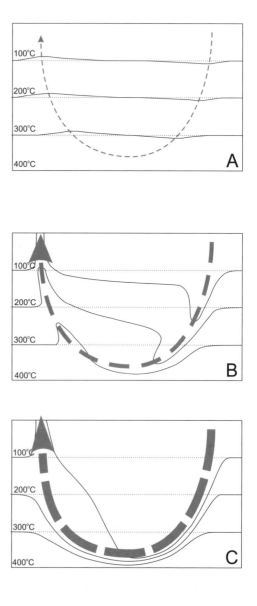

Figure 3. Thermal structure of a highly generalized convection cell. The lower boundary is assumed to represent the natural permeable/impermeable barrier near 400 °C. Heat is introduced from below by conduction. All figures are schematic representations of the hottest phase of convection. A) Very low flow velocity. Advective heat transport is negligible and the regularly spaced isotherms of a conductive temperature profile see only minor local disturbance. In the downflow zones they show slight depressions, in upflow zones they show slight bulges. In a hydrothermal setting, this would be typical for permeabilities near the conductive/advective boundary, i.e., near 10^{-16} m². B) Intermediate flow velocity. Disturbance of initial isotherms (dotted) is now much more pronounced. Fluid velocity is too high to allow static thermal equilibration. Heat loss in the upflow path is limited and a significant hydrothermal anomaly develops. Flow is slow enough for significant fluid heating to occur at depth. Typical of hydrothermal systems near 10^{-15} m² bulk permeability. C) High fluid velocity. Conductive heating at the bottom cannot keep pace with the velocity with which heat is adveted away by the fluid and the fluid never becomes really hot. Even in the upflow zone, fluid temperatures are moderate, although little heat is lost to the country rock.

(and hence value of the hot hydrostatic head) and the faster the volumetric fluid flow.

Material properties. The fluid velocity will also be controlled by the permeability of the rock type through which the fluid is flowing. According to Darcy's law, the permeability of the rock will control the velocity, for a given gradient in hydrostatic head. The velocity is also inversely proportional to the fluid's dynamic viscosity, which in turn is also a function of temperature (generally, viscosity will decrease with increasing temperature; however, low density water shows in increase in viscosity with temperature above the critical point).

Self-organization of thermal structure. The positive dependence of fluid velocity on permeability, on the inverse of fluid viscosity, and on the gradient in hydrostatic head (and therefore on fluid temperature) on the one hand, and the negative dependence of fluid temperature on the fluid velocity on the other hand, will necessarily counteract and lead to a "self-control" of fluid velocity. As viscosity is a material property of the fluid, the "geological" factors that control the temperature are the permeabilities (potentially a function of temperature as well) and thermal conductivities of the rocks involved. The variations in conductivity between various rock types are small compared to the potential orders-of-magnitude variations in permeability. Hence, the first-order parameter controlling the thermal structure and evolution of a hydrothermal system around a heat source will be the host rock permeability.

Thermal evolution patterns above a cooling pluton

The dominant role of permeability has been emphasized in essentially all numerical studies that investigated the cooling of plutons by hydrothermal fluid flow. Exploring parameter space by numerical simulation shows that the thermal evolution of hydrothermal systems above a cooling pluton follows a rather simple pattern, and these findings suggest that careful fluid inclusion studies could allow reconstruction of the large-scale permeability of a system.

General pattern from porous media simulations. The general thermal evolution due to fluid flow around a cooling pluton was established 30 years ago in the pioneering studies by Cathles (1977) and Norton and Knight (1977). Due to simplifications in the simulation algorithms, boiling was only modeled as a sharp liquid to vapor transition, and this may have contributed to the occasional misconception of sharp boiling fronts in epithermal Au-Ag deposits. This changed when Hayba and Ingebritsen (1997) included a realistic description of two-phase (liquid + vapor) flow. The general geometry of their model is similar to that of Cathles (1977) and comprises a small pluton of 2 km height and 750 m half-width (Fig. 4).

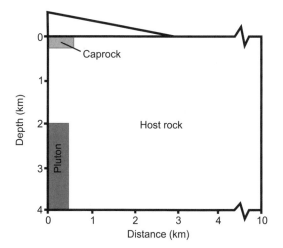

Figure 4. General model geometry used for numerical simulation of cooling plutons. Host rock permeability was usually taken as isotropic and homogenous. Cathles (1977) used the same permeability for pluton and host rock, Norton and Knight (1977) used an essentially impermeable pluton, and applied a temperature dependent permeability function to the pluton. The bottom boundary is typically assigned a constant heat flux and the left side is a no flow boundary. The top is usually a constant temperature and/or pressure boundary. Topography and caprock were studied by Hayba and Ingebritsen (1997).

Besides the two-phase treatment, Hayba and Ingebritsen (1997) included a temperature-dependent permeability for the igneous body to mimic the ductile to brittle transition between 360 °C and 400 °C, and the pluton was essentially impermeable above 400 °C. The host rock was treated as a homogenous porous medium with isotropic permeability. For these conditions, the thermal evolution is essentially a function of host rock permeability only. In Figure 5, we show how the temperature-depth distribution evolves with time in such hydrothermal systems along the upward flow path above the pluton in Figure 4 for three different host rock permeabilities. Figure 6 shows the respective flow patterns during the hottest phase of these systems.

In Figures 5A and 6A, for a host rock permeability of 10^{-14} m^2 (10 mD), the first obvious result is that the system does not reach a temperature higher than 250 °C. The evolution of the system starts with advectively dominated transport of heat into the rocks overlying the pluton and rapidly leads to a temperature profile that is largely depth-independent between 1500 m and ca. 400 m. At shallower depths, the temperature profile follows the "boiling curve with depth." The thermal plume has a narrow shape with some spreading at the top (Fig. 6A).

At 10^{-15} m^2 (1 mD) host rock permeability, the hottest system develops (Figs. 5B and 6B). After an initial phase of heating up the rock matrix in the upflow zone, the temperature-depth profile follows the water boiling curve down to almost 2 km depth. During the hottest stage, the plume is wider at the bottom and more focused at the top (Fig. 6B) when compared to the 10^{-14} m^2 case. The "boiling curve with depth" thermal profile is sustained until the plutonic heat source is depleted. Then, the system starts to cool from below, and the boiling zone is restricted to shallower and shallower depths.

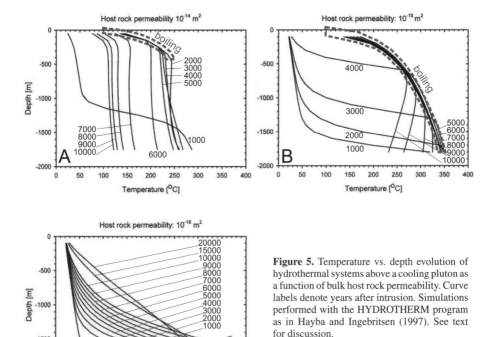

Figure 5. Temperature vs. depth evolution of hydrothermal systems above a cooling pluton as a function of bulk host rock permeability. Curve labels denote years after intrusion. Simulations performed with the HYDROTHERM program as in Hayba and Ingebritsen (1997). See text for discussion.

For a host rock permeability of 10^{-16} m^2 (0.1 mD), the advective contribution to heat transport is substantially reduced and the conductive component important. Accordingly, a thermal aureole develops, rather than a plume (Fig. 6C). The upflow zone is much more slowly heated (Fig. 5C), and with time the temperature becomes almost a linear function of depth. A straight line would be the steady state solution for purely conductive heat transport, and the slight curvature is the expression of a small advective component combined with the downward retreat of the isotherms.

The patterns in the three cases are very different, potentially allowing them to be distinguished where a fossil system is accessible over a sufficiently large vertical interval. However, hydrothermal systems—and in particular ore-forming systems—frequently use fractures as main fluid pathways and the question if flow in discrete fractures may significantly alter this conclusion has not yet sufficiently been studied. A first application to a vein-type Pb-Zn deposit will be discussed later and probably indicates that fracture flow does not alter this general picture significantly.

Influence of an impermeable cap rock. Many geothermal systems have a low-permeability cap rock at shallow depth. Simulations by Hayba and Ingebritsen (1997) show that this cap rock acts as a hydrological barrier that can either suppress or enhance the development of vertically extensive boiling zones. The statement in Hayba and Ingebritsen (1997) that this may lead to errors for pressure estimates in epithermal environments is somewhat ambiguous. If the pressure estimates are actually based on fluid inclusion boiling assemblages, the pressure determination is exact. In that case, a now eroded cap rock in a fossil system would lead to a wrong depth estimate. If, however, fluid inclusions do not indicate boiling, than the necessary pressure correction is unknown unless there is geological information on the paleo-depth.

Effect of topography. Ore-forming epithermal systems are often linked to supra-subduction magmatism which in turn typically produces large composite volcanoes with heights reaching 3 km above their surroundings. Groundwater in the volcano overlying a deeper subvolcanic intrusion can have considerable influence on the hydrothermal system evolving at depth.

Hayba and Ingebritsen (1997) simulated the effects of a rather moderate topographic elevation (mimicking a volcano) of 0.55 km height centered above the intrusion with a slope of

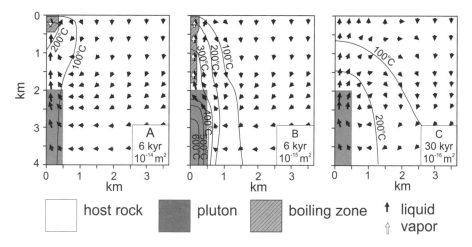

Figure 6. Flow patterns, thermal structure, and boiling zones during the hottest phases of hydrothermal systems with bulk host rock permeabilities of 10^{-14}, 10^{-15}, and 10^{-16} m^2 (left to right). Modified from Hayba and Ingebritsen (1997). Compare Figure 5.

20°. This was sufficient to suppress the hydrothermal plume reaching the top of the volcano. The downward directed flow causes the hot hydrothermal plume to escape laterally towards the base of the volcano. The development and dimensions of a boiling zone are a strong function of whether a low-permeability cap rock at the base of the volcano was present (Fig. 7B,C). Without a cap rock (Fig. 7C), boiling is restricted to a rather small zone near the pluton at temperatures higher than 300 °C. With the cap rock (Fig. 7B), a rather wide boiling zone develops between the pluton and the cap rock with high degrees of vapor saturations developing immediately below the cap rock. Due to the effects of relative permeability, the liquid phase becomes essentially stationary in this area and may locally even show a substantial downflow component, i.e., parts of the boiling zone immediately below the cap rock may act as a heat pipe.

A much more comprehensive study of the effect of topography has been presented by Hurwitz et al. (2003). Elevation in their model was significantly higher (1.5 km height with a 30° slope) and the magmatic heat source wider (up to 3 km) and deeper (5 km beneath the "volcano" top). They also investigated the effects of various permeability structures including simple anisotropic permeability models and high permeability conduits, varying heat input, and varying recharge under conditions of unconfined groundwater flow. Some key conclusions of earlier studies remained valid also in this context: again, the boundary between conduction dominated and advection dominated heat transport is found for systems with a bulk (isotropic) permeability of ca. 10^{-16} m^2, pressures are typically (close to) hydrostatic, and the hottest/most extensive hydrothermal systems develop when host rock permeability is close to 10^{-15} m^2. The hot plume can reach to the groundwater table (the iso-surface of 1 bar pressure in water) and then follows a "boiling curve with depth" temperature-pressure-distribution that is indistinguishable from the low-topography case (Hayba and Ingebritsen 1997). A major difference is that groundwater outflow in the volcano flanks may suppress the water table to hundreds of meters beneath the surface of the volcano. In such cases, the boiling zones develop beneath a considerable cover, which may have important implications for the preservation potential of hydrothermal mineral deposits.

Systems with high bulk permeabilities $\geq 10^{-14}$ m^2 differ substantially from their low-topography, shallow-intrusion equivalents. Boiling is restricted to the bottom parts of the systems, typically in the temperature range between 300 °C and 350 °C. Due to the high pressure of the overlying cold water column, the system has a strong tendency to remain in the liquid state. The 100 °C isotherm, which is located at or near the surface in most flat-topography systems, here typically lies 1-2 km beneath the groundwater table at 100-200 bar pressure. The temperature at which the deep boiling zone occurs is therefore mostly a function of the distance between intrusion and groundwater table, the latter being most strongly influenced by the near-surface permeability distribution and recharge by precipitation.

Effect of depth of emplacement and pluton size. Hot fluids moving up from deeper plutons heat a larger volume of rock on their way to the surface. Systems with bulk host rock permeability of 10^{-15} m^2 are still able to form a boiling water column in such cases (Fig. 7D). Unfortunately, Hayba and Ingebritsen (1997) provide no temperature-depth profiles for these simulations. We therefore repeated the respective simulations using their HYDROTHERM computer program and found that the boiling zone for a 10^{-15} m^2 bulk host rock permeability extended to only ~1 km below the surface, compared to 2 km in the shallower intrusion shown in Figure 7A. For a 10^{-14} m^2 permeability, the familiar profile (similar to Fig. 5A) with boiling in the upper ca. 500 m and with an underlying 250 °C isothermal leg (down to 2 km) develops.

Larger and/or hotter plutons can sustain their hydrothermal systems for longer times. However, the wider top surface influences the convection geometry and may set up multiple cells over the intrusion, each of which may produce boiling zones. Their vertical extent follows the pattern expected for the host rock permeability, provided that the respective convection is stable enough in space and time that this thermal structure can develop.

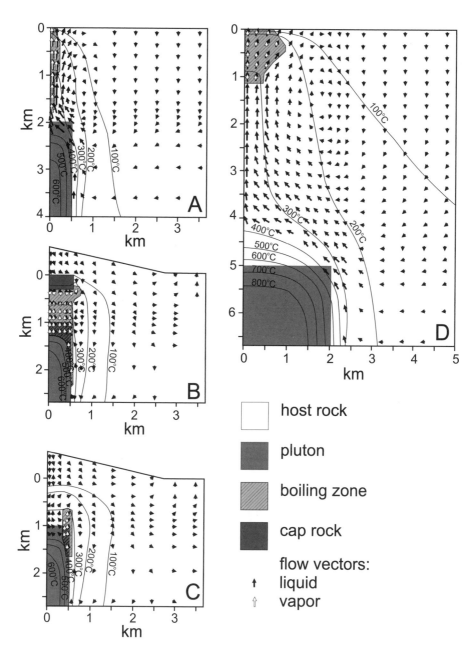

Figure 7. Influence of topography, depth and size of pluton, and presence of a cap rock on the flow pattern, thermal structure and boiling zones of a hydrothermal system with 10^{-15} m^2 host rock bulk permeability during its hottest phase. Illustrations based on Hayba and Ingebritsen (1997). Pictures have been adjusted to the same horizontal and vertical scale. See text for further explanation.

Multiple intrusions. If the country rock is preheated by an earlier intrusion, the rising fluid loses less heat to its surroundings and the maximum temperatures are reached more rapidly. The longest lasting boiling zones occur for permeabilities somewhat lower ($10^{-15.5}$ m^2) than in the non-preheated case (10^{-15} m^2). However, from the plots given by Hayba and Ingebritsen (1997), the pre-heated systems appear not to boil at shallow depths. If the rocks were preheated by an earlier advective hydrothermal system, they may cool much more rapidly. The timing and exact location of a new intrusion determines whether the new system can take advantage of the preheating or not.

The influence of narrow, high permeability zones. Hayba and Ingebritsen (1997) did not study the role of fractures and, to the best of our knowledge, no other study has been published that systematically investigated the role of fractures on the development of boiling zones in hydrothermal systems, at least not by explicitly modeling flow in networks of discrete fractures. To get at least a rough idea of the role of fracture zones, we have run a number of HYDROTHERM simulations putting a 10 m wide, vertical high-permeability (10^{-13} m^2) zone above the centre of the pluton (i.e., at the left margin of the half-space model of Fig. 4).

Surprisingly, the thermal evolution in this high permeability zone shows only minor changes compared to the homogenous porous medium case. For host rock permeability of 10^{-14} m^2, as in the homogenous case, the upflow zone does not get hotter than ~250°C. Again, boiling is restricted to upper few hundred meters and the temperature change is nearly constant down to 1500 m. The temperature in this zone is ca. 10 °C lower than in the homogeneous case.

As in the homogenous permeability case, the hottest systems evolve with a host rock permeability of 10^{-15} m^2. Again, the temperature profile follows the boiling curve for long times. However, in this case, the boiling zone never reaches deeper than approximately 1000 m. Below this depth—which is rarely of interest in the case of hydrothermal ore deposits—the temperature is slightly lower than in the pure homogeneous medium case and does not reach the boiling curve. Other differences relative to the homogenous permeability case are a shorter heating stage and less pronounced negative thermal profiles during the cooling phase.

Although these simulations may be a reasonable first proxy, we do not claim that they caught all features that would appear if discrete narrow fractures were modeled. The simulations assumed thermal equilibrium between rock matrix and fluid. If fractures were narrow but the spacing between them comparatively large, thermal equilibration of the rock matrix between them will take a time that may be substantially longer than the individual time steps used in a simulation (Pruess 1990).

Whether or not thin fractures can drain the high-temperature, low permeability zone and attain fluid flow rates high enough to act as hot hydrothermal flow paths, remains to be seen in future research.

COMPARING FLUID INCLUSION DATA AND SIMULATION PREDICTIONS

Studies of active geothermal systems in the Western Pacific area reveal numerous similarities between these mostly boiling systems and epithermal mineral deposits. Hedenquist et al. (1992) summarized a number of key features and demonstrated that temperature-depth distributions measured in geothermal wells agree favorably with data obtained from fluid inclusions in the same system. Unlike the snapshot provided by the well measurements, the fluid inclusions also record earlier stages in the thermal history.

Few fluid inclusion studies on epithermal vein type deposits have yielded data sets that are petrographically well-enough characterized and spread over a large enough vertical section to allow meaningful comparisons to temperature-depth distributions in modern systems (Vikre 1985; Simmons et al. 1988). At Fresnillo, Simmons et al. (1988) integrated fluid inclusions with other geochemical data and were able to produce a fairly detailed model of deposit formation. However, their interpretation was not quantitatively supported by numerical modeling.

How quantitative numerical simulation can aid the interpretation of fluid inclusion data has recently been shown for the Oligocene hydrothermal Pb-Zn vein type deposits of Madan, Bulgaria (Kostova et al. 2004). A set of six 10-20 km long fracture zones, located at the southwestern edge of an extensional structure interpreted to represent a metamorphic core complex (Dimov et al. 2000), host a number of vein-type ore deposits, sometimes accompanied by replacement ore bodies where a marble horizon is cut by the veins. The heat source for hydrothermal convection is unknown and may be a yet unidentified pluton at depth or a strong thermal gradient in the rapidly expanding core complex. A simplified paragenetic sequence shows an early quartz-pyrite stage followed by a main stage of dominantly quartz-galena-sphalerite and later quartz +/− carbonates +/− sulphosalts. More complex classifications have been proposed but appear to be local or quantitatively less important variations of this theme. Boiling has long been recognized as a major process during the formation of the main stage mineralization by Bonev and co-workers, including quantitative liquid and vapor geochemistry from large primary inclusions in galena (Bonev 1977, 1984; Piperov et al. 1977). The fluids are generally low-salinity with mostly 4 to 5 wt% $NaCl_{eq.}$.

Kostova et al. (2004) selected primary and pseudosecondary fluid inclusions that could be unambiguously related to individual paragenetic stages in a set of samples covering a 500 m vertical interval in the vein mineralization of one of the deposits (Yuzhna Petrovitsa) in the Madan district. Grouping the homogenization results as a function of present day elevation revealed significantly different trends for inclusions from the three main paragenetic stages (Fig. 8A), which could not readily be explained without additional information. Only the temperature-depth distribution for the main stage mineralization followed the "boiling curve with depth" profile observed in modern geothermal systems.

A powerful explanation became apparent in the context of the 10^{-15} m^2 permeability simu-

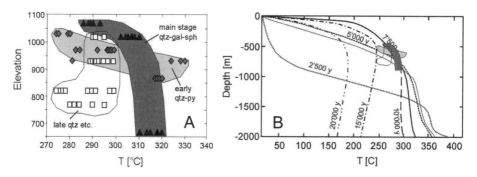

Figure 8. A) Fluid inclusion homogenization temperatures from three different paragenetic stages in the epithermal Pb-Zn veins of the Yuzhna Petrovitsa deposit of the Madan ore field as a function of present-day topographic elevation (from Kostova et al. 2004, with additional data from Driesner et al. 2005). B) The same data fields plotted onto the temperature-depth curves for the simulated thermal evolution of a hydrothermal system above a cooling pluton in a porous rock matrix with 10^{-15} m^2 bulk permeability from Hayba and Ingebritsen (1997); numbers are years after intrusion. The early quartz-pyrite stage apparently formed during the heating-up of the system in a very strong thermal gradient, while the main stage quartz-galena-sphalerite mineralization formed during a prolonged period during which the whole system was boiling to depth (solid black curve). Notice that economic mineralization started to form from the ascending fluid where this boiling-point-with-depth curve starts to show a pronounced cooling curvature near ca. 1000 m depth. Thermodynamic modeling shows that the precipitation of galena and sphalerite was mainly a response to this cooling, i.e., a physical consequence of boiling and not a direct chemical effect. The small salinity measured in the fluid inclusions would shift the actual locations towards slightly greater depth.

lation results by Hayba and Ingebritsen (1997). In fact, the fluid inclusion data could be plotted directly onto Figure 4 of Hayba and Ingebritsen (1997) and showed how the various paragenetic stages related to distinct stages in the thermal evolution of the hydrothermal system (Fig. 8B). The early quartz-pyrite stage shows the steepest geothermal gradient and represents the early heating-up of the system under single-phase liquid conditions. The quartz-galena-sphalerite stage formed during the hottest phase when the hydrothermal system was boiling over the whole vertical section. Finally, the late quartz-carbonate phase formed during the period when the system was cooling from below and the lower parts of the system were under single phase liquid conditions while the upper parts were still boiling. Since the measured fluid salinities were low, the comparison to numerical simulation obtained with pure water equations of state seemed acceptable. Over the vertical interval of interest in this context, the boiling curves with depth for pure water and for an aqueous solution of 5 wt% NaCl show little difference.

With ore-metal concentration in the fluid from LA-ICPMS analyses, pH and f_{S2} constraints from mineral buffers, and the solubility relations provided by Barret and Anderson (1988), Kostova et al. (2004) concluded that the boiling process caused the main stage mineralization mostly by imposing a temperature-depth distribution that forced galena and sphalerite to precipitate in response to cooling (Fig. 9A,B). More accurate thermodynamic modeling by Driesner et al. (2005) confirmed this conclusion. The direct chemical consequences of boiling appear to have played only a minor role in base metal precipitation. A particularly interesting result was the observation that in the late stage the fluids became more saline, and therefore the solubility of galena and sphalerite increased. LA-ICPMS analyses showed an increase in Pb and Zn from the

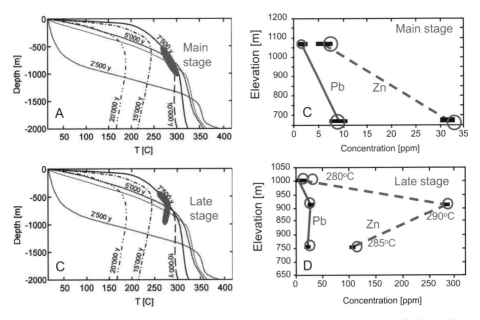

Figure 9. Ore metal concentration in Madan main stage (a+b, top) and late stage (c+d, bottom) fluids over the vertical extent of the presently accessible section. *Left*: temperature curves for hydrothermal system evolution from Hayba and Ingebritsen (1997) and fluid inclusion homogenization temperatures (grey fields). *Right*: measured LA-ICPMS fluid inclusion data (black rectangles) compared to modeled solubilities (grey curves and circles) from Driesner et al. (2005). The solubilities were modeled assuming that a fluid of the composition measured in fluid inclusions from the deepest samples was rising in the vein according to the simulated temperature curves, and was in full equilibrium with host rock and vein mineral buffers. Agreement is within a few percent and quantitatively predicts the observed redistribution of metals in the late stage.

lowest part of the system to some intermediate depth and then a strong decrease towards shallower depths (Fig. 9C,D). The reason became apparent when the temperature trends from fluid flow simulations were combined with thermodynamic modeling of ore mineral solubilities. Due to the late stage cooling from below, the incoming fluids rose through a positive thermal gradient, dissolving galena and sphalerite upon heating, and then re-precipitating them in the upper parts of the system where the fluids were forced to cool along the boiling curve. This result explains the widespread dissolution textures in main stage galena and sphalerite and the occurrence of late-stage sphalerite with fluid inclusion homogenization temperatures (Bonev and Kouzmanov 2002) in the range predicted by this modeling. The degree of redistribution of galena and sphalerite in terms of mass appears to be rather small, which is in good agreement with the prediction from the fluid flow modeling that this stage is highly transient. An additional field test for this model would be to map the exact vertical distribution of dissolution textures in main stage ore minerals and of the occurrence of late stage sphalerite.

In our view, the results of Kostova et al. (2004) demonstrate that quantitative fluid flow simulations and the study of carefully selected and characterized fluid inclusions can go hand in hand to achieve a meaningful quantitative reconstruction of a fossil, ore-forming hydrothermal system including dynamic aspects of its thermal evolution. Using a comparatively easy-to-learn simulation program such as HYDROTHERM should become a routine tool in future work of this kind.

The next step in these investigations would be to use a full reactive transport computer code that allows evaluation of the chemical effects of vapor-liquid partitioning of CO_2, H_2S, and other gases in a flowing multiphase system. The Madan deposits would be a good first test case for such models.

FLOW OF SALINE FLUIDS IN MAGMATIC-HYDROTHERMAL SYSTEMS

Numerous fluid inclusion studies have shown that fluids in porphyry copper-type magmatic-hydrothermal system are generally saline and there is now general agreement that the saline fluids are largely of magmatic origin. Adding salt to water greatly extends the pressure and temperature range over which fluid phase separation can occur. The density contrasts between liquid and vapor in saline systems can be substantially larger than in the pure water case, even at high temperatures (Heinrich 2007; this volume). For both reasons, vapor columns can have greater total buoyancy than in the pure water case. We can expect that the hydrodynamics of saline systems can differ considerably from low-salinity geothermal systems and also be more complex because vapor can become physically separated from liquid. Furthermore, the exact *P-T-X* paths that vapor and/or brine take can be crucial for the formation of various types of ore deposits (see Heinrich 2007; this volume).

Geiger et al. (2006a,b) have recently presented the theory and first results for a software system that simulates multiphase fluid flow in saline systems under conditions relevant to magmatic-hydrothermal systems. The software includes a new accurate model of the phase relations, volumetric properties and enthalpies of H_2O-NaCl fluids (valid between 0-1000 °C, 1-5000 bar, and 0-100 wt% NaCl; Driesner and Heinrich 2007; Driesner 2007), and a novel combined finite element and finite volume scheme to efficiently handle multiphase fluid flow in this two-component system. Using this software, Geiger et al. (2005a) showed a general overview of possible flow patterns under common crustal conditions. More recent, unpublished tests have shown that vapor saturations in the two-phase case are over-predicted, and a modified scheme is currently in the test phase. The simulations shown below (Geiger et al. 2005b; Driesner et al. 2006) were done with the original scheme and may not represent the natural situation in all respects. Nevertheless, we expect that the thermal evolution and the sequence of phase states are generally predicted correctly. We show the results of two reconnaissance simulations to illustrate the current state of development and the potential results that can emerge from this kind of research.

Kissling (2005a,b) presented another simulation scheme based on the TOUGH2 simulation program but used a model for the properties of NaCl-H$_2$O fluids by Palliser and McKibbin (1998a,b,c) that is known to show unphysical behavior, such as negative heat capacities, under certain *T-P-X* conditions. First applications included 1-D test scenarios including phase separation and convection under single phase conditions.

Porphyry case study

Geiger et al. (2005b) investigated the role of host rock permeability and topography on the development of a hydrothermal system around a rather deep-seated pluton (at 5-8 km depth) with a porphyry "finger" extending 2 km towards the surface from the centre of the pluton (Fig. 10). We assumed that the model is symmetrically repeated to the left. Accordingly, the left model boundary is taken as no flow boundary. The horizontal dimension is 20 km and the maximum depth 12 km. A magma chamber of 5 km half-width is emplaced between 5 and 8 km depth, and a magma finger of 200 m half-width protrudes 2 km upwards. A 1 km high volcano with a 20° slope is present in one simulation. The model domain is spatially discretized by an unstructured mesh of ~12000 triangular finite elements that has an enhanced resolution in the region of the magma chamber and above it. The country rock is subdivided into two domains: from the middle of the magma chamber (6.5 km) to the bottom, it is impermeable. Preliminary studies with lower resolution had shown that this has no influence on the overall convection patterns, and excluding this region from the flow computations saves substantial computing time. The full magma chamber geometry was nevertheless incorporated into the thermal model to capture its conductive cooling directly rather than assuming a time-dependent heat flux.

The host rock in the upper domain has a homogenous isotropic permeability. For the magma chamber, preliminary studies (Geiger et al. 2005b) had shown that including a temperature-pressure dependent permeability that mimics hydrofracturing and fracture plugging by mineral precipitation was needed to maintain two-phase conditions in the porphyry finger. This was implemented as follows: we interpolate the permeability log-linearly between 10^{-20} m^2 at the solidus temperature at 700 °C and 10^{-15} m^2 at the brittle-ductile transition at 360 °C. The latter value also applies below 360 °C. Hydrofracturing occurs if, after solving for fluid pressure, the fluid pressure exceeds the lithostatic pressure plus the tensile strength of the rock at a node

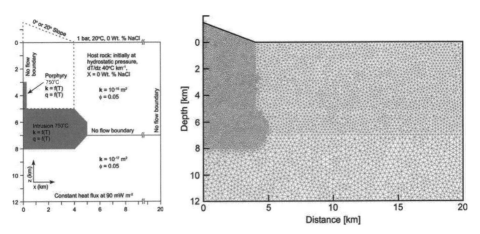

Figure 10. Model geometry and parameters (*left*; dashed box marks area shown in Figs. 11 and 12) and finite element mesh (*right*) used in the simulation of magmatic-hydrothermal systems by Geiger et al. (2005).

within the finite element mesh. In this case, we calculate the fluid volume that has been stored in each finite element that is in excess of that which could be stored at lithostatic pressure. The permeability is then adjusted by the amount necessary to release this excess volume. When the excess fluid pressure has decayed and is less than the lithostatic pressure, we return to the permeability of the element to the normal temperature-dependent value. In regions with temperatures above the brittle-ductile transition where the permeability is low, hydrofracturing often occurs many times, mimicking repeated generation of fractures and sealing of fractures due to mineral precipitation and ductile deformation.

Intrusion temperature was set to 750 °C, and the intrusion was allowed to cool due to conduction and hydrothermal convection. Once a finite element reached the solidus temperature, 5 wt% of the melt's mass in that element were liberated as 10 wt% NaCl fluid and fed into the base of the porphyry finger.

Magmatic hydrothermal system with a flat topography and homogenous isotropic host rock permeability of 10^{-15} m^2. The simplest simulation uses a homogenous isotropic host rock permeability of 10^{-15} m^2 under a flat topography. From experience with pure water systems this shall produce the hottest system. Figure 11 shows a series of snapshots of the region above the magma chamber (dashed box in Fig. 10). We plotted bulk fluid density since this reflects best the effects that we will discuss in the following. The right column in Figure 11 shows the paths that fluids will take in the phase diagram when rising along the left edge of the respective model shown in the left column.

At 12,000 years after intrusion (Fig. 11A), two thermal plumes with similar temperature-depth profiles have developed. The plume to the right formed originally at the edge of the pluton where strong thermal gradients along an inclined interface strongly favor the rapid rise of hot, buoyant fluids. Boiling of the rising liquid-like fluid occurs only in its upper ~1000 m. This plume gradually moves inward to eventually join the central plume. The porphyry finger has crystallized. Its upper two thirds are at hydrostatic pressure and the thermal anomaly related to it has shrunk to a small bump on top of the magma chamber.

Along the left edge of the model, fluid flows along the *T-P-X* path depicted in Figure 11B. Within the magma chamber, fluid has the same lithostatic pressure as the magma and any free fluid will be a single phase with a density of ~400 kg m^{-3}. When released through the hydrofracturing zone, the fluid experiences a transition from near-lithostatic to near-hydrostatic pressures (from point 1 to point 2 in the phase diagram) over a distance of tens to a few hundred meters. In the phase diagram the 10 wt% NaCl fluid will hit the vapor+liquid coexistence surface at temperatures around ~550 °C (point 2). This phase separation is not boiling in the strict sense but rather the condensation of brine from a low density, vapor-like fluid. The fluid in the thermal plume stays in the vapor+liquid region of the phase diagram (between points 2 and 3) to almost the top of the model. Saturations shown in the figure indicate a very strong volumetric dominance of vapor. This can be explained by the fact that, at these pressure, temperature, and composition conditions, the volume fraction of the vapor is intrinsically very large (i.e., almost at unity). The brine probably forms isolated droplets that rarely move. In other words, due to the thermodynamics of the NaCl-H$_2$O system, fluid flow at these conditions is dominated by vapor flow. However, since these simulations were performed with an earlier algorithm that tends to overestimate vapor saturations, this result needs to be treated with some caution. Near the top of the model, the plume condenses into a thin layer of groundwater (between point 3 and 4), which is likely an artifact of the type of boundary condition applied at the top of the model. In reality, a field of hot springs is expected as surface expression.

At 30,000 years (Fig. 11C,D), the plume originally located on the right edge of the pluton has merged with the central plume and only a ghost thermal anomaly remains. A small pocket of boiling remains in the uppermost part of this remnant anomaly. The path that the ascending fluid takes along the left edge resembles the situation at 12,000 years except for an excursion

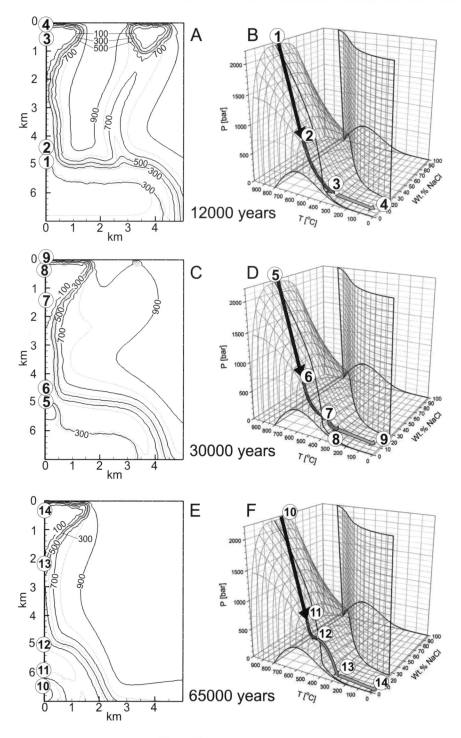

Figure 11. *caption on facing page*

into the vapor+halite field at intermediate depths (between points 7 and 8 in Fig. 11D). This implies a very low pressure and may be an artifact of overestimated vapor saturations in the plume. Whether or not this feature will appear in future simulations with improved algorithms remains to be seen.

A significant change in the fluid path in the phase diagram occurs in the late stages, illustrated here at 65,000 years (Fig. 11E,F). The remaining magma chamber is small. At about one third the distance between magma chamber and surface the rising fluid leaves the vapor+liquid coexistence surface and travels as a low density, single phase fluid that "contracts" and becomes denser as it rises and cools between points 12 and 13. Pressure decreases at shallow depth which causes the fluid to enter a second boiling zone (point 13). Such a path could provide a good way to transport gold from the magmatic body to the shallow epithermal environment (Heinrich et al. 2004). The second boiling zone would favor the precipitation of gold. However, since these conditions persist only for a short period of a few thousand years, we doubt that this scenario is likely to provide a good epithermal gold deposit.

Magmatic hydrothermal system with volcano topography and homogenous isotropic host rock permeability of 10^{-15} m^2. A volcano topography (Fig. 12) depresses the top of the central plume at 12,000 years. The second plume is almost eliminated. The evolution of the "vapor contraction" path favorable for gold transport developed here already at 30,000 years (between points 7 and 8 in Fig. 12C,D) and remains rather stable to at least 65,000 years (Fig. 12E,F, between points 13 and 14). The fluid does not hit the vapor+halite phase region and a spatially rather stable yet fluctuating interface of mixing with cold groundwater forms beneath the volcano. As pointed out by Hayba and Ingebritsen (1997), this is a favorable place for ore precipitation by fluid mixing such as suggested for the epithermal vein deposits at Creede, Colorado (Hayba 1997).

The results from these two simulations show how a NaCl-H_2O flow model can identify the impact of single changing conditions such as the presence or absence of a volcanic edifice. Key features, such as boiling at high temperature and pressure or vapor contraction, would not occur in a pure water model.

OUTLOOK

In our opinion, the emerging magmatic-hydrothermal simulations look very promising in the sense that they appear to reproduce first order phenomena observed in natural examples. Moreover, the parameter space is strongly dominated by very few key variables, in particular the permeability structure of the system. This limits the possible scenarios for the formation of certain ore deposit types to a rather small set. Understanding these on a generic level is likely to be the biggest contribution that simulations can provide in the near future. More practically, the results obtained so far indicate that at least the system-scale permeability structure will leave unique fingerprints in the form of a specific thermal evolution which potentially can be reconstructed with careful fluid inclusion studies.

Ideally, this type of simulation might become a routine tool to study transient processes and play a role similar to that phase diagrams played in the understanding of static problems in petrology over the last decades. In particular, the availability of relatively easy-to-learn

---

Figure 11 (*on facing page*). Evolution of a magmatic-hydrothermal system with a constant host rock permeability of 10^{-15} m^2 and including the full NaCl-H_2O phase diagram and fluid properties. Pictures on the left show the fluid density distribution (in kg m^{-3}, in two-phase zones the density of the mixture is given) in the dashed box of Figure 10. Pictures on the right show the path of fluid through the NaCl-H_2O phase diagram upon rising along the left side of the left picture. See text for further explanation.

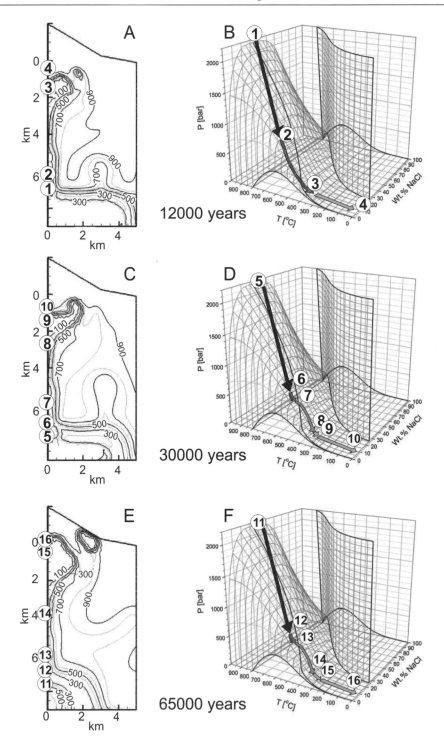

Figure 12. *caption on facing page*

computer programs with graphical user interfaces is of great help. We strongly recommend to the reader to acquire the USGS HYDROTHERM computer program (Hayba and Ingebritsen 1994; available under *http://water.usgs.gov/software/hydrotherm.html*). A more easy-to-use version with a graphical user interface that can be run on standard PCs ("HYDROTHERM 2D interactive") awaits official release. Many of the most fundamental results presented in this review have been computed this way. HYDROTHERM has a sufficient number of analysis possibilities and, although restricted to pure water applications, provides an excellent way to evaluate possible scenarios for many cases in hydrothermal research. We have used it successfully in teaching. When combined with careful fluid inclusion studies, these simulations can help to substantiate the physical reality of inferred flow paths, etc.

In the low-temperature domain to ~300 °C, more complicated fluid systems, and in some cases also reactive transport, can be simulated with programs from the TOUGH family from LBNL (Pruess 1991). These programs, however, are rather expensive and have a rather steep learning curve. They do, however, have a large set of capabilities beyond those of HYDROTHERM and are well-maintained.

In spite of the first encouraging results and the increasing number and capabilities of available programs, an important problem in the near future will be the definition of benchmarks to verify that the codes do indeed simulate hydrothermal processes correctly. In addition, a number of pressing research needs for the near future is apparent:

- The accuracy with which material properties are known plays a crucial role in the accuracy of the simulation results. Hence, experimental studies on fluid phase equilibria (in particular, multicomponent systems such as H_2O-$NaCl$-CO_2) are needed.

- Of similar importance is the fact that we have hardly any idea about the actual behavior of the rock's intrinsic permeability at high temperature, or of the relative permeability functions for multiphase fluid flow under hydrothermal conditions. While good simulation results can be obtained for a given permeability function at high temperatures, we simply do not know, quantitatively, how various rocks behave under those conditions, which hinders true predictive simulations (although the rather good agreement between simulation and natural observations seems to indicate that the numbers are broadly in the right range when applied to "postdictive" cases).

- High resolution studies of the influence of discrete faults and how they influence the flow models are needed.

- The feedback between fluid pressurization and rock mechanics is lacking from the current simulation codes.

- In order to quantitatively understand the formation of ore grades, reactive transport simulation schemes need to be coupled to the most advanced flow simulation programs.

- Finally, the most important task is probably to identify testable predictions in the simulations that, in ideal cases, might even allow "field calibration" of the unknowns just mentioned.

Figure 12 (*on facing page*). Evolution of a magmatic-hydrothermal system with a constant host rock permeability of 10^{-15} m^2 beneath a volcano, including the full $NaCl$-H_2O phase diagram and fluid properties. Pictures on the left show the fluid density distribution (in kg m^{-3}, in two-phase zones the density of the mixture is given) in the dashed box of Figure 10. Pictures on the right show the path of fluid through the $NaCl$-H_2O phase diagram upon rising along the left side of the left picture. See text for further explanation.

ACKNOWLEDGMENTS

We thank Larry Cathles, James Cleverly, and Steve Ingebritsen for their detailed and insightful reviews. Dim Coumou's help in the preparation of Figures 11 and 12 is greatly appreciated. Finally, we would like to thank Axel Liebscher and Chris Heinrich for inviting us to write this paper.

REFERENCES

Barret TJ, Anderson GM (1988) The solubility of sphalerite and galena in 1-5 m NaCl solutions to 300 °C. Geochim Cosmochim Acta 52:813-820

Bonev I (1977) Primary fluid inclusions in galena crystals. 1. Morphology and origin. Mineral Deposita 12:64-76

Bonev I (1984) Mechanisms of the hydrothermal ore deposition in the Madan lead-zinc deposits, central Rhodopes, Bulgaria. Proceedings of the 6th quadrennial IAGOD symposium: 69-73

Bonev I, Kouzmanov K (2002) Fluid inclusions in sphalerite a negative crystals: a case study. Eur J Mineral 14:607-620

Cathles LM (1977) An analysis of the cooling of intrusives by ground-water convection which includes boiling. Econ Geol 72:804-826

Cathles LM (1983) An analysis of the hydrothermal system responsible for massive sulfide deposition in the Hokuroko Basin of Japan. Econ Geol Monograph 5:439-487

Cathles LM (1990) Scales and effects of fluid-flow in the upper crust. Science 248:323-329

Cathles LM (1993) A capless 350 °C flow zone model to explain megaplumes, salinity variations, and high-temperature veins in ridge axis hydrothermal systems. Econ Geol 88:1977-1988

Dimov D, Dobrev S, Ivanov Z, Kolkovski B, Sarov S (2000) Structure, alpine evolution and mineralizations of the Central Rhodopes area (South Bulgaria). *In*: Geodynamics and Ore Deposits Evolution of the Alpine-Balkan-Carpathian-Dinaride Province. Guide to Excursion B. Ivanov Z (ed) University Press "St. Kliment Ohridski", Sofia

Driesner T (2007) The System H_2O-NaCl. II. Correlations for molar volume, enthalpy, and isobaric heat capacity from 0 to 1000 °C, 1 to 5000 bar, and 0 to 1 X_{NaCl}. Geochim Cosmochim Acta. (in press)

Driesner T, Geiger S, Heinrich CA (2006) Modeling multiphase fluid flow of H_2O-NaCl fluids by combining CSP5.0 with SoWat2.0. Geochim Cosmochim Acta 70/18S:A147

Driesner T, Heinrich CA (2007) The System H_2O-NaCl. I. Correlation formulae for phase relations in pressure-temperature-composition space from 0 to 1000 °C, 0 to 5000 bar, and 0 to 1 X_{NaCl}. Geochim Cosmochim Acta (in press)

Driesner T, Kostova B, Heinrich CA (2005) Hydrodynamic and thermodynamic modeling of the formation of the Yuzhna Petrovitsa hydrothermal Pb-Zn ore deposit, Madan, Bulgaria. Geochim Cosmochim Acta 69/10S:A157

Faust CR, Mercer JW (1979a) Geothermal reservoir simulation 1. Mathematical models for liquid- and vapor-dominated hydrothermal systems. Water Resour Res 15:23-30

Faust CR, Mercer JW (1979b) Geothermal reservoir simulation 2. Numerical solution techniques for liquid- and vapor-dominated hydrothermal systems. Water Resour Res 15:31-46

Fournier RO (1999) Hydrothermal processes related to movement of fluid from plastic into brittle rock in the magmatic-epithermal environment. Econ Geol 94:1193-1211

Geiger S, Driesner T, Heinrich CA, Matthai SK (2005a) On the dynamics of NaCl-H_2O fluid convection in the Earth's crust. J Geophys Res 110:B07101

Geiger S, Driesner T, Heinrich CA, Matthai SK (2006a) Multiphase thermohaline convection in the earth's crust: I. A new finite element - Finite volume solution technique combined with a new equation of state for NaCl-H2O. Transport Porous Media 63:399-434

Geiger S, Driesner T, Heinrich CA, Matthai SK (2006b) Multiphase thermohaline convection in the earth's crust: II. Benchmarking and application of a finite element - Finite volume solution technique with a NaCl-H2O equation of state. Transport Porous Media 63:435-461

Geiger S, Driesner T, Heinrich CA, Mattthai SK (2005b) Coupled heat and salt transport around cooling magmatic intrusions. Geochim Cosmochim Acta 69:A739

Geiger S, Robert S, Matthai SK, Zoppou C, Burri A (2004) Combining finite element and finite volume methods for efficient multiphase flow simulations in highly heterogeneous and structurally complex geologic media. Geofluids 4:284-299

Gerritsen MG, Durlofsky LJ (2005) Modeling fluid flow in oil reservoirs. Annu Rev Fluid Mech 37:211-238

Grant MS, Sorey ML (1979) The compressibility and hydraulic diffusivity of a water-steam flow. Water Resour Res 15:684-686

Hanson RB (1995) The hydrodynamics of contact metamorphism. Geol Soc Amer Bull 107:595-611

Hanson RB (1996) Hydrodynamics of magmatic and meteoric fluids in the vicinity of granitic intrusions. Trans Royal Soc Edinburgh: Earth Sci 87:251-259

Hayba DO (1997) Environment of ore deposition in the Creede mining district, San Juan Mountains, Colorado. 5. Epithermal mineralization from fluid mixing in the OH vein. Econ Geol 92:29-44

Hayba DO, Ingebritsen SE (1994) The computer model HYDROTHERM, a three-dimensional finite-difference model to simulate ground-water flow and heat transport in the temperature range of 0 to 1200 °C. U.S. Geol Survey Water-Res Invest Report 94-12252

Hayba DO, Ingebritsen SE (1997) Multiphase groundwater flow near cooling plutons. J Geophys Res 102:12235-12252

Hedenquist JW, Reyes AG, Simmons SF, and Taguchi S (1992) The thermal and geochemical structure of geothermal and epithermal systems - a framework for interpreting fluid inclusion data. Eur J Mineral 4:989-1015

Heinrich CA (2007) Fluid – fluid interactions in magmatic-hydrothermal ore formation. Rev Mineral Geochem 65:363-387

Heinrich CA, Driesner T, Stefánsson A, Seward TM (2004) Magmatic vapor contraction and the transport of gold from the porphyry environment to epithermal ore deposits. Geology 32:761-764

Hurwitz S, Kipp KL, Ingebritsen SE, Reid ME (2003) Groundwater flow, heat transport, and water table position within volcanic edifices: implications for volcanic processes in the Cascades Range. J Geophys Res 108:2557

Huyakorn PS, Pinder GF (1983) Computational Methods in Subsurface Flow. Academic Press

Ingebritsen SE, Sanford W, Neuzil C (2006) Groundwater in Geologic Processes. Cambridge University Press

Jupp T, Schultz A (2000) A thermodynamic explanation for black smoker temperatures. Nature 403:880-883

Jupp T, Schultz A (2004) Physical balances in subseafloor hydrothermal convection cells. J Geophys Res - Solid Earth 109:B05101

Kissling WM (2005a) Transport of three-phase hypersaline brines in porous media: Theory and code implementation. Transport Porous Media 61:25-44

Kissling WM (2005b) Transport of three-phase hypersaline brines in porous media: Examples. Transport Porous Media 61:141-157

Kostova B, Pettke T, Driesner T, Petrov P, Heinrich CA (2004) LA ICP-MS study of fluid inclusions in quartz from the Yuzhna Petrovitsa deposit, Madan ore field, Bulgaria. Schweiz Mineral Petrograph Mitt 84:25-36

Narasimhan TN, Witherspoon PA (1976) An integrated finite difference method for analyzing fluid flow in porous media. Water Resour Res 12:57-65

Norton D, Knight J (1977) Transport phenomena in hydrothermal systems: cooling plutons. Am J Sci 277:937-981

Oliver NHS, Rubenach MJ, Fu B, Baker T, Blenkinson TG, Cleverley JS, Marshall LJ, Ridd PJ (2006) Granite-related overpressure and volatile release in the mid crust: fluidized breccias from the Cloncurry District, Australia. Geofluids 6:346-358

O'Sullivan MJ, Pruess K, Lippmann MJ (2001) State of the art of geothermal reservoir simulation. Geothermics 30:395-429

Palliser C, McKibbin R (1998a) A model for deep geothermal brines, I: T-P-X state-space description. Transport Porous Media 33:65-80

Palliser C, McKibbin R (1998b) A model for deep geothermal brines, II: thermodynamic properties - density. Transport Porous Media 33:129-154

Palliser C, McKibbin R (1998c) A model for deep geothermal brines, III: thermodynamic properties - enthalpy and viscosity. Transport Porous Media 33:155-171

Piperov NB, Penchev NB, Bonev I (1977) Primary fluid inclusions in galena crystals. 2. Chemical composition of the liquid and gas phase. Mineral Deposita 12:77-89

Pruess K (1990) Modeling geothermal reservoirs: fundamental processes, computer simulation and field applications. Geothermics 19:3-25

Pruess K (1991) TOUGH2 - a general purpose numerical simulator for multiphase fluid and heat flow. Lawrence Berkeley Laboratory Report LBL 29400

Shewchuk JR (1996) Triangle: Engineering a 2D quality mesh generator and Delaunay triangulator. *In*: Applied Computational Geometry. Towards Geometric Engineering, Vol. 1148. Lin MC, Mocha D (eds) Springer, p 203-222

Simmons SF, Gemmell JB, and Sawkins FJ (1988) The Santo Nino silver-lead-zinc vein, Fresnillo district, Zacatecas, Mexico: Part II. Physical and chemical nature of ore-forming solutions. Econ Geol 83:1619-1641

Stueben K (2001) A review of algebraic multigrid. J Compu Appl Mathematics 128:281-309

Vikre PG (1985) Precious metal vein system in the National Disctrict, Humboldt County, Nevada. Econ Geol 80:360-393

Young R (1993) Two-phase geothermal flows with conduction and the connection with Buckley-Leverett theory. Transport Porous Media 12:231-278

Reviews in Mineralogy & Geochemistry
Vol. 65, pp. 213-239, 2007
Copyright © Mineralogical Society of America

7

Fluid Phase Separation Processes in Submarine Hydrothermal Systems

Dionysis I. Foustoukos

Geophysical Laboratory
Carnegie Institution of Washington
Washington, DC 20015, U.S.A.
dfoustoukos@ciw.edu

William E. Seyfried, Jr.

Department of Geology and Geophysics
University of Minnesota
Minneapolis, Minnesota 55455, U.S.A.
wes@umn.edu

INTRODUCTION

Mineral-fluid equilibria play a key role in governing the chemical evolution of modern submarine hydrothermal systems (Seyfried et al. 1999, 2004; Foustoukos and Seyfried 2005). Seawater circulating through high permeability zones of the upper oceanic crust, reaches areas close to the brittle-ductile boundary, where high temperature and pressure conditions enhance water/rock interactions, characterized by formation of hydrous alteration minerals and compositional changes in seawater chemistry, particularly enrichment of dissolved volatiles and transition metals (Fig. 1) (Kelley et al. 2002; German and Von Damm 2003). The buoyant hydrothermal fluids ultimately vent at the seafloor with exit temperatures reaching values higher than 350 °C. Mixing between ambient seawater and hydrothermal vent fluid induces metal sulfide precipitation, accounting for the common reference to the vent fluids as "black smokers." The large spatial compositional variability observed even between adjacent black smoker vent fluids, however, is mainly the result of phase separation and boiling processes intrinsic to the NaCl-H_2O system at elevated temperature and pressure conditions (Bischoff and Pitzer 1985; Bischoff and Rosenbauer 1987; Butterfield et al. 1994; Von Damm 1995, 2000; Berndt and Seyfried 1997; Lilley et al. 2003). Although this compositional variability is best indicated by the wide range of dissolved chloride concentrations of vent fluids that greatly deviate from seawater values (Von Damm et al. 1997; Von Damm 2000, 2004; Seyfried et al. 2003), confirming evidence of phase separation processes can also be traced to the distribution and abundance of dissolved volatiles and other aqueous species known to fractionate between vapor and liquid when seawater phase separation and segregation occurs (Oosting and Von Damm 1996; Berndt and Seyfried 1997; Fornari et al. 1998; Butterfield et al. 1999; Lilley et al. 2003; Von Damm et al. 2003). Given the nature of magmatic and tectonic processes associated with the formation of the ocean crust at mid-ocean ridges, which can result in high temperatures at low to moderate pressure conditions (Wilcock and McNabb 1996; Bohnenstiehl et al. 2004; Van Ark et al. 2007), phase separation processes are not surprising, although the rate of change from vapor to liquid dominated systems and the compositional diversity expressed by ridge vent fluids, remain enigmatic (Sohn et al. 1999; Lilley et al. 2003; Von Damm et al. 2003; Larson et al. 2007).

1529-6466/07/0065-0007$05.00

DOI: 10.2138/rmg.2007.65.7

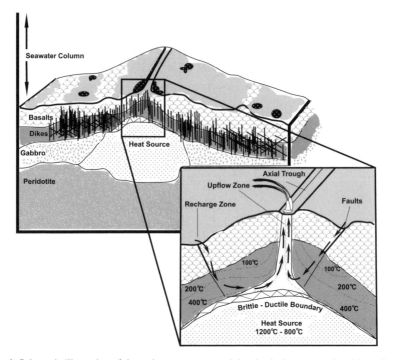

Figure 1. Schematic illustration of the main components and the physical processes that drive submarine hydrothermal circulation in mid-ocean ridge spreading centers (modified after Kelley et al. 2002). Seawater penetrates the basaltic substrate through faulting, reaching high temperature-pressure alteration zones very near the brittle-ductile boundary of partially molten with crystallized mantle material. Mineral-fluid equilibria enhanced by the elevated temperature and pressure conditions established in the subseafloor reaction zone, alters the chemical composition of the circulating seawater, while buoyant fluid conditions fuel hydrothermal venting to the seafloor. Most importantly, however, the physical conditions of hydrothermal circulation can intersect the vapor-liquid boundary of the NaCl-H$_2$O system (Bischoff 1991; Berndt et al. 2001), resulting to fluid phase separation and formation of a low and high salinity/density vapor and liquid fractions, respectively.

The results of recent time series investigations of hydrothermal vent systems at mid-ocean ridges have provided important insight on the direction, magnitude and rate of change of chemical components during phase separation and segregation processes. The relatively high temperatures and pressures and often extreme chloride variability between coexisting vapors and liquids, which is manifest as well by sharp contrasts in density (Bischoff 1991; Berndt et al. 2001), however, has presented serious challenges for obtaining a more quantitative understanding of the geochemical controls regulating the distribution of species other than dissolved chloride. These difficulties largely result from a lack of thermodynamic and kinetic data applicable to conditions at sites inferred for phase separation in the oceanic crust. Indeed, the low-density conditions for the vapor established during phase separation impose significant limitations to the estimated dissociation equilibria of 1:1 aqueous complexes (e.g., NaCl$_{(aq)}$, HCl$_{(aq)}$) (Ho et al. 1994, 2000, 2001), while aqueous speciation, especially for the highly ionic saline phases, is strongly obscured by the possible existence of polyatomic ionic species (Pitzer and Schreiber 1987; Oelkers and Helgeson 1990, 1993a,b; Driesner et al. 1998; Sharygin et al. 2002). As a consequence of this, mineral-fluid buffering reactions, acid generation, redox processes, and mass transfer induced by vapor-liquid partitioning are largely unconstrained. This is particularly true for the high-density liquid phase, which tends to segregate from a compositionally homogeneous fluid during phase separation and accumulate above the magma-hydrothermal interface (Fontaine and Wilcock 2006), while facilitating mineral dissolution and re-precipitation of more stable phases, as often inferred from the chemical, mineralogical and fluid inclusion data from rocks recovered from fossil hydrothermal systems (Richardson et al. 1987; Nehlig et al. 1994; Alt et al. 1999; Gillis and Roberts 1999; Alt and Teagle 2000; Gillis 2002).

To address the lack of proper thermodynamic data to accurately model dissolved aqueous species partitioning between the low and high salinity fluids, a number of experimental studies have been performed to constrain fractionation patterns in phase separated homogeneous and heterogeneous systems. Accordingly, experimental data that exist describe vapor-liquid partitioning of highly mobile trace elements (e.g., Br, Li, Rb, Cs, B) (Berndt and Seyfried 1990, 1997; Liebscher et al. 2005, 2006b; Foustoukos and Seyfried 2007a), and metals (Pokrovski et al. 2002, 2005), as well as the vapor-liquid isotopic fractionation of several stable isotopes (δD, $\delta^{18}O$, $\delta^{11}B$, δ^7Li, $\delta^{37}C$) (Spivack et al. 1990; Berndt et al. 1996; Shmulovich et al. 1999; Driesner et al. 2000; Foustoukos et al. 2004; Liebscher et al. 2005, 2006a, 2007) at a wide range of temperatures and pressures. The strong contrast in density and composition between coexisting vapors and liquids approximating physical conditions deep in the 2-phase region of the NaCl-H$_2$O system (Bischoff 1991; Berndt et al. 2001), however, inhibits sampling at conditions very near the vapor-liquid-halite boundary (Seyfried et al. 1987; Shmulovich et al. 1999; Pokrovski et al. 2002). Recent advances in experimental geochemistry, however, have permitted such volume limitations to be partially overcome following an open-system approach that allows the halite-vapor stability field to be reached (Foustoukos and Seyfried 2007a).

Experimental challenges introduced during the formation of low-density vapors, are also linked to the dearth of experimental data on mineral-fluid equilibria for heterogeneous phase separated systems. Under these conditions, the activity and chemical potential of aqueous species dissolved in the vapor and liquid fractions are controlled by the imposed mineral-fluid equilibria as dictated by the composition of the coexisting mineral phases. Despite the fact that such experiments have the potential to enhance our understanding of phase separation effects on water/rock interactions, existing experimental data are limited to the quartz-NaCl-KCl-H$_2$O and basalt-seawater systems (Bischoff and Rosenbauer 1987; Foustoukos and Seyfried 2007b). Fluid immiscibility conditions in the presence of redox buffering mineral assemblages, however, are of particular significance, as phase separation might be a major mechanism for extended hydrothermal alteration and migration of subseafloor reaction zones towards higher oxidation states. This phase-separation driven oxidation process has been envisioned to occur during release of H$_{2(aq)}$ and H$_2$S$_{(aq)}$ into the low-density and buoyant vapor (Ridge 1974; Drummond 1981; Drummond and Ohmoto 1985; Bischoff and Rosenbauer 1987). Accordingly, in order to evaluate the extent of redox gradients between vapor and liquid and the associated effects on metal transport, we also present new, unpublished experimental data from a series of phase separation experiments performed at 390 °C and pressures ranging from 290 to 240 bars, involving NaCl-KCl-bearing aqueous solutions and a pyrite-pyrrhotite-magnetite mineral assemblage.

PHASE RELATIONS IN THE NaCl-H$_2$O SYSTEM

Boiling of pure H$_2$O solution results in the formation of vapor ("steam") coexisting with liquid H$_2$O, while at conditions beyond the critical point of H$_2$O (374 °C – 220 bars) only one phase exists. By introducing NaCl into the system, however, the critical point moves to higher temperature and pressure, allowing separation of compositionally distinct vapor and liquid (brine) over a wide range of temperature and pressure conditions (Keevil 1942; Sourirajan and Kennedy 1962; Bischoff 1991). For these NaCl-bearing fluids, phase separation is primarily driven by the long-range interactions (Coulombic forces) developed between dissolved ions

(Pitzer 1995). The establishment of strong ionic forces capable of enriching vapor with electrolytes is only feasible during a corresponding decrease of dielectric constant of the solutions, occuring with progressive decrease of solution density at elevated temperatures and pressures (Oelkers and Helgeson 1990). Upon phase separation of the homogeneous NaCl-bearing fluid, two distinctively different liquid fractions are formed, assembling a conjugate pair of low ("vapors") and high ("liquids") salinity/density aqueous solutions.

The corresponding vapor-liquid compositional variability can be quantitatively described by the now well-established phase relations in the NaCl-H_2O system (Sourirajan and Kennedy 1962; Bischoff and Rosenbauer 1984, 1988; Bischoff 1991; Berndt et al. 2001; Driesner and Heinrich 2007). Numerous experimental studies conducted to measure vapor/liquid composition and density variability (Keevil 1942; Olander and Liander 1950; Copeland et al. 1953; Sourirajan and Kennedy 1962; Khaibullin and Borisov 1966; Liu and Lindsay 1972; Mashovets et al. 1973; Urusova 1974, 1975; Parisod and Plattner 1981; Wood et al. 1984; Bischoff et al. 1986; Chou 1987; Pitzer et al. 1987; Rosenbauer and Bischoff 1987; Bischoff and Rosenbauer 1988; see Liebscher 2007, this volume, for a review of experimental studies) resulted in the development of vapor-liquid-halite equilibrium tables allowing an accurate estimation of the composition and density of the different phases (Sourirajan and Kennedy 1962; Bischoff 1991). Data reveal the complex changes of the two-phase surface of the NaCl-H_2O system (Fig. 2a), with some uncertainty related to accurately identifying the vapor-liquid-halite curve and the vapor-liquid boundary at conditions very close to the critical point of H_2O (Bischoff et al. 1986). Recent advances in thermodynamic modeling, however, allow a more detailed analysis of these data. The mathematical model *Salt-Therm 1.0* (Berndt et al. 2001), e.g., accurately estimates the dissolved Cl concentrations in conjugate vapor-liquid pairs over a wide range of pressure/temperature conditions. This code includes an integrated compilation of compositional and density data retrieved from NaCl-H_2O experiments and theoretical studies (Haas 1976a,b; Pitzer et al. 1984; Tanger and Pitzer 1989; Bischoff 1991; Anderko and Pitzer 1993). Density, temperature, and the corresponding phase relations are stored as multispline functions of enthalpy (0 – 5000 J/g), pressure (1 – 1000 bars) and NaCl composition (1 – 100%), allowing for fitting of the thermodynamic properties across the *P-T* space. Expressing the physical properties as enthalpy-based functions provides detailed estimations of the NaCl-H_2O phase relations over a wide range of pressure and temperature conditions (Mercer and Faust 1979; Hayba and Ingebritsen 1994). The code has been successfully implemented in field and theoretical studies assessing the effect of phase separation not only during the chemical and thermal evolution of the low-salinity vent fluids at the Main Endeavour Field (JDF) (Seyfried et al. 2003), but also by imposing important physical constraints on the range of maximum dissolved Cl concentrations of vent fluids ascending from subseafloor reaction zones (Fontaine et al. 2007). Such compilation efforts of experimentally determined NaCl-H_2O phase relations have proven essential to study two-phase (fluid) phenomena observed in vent fluids issuing from mid-ocean ridge and back-arc hydrothermal systems (Berndt and Seyfried 1997; Charlou et al. 2002; Seyfried et al. 2003; Stoffers et al. 2006; Von Damm 2000).

Utilizing *Salt-Therm 1.0* (Berndt et al. 2001) together with results of previous studies (Bischoff 1991; Sourirajan and Kennedy 1962), the P-T-x_{NaCl} salinity isopleths can be constructed, revealing a "bell-like" shape (Fig. 2a), with the wt% NaCl composition of conjugate vapor-liquid pairs arranged around the critical curve of the NaCl-H_2O system. The topology of the *P-T-x* surface shows that the vapor-liquid region occupies a large portion of the system, while under more extreme P-T (i.e., low H_2O density conditions) the halite stability field is encountered. The solubility of halite with coexisting vapor and liquid, however, does not intersect the critical curve. The extreme *P-T-x* conditions needed for halite-saturated phases is reflected by the dearth of experimentally determined densities for the halite-saturated liquids, encouraging extrapolations from measurements conducted in undersaturated liquids to constrain phase relations (Bischoff 1991). A similar "bell-shaped" topology is also indicated when the enthalpy of the low and high

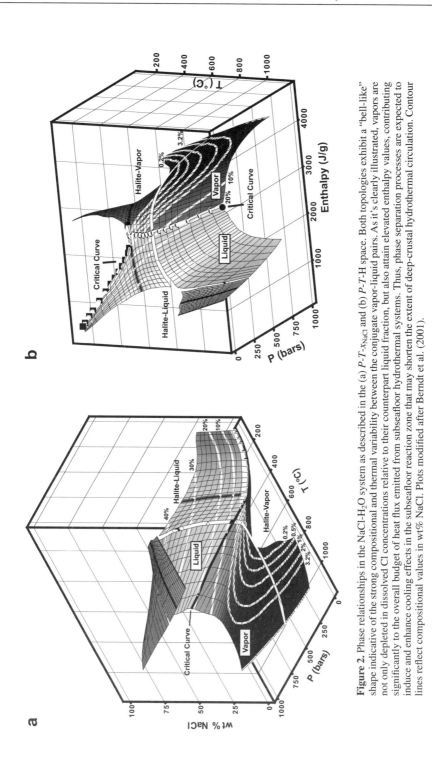

Figure 2. Phase relationships in the NaCl-H_2O system as described in the (a) P-T-x_{NaCl} and (b) P-T-H space. Both topologies exhibit a "bell-like" shape indicative of the strong compositional and thermal variability between the conjugate vapor-liquid pairs. As it's clearly illustrated, vapors are not only depleted in dissolved Cl concentrations relative to their counterpart liquid fraction, but also attain elevated enthalpy values, contributing significantly to the overall budget of heat flux emitted from subseafloor hydrothermal systems. Thus, phase separation processes are expected to induce and enhance cooling effects in the subseafloor reaction zone that may shorten the extent of deep-crustal hydrothermal circulation. Contour lines reflect compositional values in wt% NaCl. Plots modified after Berndt et al. (2001).

salinity phases are plotted in the *P-T*-H space (Fig. 2b). The computational ability to determine the enthalpy of the vapor-liquid conjugate pairs is a unique characteristic of the *SalTherm 1.0* algorithm that distinguishes it from previous NaCl-H$_2$O composition/density compilations. The topology of the *P-T*-H surface reveals relatively high enthalpy for the low-density vapors, supporting the hypothesis that vapor addition provides an effective means of heat transfer from the lower to upper parts of the subseafloor hydrothermal system (Fig. 1). On the other hand, the denser liquids will descend and accumulate at depth (Bischoff and Rosenbauer 1989; Fontaine and Wilcock 2006) establishing a low-enthalpy regime. Once more, extreme phase separation conditions associated with vapor-halite coexistence require elevated enthalpy values (~3000 - 4000 J/g). Accordingly, hydrothermal systems characterized by halite formation can be expected to be relatively short-lived owing to the effects of cooling of the subseafloor reaction zone by the formation and ascent (isoenthalpic adiabatic decompression) of high-enthalpy vapors (Bischoff and Pitzer 1985; Drummond and Ohmoto 1985).

Phase separation processes in submarine hydrothermal systems are generally defined as "supercritical" or "subcritical" in reference to the critical point of a 3.2 wt% NaCl solution (407 °C – 298 bars), a composition that approximates seawater. This classification was first used by Welham and Craig (1979) to describe phase separation inferred to contribute to the observed composition of high-temperature (400 °C) vent fluids sampled at 21°N East Pacific Rise (EPR). These investigators linked variability of dissolved chloride to phase relations determined earlier by Sourirajan and Kennedy (1962), pointing out that below the critical point of seawater phase separation can be equated to "subcritical" *boiling* producing ~3.2 wt% NaCl liquid coexisting with a low-salinity vapor, while during "supercritical" phase separation, a high-salinity vapor (≥3.2 wt% NaCl) is formed by *condensation* of a small amount of liquid (see figure 15 in Bischoff and Pitzer 1985; cf. introductory discussion in Liebscher and Heinrich 2007). These definitions, however, reflect vapor and liquid abundances at conditions very close to the critical curve of the NaCl-H$_2$O system for a fluid with seawater bulk composition. This can be illustrated more clearly, when the relative fractions of vapor (x_{vapor}) and liquid phases (x_{liquid}) are determined for "subcritical" and "supercritical" conditions following simple mass balance constraints:

$$x_{liquid} Salinity_{liquid} + x_{vapor} Salinity_{vapor} = Salinity_{bulk} \tag{1}$$

$$x_{liquid} + x_{vapor} = 1 \tag{2}$$

Data presented in Figure 3 indicate that differences between the relative fraction of liquid under "subcritical" and "supercritical" conditions are noticeable only near the two-phase boundary. As pressures and temperatures depart from the critical curve, however, vapor is volumetrically dominant and attains low dissolved Cl concentrations, reducing in this way the distinction between "supercritical" and "subcritical" conditions. This can be further supported by recent experimental results that reveal identical trace element partitioning between vapor and liquids under both "subcritical" and "supercritical" conditions (Foustoukos and Seyfried 2007a). Thus, in some cases, classifying phase separation as "sub-" or "supercritical" based on pressure and temperature conditions relative to the critical point of seawater might not be particularly meaningful, especially in the light of segregation effects that can radically change the bulk composition of the fluid.

FIELD OBSERVATIONS OF PHASE SEPARATION IN SUBMARINE HYDROTHERMAL SYSTEMS

Since the first discovery of hot springs on the Galapagos Spreading Center (Corliss et al. 1979; Edmond et al. 1979), the overwhelming effect of phase separation on the chemical evolution of submarine hydrothermal systems has become obvious. As noted earlier, the most

Figure 3. Diagram depicting the evolution of liquid fraction (x_{liquid}) during "subcritical" (■) and "supercritical" (□) phase separation. These conditions are directly linked to the relative abundance of the two phases, with "subcritical" boiling inducing formation of small amounts of vapor, while "supercritical" conditions are characterized by condensation of diminutive fractions of liquid. It appears, however, that these definitions can only be applied at conditions very close to the two-phase boundary, where liquid may indeed attain large fraction values. Accordingly, even when P-T conditions do not greatly deviate from the critical curve, low-salinity vapors are the dominant phases resulting in indistinctness between "sub-" and "supercritical" phase separation processes. Phase relationships in the NaCl-H_2O were obtained from Berndt et al. (2001).

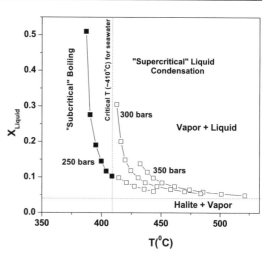

compelling evidence for this comes from sharp differences in dissolved chloride concentration between high-temperature vent fluids (~30 - 1250 mmol/kg$_{\text{solution}}$) and seawater (540 mmol/kg$_{\text{solution}}$) (see Table 2 in German and Von Damm, 2003). Furthermore, the enrichment of volatile species in low-chloride vent fluids recognized from the earliest studies at 21°N East Pacific Rise and South Juan de Fuca Ridge (Welhan and Craig 1979; Bischoff 1980; Von Damm and Bischoff 1987) is difficult or impossible to explain in the absence of phase separation effects. Interestingly, these first manifestations of phase separation phenomena were also linked to enhanced metal transport and elevated water/rock ratios (Berndt and Seyfried 1990), while processes like adiabatic decompression were introduced to describe the thermal evolution of vent fluids (Bischoff and Rosenbauer 1989). Subsequent field studies in a number of seafloor hydrothermal vent sites further supported the significance of phase separation and segregation on heat and mass transport throughout the crust at mid-ocean ridges. Indeed, vent fluids with dissolved chloride that differs significantly from seawater values have now been sampled from virtually all ridge segments, regardless of spreading rate or the bedrock composition that host the hydrothermal system, indicating the universality of the phenomena (Campbell et al. 1988; Butterfield et al. 1990, 1994; Jean-Baptiste et al. 1991; Lilley et al. 1993, 2003; Edmond et al. 1995; James et al. 1995; Von Damm et al. 1995, 1997, 1998, 2003; Charlou et al. 1996, 2000, 2002; Edmonds et al. 1996; Gamo et al. 1996; Oosting and Von Damm 1996; Lein et al. 2000; Von Damm 2000, 2004; Chiba et al. 2001; Douville et al. 2002; Seewald et al. 2003; Seyfried et al. 2003; Foustoukos et al. 2004; Koschinsky et al. 2006). Recent discoveries have also revealed the existence of high-chloride fluids issuing from chimney structures at the Kairei and Edmond vent fields along the Central Indian ridge (Gallant and Von Damm 2006). In general, vent fluids sampled from hydrothermal systems located at slow spreading ridges appear to attain relatively constant vapor- or liquid-like dissolved chloride concentrations for relatively long intervals, while for hot springs at fast and intermediate spreading ridges, changes in the spatial and temporal variability of vent fluid compositions are more obvious due to the inherent frequency of magmatic and tectonic events associated with these systems (Detrick et al. 1987; Kent et al. 1993; Van Ark et al. 2007). Of particular interest, however, are the back-arc hydrothermal systems, like those situated at the seamounts of the Tonga arc, where hydrothermal circulation takes place within shallow volcanoes (<500 m below seafloor), which moderates temperatures (~250 °C) owing to phase separation effects, while producing vent fluids depleted in transition metals, but enriched in dissolved volatile species (e.g., $CO_{2(aq)}$). Due to the lower pressure of these relatively shallow hydrothermal systems, however, fluids vent in the form of "flame-like" discharges with robust steam separation phenomena occurring at chimney orifices (Stoffers et al. 2006).

Perhaps the most intensively and continuously studied hydrothermal vent systems where phase separation effects have long been recognized, involve the hot springs located at 9-10°N and 21°S East Pacific Rise (Von Damm et al. 1995, 1997, 2003; Oosting and Von Damm 1996; Von Damm 2000, 2004). In addition to the relatively high vent fluid temperatures (>390 °C), resulting from the high seafloor pressures (~250 bars), these Cl-depleted fluids have revealed conspicuous deviations in Br/Cl ratios from seawater values (Fig. 4). The Br/Cl ratio, however, is generally accepted as one of the most conservative chemical parameters during water/ rock interaction, so the observed departures likely indicate fluid phase separation effects. Furthermore, early hydrothermal experiments suggested minimal fractionation of these species between vapor and liquid (Berndt and Seyfried 1990; Berndt and Seyfried 1997). Accordingly, it was proposed that more extreme phase separation conditions involving coexistence of halite might be a plausible mechanism to affect Br/Cl ratios. Recent experimental data confirmed the strongly incompatible nature of Br in the halite crystal structure at high temperature, that enhaces Br partitioning into the vapor for all conditions in which vapor-halite coexist (Foustoukos and Seyfried 2007a). Thus, for vent fluids indicating Br/Cl ratios less than seawater such as at 9-10°N EPR, dissolution of 5-25 mmol/kg$_{solution}$ halite is considered an additional source of dissolved Cl (Fig. 4) (Berndt and Seyfried 1997). Recent studies have also proposed that Rayleigh distillation processes developed during vapor-liquid segregation might contribute to the Br-depletion of the escaping vapors (Liebscher et al. 2006b). In the case of the Brandon vent fluids at 21°S EPR, chloride-normalized Br concentrations show a

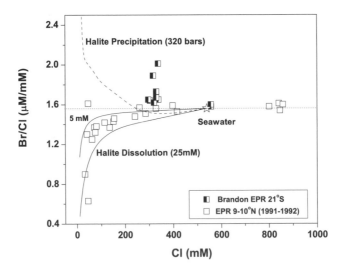

Figure 4. The effect of extreme phase separation and corresponding halite precipitation/dissolution processes on the Br/Cl ratios of vapor-rich vent fluids. Depletion on Br/Cl ratios relative to seawater values is closely related to dissolution of halite, a process that has been suggested to control the distribution of Br/Cl in the Cl-depleted vent fluids at 9-10°N EPR (Berndt and Seyfried 1997). In contrast, the Brandon 21°S EPR vent fluids (Von Damm et al. 2003) exhibit a sharp increase of Br/Cl ratios indicative of halite precipitation. Intersection of the halite stability field, however, is precluded at these moderate dissolved Cl concentrations (Bischoff 1991; Berndt et al. 2001), while predicted Br/Cl ratios (Eqn. 5) are still lower than those encountered at the Brandon vapor-rich vent fluids. Accordingly, a subseafloor mixing process is proposed, involving inputs of significantly Cl-depleted but Br-enriched fluids evolved under extreme phase separation conditions that enhance *salting-out* and Br-halite immiscibility effects (Foustoukos and Seyfried 2007a). Concentration units are in mmol/kg$_{solution}$ (mM) and μmol/kg$_{solution}$ (μM).

clear increase relative to seawater (Fig. 4). The Br/Cl enrichment is indeed significant, being comparable in magnitude with the Br/Cl depletion observed in the vent fluids at 9-10°N EPR, indicative of halite precipitation/dissolution processes. Along the halite-liquid-vapor phase boundary, however, NaCl-H$_2$O phase relations (Bischoff 1991; Berndt et al. 2001) suggest that vapors should attain significantly lower Cl concentrations than the case for Brandon vent fluids; a conclusion that is counterintuitive in light of the Br/Cl data. These somewhat contradictory lines of evidence can be resolved by mixing a Cl-depleted seawater source fluid at approximately 420 °C, 320 bars with a high Br/Cl vapor formed at temperatures sufficiently high to cause halite precipitation (Foustoukos and Seyfried 2007a). Owing to the exponential increase in Br/Cl with decrease in dissolved chloride of the vapor-rich fluid coexisting with halite (Fig. 4), even modest amounts of this fluid could affect greatly the Br/Cl ratio of the mixture. If this supercritical mixing scenario proposed for the chemical evolution of the Brandon vapor vent fluids is correct, then it provides evidence of the complex nature of phase-separation related chemical and physical phenomena occurring deep in the oceanic crust.

EXPERIMENTAL STUDIES OF PHASE SEPARATION IN THE NaCl-H$_2$O SYSTEM

Empirical expressions and theoretical modeling

Although considerable progress has been made in the development of theoretical models of concentrated NaCl fluids, such as the "ion interaction model" (Pitzer 1990, 1995) and "scaled particle theory" (Masterton and Lee 1970), the effective working range of these models is largely limited to near-critical conditions and also restricted in compositional space. As a consequence of this, often tenuous assumptions are necessary to approximate chemical speciation in vapor and liquid at elevated temperatures and pressures. One particularly critical assumption often made involves assigning a value of unity for the activity coefficient for neutral aqueous species. For some time, however, it has been recognized that this assumption is inconsistent with experimental data depicting the dissociation of 1:1 neutral complexes (Ho et al. 1994, 2000, 2001). In fact, theoretically based regression calculations indicate that activity coefficients of neutral ion pairs in high temperature liquids may be large positive numbers (Oelkers and Helgeson 1991, 1993a). In effect, the larger that activity coefficient, the more likely it is that the neutral species (e.g., HCl$_{(aq)}$, NaCl$_{(aq)}$), will be "salted-out" of the liquid and effectively replaced by polyatomic ionic species (Pitzer and Schreiber 1987; Oelkers and Helgeson 1990). At pressures and temperatures within the 2-phase region of the NaCl-H$_2$O system, establishment of a low H$_2$O dielectric constant ($\varepsilon < 15$) has been shown to favor formation of higher order ionic complexes in electrolyte-concentrated aqueous solutions (Oelkers and Helgeson 1993b; Driesner et al. 1998; Sharygin et al. 2002) with vapors becoming more molecular (nonionic), while liquids get even more ionic. This enhances *salting-out* effects imposed on neutral aqueous species dissolved in the increasingly concentrated liquid (Pitzer and Schreiber 1987; Oelkers and Helgeson 1990, 1993a,b; Sharygin et al. 2002; Grover and Ryall 2004), resulting in a significant enhancement of elemental partitioning into the low salinity vapors.

To better understand the role of phase separation on the chemical composition of vapor and liquid, inherent difficulties related to the complexity of aqueous speciation at low density and dielectric constant conditions of H$_2$O need to be overcome, which is typically accomplished by use of empirical expressions. One such formulation is based on correlating apparent partitioning coefficients ($K_m = (M_{el})_{vapor}/(M_{el})_{brine}$[1]) with the density contrast between coexisting phases. This is a well-known empirical expression in studies involving mass transport in the steam-

---

[1] $(M_{el})_{vapor}$ and $(M_{el})_{brine}$ designates the total dissolved concentrations of each element in the corresponding phase

liquid system (Styrikovich et al. 1955; Smith et al. 1987; Simonson et al. 2000; Pokrovski et al. 2002; Palmer et al. 2004), and allows K_m to be described as a linear function of the density of vapor and coexisting liquid:

$$\log K_m = n_i \log\left(density_{vapor} / density_{liquid}\right) \tag{3}$$

where (n_i) is an empirical fitting coefficient linked to the vapor-liquid hydration and the volatility of the aqueous species (Simonson et al. 2000; Palmer et al. 2004). Elevated positive values of (n_i) indicate less affinity of non-volatile components to fractionate into the vapor phase, while negative values characterize species with greater volatility (e.g., dissolved gases; Smith et al. 1987), which also tend to exhibit negative hydration behavior. The different (n) values between volatile and non-volatile components has been estimated from vapor-liquid experimental and theoretical studies on vapor-liquid partitioning for a number of dissolved trace elements, metals and dissolved gases (Table 1) (Drummond 1981; Pokrovski et al. 2005; Foustoukos and Seyfried 2007a).

Vapor-liquid fractionation patterns can be also expressed in thermodynamic terms involving the variability of H_2O density in the P-T space (Foustoukos and Seyfried 2007a). For this to be effective, however, partition coefficients between vapor and liquid/halite phases must be expressed as ratios with Cl:

$$D_{el/Cl} = \frac{\left(M_{el}/M_{Cl}\right)_{vapor}}{\left(M_{el}/M_{Cl}\right)_{liquid}} \tag{4}$$

where $(M_{el}/M_{Cl})_{vapor}$ and $(M_{el}/M_{Cl})_{liquid}$ are the element to chloride ratios for vapor and liquid/halite, respectively. $D_{el/Cl}$ values greater than unity indicate the preferential fractionation into the vapor phase, while the opposite is true (liquid enrichment) when $D_{el/Cl}$ is less than 1. This formulation, when first introduced (Berndt and Seyfried 1990), was based on the assumption that the relative tendency of elemental partitioning should be governed by the type and strength of bond that forms the dominant aqueous species, which for most alkali and alkali earth elements is described by neutral chloro-complexes (Oelkers and Helgeson 1990; Sharygin et al. 2002; Sue and Arai 2004). Thus, an important advantage of using element/Cl ratios to assess trace element partitioning in multi-phase systems is the minimal effects of bulk fluid composition on the apparent partitioning coefficients (K_m) of elements that exist largely if not entirely complexed with chloride or other elements (e.g., Na), which are linked to Cl through electrical neutrality constraints (Shock et al. 1997; Sverjensky et al. 1997).

Table 1. Vapor-liquid hydration parameters (see Eqn. 3).

Element	$n(i)$	Ref.
$H_{2(aq)}$	−1.66 (±0.01)	[1]
$H_2S_{(aq)}$	−1.05 (±0.01)	[1]
$B(OH)_{3(aq)}$	0.52 (±0.1)	[2]
As (300-400 °C)	1.11 (±0.003)	[3]
As (450 °C)	0.69 (±0.05)	[3]
$SiO_{2(aq)}$	2.01 (±0.07)	[3]
Sb	2.29 (±0.08)	[3]
Au	2.73 (±0.31)	[3]
Br	2.99 (±0.03)	[2]
Li	3.01 (±0.05)	[2]
Rb	3.44 (±0.05)	[2]
Cs	3.46 (±0.05)	[2]
Na	3.83 (±0.06)	[3]
Fe	3.84 (±0.07)	[3]
Cu	3.86 (±0.09)	[3]
Zn	4.27 (±0.11)	[3]
Ag	4.69 (±0.10)	[3]

References: [1] This study and Drummond (1981); [2] Foustoukos and Seyfried (2007a); [3] Pokrovksi et al. (2005)

Elemental partitioning between vapor-liquid and vapor-halite

Early experimental studies conducted within the two-phase region of the NaCl-H_2O system were motivated by the need to better understand the relative effect of phase separation and mineral-fluid equilibria on the distribution of mobile elements (e.g., B, Li, Sr, Rb; Berndt and Seyfried 1990). Results revealed slight enrichment of Sr, Ba, Ca, and K into liquids, while Li and Br appeared to favor the low-salinity vapor fluids as noted previously. This was also the case for dissolved B, which most likely is controlled by the polar and gas-like nature of the dominant B-bearing hydroxyl aqueous species ($B(OH)_{3(aq)}$). In addition, the relative partitioning of the different trace elements between Cl-bearing vapor and liquid appears to be largely controlled by the electronegativity differences between the cations and anions that constitute the predominant neutral aqueous species. The lower the electronegativity values of the neutral chloro-complexes the bonding is more ionic, resulting in stronger partitioning into the liquid phase as expressed by the relative order of vapor enrichment Br > Li > Na,Cl > K > Ca > Sr > Ba derived from experimental results (Berndt and Seyfried 1990). This conclusion is in general agreement with recent studies (Table 1; Pokrovski et al. 2005; Foustoukos and Seyfried 2007a), although, at extremely low H_2O density and dielectric constant conditions, *salting-out* effects and mineral-fluid immiscibility phenomena, involving halite formation under phase separation conditions, will eventually prevail over tendencies attributed to electronegativity differences (Foustoukos and Seyfried 2007a).

Advances in understanding of element fractionation during phase separation resulted from a series of hydrothermal experiments conducted by Pokrovksi et al. (2005). Despite the fact that this study was mainly conducted to evaluate metal transport in magmatic-hydrothermal systems, results are applicable to submarine hydrothermal systems and relate well to experimental studies where density effects are similarly emphasized (Foustoukos and Seyfried 2007a). A density model involving apparent partitioning coefficients and density ratios between coexisting vapor and liquid (Eqn. 3) provides an effective first-approximation that can be used to describe vapor-liquid partitioning of aqueous species without considering complex speciation models. Indeed, predictions of phase separation effects based on the density model (Pokrovski et al. 2005) are generally in good agreement with fluid inclusion data (Audetat et al. 1998; Heinrich et al. 1999; Ulrich et al. 1999; Baker et al. 2004) from magmatic-hydrothermal systems. In general, the experimental and field data show a decrease of species volatility in the order of: As > Sb > Au > Fe > Cu > Zn > Ag (Table 1). Limitations, however, arise if metal transport is affected by abundant S-bearing volatile species (e.g., Cu, Au), a processes not described in the sulfur-free vapor-liquid experiments of Pokrovksi et al. (2005). Nevertheless, this study has shown the linkage between hydration, volatility and vapor enrichment for a wide compositional range; phenomena directly related with *salting-out* effects expected to be imposed with the progressive increase of ionic difference between vapor and liquid.

Building upon the previous experimental studies, recent phase separation experiments describe trace element fractionation patterns (Li, Br, Rb, Cs, B) not only within the vapor-liquid but also within the halite-vapor ± liquid region (Foustoukos and Seyfried 2007a). These experiments used an "open-system" experimental approach, utilizing flow-through reactors that provide sufficient volume to accommodate the low-density vapor phases. Experiments involved formation of conjugate vapor/liquid pairs at pressures ranging from 250 to 350 bars and temperatures from 388 °C to 550 °C. Results revealed a decrease in vapor enrichment in the order of Li ≅ Br > Rb ≅ Cs, which relates well to predictions based on electronegativity differences between the charged species constituting the dominant aqueous complexes. This is best illustrated by the distribution of apparent (K_m) (Table 1), as well as, by the Cl-normalized ($D_{el/Cl}$) partition coefficients, which are functions of the water density (ρ_w) across the *P-T* space of the vapor-liquid and halite-vapor region:

$$\log(D_{el/Cl}) = a + b\log(\rho_w) + c\log(\rho_w)^2 + d\log(\rho_w)^3 \tag{5}$$

where *a, b, c, d* are coefficients unique to specific trace elements (Table 2), and are valid for H_2O density range from 0.094 to 0.28 g/cm^3. These data are in good agreement with partitioning coefficients derived from previous phase separation experiments (Berndt and Seyfried 1990, 1997; Liebscher et al. 2006b) (Fig. 5), although these earlier studies were conducted at conditions distant to the halite-vapor-liquid boundary (NaCl-H_2O). Most importantly, results indicate an abrupt change in par-

Table 2. Fit parameters for the polynomial expression of Cl-normalized partitioning coefficients (see Eqn. 5).

	Br	Li	Rb	Cs
a	−13.04	−10.29	−23.43	−23.94
b	−54.95	−44.36	−102.95	−105.20
c	−76.93	−64.05	−148.98	−152.28
d	−36.09	−31.14	−70.71	−72.31

titioning as the halite field of stability is approached, characterized by highly non-linear enrichment in vapor for all species, effectively caused by strong *salting-out* effects. Upon intersection of the halite-vapor ± liquid field (Fig. 5), however, vapor enrichment is further enhanced, likely encouraged by trace element exclusion from the halite structure. Apparently, not only Br is excluded from NaCl mineral phases (Berndt and Seyfried 1997), but also other elements that are largely transported as aqueous chloro-complexes.

In the case of dissolved boron, compositional variability during vapor-liquid equilibria primarily relates to *salting-out* effects and the polar and gas-like nature of the dominant B-bearing hydroxyl aqueous species ($B(OH)_{3(aq)}$) (Styrikovich et al. 1960; Berndt and Seyfried 1990; Schmidt et al. 2005). Apparent partition coefficients (K_m) data describing transfer of boron between vapor and coexisting liquid support a strong volatile behavior and a clear enhanced affinity to participate in the low-density vapor phase (hydration (n) value = 0.52) (Table 1), although previous experimental studies indicate otherwise for some conditions

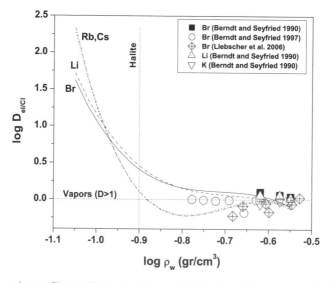

Figure 5. Trace element Cl-normalized partitioning coefficients ($D_{el/Cl}$) between vapor-liquid and halite-vapor expressed as function of the density of H_2O (Eqn. 5). This empirical expression is based on after data retrieved from experiments conducted under phase separation conditions that included the vapor-liquid and vapor-halite regions (Foustoukos and Seyfried 2007a). In general, predicted $D_{el/Cl}$ values are in agreement with previous experimental studies. Within the vapor-liquid region, Li-Br and Rb-Cs pairs appear to follow an inverse fractionation, with the latter being enriched in liquids, while upon intersection of the halite stability field, elemental partitioning strongly favors vapor-enrichment apparently affected by strong *salting-out* and elemental halite exclusion effects, similar to those initially suggested for Br (Berndt and Seyfried 1997).

(Berndt and Seyfried 1990; Liebscher et al. 2005). More specifically, Liebscher et al. (2005) suggested that boron vapor-liquid fractionation is weak and slightly more favorable towards liquid enrichment. At lower bulk chloride composition, boron does indeed appear to partition weakly into the liquid, however, when the salinity of the coexisting liquid exceeds ~40-50 wt% NaCl and halite stability is achieved (Fig. 2a), then vapor enrichment becomes clear (Foustoukos and Seyfried 2007a). Thus, following patterns observed for halide-bonded alkaline and alkaline earth elements, boron tends to be strongly incompatible in high salinity fluid and mineral phases, as has been previously inferred from fluid (vapor/liquid)-saturated melt experiments (Schatz et al. 2004). Apparently, it is the inherent concentration of the liquid phase that governs the extent of vapor-liquid B fractionation, a conclusion consistent again with *salting-out* phenomena. Moreover, boron Cl-normalized partitioning coefficients appear to be a logarithmic function of the reduced density conditions (ρ_{rd}):

$$\log\left(D_{B/Cl}\right) = 0.570335 - 6.98584\log\left(\rho_{rd}\right) \tag{6}$$

where reduced density is defined as the ratio of the density of H_2O at experimental conditions with that along the critical curve (ρ/ρ_{cr}). This empirical expression is similar to that utilized to describe the thermodynamic behavior of dissolved boron in the water-steam system (Kukuljan and Alvarez 1999). The conservative behavior of boron during phase separation is reflected in the $D_{B/Cl} - \rho_{rd}$ relationship, but can also be shown by the linear 1:1 correlation between B/Cl ratio and Cl over a wide range of chemical and physical conditions, although these data are dependent on fluid compositional constraints induced prior to phase separation processes (Foustoukos and Seyfried 2007a). This conservative behavior of boron during phase separation has important implications for distinguishing fluid/rock interaction from phase separation effects based on dissolved boron concentrations in submarine and epithermal hydrothermal systems. Boron derived from leaching of basalt followed by phase separation would follow the described linear correlation, while addition of excess boron to the hydrothermal system from magmatic degassing might result in a steepening of the line between B/Cl ratio and total dissolved Cl.

Stable isotope fractionation in the two-phase region of the NaCl-H_2O system

In order to better distinguish mineral-fluid equilibria from phase separation effects, a number of experimental studies have been performed to explore stable isotope vapor-liquid fractionation of highly mobile elements (e.g., Li, B). Once again, these studies have been limited to *P-T-x* space, where salinity and density contrast between the vapor-liquid phases is modest (Fig. 2a). Thus, for the majority of stable isotope systems investigated only slight isotopic fractionation is observed (Table 3). For example, vapor-liquid fractionation for $\delta^{11}B$ is less than 1‰ based on data reported by Spivack et al (1990) and Liebscher et al. (2005), with experimental conditions ranging from 400 °C to 450 °C and 234 bars to 379 bars. Both studies concluded that trigonal speciation of boron might be similar in vapor/liquid phases with the highest fractionation of 2.2‰ predicted to occur along the vapor-liquid-halite boundary at 450 °C (Liebscher et al. 2005). Furthermore, $\delta^{11}B$ fractionation was purported to alter the boron isotopic composition of ascending vapor vent fluids, if open system, Rayleigh distillation processes prevail. These models, however, predict a decrease in dissolved boron content for low-salinity vapor fluids, a conclusion that departs from recent experimental data (Foustoukos and Seyfried 2007a). Thus, for the range of pressures and temperatures where $\delta^{11}B$ isotopic fractionation has been studied, it appears that the effect of phase separation on the isotopic variability of $\delta^{11}B$ measured in high temperature vent fluids is minimal. Similarly, δ^7Li fractionation is not affected by vapor-liquid segregation (Table 3), in a manner consistent with the chemical composition of the vapor-rich vent fluids collected from the Main Endeavour Field (Juan De Fuca ridge) (Foustoukos et al. 2004), as well as by experimental data retrieved under subcritical and supercritical phase separation conditions (Berndt et al. 1996; Foustoukos et al. 2004; Liebscher et al. 2007). Finally, the same can be also inferred for Cl isotope fractionation between vapor and liquid, with such effects being less than 0.4‰ (Bonifacie et al. 2005; Liebscher et al. 2006a).

Table 3. Experimental data on vapor-liquid fractionation in
the NaCl-H$_2$O system for selected stable isotopes.

Isotope	1000lna_{v-l}	T (°C)	P (bars)	wt% NaCl (vapor)	wt% NaCl (liquid)	Ref.
δD	2.0-10.4	399-450	277-397	0.33-2.44	5.11-24.67	[1]
δD	2.3-14.5	350-600	152-910	0.003-4	8-33	[2]
δ^{18}O	−1.84-(−0.16)					[2]
δD	7.0-10.5	301-396	75-230	0.0002-0.058	18.52-19.66	[3]
δ^{18}O	−1.39-(−0.74)					[3]
δ^{11}B	(−2.33)-0.38	425-450	346-379	0.702-1.7	9.83-18.61	[4]
δ^{11}B	0.1-0.9	400-450	234-418	0.064-7.38	4.60-26.2	[5]
δ^{7}Li	(−0.5)-0.2	431-447	322-397	0.38-1.72	14.6-18.58	[1],[6]
δ^{7}Li	(−0.1)-1.0	400	204-282	0.035-24.6*	14.6-18.58	[7]
δ^{37}Cl	(−0.37)-0.32	400-450	234-418	0.064-7.38	4.60-26.2	[8]

* Concentration reflects wt% LiCl.
References: [1] Berndt et al. 1996; [2] Shmulovich et al. 1999; [3] Driesner and Seward 2000; [4]Spivack et al. 1990;
[5] Liebscher et al. 2005; [6] Foustoukos et al. 2004; [7] Liebscher et al. 2007; [8] Liebscher et al. 2006a

Hydrogen isotope fractionation, however, has an important impact on the D/H composition of coexisting vapor and liquid. Experiments conducted at temperatures below 350 °C and along the steam saturation curve (Horita et al. 1993a,b, 1995) recognized that δD fractionation between liquid water and steam is not only temperature dependent, but is also affected by the NaCl content of the aqueous solutions ("isotope salt effect"). Phase separation experiments performed by Berndt et al. (1996), Shmulovich et al. (1999) and Driesner and Seward (2000) did show that low-salinity vapors are D-enriched relative to liquids in the two-phase region of the NaCl-H$_2$O system, with fractionation generally being between 2 and 15‰ (Table 3). An empirical expression in the form of: 1000 ln$a_{(vapor-liquid)} = a + b$ log ($P_{cr} - P + c$) was developed to estimate the relative extent of vapor-liquid fractionation effects as a function of the deviation from the critical curve of the NaCl-H$_2$O system (Berndt et al. 1996). More recently, theoretical and experimental studies supported the greater effect of vapor density in the observed D/H fractionations (Driesner 1997; Shmulovich et al. 1999). Accordingly, an expression analogous to that developed earlier for boron (Eqn. 6), can be applied to D/H vapor-liquid isotopic fractionation, incorporating experimental data in the NaCl-H$_2$O system and at temperatures greater than 350 °C (Berndt et al. 1996; Shmulovich et al. 1999; Driesner et al. 2000):

$$1000\ln a_{(vapor-liquid)} = 4.1(\pm 0.4) - 31.2(\pm 3.8)\log(\rho_{rd}) \qquad (7)$$

Again reduced density (ρ_{rd}) is defined as the ratio of the density of H$_2$O at experimental conditions to that along the critical curve (ρ/ρ_{cr}). Equation (7)^2 is valid for the density range of 0.14-0.29 g/cm^3 (Fig. 6a), which significantly extends the range of pressure-temperature conditions applicable to seafloor hydrothermal systems. However, no correction factors are added to account for fractionation behavior very near to the critical point, as was done by Berndt et al. (1996) (parameter (*c*)).

Thus, a simple equilibrium process can produce vapors with δD values 2-15‰ higher than for the residual liquid (Fig. 6a). Clearly this does not explain the negative values from the East Pacific Rise (Fig. 6b), all of which are associated with low chloride vent fluids. However, phase

---

^{2} Correlation coefficient $r^2 = 0.62$. Estimated standard deviation of the fitting curve (Root-MSE) = 2.4

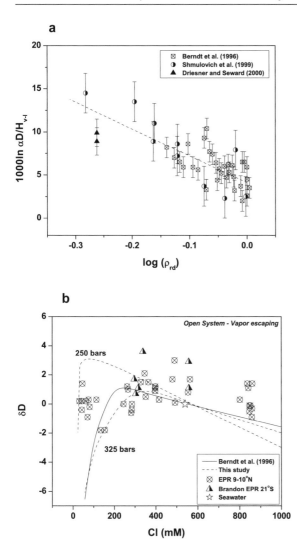

Figure 6. Hydrogen isotope fractionation during fluid phase separation. Experimental data in the NaCl-H$_2$O system reveal the tendency of low-salinity vapors to be enriched in D/H relative to coexisting liquids with corresponding isotope fractionation values ranging from 2 to 15‰ (a). Formulations involving the concept of reduced density (ρ_{rd}), which describes the ratio of water density at supercritical and critical conditions, respectively, allowed for D/H vapor-liquid isotopic fractionation to be predicted at a wide range of temperature and pressure conditions in the two-phase region of the NaCl-H$_2$O system (Eqn. 7). For the Cl-depleted vent fluids at 9-10°N EPR (Shanks 2001), however, open-system and Rayleigh distillation during vapor escaping, at pressures between 250 and 325 bars, should be accounted for, permitting not only liquids but also vapors to attain negative δD values (b). The hydrogen isotopic composition of vent fluids issuing from the Brandon hydrothermal system (Von Damm et al. 2003), even though might be affected by complex physical and chemical processes likely involving supercritical mixing, appears to be within the compositional range of the 9-10°N EPR vent fluids.

separation is a dynamic process, so Rayleigh distillation effects need to be considered. Berndt et al. (1996), for example, described a possible process that involves open system escape of vapor produced during incipient phase separation, as might occur during isobaric heating events related to dike emplacement into the near surface of the mid-ocean ridge. Under these conditions, the residual liquid is predicted to attain very negative δD values and the vapor evolved also may have negative δD values. Figure 6b shows modeling results based on the empirical expressions of D/H fractionation presented above, in comparison with the δD composition of selected vapor and liquid samples from 9-10°N (Shanks 2001) and 21°S EPR (Von Damm et al. 2003). Although the model proposed by Berndt et al. (1996) to account for the negative δD values of the vent fluids has merit at East Pacific Rise, the recognized concern about the level of distillation necessary to accomplish this is a significant obstacle, as is the relative lack of evidence of Cl-enriched vent fluids also with strongly negative δD values. These observations suggest that perhaps phase separation alone is not the answer, but that this in combination with rock-fluid interaction may be. Indeed, some of the more isotopically negative (D/H) vapors at 9-10°N EPR have relatively high $\delta^{18}O$ values, confirming significant rock-fluid interaction at low water/rock ratios likely involving formation of chlorite-bearing alteration mineral phases (Shanks et al. 1995; Berndt et al. 1996). The effect of mineral-fluid equilibria can be further supported by the reverse $\delta^{18}O$ isotope fractionation patterns observed in phase separation experiments, where vapors attain lower $\delta^{18}O$ values than the coexisting liquids (Table 3) (Horita et al. 1995; Shmulovich et al. 1999; Driesner et al. 2000); a trend not consistent with the isotopic composition of vent fluids issuing from the submarine hydrothermal systems at East Pacific Rise. Additional discussions on D/H and more specifically on $^{18}O/^{16}O$ isotopic fractionation during phase separation processes can be found in Liebscher (2007) and Webster and Mandeville (2007).

Phase separation and mineral-fluid equilibria

Experimental studies investigating phase separation effects on NaCl fluids coexisting with mineral phases are very scare and largely limited to quartz-NaCl-KCl-H$_2$O equilibria and basalt-seawater systems (Bischoff and Rosenbauer 1987; Foustoukos and Seyfried 2007b). The quartz-fluid equilibria phase separation experiments were conducted at a wide range of temperatures and pressures (389-430 °C, 241-352 bars). Experimental data allowed for the non-ideal activity-concentration relations of dissolved SiO$_2$ in high-temperature NaCl-bearing vapor fluids to be described at conditions where *salting-in* effects significantly alter well-known quartz-NaCl-H$_2$O thermodynamic equilibrium patterns (Fournier and Potter 1983; Von Damm et al. 1991). Results revealed that dissolved silica concentrations in vapor phase for the P-T range investigated, can be expressed as functions of temperature (°C), density of H$_2$O (ρ_w) (g/cm^3), and molality of dissolved chloride (m_{cl}):

$$\log m_{SiO_2(aq)} = -5.10347 + 0.00883565T_c + 0.105367\log m_{Cl} + 1.03419\log\rho_w \qquad (8)$$

Accordingly, assuming quartz-bearing alteration assemblages, dissolved silica concentrations measured in moderately low-salinity vent fluids can be used to constrain subseafloor depth (pressure) and temperature of the water/rock interaction, providing a geochemical proxy to identify the physical conditions at the point of last quartz-fluid equilibration in subseafloor reaction zones. Applying this approach to fluids issuing from vents at 9-10°N and 21°S EPR suggests relatively shallow subseafloor depth, but still supercritical conditions with temperatures in excess of 420 °C (Foustoukos and Seyfried 2007a,b).

Experimental studies involving redox-buffering mineral assemblages are limited to a single investigation conducted by Bischoff and Rosenbauer (1987), where basalt-seawater reactions were evaluated at 390 °C and 240-250 bars (Table 4). Results suggest that phase separation could significantly contribute to enhancing volatile and metal vapor enrichment, a hypothesis that has been previously presented in a series of theoretical studies (Ridge 1974; Drummond 1981; Drummond and Ohmoto 1985). Such vapor-enrichment could lead to a severe depletion of metals and progress oxidation processes as vapor and liquid segregate due to density contrast. To better describe the effect of phase separation on Fe and volatile species (e.g., H$_{2(aq)}$, H$_2$S$_{(aq)}$) transfer during heterogeneous phase relations, a series of experiments were performed at 390 °C and pressures ranging from 290 to 240 bars (unpublished data); conditions broadly analogous to those used by Bischoff and Rosenbauer (1987) for the basalt-bearing system. For these experiments, however, a pyrite-pyrrhotite-magnetite (PPM) mineral assemblage was used to provide requisite buffering of key aqueous species. Experimental methodologies followed those reported by Foustoukos and Seyfried (2005, 2007b). In particular, a flexible gold reaction cell (Seyfried et al. 1987) was used, while phase separation was induced by decreasing pressure along the 390 °C isotherm. Vapor and liquid endmembers were sampled by changing the orientation of the vessel, such that gravity effects facilitate segregation of vapor and liquid-rich components. In addition to the redox buffering assemblage, mineral

Table 4. Selected chemical composition of fluid samples collected during heterogeneous (mineral-fluid) phase separation experiments.

Mineral Reactants	Phase Region		T (°C)	P (bars)	ΔT^* (°C)	Cl (mm)§	Fe (mm)	$H_{2(aq)}$ (mm)	$H_2S_{(aq)}$ (mm)	pH (25 °C)	Ref.
Po - Py - Mgt											
Experiment 1											[1]
	single	1	390	281		652	3.72	0.75		3.4	
	single	2	390	280		633	4.30	0.7		3.5	
	single	3	390	279		656	4.18	0.69		3.5	
	vapor	4	390	254	1	43	0.23	6.19		3.2	
	liquid	5	390	254	1	1002	8.28	0.84		3.4	
	single	6	390	286		704	6.74	0.76		3.3	
	single	7	390	281		696	6.91	0.91		3.8	
	vapor	8	390	243	5	17	0.29	11.2			
	single	9	390	286		644	8.78	0.79		3.6	
	single	10	390	286		659	9.34	0.93		3.7	
	vapor	11	390	241	5	15	0.26	6		3.7	
Experiment 2	single	1	390	293		611	4.16	0.87	13.4	4.6	
	single	2	390	288		613	4.57	1.17	13.3		
	vapor	3	390	251	−2	46	0.39	4.70	48		
	single	4	390	282	0	621	2.4	0.68	13.2		
	vapor	5	390	251	4	30	0.22	3.75	37.7	3.9	
	liquid	6	390	251	4	1742	12.0	0.46	8.2	3.7	
	single	7	390	283		697	2.84	0.91	12.5	3.0	
MORB											
Experiment 1	single	1	406	330		552	11.85	0.6	3.3	3.1	[2]
	liquid	2	392	251	2	1710	35.27	<0.1	1	2.9	
	vapor	3	392	251	0	33	0.664	1.1	6	3.4	
Experiment 2	single	1	406	330	−1	527	4.7	<0.1		3.7	
	liquid	2	388	240		1095	8.4			3.5	
	vapor	3	388	240	−1	23	0.133			4	

* Difference between measured and estimated temperatures at given pressure based on NaCl-H_2O phase relationships (Berndt et al. 2001).
§ mm : mmolal.
References: [1] This study - unpublished data; [2] Bischoff and Rosenbauer (1987)

reactants[3] included K-feldspar, muscovite and quartz, which together with constraints imposed by the addition of dissolved KCl to the NaCl dominated fluid (see Foustoukos and Seyfried 2005) fixed $pH_{in\ situ}$ at values approximately 5. During and prior to phase separation, sufficient reaction time was allocated (3 to 5 days) to permit for equilibrium to be achieved. This can be clearly illustrated by the good agreement in dissolved volatiles ($H_{2(aq)}$, $H_2S_{(aq)}$) and metals (Fe) as thermodynamic equilibrium was approached from different pressure and temperature conditions (Table 4).

Despite the fact that these experiments were conducted at the *P-T-x* conditions near the two-phase boundary of the $NaCl$-H_2O system, distinct vapor-liquid partitioning for the dissolved volatile species was observed (Fig. 7a). For example, $H_{2(aq)}$ concentrations in vapors in equilibrium with PPM attained values nearly 10 times higher than those measured in the single phase fluid, while in the case of $H_2S_{(aq)}$ the vapor enrichment was less, but still significant. Dissolved volatile concentrations in the liquid, however, did not depart significantly from values typical of single-phase conditions prior to initiation of phase separation. Thus, the intense oxidation conditions anticipated was not realized at the specific pressure and temperature conditions investigated. Apparently, when phase separation occurs at conditions deeper in the two-phase region, *salting-out* effects are likely more extreme, which would then trigger stronger oxidation effects due to more distinct differences in the density of coexisting vapors and liquids. Theoretical modeling calculations based on extrapolated values of fugacity-concentration (Kishima and Sakai 1984) and activity-concentration relationships (Ding and Seyfried 1990) for dissolved hydrogen, support this inference (Fig. 7b). In effect, as the system migrates towards lower water density, hydrogen tends to be more soluble in the low-salinity vapors while strong *salting-out* phenomena imposed on neutral species enhance the incompatibility of dissolved H_2 in the highly ionic liquid. Accordingly, volatile release from an initially reducing chemical system could initiate severe oxidation in a manner consistent with earlier suggestions (Ridge 1974; Drummond 1981; Drummond and Ohmoto 1985; Seyfried et al. 1988).

Early studies were able to suggest such behavior from available Henry's law constant data along the steam-saturation curve, following the formulation of volatility ratios, which can be defined as follows: $VR = (n_{gas}/n_{water})_{vapor}/(n_{gas}/n_{water})_{liquid}$ (Ellis and Golding 1963; Drummond 1981)[4]. The different values of (*VR*) in the vapor-liquid system were then regressed as a linear logarithmic function of temperature based on a series of gas solubility experimental studies (Drummond 1981 and references therein). Volatility ratios, however, are directly linked to the apparent partitioning coefficients (K_m) (Smith et al. 1987) encouraging development of largely equivalent, but more robust density-based models (Eqn. 3), especially for dissolved $H_{2(aq)}$ and $H_2S_{(aq)}$ (Table 1). Even though the models are strictly applicable to the pure water/steam system, data extrapolated to other conditions (Fig. 8) suggest tendencies for vapor enrichment in a manner generally consistent with the limited data that do exist (Fig. 7a).

In contrast, Fe/Cl molar ratios between coexisting phases do not reveal strong vapor-liquid fractionation (Fig. 7a). These results are in agreement with previous phase separations experiments (400-450 °C) (Berndt et al. 1993) that indicate slightly higher Fe/Cl ratios in vapors than in coexisting liquids. Apparently, iron speciation effects are dominated by the neutral $FeCl_{(aq)}$, favoring vapor-enrichment. It can be predicted, however, that dissolved Fe concentrations and Fe/Cl ratios in the vapor will increase with more extreme phase separation effects, especially when halite saturation is achieved. At this point, *salting-out* effects are of sufficient magnitude to likely follow the general trends reported earlier for the alkali elements and dissolved boron (Foustoukos and Seyfried 2007a). Iron enrichment in the vapor, however,

---

[3]   Mineral composition in Foustoukos and Seyfried (2005, 2007b).

[4] n_x = moles of gas or water per unit volume of vapor/liquid phase.

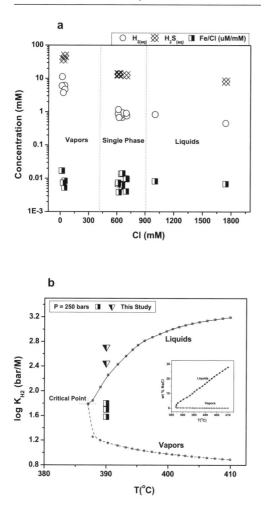

Figure 7. Experimental data on heterogeneous mineral-fluid equilibria under phase separation conditions (390 °C, 240-250 bars) involving a redox-buffering mineral assemblage of pyrite, pyrrhotite and magnetite (PPM). As it is clearly illustrated, dissolved volatile species (H_2, H_2S) have a strong tendency to partition in the vapor, while liquid fractions attain redox conditions somewhat similar to the homogeneous fluid system (single phase) (a). This can be best explained by the minimal *salting-out* effects expected at the specific experimental *P-T* conditions that do not depart significantly from the critical curve of the $NaCl$-H_2O system. (b) Estimated Henry's law constant values for the conjugated vapor-liquid pairs along the 250 isobar are in general agreement with experimental data, further suggesting, elevated hydrogen solubility in the low-chloride vapors when conditions deeper in the two-phase region are approached. Vapor-liquid fractionation is also minimal for the Fe/Cl molar ratios, most likely restrained by the weak *salting-out* effects at these *P-T* conditions.

might be lower than for alkali and monovalent elements due to the predictably lower degree of volatility suggested by the density-based partitioning (K_m) model (Pokrovski et al. 2005). Most importantly, however, the significantly different partitioning between Fe and dissolved volatile species (H_2, H_2S) indicate the key role of phase separation in delivering highly reducing but metal-depleted vapor fluids (Fig. 8), while the low Fe/$H_2S_{(aq)}$ ratios attained could eventually impose serious constraints on sulfide precipitation and exhalative ore deposit formation. Even though such a process might not be of limited significance for deep-sea submarine hydrothermal systems where high pressure contributes to formation of Cl-bearing vapors, which enhances Fe solubility (Fig. 7b; Berndt et al. 1993), while the opposite may be true for the recently discovered seamount-hosted hydrothermal systems (Stoffers et al. 2006), where steam-like discharge is common place. In any event, phase separation/segregation in conjunction with the water/rock interaction effects might be an effective mechanism to transport metals and dissolved volatiles from the deeper parts of the oceanic crust, than what can be predicted based on mineral-fluid equilibria constraints alone.

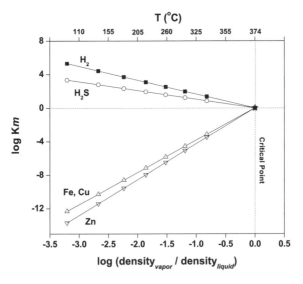

Figure 8. Apparent partitioning coefficients (K_m) of dissolved H_2 and H_2S estimated based on gas solubility data (Drummond 1981) and expressed as function of the vapor-liquid density along the steam-saturation curve. In agreement with experimental data presented earlier (Table 4, Fig. 6a), dissolved H_2 appears to have a higher tendency for vapor enrichment relative to $H_2S_{(aq)}$. Not surprisingly, dissolved volatile species also exhibit distinctly different partitioning in the vapor fraction relative to transition metals (e.g., Fe, Cu), indicative of the important role of phase separation in inducing great oxidation events in the subseafloor reaction zone, while delivering volatile-enriched but metal-depleted vapor fluids to the seafloor.

FINAL REMARKS – NUMERICAL SIMULATIONS

In addition to the progress that has been made on the chemical effects intrinsic to multi-phase NaCl-H_2O system at elevated temperatures and pressures, there have also been important and vigorous efforts to describe the physical effects of two-phase flow and thermohaline convection patterns in submarine hydrothermal systems. Initial attempts to do this involved multiphase, near critical and supercritical fluid flow, although typically restricted to a single component system (Ingebritsen and Hayba 1994; Hayba and Ingebritsen 1997; Lowell and Xu 2000), and in the absence explicit consideration of terms describing dissolved chloride effects on physical parameters in the governing equations. Accordingly, these models were mainly based on treating the physical properties of aqueous solutions as simple functions of enthalpy, while establishing mass and heat conservation during flow following methodologies introduced by Faust and Mercer (1979) and Hayba and Ingebritsen (1994). Recent theoretical developments, now express temperature, pressure and salinity changes as a function of density permitting in this way multiphase, multicomponent fluid flow models to be developed for a wide range of temperature and pressure conditions (Bai et al. 2003; Kawada et al. 2004; Lewis and Lowell 2004; Geiger et al. 2005; Kissling 2005; Geiger et al. 2006; Fontaine et al. 2007). Driesner and Geiger (2007) present a more detailed description of the relative merits of the different multicomponent fluid flow models.

Numerical simulations incorporate phase separation by utilizing two different fluid flow patterns: one-dimensional pipe flow and two-dimensional convection in porous media, with the later being the most stable due to the large effect of vapor/liquid salinity and density oscillations on the stability of the 1D-pipe flow models (Xu and Lowell 1998; Berndt et al. 2001). Results and flow patterns originating from these theoretical studies have been applied not only to describe the thermal and chemical evolution of high-temperature vapor vent fluids at mid-ocean ridges (Seyfried et al. 2003), but also to better understand formation and consequences of thermal convection of a putative brine (liquid) layer during phase separation/segregation processes (Bischoff and Rosenbauer 1989; Lowell and Germanovich 1997; Schoofs and Hansen 2000; Kawada et al. 2004; Geiger et al. 2005; Fontaine and Wilcock 2006). Brine storage is an important component of the subseafloor reaction zone, with brine layers appearing to be stably stratified but unable to convect (Kawada et al. 2004; Fontaine and Wilcock 2006). Thus, hot subsurface brines stored deep in the crust would eventually mix with evolved seawater fluids, resulting in a possible positive correlation between dissolved Cl concentrations and temperature increase (Larson et al. 2007), although little evidence exists for this at present. Accordingly, the process controlling variables have yet to be fully established, but even with the limited information on these systems that is available, the effect of phase separation cannot be ignored. The challenge that is before us, however, is to link the field observations with quantitative models involving homogeneous and heterogeneous phase relations.

ACKNOWLEDGMENTS

We would like to thank Mike Berndt, Fabrice Fontaine, Gleb Pokrovski, Chris Heinrich, Axel Liebscher and Thomas Driesner for valuable discussions and constructive suggestions. This study has been supported by the NSF OCE-0351069 grant and a postdoctoral fellowship (DIF) from the Carnegie Institution of Washington.

REFERENCES

Alt JC, Barrie CT, Hannington MD (1999) Hydrothermal alteration and mineralization of oceanic crust; mineralogy, geochemistry, and processes. *In:* Volcanic-associated Massive Sulfide Deposits; Processes and Examples in Modern and Ancient Settings. Barrie CT, Hannington MD (eds) Rev Econ Geol 8:133-155

Alt JC, Teagle DAH (2000) Hydrothermal alteration and fluid fluxes in ophiolites and oceanic crust. *In:* Ophiolites and Oceanic Crust: New Insights from Field Studies and the Ocean Drilling Program. Dilek Y, Moores EM, Elthon D, Nicolas A (eds) Geol Soc Am Special Paper 349:273-282

Anderko A, Pitzer KS (1993) Equation of state representation of phase equilibria and volumetric properties of NaCl-H_2O above 573 °C. Geochim Cosmochim Acta 57:1657-1680

Audetat A, Gunther D, Heinrich CA (1998) Formation of a magmatic-hydrothermal ore deposit: insights with LA-ICP-MS analysis of fluid inclusions. Science 279:2091-2093

Bai W, Xu W, Lowell RP (2003) The dynamics of submarine geothermal heat pipes. Geophys Res Lett 30:1108

Baker T, Achterberg E, Ryan CG, Lang JR (2004) Composition and evolution of ore fluids in a magmatic-hydrothermal skarn deposits. Geology 32:117-120

Berndt ME, Person ME, Seyfried WE (2001) Phase separation and two-phase flow in seafloor hydrothermal systems: Geophysical modeling in the NaCl-H_2O system. 11th Annual V. M. Goldschmidt Conference, Geochemical Society, p 3692-3693

Berndt ME, Seal RR, Shanks WC, Seyfried WE (1996) Hydrogen isotope systematics of phase separation in submarine hydrothermal systems: experimental calibration and theoretical models. Geochim Cosmochim Acta 60:1595-1604

Berndt ME, Seyfried WE (1990) Boron, bromine, and other trace elements as clues to the fate of chlorine in mid-ocean ridge vent fluids. Geochim Cosmochim Acta 54:2235-2245

Berndt ME, Seyfried WE (1993) Experimental investigation of phase-separation and the behavior of Fe in subseafloor hydrothermal systems. EOS Trans, Am Geophys Union 74:666

Berndt ME, Seyfried WE (1997) Calibration of Br/Cl fractionation during subcritical phase separation of seawater; possible halite at 9 to 10° N East Pacific Rise. Geochim Cosmochim Acta 61:2849-2854

Bischoff JL (1980) Geothermal system at 21° N, east pacific rise: physical limits on geothermal fluid and role of adiabatic expansion. Science 207:1465-1469

Bischoff JL (1991) Densities of liquids and vapors in boiling NaCl-H_2O solutions; a PVTx summary from 300° to 500 °C. Am J Sci 291:309-338

Bischoff JL, Pitzer KS (1985) Phase relations and adiabats in boiling seafloor geothermal systems. Earth Planet Sci Lett 75:327-338

Bischoff JL, Rosenbauer RJ (1984) The critical point and two-phase boundary of seawater, 200-500 °C. Earth Planet Sci Lett 68:172-180

Bischoff JL, Rosenbauer RJ (1987) Phase separation in seafloor geothermal systems: An experimental study of the effects on metal transport. Am J Sci 287:953-978

Bischoff JL, Rosenbauer RJ (1988) Liquid-vapor relations in the critical region of the system sodium chloride-water from 380 to 415 °C: A refined determination of the critical point and two-phase boundary of seawater. Geochim Cosmochim Acta 52:2121-2126

Bischoff JL, Rosenbauer RJ (1989) Salinity variations in submarine hydrothermal systems by layered double-diffusive convection. J Geol 97:613-623

Bischoff JL, Rosenbauer RJ, Pitzer KS (1986) The system sodium chloride-water: Relations of vapor-liquid near the critical temperature of water and vapor-liquid-halite from 300 to 500 °C. Geochim Cosmochim Acta 50:1437-1444

Bohnenstiehl DR, Dziak RP, Tolstoy M, Fox CG, Fowler M (2004) Temporal and spatial history of the 1999–2000 Endeavour Segment seismic series, Juan de Fuca Ridge. Geochem Geophys Geosy 5:doi:10.1029/2004GC000735

Bonifacie M, Charlou JL, Jendrzejewski N, Agrinier P, Donval JP (2005) Chlorine isotopic composition of high temperature hydrothermal vent fluids over ridge axes. Chem Geol 221:279-288

Butterfield DA, Jonasson IR, Massoth GJ, Feely RA, Roe KK, Embley RE, Holden JF, McDuff RE, Lilley MD, Delaney JR (1999) Seafloor eruptions and evolution of hydrothermal fluid chemistry. *In*: Mid-ocean Ridges; Dynamics of Processes Associated with Creation of New Ocean Crust. Cann JR, Elderfield H, Laughton A (eds) Cambridge University Press, p 153-170

Butterfield DA, Massoth GJ, McDuff RE, Lupton JE, Lilley MD (1990) Geochemistry of hydrothermal fluids from Axial Seamount Hydrothermal Emissions Study Vent Field, Juan de Fuca Ridge: Subseafloor boiling and subsequent fluid-rock interaction. J Geophys Res 95:12,895-12,921

Butterfield DA, McDuff RE, Mottl MJ, Lilley MD, Lupton JE, Massoth GJ (1994) Gradients in the composition of hydrothermal fluids from the Endeavour segment vent field: phase separation and brine loss. J Geophys Res 99:9561-9583

Campbell AC, Palmer MR, Klinkhammer GP, Bowers TS, Edmond JM, Lawrence JR, Casey JF, Thompson G, Humphris S, Rona P, Karson JA (1988) Chemistry of hot springs on the Mid-Atlantic Ridge. Nature 335:514-519

Charlou JL, Donval JP, Douville E, Jean-Baptiste P, Radford-Knoery J, Fouquet Y, Dapoigny A, Stievenard M (2000) Compared geochemical signatures and the evolution of Menez Gwen (37°50'N) and Lucky Strike (37°17'N) hydrothermal fluids, south of the Azores triple junction on the Mid-Atlantic Ridge. Chem Geol 171:49-75

Charlou JL, Donval JP, Fouquet Y, Jean-Baptiste P, Holm N (2002) Geochemistry of high H_2 and CH_4 vent fluids issuing from ultramafic rocks at the Rainbow hydrothermal field (36°14'N, MAR). Chem Geol 191:345-359

Charlou JL, Donval JP, Jean-Baptiste P, Dapoigny A, Rona PA (1996) Gases and helium isotopes in high temperature solutions sampled before and after ODP Leg 158 drilling at TAG hydrothermal field (26° N, MAR). Geophys Res Lett 23:3491-3494

Chiba H, Masuda H, Fujioka K (2001) Chemistry of hydrothermal fluids at the TAG active mound, MAR 26° N, in 1998. Geophys Res Lett 28:2919-2922

Chou IM (1987) Phase relations in the system NaCl-KCl-H_2O. III: solubilities of halite in vapor-saturated liquids above 445 °C and redetermination of phase equilibrium properties in the system NaCl-H_2O to 1000 °C and 1500 bars. Geochim Cosmochim Acta 51:1965-1975

Copeland CS, Silverman J, Benson SW (1953) The system NaCl-H_2O at supercritical temperatures and pressures. J Chem Phys 21:12-16

Corliss JB, Dymond J, Gordon LI, Edmond JM, Von Herzen RP, Ballard RD, Green K, Williams D, Bainbridge A, and et al. (1979) Submarine thermal springs on the Galapagos Rift. Science 203:1073-83

Detrick RS, Buhl B, Vera E, Mutter J, Orcutt J, Madsen J, Brocher T (1987) Multi-channel seismic imaging of a crustal magma chamber along the East Pacific Rise. Nature 326:35-41

Ding, K, Seyfried WE (1990) Activity coefficients of H_2 and H_2S in NaCl solutions at 300-425 °C, 300-500 bars with application to ridge crest hydrothermal systems. EOS Trans, Am Geophys Union 71:1680

Douville E, Charlou JL, Oelkers EH, Bienvenu P, Jove Colon CF, Donval JP, Fouquet Y, Prieur D, Appriou P (2002) The Rainbow vent fluids (36° 14'N, MAR): the influence of ultramafic rocks and phase separation on trace metal content in Mid-Atlantic Ridge hydrothermal fluids. Chem Geol 184:37-48

Driesner T (1997) The effect of pressure on deuterium-hydrogen fractionation in high-temperature water. Science 277:791-794

Driesner T, Geiger S (2007) Numerical simulation of multiphase fluid flow in hydrothermal systems. Rev Mineral Geochem 65:187-212

Driesner T, Heinrich CA (2007) The system H_2O-NaCl. I. Correlation formulae for phase relations in temperature-pressure-composition space from 0 to 1000 °C, 0 to 5000 bar, and 0 to 1 X_{NaCl}. Geochim Cosmochim Acta (in press)

Driesner T, Ha TK, Seward TM (2000) Oxygen and hydrogen isotope fractionation by hydration complexes of Li^+, Na^+, K^+, Mg^{2+}, F^-, Cl^-, and Br^-: a theoretical study. Geochim Cosmochim Acta 64:3007-3033

Driesner T, Seward TJ, Tironi IG (1998) Molecular dynamics simulation study of ionic hydration and ion association in dilute and 1 molal aqueous solutions from ambient to supercritical conditions. Geochim Cosmochim Acta 62:3095-3107

Drummond SE (1981) Boiling and mixing of hydrothermal fluids: chemical effects on mineral precipitation. PhD Dissertation. The Pennsylvania State University, State College, Pennsylvania

Drummond SE, Ohmoto H (1985) Chemical evolution and mineral deposition in boiling hydrothermal systems. Econ Geol 80:126-147

Edmond JM, Campbell AC, Palmer M, Klinkhammer GP, German CR, Edmonds HN, Elderfield H, Thompson G, Rona P (1995) Time series studies of vent fluids from the TAG and MARK sites (1986, 1990) Mid-Atlantic Ridge: a new solution chemistry model and mechanism for Cu/Zn zonation in massive sulfide ore deposits. *In*: Hydrothermal Vents and Processes. Parson LM, Walker CL, Dixon DR (eds) Geological Society Special Publication No. 87, p 77-86

Edmond JM, Measures C, McDuff RE, Chan LH, Collier R, Grant B, Gordon LI, Corliss JB (1979) Ridge crest hydrothermal activity and the balances of the major and minor elements in the ocean: the Galapagos data. Earth Planet Sci Lett 46:1-18

Edmonds HN, German CR, Green DRH, Huh Y, Gamo T, Edmond JM (1996) Continuation of the hydrothermal fluid chemistry time series at TAG, and the effects of ODP drilling. Geophys Res Lett 23:3487-3489

Ellis AJ, Golding RM (1963) The solubility of carbon dioxide above 100 °C in water and in sodium chloride solutions. Am J Sci 261:47-60

Faust CR, Mercer JW (1979) Geothermal reservoir simulation. 1. Mathematical models for liquid- and vapor-dominated hydrothermal systems. Water Resour Res 15:23-30

Fontaine FJ, Wilcock WS (2006) Dynamics and storage of brine in mid-ocean ridge hydrothermal systems. J Geophys Res 111:doi:10.1029/2005JB003866

Fontaine FJ, Wilcock WS, Butterfield DA (2007) Physical controls on the salinity of mid-ocean ridge hydrothermal vent fluids. Earth Planet Sci Lett 257:132-145

Fornari DJ, Shank T, Von Damm KL, Gregg TKP, Lilley M, Levai G, Bray A, Haymon RM, Perfit MR, Lutz R (1998) Time-series temperature measurements at high-temperature hydrothermal vents, East Pacific Rise 9°49'-51'N: evidence for monitoring a crustal cracking event. Earth Planet Sci Lett 160:419-431

Fournier RO, Potter PW (1983) An equation correlating the solubility of quartz in water from 25° to 900 °C at pressures up to 10000 bars. Geochim Cosmochim Acta 46:1969-1973

Foustoukos DI, James RH, Berndt ME, Seyfried WE (2004) Lithium isotopic systematics of hydrothermal vent fluids at the Main Endeavour Field, Northern Juan de Fuca Ridge. Chem Geol 212:17-26

Foustoukos DI, Seyfried WE (2005) Redox and pH constraints in the subseafloor root zone of the TAG hydrothermal system, 26°N Mid-Atlantic Ridge. Earth Planet Sci Lett 235:497-510

Foustoukos DI, Seyfried WE (2007a) Trace element partitioning between vapor, brine and halite under extreme phase separation conditions. Geochim Cosmochim Acta 71:2056-2071

Foustoukos DI, Seyfried WE (2007b) Quartz solubility in the two-phase and critical region of seawater fluids: Implications for submarine hydrothermal vent systems at 9°50'N East Pacific Rise. Geochim Cosmochim Acta 71:186-201

Gallant RM, Von Damm KL (2006) Geochemical controls on hydrothermal fluids from the Kairei and Edmond vent fields, 23°-25°S, Central Indian Ridge. Geochem Geophys Geosy 7: doi:10.1029/2005GC001067

Gamo T, Chiba H, Masuda H, Edmonds HN, Fujioka K, Kodama Y, Nanba H, Sano Y (1996) Chemical characteristics of hydrothermal fluids from the TAG mound of the mid- Atlantic ridge in August 1994: implications for spatial and temporal variability of hydrothermal activity. Geophys Res Lett 23:3483-3486

Geiger S, Driesner T, Heinrich CA (2005) On the dynamics of $NaCl$-H_2O fluid convection in the Earth's crust. J Geophys Res 110:doi:10.1029/2004JB003362

Geiger S, Driesner T, Heinrich CA, Matthäi SK (2006) Multiphase thermohaline convection in the earth's crust: II. Benchmarking and application of a finite element - Finite volume solution technique with a $NaCl$-H_2O equation of state. Transport Porous Med 63:435-461

German CR, Von Damm KL (2003) Hydrothermal processes. Treatise on Geochemistry 6:181-222

Gillis KM (2002) The root zone of an ancient hydrothermal system exposed in the Troodos ophiolite, Cyprus. J Geol 110:57-74

Gillis KM, Roberts MD (1999) Cracking at the magma-hydrothermal transition: evidence from the Troodos ophiolite, Cyprus. Earth Planet Sci Lett 169:227-244

Grover PK, Ryall RL (2004) Critical appraisal of salting-out and it's implications for chemical and biological sciences. Chem Rev 105:10.1021/cr030454p

Haas JL (1976a) Physical properties of the coexisting phases and thermochemical properties of the H_2O component in boiling $NaCl$ solutions, preliminary steam tables for $NaCl$ solutions. Geol Survey Bull 1421-A

Haas JL (1976b) Thermodynamic properties of the coexisting phases and thermochemical properties of the NaCl component in boiling NaCl solutions. Geol Survey Bull 1421-B

Hayba DO, Ingebritsen SE (1994) The computer model Hydrotherm, a three-dimensional finite-difference model to simulate ground-water flow and heat transport in the temperature range of 0 to 1,200 °C. U.S. Geological Survey, Water-Resources Investigations Report # 94-4045

Hayba DO, Ingebritsen SE (1997) Multiphase groundwater flow near cooling plutons. J Geophys Res 102:12235-12252

Heinrich CA, Gunther D, Audetat A, Ulrich T, Frischknecht R (1999) Metal fractionation between magmatic brine and vapor, and the link between porphyry-style and epithermal Cu-Au deposits. Geology 27:755-758

Ho PC, Palmer DA, Bianchi H, Wood RH (2000) A flow-through high-pressure electrical conductance cell for determination of ion association of aqueous electrolyte solutions at high temperature and pressure. Steam, Water, and Hydrothermal Systems: 13th Physics and Chemistry Meeting the Needs of Industry, Proc Int Conf on the Properties of Water and Steam, p 661-668

Ho PC, Palmer DA, Gruszkiewicz MS (2001) Conductivity measurements of dilute aqueous HCl solutions to high temperatures and pressures using a flow-through cell. J Phys Chem B 105:1260-1266

Ho PC, Palmer DA, Mesmer RE (1994) Electrical conductivity measurements of aqueous sodium chloride solutions to 600 °C and 300 MPa. J Solution Chem 23:997-1018

Horita J, Cole DR, Wesolowski DJ (1995) The activity-composition relationship of oxygen and hydrogen isotopes in aqueous salt solutions; III, Vapor-liquid water equilibration of NaCl solutions to 350 °C. Geochim Cosmochim Acta 59:1139-1151

Horita J, Wesolowski D, Cole DR (1993a) The activity-composition relationship of oxygen and hydrogen isotopes in aqueous salt solutions: I. vapor-liquid water equilibration of single salt solutions from 50 to 100 °C. Geochim Cosmochim Acta 57:2797-2817

Horita J, Wesolowski D, Cole DR (1993b) The activity-composition relationship of oxygen and hydrogen isotopes in aqueous salt solutions: II. vapor-liquid water equilibration of mixed salt solutions from 50 to 100 °C and geochemical implications. Geochim Cosmochim Acta 57:4703-4711

Ingebritsen SE, Hayba DO (1994) Fluid flow and heat transport near the critical point of H_2O. Geophys Res Lett 21:2199-2202

James RH, Elderfield H, Palmer MR (1995) The chemistry of hydrothermal fluids from the Broken Spur site 29°N Mid Atlantic Ridge. Geochim Cosmochim Acta 59:651-661

Jean-Baptiste P, Charlou JL, Stievenard M, Donval JP, Bougault H, Mevel C (1991) Helium and methane measurements in hydrothermal fluids from the Mid-Atlantic Ridge: the Snake Pit site at 23°N. Earth Planet Sci Lett 106:17-28

Kawada Y, Yoshiba S, Watanabe S (2004) Numerical simulations of mid-ocean ridge hydrothermal circulation including the phase separation of seawater. Earth Planet Space 56:193-215

Keevil NB (1942) Vapor pressures of aqueous solutions at high temperatures. J Am Chem Soc 64: 841-850

Kelley DS, Baross JA, Delaney JR (2002) Volcanoes, fluids and life at mid-ocean ridge spreading centers. Annu Rev Earth Planet Sci 30:385-491

Kent GM, Harding A, Orcutt JA (1993) Distribution of magma beneath the East Pacific Rise between the Clipperton Transform and the 9°17'N overlapping spreading center from forward modeling of common depth point data. J Geophys Res 98:13971-13996

Khaibullin IK, Borisov NM (1966) Experimental investigation of the thermal properties of aqueous and vapor solutions of sodium and potassium chlorides at phase equilibrium. Teplofizika Vysokikh Temperatur 4:518-523

Kishima N, Sakai H (1984) Fugacity-concentration relationship of dilute hydrogen in water at elevated temperature and pressure. Earth Planet Sci Lett 67:79-86

Kissling WM (2005) Transport of three-phase hyper-saline brines in porous media: Theory and code implementation. Transport Porous Med 61:25-44

Koschinsky A, Garbe-Schoenberg D, Seifert HJ, Strauss H, Weber S, Marbler H (2006) 400 °C hot boiling fluids from a hydrothermal field at 5°S on the mid-atlantic ridge - Results of Meteor Cruise M64/1. AGU Fall Meeting p QS33A-1456

Kukuljan JA, Alvarez JL (1999) Distribution of $B(OH)_3$ between water and steam at high temperatures. J Chem Thermodyn 31:1511-1521

Larson BI, Olson EJ, Lilley MD (2007) In-situ measurements of dissolved chloride in high temperature hydrothermal fluids. Geochim Cosmochim Acta 71: 2510-2523

Lein AY, Grichuk DV, Gurvich EG, Bogdanov YA (2000) New type of the hydrothermal fluid enriched with H_2 and CH_4 in the rift zone of the Mid-Atlantic Ridge. Dokl Akad Nauk 375:380-383

Lewis K, Lowell RP (2004) Mathematical modeling of phase separation of seawater near an igneous dike. Geofluids 4:197-209

Liebscher A (2007) Experimental studies in model fluid systems. Rev Mineral Geochem 65:15-47

Liebscher A, Heinrich CA (2007) Fluid–fluid interactions in the Earth's lithosphere. Rev Mineral Geochem 65:1-13

Liebscher A, Meixner A, Romer RL, Heinrich W (2007) Experimental calibration of the vapour-liquid phase relations and lithium isotope fractionation in the system H_2O-LiCl at 400 °C/20 to 28 MPa. Geofluids (in press)

Liebscher A, Barnes J, Sharp Z (2006a) Chlorine isotope vapor-liquid fractionation during experimental fluid-phase separation at 400 °C/23 MPa to 450 °C/42 MPa. Chem Geol 234:340-345

Liebscher A, Luders V, Heinrich W, Schettler G (2006b) Br/Cl signature of hydrothermal fluids: liquid-vapour fractionation of bromine revisited. Geofluids 6(2):113-121, doi: 10.1111/j.1468-8123.2006.00135.x

Liebscher A, Meixner A, Romer RL, Heinrich W (2005) Liquid-vapor fractionation of boron and boron isotopes: Experimental calibration at 400 °C/23 MPa to 450 °C/42 MPa. Geochim Cosmochim Acta 69:5693-5704

Lilley MD, Butterfield DA, Lupton JE, Olson EJ (2003) Magmatic events can produce rapid changes in hydrothermal vent chemistry. Nature 422:878-881

Lilley MD, Butterfield DA, Olson EJ, Lupton JE, Macko SA, McDuff RE (1993) Anomalous methane and ammonium concentrations at an unsedimented mid-ocean-ridge hydrothermal system. Nature 364:45-47

Liu C-t, Lindsay WT (1972) Thermodynamics of sodium chloride solutions at high temperatures. J Solution Chem 1:45-69

Lowell RP, Germanovich LN (1997) Evolution of a brine-saturated layer at the base of a ridge-crest hydrothermal system. J Geophys Res 102:10245-10255

Lowell RP, Xu W (2000) Subcritical two-phase seawater convection near a dike. Earth Planet Sci Lett 174:385-396

Mashovets VP, Zarembo VI, Fedorov MK (1973) Vapor pressures of aqueous solutions of NaCl, NaBr, and NaI at temperatures in the range of 150-350 °C. J Appl Chem-USSR 46:684-686

Masterton WL, Lee TP (1970) Salting coefficients from scaled particle theory. J Phys Chem 74:1776-1782

Mercer JW, Faust CR (1979) Review of numerical simulation of hydrothermal systems. Hydr Sci Bulletin 24:335-344

Nehlig P, Juteau T, Bendel V, Cotten J (1994) The root zones of oceanic hydrothermal systems - Constraints from the Samail ophiolite (Oman). J Geophys Res 99:4703-4713

Oelkers EH, Helgeson HC (1990) Triple-ion anions and polynuclear complexing in supercritical electrolyte solutions. Geochim Cosmochim Acta 54:727-738

Oelkers EH, Helgeson HC (1991) Calculation of activity coefficients and degrees of formation of neutral ion pairs in supercritical electrolyte solutions. Geochim Cosmochim Acta 55:1235-1251

Oelkers EH, Helgeson HC (1993a) Calculation of dissociation constants and the relative stabilities of polynuclear clusters of 1:1 electrolytes in hydrothermal solutions at supercritical pressures and temperatures. Geochim Cosmochim Acta 57:2673-2697

Oelkers EH, Helgeson HC (1993b) Multiple ion association in supercritical aqueous solutions of single electrolytes. Science 261: 888-891

Olander A, Liander H (1950) The phase diagram of sodium chloride and steam above the critical point. Acta Chim Scand 4:1437-1445

Oosting SE, Von Damm KL (1996) Bromide/chloride fractionation in seafloor hydrothermal fluids from 9-10° N East Pacific Rise. Earth Planet Sci Lett 144:133-145

Palmer DA, Simonson JM, Jensen JP (2004) Partitioning of electrolytes to steam and their solubilities in steam. *In*: Aqueous Systems at Elevated Temperatures and Pressures. Palmer DA, Fernandez-Prini R, Harvey AH (eds) Academic Press, p. 409-439

Parisod CJ, Plattner E (1981) Vapor-liquid equilibria of the NaCl-H_2O system in the temperature range 300-440 °C. J Chem Eng Data 26:16-20

Pitzer KS (1990) Ion interaction approach theory and data correlation. *In*: Activity Coefficients in Electrolyte Solutions. Pitzer KS (ed) CRC Press Ltd, p 75-153

Pitzer KS (1995) Ionic fluids - near critical and related properties. J Phys Chem 99:13070-13077

Pitzer KS, Bischoff JL, Rosenbauer RJ (1987) Critical behavior of dilute sodium chloride in water. Chem Phys Lett 134:60-63

Pitzer KS, Peiper JC, Busey RH (1984) Thermodynamic properties of aqueous sodium chloride solutions. J Phys Chem Ref Data 13:1-102

Pitzer KS, Schreiber DR (1987) The restricted primitive model for ionic fluids: properties of the vapor and the critical region. Mol Phys 60:1067-1078

Pokrovski GS, Roux J, Harrichourry JC (2005) Fluid density control on vapor-liquid partitioning of metals in hydrothermal systems. Geology 33:657-660

Pokrovski GS, Zakirov IV, Roux J, Testemale D, Hazemann JL, Bychkov AY, Golikova GV (2002) Experimental study of arsenic speciation in vapor phase to 500 °C: Implications for As transport and fractionation in low-density fluids and volcanic gases. Geochim Cosmochim Acta 66:3453-3480

Ridge JD (1974) Note on boiling of ascending ore fluids and the position of volcanic-exhalative deposits in the modified Lindgren classification. Geology 2:287-288

Richardson CJ, Cann JR, Richards HG, Cowan JG (1987) Metal-depleted root zones of the Troodos ore-forming hydrothermal systems, Cyprus. Earth Planet Sci Lett 84:243-253

Rosenbauer RJ, Bischoff JL (1987) Pressure-composition relations for coexisting gases and liquids and the critical points in the system sodium chloride-water at 450, 475, and 500 °C. Geochim Cosmochim Acta 51:2349-54

Schatz OJ, Dolejs D, Stix J, Williams-Jones AE, Layne GD (2004) Partitioning of boron among melt, brine and vapor in the system haplogranite-H$_2$O-NaCl at 800 °C and 100 MPa. Chem Geol 210:135-147

Schmidt C, Thomas R, Heinrich W (2005) Boron speciation in aqueous solutions at 22 to 600 °C and 0.1 MPa to 2 GPa. Geochim Cosmochim Acta 69:275-281

Schoofs S, Hansen U (2000) Depletion of a brine layer at the base of ridge-crest hydrothermal systems. Earth Planet Sci Lett 180:341-353

Seewald J, Cruse A, Saccocia P (2003) Aqueous volatiles in hydrothermal fluids from the Main Endeavour Field, northern Juan de Fuca Ridge: temporal variability following earthquake activity. Earth Planet Sci Lett 216:575-590

Seyfried WE, Berndt ME, Seewald JS (1988) Hydrothermal alteration processes at mid-ocean ridges: Constraints from diabase alteration experiments, hot-spring fluids and composition of the oceanic crust. Can Mineral 26:787-804

Seyfried WE, Ding K, Berndt ME, Chen X (1999) Experimental and theoretical controls on the composition of mid-ocean ridge hydrothermal fluids In: Volcanic-Associated Massive Sulfide Deposits; Processes and Examples in Modern and Ancient Settings. Barrie CT, Hannington MD (eds) Rev Econ Geol 8:181-200

Seyfried WE, Foustoukos DI, Allen DE (2004) Ultramafic-hosted hydrothermal systems at mid-ocean ridges: chemical and physical controls on pH, redox and carbon reduction reactions. In: Mid-Ocean Ridges: Hydrothermal Interactions Between the Lithosphere and Oceans. German CR, Lin J, Parson LM (ed) Geophys Monogr Ser 148:267-284

Seyfried WE, Seewald JS, Berndt ME, Ding K, Foustoukos DI (2003) Chemistry of hydrothermal vent fluids from the Main Endeavour Field, Northern Juan de Fuca Ridge: Geochemical controls in the aftermath of June 1999 seismic events. J Geophys Res 108:2429

Seyfried WE, Janecky DR, Berndt ME (1987) Rocking autoclaves for hydrothermal experiments: II. The flexible reaction-cell system. In: Hydrothermal Experimental Techniques. Ulmer G, Barnes H (eds) Wiley-Interscience, p 216-239

Shanks WC (2001) Stable isotopes in seafloor hydrothermal systems: Vent fluids, hydrothermal deposits, hydrothermal alteration, and microbial processes. Rev Mineral Geochem 43:469-525

Shanks WC, Bohlke JK, Seal RR (1995) Stable isotopes in mid-ocean ridge hydrothermal systems; interactions between fluids, minerals, and organisms. In: Seafloor Hydrothermal Systems; Physical, Chemical, Biological, and Geological Interactions. Humphris SE, Zierenberg RA, Mullineaux LS, Thomson RE (eds) Geophys Monogr 91:194-221

Sharygin AV, Wood RH, Zimmerman GH, Balashov VN (2002) Multiple ion association versus redissociation in aqueous NaCl and KCl at high temperatures. J Phys Chem B 106:7121-7134

Shmulovich K, Landwehr D, Simon K, Heinrich W (1999) Stable isotope fractionation between liquid and vapour in water-salt systems up to 600 °C. Chem Geol 157:343-354

Shock EL, Sassani DC, Willis M, Sverjensky DA (1997) Inorganic species in geologic fluids: Correlations among standard molal thermodynamic properties of aqueous ions and hydroxide complexes. Geochim Cosmochim Acta 61:907-950

Simonson JM, Palmer DA, Gruszkiewicz MS (2000) Liquid-vapor partitioning of aqueous electrolytes at temperatures to the solvent critical point. 14th Symposium on Thermophysical Properties. National Institute of Standards and Technology, Boulder, CO USA, *http://www.symp14.nist.gov/PDF/AQU04SIM.PDF*

Smith CL, Ficklin WH, Thompson JM (1987) Concentrations of arsenic, antimony, and boron in steam and steam condensate at the geysers, California. J Volcanol Geotherm Res 32:329-341

Sohn RA, Hildebrand JA, Webb SC (1999) A microearthquake survey of the high-temperature vent fields on the volcanically active East Pacific Rise (9°50'N). J Geophys Res 104:25367-25377

Sourirajan S, Kennedy GC (1962) The system NaCl-H$_2$O at elevated temperatures and pressures. Am J Sci 260:115-242

Spivack AJ, Berndt ME, Seyfried WE (1990) Boron isotope fractionation during supercritical phase separation. Geochim Cosmochim Acta 54:2337-2339

Stoffers P, Worthington TJ, Schwarz-Schampera U, Hannington M, Massoth GJ, Hekinian R, Schmidt M, Lundsten LJ, Evans LJ, Vaiomo'unga R, Kerby T (2006) Submarine volcanoes and high-temperature hydrothermal venting on the Tonga arc, southwest Pacific. Geology 34:453-456

Styrikovich MA, Khaibullin IK, Tskhvirashvili DG (1955) Solubility of salts in high-pressure steam. Dokl Akad Nauk 100:1123-6

Styrikovich MA, Tskhvirashvili DG, Hebieridze DP (1960) A study of the solubility of boric acid in saturated water vapor. Dokl Akad Nauk 134:615-617

Sue K, Arai K (2004) Specific behavior of acid-base and neutralization reactions in supercritical water. J Supercrit Fluid 28:57-68

Sverjensky DA, Shock EL, Helgeson HC (1997) Prediction of the thermodynamic properties of aqueous metal complexes to 1000 °C and 5kbar. Geochim Cosmochim Acta 61:1359-1412

Tanger JC, Pitzer KS (1989) Thermodynamics of NaCl-H_2O: a new equation of state for the near critical region and comparisons with other equations for adjoining regions. Geochim Cosmochim Acta 53:973-987

Ulrich T, Gunther D, Heinrich CA (1999) Gold concentrations of magmatic brines and the metal budget of porphyry copper deposits. Nature 399:676-679

Urusova MA (1974) Phase equilibria in the sodium-hydroxide-water and sodium chloride-water systems at 350-550 °C. Russ J Inorg Chem 19:450-454

Urusova MA (1975) Volume properties of aqueous solutions of sodium chloride at elevated temperatures and pressures. Russ J Inorg Chem 20:1717-1721

Van Ark EM, Detrick RS, Canales JP, Carbotte SM, Harding AJ, Kent GM, Nedimovic MR, Wilcock WSD, Diebold JB, Babcock JM (2007) Seismic structure of the Endeavour Segment, Juan de Fuca Ridge: Correlations with seismicity and hydrothermal activity. J Geophys Res 112:doi:10.1029/2005JB004210

Von Damm KL (1995) Controls on the chemistry and temporal variability of seafloor hydrothermal fluids. *In:* Seafloor Hydrothermal Systems: Physical, Chemical, Biologic and Geologic Interactions, Humphris SE, Zierenberg RA, Mullineaux LS, Thompson RE (eds) Geophys Monogr 91:222-248

Von Damm KL (2000) Chemistry of hydrothermal vent fluids from 9°-10° N, East Pacific Rise; "time zero," the immediate posteruptive period. J Geophys Res 105:11203-11222

Von Damm KL (2004) Evolution of the hydrothermal system at East Pacific Rise 9°50´N: Geochemical evidence for changes in the upper oceanic crust. *In:* Mid-Ocean Ridges: Hydrothermal Interactions Between the Lithosphere and Oceans, German CR, Lin J, Parson LM (eds), Geophys Monogr Ser 148:285-304

Von Damm KL, Bischoff JL (1987) Chemistry of hydrothermal solutions from the southern Juan de Fuca Ridge. J Geophys Res 92:11334-46

Von Damm KL, Bischoff JL, Rosenbauer RJ (1991) Quartz solubility in hydrothermal seawater: an experimental study and equation describing quartz solubility for up to 0.5 M NaCl solutions. Am J Sci 291:977-1007

Von Damm KL, Bray AM, Buttermore LG, Oosting SE (1998) The geochemical controls on vent fluids from the Lucky Strike vent field, Mid-Atlantic Ridge. Earth Planet Sci Lett 160: 521-536

Von Damm KL, Buttermore LG, Oosting SE, Bray AM, Fornari DJ, Lilley MD, Shanks WC (1997) Direct observation of the evolution of a seafloor 'black smoker' from vapor to brine. Earth Planet Sci Lett 149:101-111

Von Damm KL, Lilley MD, Shanks WC, Brockington M, Bray AM, O'Grady KM, Olson E, Graham A, Proskurowski G (2003) Extraordinary phase separation and segregation in vent fluids from the southern East Pacific Rise. Earth Planet Sci Lett 206:365-378

Von Damm KL, Oosting SE, Kozlowski R, Buttermore LG, Colodner DC, Edmonds HN, Edmond JM, Grebmeier JM (1995) Evolution of East Pacific Rise hydrothermal vent fluids following a volcanic eruption. Nature 375:47-50

Webster J, Mandeville C (2007) Fluid immiscibility in volcanic environments. Rev Mineral Geochem 65:313-362

Welhan JA, Craig H (1979) Methane and hydrogen in East Pacific Rise hydrothermal fluids. Geophys Res Lett 6:829-831

Wilcock WSD, McNabb A (1996) Estimates of crustal permeability on the Endeavour segment of the Juan de Fuca mid-ocean ridge. Earth Planet Sci Lett 138:83-91

Wood SA, Crerar D, Brantley SL, Borscsik M (1984) Mean molar stoichiometric activity coefficients of alkali halides and related electrolytes in hydrothermal solutions. Am J Sci 284:668-705

Xu WY, Lowell RP (1998) Oscillatory instability of one-dimensional two-phase hydrothermal flow in heterogeneous porous media. J Geophys Res 103:20859-20868

Reviews in Mineralogy & Geochemistry
Vol. 65, pp. 241-258, 2007
Copyright © Mineralogical Society of America

8

Fluids in Hydrocarbon Basins

Karen S. Pedersen and Peter L. Christensen

Calsep A/S
Gl. Lundtoftevej 1C
DK-2800 Kongens Lyngby, Denmark
ksp@calsep.com

INTRODUCTION

A petroleum reservoir consists of a zone with hydrocarbons that may potentially be produced and used as an energy resource. The hydrocarbons may have liquid (oil) properties, gas properties or the reservoir may consist of an oil zone with a gas cap on top. Often a water-rich zone is found beneath the hydrocarbon zone. The petroleum reservoir fluids found in the hydrocarbon zone may be divided into

- Natural gas mixtures
- Gas condensate mixtures
- Near-critical mixtures
- Volatile oils
- Black oil (or "ordinary" oils)
- Heavy oils

The various fluid types can be distinguished by the content of heavy hydrocarbons and by the phase behavior (number of phases, phase compositions and phase properties). Table 1 shows examples of each type of fluid composition. A natural gas mixture only contains trace amounts of components heavier than C_6 while at the other end of the fluid scale, a heavy oil is

Table 1. Examples of reservoir fluid compositions in mol%. *M* stands for molecular weight and the C_{7+} density is a 1.01 bar and 15 °C.

Component	Natural Gas	Gas Condensate	Near Critical Fluid	Volatile Oil	Black Oil	Heavy Oil
N_2	0.340	0.53	0.46	0.55	0.59	0.00
CO_2	0.840	3.30	3.36	2.83	0.36	1.44
C_1	90.400	72.98	62.36	55.57	40.81	18.72
C_2	5.199	7.68	8.90	8.59	7.38	0.14
C_3	2.060	4.10	5.31	5.75	7.88	0.03
iC_4	0.360	0.70	0.92	1.01	1.20	0.01
nC_4	0.550	1.42	2.08	2.44	3.96	0.01
iC_5	0.140	0.54	0.73	0.90	1.33	0.01
nC_5	0.097	0.67	0.85	1.24	2.09	0.27
C_6	0.014	0.85	1.05	1.58	2.84	0.41
C_{7+}		7.24	9.58	19.56	31.56	76.70
$C_{7+} M$		167.4	179.0	201.1	213.2	540.0
C_{7+} density (g/cm^3)		0.816	0.821	0.845	0.843	1.006

1529-6466/07/0065-0008$05.00

DOI: 10.2138/rmg.2007.65.8

dominated by components heavier than C_6. The properties of heavy oil may in fact be influenced by components as heavy as C_{200}.

The phase behavior of petroleum reservoir fluids is often referred to as *PVT* properties where *P* stands for pressure, *V* for volume (or more correctly molar volume) and *T* for temperature. Because production takes a petroleum reservoir fluid through significant changes in pressure and temperature, it is of great practical interest to know the volumetric changes occurring as a result of changed *T* and *P* conditions. It is also of interest to know how the total volume will split between gas and liquid as a function of *T* and *P* and how changes in *T* and *P* will affect the densities and other physical properties of the gas and liquid phases. This is why *PVT* studies and *PVT* simulations are important disciplines in the oil industry. Comprehensive descriptions of reservoir fluid properties including both hydrocarbons and water are given by Pedersen and Christensen (2006) and Dandekar (2006).

PHASE DIAGRAMS

Pure components: methane (CH_4) and ethane (C_2H_6)

All hydrocarbon fluids in petroleum reservoirs contain methane (C_1) and ethane (C_2) as a more or less abundant component. In their pure forms, methane and ethane are gases at both reservoir and ambient conditions. Mixed with heavier hydrocarbons, however, they may dissolve in a liquid oil phase, and they will also have some, although limited solubility in aqueous systems. Figure 1 shows the vapor pressure curves of pure methane and pure ethane, along which liquid and gas coexist. At temperature below the vapor pressure curve, pure methane and ethane will be in a liquid state and at higher temperature in a gaseous state. It is seen that liquefaction of pure methane will not take place at any pressure unless the temperature is −82 °C or lower. A vapor pressure curve ends in the critical point. The temperature at the critical point is called the critical temperature (T_c) and the pressure at the critical point is the critical pressure (P_c). No phase transition can take place at temperatures above T_c and pressures above P_c, and the fluid state at higher temperatures and pressures is often referred to as a supercritical or dense phase state (for phase diagrams of H_2O see Liebscher and Heinrich 2007, this volume).

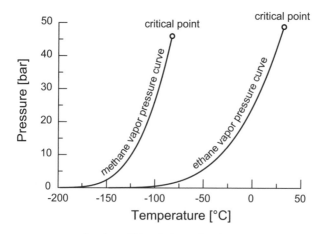

Figure 1. Vapor pressure curves of methane (C_1) and ethane (C_2). Vapor and liquid coexist along the vapor pressure curve. At lower temperatures the actual component is in liquid form and at higher temperatures it is in vapor form. At pressures above the critical point the component is in a supercritical form (Pedersen and Christensen 2006).

How the vapor pressure (P^{vap}) varies with temperature is defined by the acentric factor, ω (Pitzer 1955)

$$\omega = -1 - \log_{10} \left(\frac{P^{vap}}{P_c} \right)_{T=0.7T_c} \tag{1}$$

A plot of the logarithm of the reduced pure component vapor pressure, $P_r^{vap} = P^{svap} / P_c$, against the reciprocal of the reduced temperature $T_r = T/T_c$ will for most pure substances give an approximately straight line. Argon (Ar) by definition has an acentric factor of zero. From Equation (1) it can be seen that argon at a reduced temperature, T_r, of 0.7 ($T = 0.7\ T_c$) will have a vapor pressure of one tenth of the critical pressure (P_c). Most hydrocarbons will for the same reduced temperature have a vapor pressure lower than one tenth of P_c and therefore an acentric factor > 0. T_c, P_c and ω are key component parameters when a cubic equation of state is used to represent the phase behavior of petroleum reservoir fluids. Pedersen and Christensen (2006) have given several examples of the application of cubic equations of state to simulate gas and oil properties. Table 2 gives examples of T_c, P_c and ω for some reservoir fluid constituents.

Table 2. Critical temperature (T_c), critical pressure (P_c) and acentric factor (ω) of some common petroleum mixture constituents (Reid et al. 1987).

Component	T_c (Kelvin)	P_c (bar)	ω
Nitrogen (N$_2$)	126.2	33.9	0.040
Carbon dioxide (CO$_2$)	304.2	73.8	0.225
Hydrogen Sulfide (H$_2$S)	373.2	89.4	0.100
Methane (C$_1$)	190.6	46.0	0.008
Ethane (C$_2$)	305.4	48.8	0.098
Propane (C$_3$)	369.8	42.5	0.152
Iso-butane (iC$_4$)	408.1	36.5	0.176
N-butane (nC$_4$)	425.2	38.0	0.193
Iso-pentane (iC$_5$)	460.4	33.8	0.227
N-pentane (nC$_5$)	469.5	33.7	0.251
N-hexane (nC$_6$)	507.4	29.7	0.296
N-heptane (nC$_7$)	267.1	27.4	0.351
Benzene	289.0	48.9	0.212
Cyclohexane (cC$_6$)	280.3	40.73	0.213
Ethylbenzene	344.0	36.1	0.301
N-decane (nC$_{10}$)	344.5	21.1	0.490
N-eicosane (nC$_{20}$)	493.9	11.15	0.907

Multicomponent hydrocarbon fluids

Although methane (C$_1$) and ethane (C$_2$) are abundant components, petroleum reservoir fluids are generally multicomponent mixtures. Unlike for pure components, the two-phase region of a mixture is not restricted to a single line in a *P-T* diagram, but is a *P-T* area surrounded by the *P-T* phase envelope of the fluid mixture. Figure 2a shows the phase envelope of the volatile oil mixture from Table 1. The phase envelope consists of a dew point branch and a bubble point branch meeting in the critical point of the mixture. At the dew point branch the mixture is in gaseous form in equilibrium with an incipient amount of liquid and is said to be saturated. At higher temperatures but same pressure there is no liquid present. On the contrary, the gas may take up liquid components without liquid precipitation taking place. The term undersaturated gas is used to describe this state. At the bubble point branch, the mixture is in liquid form in equilibrium with an incipient amount of gas and the liquid is said to be saturated. At lower temperatures the liquid is undersaturated.

At the critical point, the two phases attain identical properties and the composition of the bulk system. At temperatures near the T_c and pressures above P_c there is only one phase present and the distinction between gas or liquid vanishes. The terms super-critical fluid or dense phase are often used. The highest pressure, at which two phases can coexist, is called the cricondenbar, whereas the highest temperature, at which two phases coexist, is called the cricondentherm. Figure

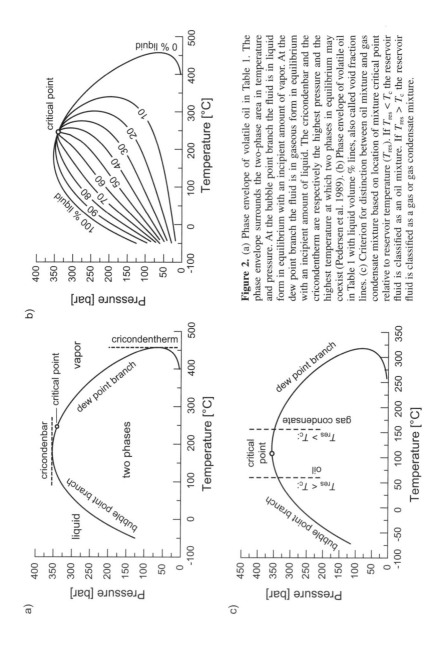

Figure 2. (a) Phase envelope of volatile oil in Table 1. The phase envelope surrounds the two-phase area in temperature and pressure. At the bubble point branch the fluid is in liquid form in equilibrium with an incipient amount of vapor. At the dew point branch the fluid is in gaseous form in equilibrium with an incipient amount of liquid. The cricondenbar and the cricondentherm are respectively the highest pressure and the highest temperature at which two phases in equilibrium may coexist (Pedersen et al. 1989). (b) Phase envelope of volatile oil in Table 1 with liquid volume % lines, also called void fraction lines. (c) Criterion for distinction between oil mixture and gas condensate mixture based on location of mixture critical point relative to reservoir temperature (T_{res}). If $T_{res} < T_c$ the reservoir fluid is classified as an oil mixture. If $T_{res} > T_c$ the reservoir fluid is classified as a gas or gas condensate mixture.

2b shows the phase envelope of the volatile oil from Table 1, contoured with lines of constant phase volume % (void fraction). The line for 50% liquid, for example, shows the *P-T* values for which 50 volume % of the fluid is in liquid form and 50 volume % in vapor (or gaseous) form. All iso-volume % lines end in the critical point at which gas and liquid are indistinguishable.

By definition, an oil mixture is in liquid form at reservoir conditions, while a gas or gas condensate mixture is in gaseous form at reservoir conditions. For a given fluid composition, the phase state depends on the reservoir temperature relative to the critical temperature of the mixture, as is shown in Figure 2c. When production from a reservoir takes place, the temperature remains approximately constant at the initial reservoir temperature, T_{res}, whereas the pressure decreases as a result of material being removed from the reservoir. The decreasing pressure will, at some stage, lead to the formation of a second phase. For an initially liquid (oil) mixture, this happens when the pressure reaches the bubble point branch at the temperature T_{res}, and the second phase forming will be a gas phase of lower density than the liquid (oil) phase from which it was formed. If a liquid phase condenses from an originally gaseous reservoir fluid at reservoir temperature, the term gas condensate is used. For a gas condensate, the saturation pressure at T_{res} will be a dew point, and the second phase forming at the saturation point will be denser than the gas from which it evolved. When the cricondentherm (defined in Fig. 2a) is lower than the reservoir temperature, the fluid is classified as a natural gas, from which no liquid precipitation can take place at the reservoir temperature at any pressure. The oil industry has settled on a series of standardized *PVT* experiments to experimentally investigate how hydrocarbon reservoir fluids behave at reservoir temperature and decreasing pressure (see Pedersen and Christensen 2006 for details).

PHYSICAL AND TRANSPORT PROPERTIES OF HYDROCARBON FLUIDS

Density and viscosity are key properties of hydrocarbon fluids. Figure 3a shows the liquid density of the black oil from Table 1 as a function of pressure at temperatures between 15 °C and 115 °C. Above the bubble point (shown as a dashed line in Fig. 3a), the density at a given temperature decreases with decreasing pressure as more space becomes available for the fluid. For lower pressures, the density is seen to increase with decreasing pressure as a result of low-density gas components evaporating from the liquid phase and leaving heavy high-density hydrocarbons in the liquid phase. Oil densities will usually be in the range from 0.5 and 1.0 g/cm^3 depending on pressure, temperature and chemical composition.

The variation in total volume (V^{tot}) with pressure (P) for a mixture of a constant composition (e.g., oil mixture above bubble point) is expressed through the compressibility

$$c_o = -\frac{1}{V^{tot}} \left(\frac{\partial V^{tot}}{\partial P} \right)_T \tag{2}$$

Figure 3b shows the Z factor of a gas evaporating from the black oil from Table 1 as a function of pressure at temperatures between 15 °C and 115 °C. Z is defined as

$$Z = \frac{PV}{RT} \tag{3}$$

where P stands for pressure, V for molar volume (volume of a weight amount equal to the molecular weight), R is the gas constant and T the absolute temperature. Z will approach 1 for pressure approaching zero as the gas approaches the properties of an ideal gas. It can be seen from Figure 3b that for a given pressure the Z factor of the gas evaporated from the black oil decreases with increasing temperature indicating that the gas properties deviate increasingly from ideal gas behavior with increasing temperature.

a)

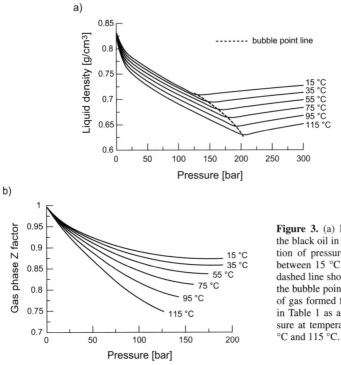

b)

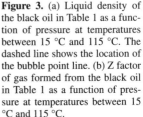

Figure 3. (a) Liquid density of the black oil in Table 1 as a function of pressure at temperatures between 15 °C and 115 °C. The dashed line shows the location of the bubble point line. (b) Z factor of gas formed from the black oil in Table 1 as a function of pressure at temperatures between 15 °C and 115 °C.

Viscosity is a flow property and of key importance to hydrocarbon fluid extraction from a reservoir rock. The dynamic viscosity, η, is defined as (e.g., Pedersen and Christensen 2006)

$$\eta = \left(\tau_{xy}\right) \bigg/ \left(\frac{\partial v_x}{\partial y}\right) \qquad (4)$$

where τ_{xy} is the applied shear stress (force per area unit), v_x is the flow velocity of fluid in x-direction, and $\partial v_x/\partial y$ is the shear rate

If the viscosity of a fluid is independent of shear rate, the fluid is said to behave Newtonian, and this flow behavior is seen for most petroleum reservoir fluids at reservoir conditions. At lower temperatures, solids as for example wax, may precipitate from a reservoir fluid, and the solid-liquid suspension will behave in a non-Newtonian manner with the viscosity generally increasing with decreasing shear rate. Wax precipitation is common in subsea pipelines used to transport reservoir fluids to nearby process plants.

The SI unit for viscosity is N s/m². It is related to other commonly used viscosity units as follows: 1 N s/m² = 1 kg/(m s) = 1 Pa s = 1000 mPa s = 1000 centipoise (cP) = 10 poise (P). The ratio of the viscosity to the density is commonly called the kinematic viscosity. With the viscosity in cP and the density in grams per cubic centimeter, the unit of the kinematic viscosity is cm²/s or Stokes.

Figure 4a shows the liquid viscosity of the black oil from Table 1 as a function of pressure at three temperatures between 15 °C and 115 °C. Above the bubble point (shown as a dashed line in Figure 4a) the viscosity at a given temperature decreases with decreasing pressure. Below the bubble point pressure the liquid viscosity increases with decreasing pressure, which is a result of gas components evaporating from the liquid phase. The viscosity of the

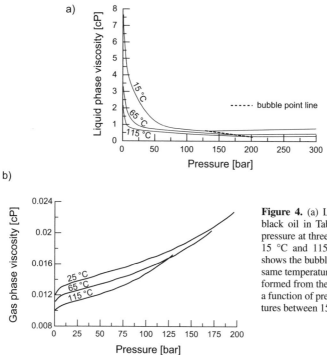

Figure 4. (a) Liquid viscosity of the black oil in Table 1 as a function of pressure at three temperatures between 15 °C and 115 °C. The dashed line shows the bubble point pressures at the same temperatures. (b) Viscosity of gas formed from the black oil in Table 1 as a function of pressure at three temperatures between 15 °C and 115 °C.

remaining high molecular weight components is high and even though a pressure reduction for a liquid of a constant composition would make the viscosity decrease, the net result below the bubble point pressure is an increasing viscosity with decreasing pressure. Figure 4b shows the viscosity of the gas phase forming from the black oil in Table 1 as a function of pressure for three different temperatures between 15 °C and 115 °C.

The viscosity of reservoir fluids can vary significantly as a function of composition (Pedersen et al. 1984). A gas condensate has a typical viscosity at reservoir conditions of 0.03 cP. Reservoir oil viscosities may vary from around 0.3 cP to 10,000 cP. Viscosities of the order of several thousands cP are seen for high-density oils with a high content of heavy aromatic compounds (Lindeloff et al. 2004).

Figure 5a shows the liquid volume % of the black oil in Table 1 as a function of pressure for a temperature of 100 °C. At high pressure the fluid is in liquid form. At the bubble point pressure, gas starts to form and the liquid volume % will decrease monotonically with decreasing pressure. Since the gas phase has a lower viscosity than the oil phase (compare Fig. 4a and 4b), the gas will be more mobile. The gas will therefore be easier to produce and unless some enhanced oil recovery technique is applied, the oil may be left behind in the reservoir and never produced.

Figure 5b shows the liquid volume % of the gas condensate mixture in Table 1 as a function of pressure for a temperature of 100 °C. At high pressure all fluid is in gaseous form. At the dew point pressure, liquid starts to form and the liquid volume % will initially increase with decreasing pressure. It will then pass through a maximum and decrease if the pressure is further decreased. A plot like the one in Figure 5b is often called a liquid dropout curve. In *PVT* studies on petroleum reservoir fluids the liquid is standard shown in percent of the dew point volume. With natural depletion from a gas condensate reservoir, the liquid formed will only to a minor

a)   b)

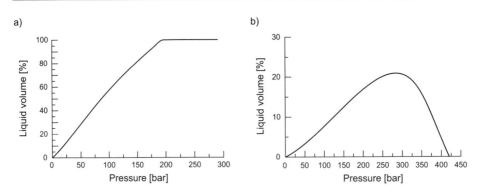

Figure 5. (a) Liquid volume % of the black oil in Table 1 as a function of pressure at a temperature of 100 °C. (b) Liquid volume % of the gas condensate in Table 1 as a function of pressure at a temperature of 100 °C.

extent contribute to the production but remain in the reservoir. As the price of hydrocarbon liquid per mass unit is higher than that of hydrocarbon gas, the value of the produced fluid will therefore decrease when the reservoir pressure gets below the dew point pressure.

PROPERTIES OF HYDROCARBON SYSTEMS EXPRESSED IN BLACK OIL TERMS

Reservoir fluids are often classified in terms of the following so-called black oil parameters

- Gas/Oil Ratio (or GOR)
- Oil formation volume factor (or B_o)
- Gas formation volume factor (or B_g)
- API gravity

The definition of the GOR is illustrated in Figure 6a. It is the relative volumes of produced gas and oil at standard conditions (1.01 bar and 15 °C). The definition of the oil formation factor (B_o) may be seen from Figure 6b. It is also called shrinkage factor and expresses how much the oil volume decreases from reservoir conditions to standard conditions. This volume decrease is the result of gas components evaporating that were kept in solution in the liquid phase at reservoir conditions. The gas formation volume factor (B_g), as is illustrated in Figure 6c, expresses the volume increase of gas from reservoir conditions to atmospheric conditions. Finally the term API gravity is defined as

$$API = \frac{141.5}{SG} - 131.5 \qquad (5)$$

where SG is the 60 °F specific gravity of the liquid produced at standard conditions. The specific gravity is defined as the mass ratio of equal volumes of oil and water at the appropriate temperatures. As the density of water at 60 °F is close to 1 g/cm³, the specific gravity of an oil sample will take approximately the same value as the density of the oil sample in g/cm³.

COMPOSITIONAL VARIATIONS WITH DEPTH

Petroleum reservoirs show variations in pressure, temperature and composition with depth. Table 3 shows typical changes in pressure, temperature and composition per 100 m vertical depth in a petroleum reservoir.

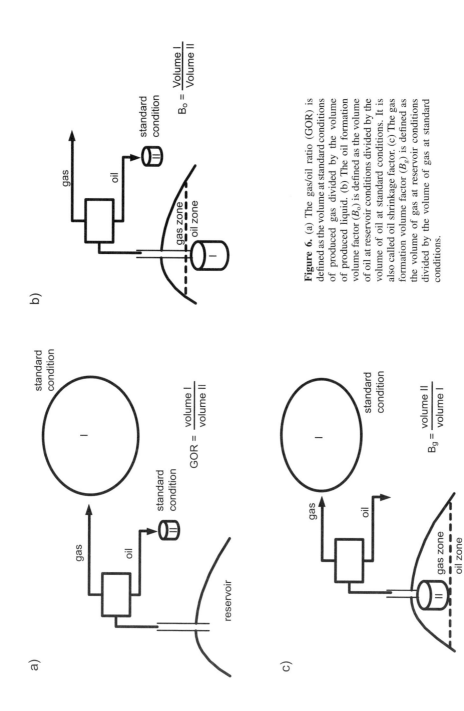

Figure 6. (a) The gas/oil ratio (GOR) is defined as the volume at standard conditions of produced gas divided by the volume of produced liquid. (b) The oil formation volume factor (B_o) is defined as the volume of oil at reservoir conditions divided by the volume of oil at standard conditions. It is also called oil shrinkage factor. (c) The gas formation volume factor (B_g) is defined as the volume of gas at reservoir conditions divided by the volume of gas at standard conditions.

Table 3. Typical variations with depth in a petroleum reservoir
(Pedersen and Christensen 2006).

Effect per 100 m downwards	Oil	Gas
P (bar)	4-6	2
T (°C)	2	2
Saturation point (bar)	−(6-11)	8-9
Molecular weight	8-21	0.3-0.4
Density (g/cm³)	0.02-0.15	0.005-0.007
C_1 mol%	−(1.7-1.5)	−0.5
C_2 - C_6 mol%	−0.2	−0.2
C_{7+} mol%	1.8-1.9	0.2-0.3

At equilibrium in a closed system with negligible depth differences, the chemical potential (μ) of a component i is the same at all positions in the system. The criterion of equal chemical potentials is however not valid for a system with considerable depth differences. For such system the depth potential must also be considered. For an isothermal system the equilibrium relation for component i becomes

$$\mu_i(h) - \mu_i(h^\circ) = M_i g(h - h^\circ)$$

(6)

where h stands for depth, M for molecular weight, g for gravitational acceleration and h° is a reference depth. The chemical potential is related to the fugacity (f) through the following relation

$$d\mu_i = RT \, d\ln f_i$$

(7)

where T is temperature and R is the gas constant. For an isothermal reservoir, Equation (7) may be rewritten to

$$\ln f_i^\circ - \ln f_i^{h^\circ} = \frac{M_i g(h - h^\circ)}{RT}$$

(8)

The fugacity (f) of component i is related to the fugacity coefficient (φ) of component i through

$$f_i = \varphi_i z_i P$$

(9)

where z is mole fraction. Making use of Equation (9), Equation (8) may for an N-component system be written as the following N equations

$$\ln\left(\varphi_i^h z_i^h P^h\right) - \ln\left(\varphi_i^{h\circ} z_i^{h\circ} P^{h\circ}\right) = \frac{M_i g(h - h^\circ)}{RT}; \quad i = 1, 2, \ldots, N$$

(10)

The mole fractions of the components must sum to 1.0 giving one additional equation

$$\sum_{i=1}^{N} z_i = 1$$

(11)

If the pressure P^{h° and the composition ($z_i^{h^\circ}$, $i = 1,...,N$) are known in the reference depth h°, the number of variables for a given depth is $N+1$, these being (z_i^h, $i = 1,...,N$) and P^h. A set of $N+1$ equations with $N+1$ variables may be solved to give the molar composition and the pressure as a function of depth. Schulte (1980) has outlined a procedure for solving these equations and given examples of compositional gradients calculated using an equation of state.

Figure 7a shows how reservoir pressure (P_{res}) and saturation pressure (P_{sat}) vary with depth for the black oil reservoir fluid in Table 1 assuming a reservoir temperature of 102 °C and a reservoir pressure (P_{res}) of 200 bar at a sampling depth of 760 m. Starting in the sampling depth the saturation pressure will decrease with further depth because the fluid will become richer in high molecular weight components while on the other hand the content of lighter components will decrease. Moving upwards, the saturation point will increase and the reservoir pressure will decrease. At a depth of 722 m the saturation pressure and the reservoir pressure coincide and the reservoir fluid is at its saturation point. A distinct change in fluid composition is seen at this position with an oil phase beneath and a gas phase above. The reservoir fluid is said to have a gas-oil contact at this depth. Above the gas-oil contact the saturation pressure will decrease with increasing distance from gas-oil contact as the concentration of high molecular weight components in the fluid decreases. The plot in Figure 7a is typical for a reservoir consisting of an oil zone (oil leg) with a gas cap on top.

The distinct change in fluid properties at the gas-oil contact is further illustrated in Figure 7b, which shows the gas/oil ratio (defined above) with depth. Coming from beneath, the gas/oil ratio increases sharply at the gas-oil contact where the fluid type changes from oil to gas condensate.

The change from an oil type of fluid to a gas condensate will not always take place through a gas-oil contact (with $P_{res} = P_{sat}$) as is shown in Figure 7b. Figure 7c shows how reservoir pressure (P_{res}) and saturation pressure (P_{sat}) vary with depth for the near critical reservoir fluid in Table 1 assuming a reservoir temperature of 150 °C and a reservoir pressure of 420 bar at a sampling depth of 960 m. At the saturation pressure at 150 °C, the fluid is close to its critical point. Further down in the reservoir the fluid becomes more oil-like and further upwards it becomes more gas-like. In both directions the saturation point decreases as compared with the saturation point at the sampling depth. At all depths, the saturation pressure is therefore lower than the reservoir pressure for which reason there is no gas-oil contact as that seen in Figure 7a.

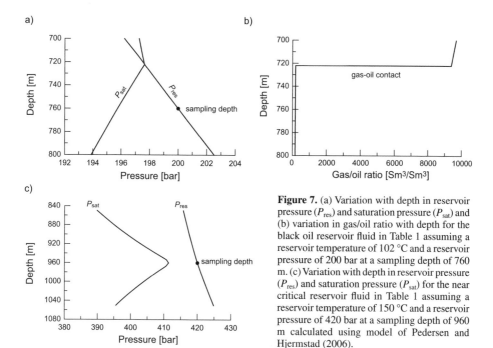

Figure 7. (a) Variation with depth in reservoir pressure (P_{res}) and saturation pressure (P_{sat}) and (b) variation in gas/oil ratio with depth for the black oil reservoir fluid in Table 1 assuming a reservoir temperature of 102 °C and a reservoir pressure of 200 bar at a sampling depth of 760 m. (c) Variation with depth in reservoir pressure (P_{res}) and saturation pressure (P_{sat}) for the near critical reservoir fluid in Table 1 assuming a reservoir temperature of 150 °C and a reservoir pressure of 420 bar at a sampling depth of 960 m calculated using model of Pedersen and Hjermstad (2006).

The compositional variation with depth is further influenced by the vertical temperature gradient seen in most reservoirs. An increasing temperature with depth has been observed to strengthen the compositional gradient (Montel and Gouel 1985; Pedersen and Lindeloff 2003), which strengthening as shown by Pedersen and Hjermstad (2006) can be explained and modeled using the theory of irreversible thermodynamics.

HYDROCARBON-WATER PHASE EQUILIBRIA

As mentioned above, a zone of formation water is often found beneath the hydrocarbon zone(s). This water is called formation water. Water deviates from hydrocarbons by having more polar forces acting between the molecules.

Water has a molecular weight (M) of 18.0, which is similar to that of methane ($M = 16.0$), but the absolute critical temperature of water is 647.3 K, more than 3 times higher than that of methane (190.6 K). Similarly, the critical pressure of water is almost 5 times higher than that of methane (221 bar for water versus 46 bar for methane). Water has these, for a low molecular weight substance, extreme properties because of attractive (association) forces acting between the water molecules.

Table 4. Mutual solubility of propane (C_3) and water (H_2O) (Kobayashi and Katz 1953).

T (K)	P (bar)	mol% C_3 in water	mol% H_2O in C_3
369.3	27.6	0.0213	3.034
369.3	48.2	0.0277	0.815
369.3	68.9	0.0287	0.752
369.3	103.5	0.0296	0.703
369.3	137.9	0.0304	0.665
369.3	206.9	0.0316	0.619
394.3	27.6	0.0231	7.260
394.3	48.2	0.0338	3.622
394.3	68.9	0.0379	1.897
394.3	103.5	0.0400	1.455
394.3	137.9	0.0414	1.370
394.3	206.9	0.0444	1.265

Table 4 shows the mutual solubility of propane (C_3) and water (H_2O) at pressures ranging from 27.6 to 206.9 bar and temperatures of 369.6 K and 394.3 K (Kobayashi and Katz 1953). The table illustrates the fact that the miscibility between water and hydrocarbons is quite limited. The polar forces acting between the water molecules make it difficult for hydrocarbon molecules to penetrate into the water phase. On the other hand it is also difficult for the water clusters (several water molecules attached through polar forces) to find room in a hydrocarbon liquid phase. The situation is different in a vapor phase, in which the distance between the molecules is much higher and the intermolecular forces therefore lower.

Formation water is not pure water but contains dissolved salts, most of which are highly soluble in water as for example sodium chloride (NaCl), potassium chloride (KCl) and calcium chloride ($CaCl_2$). The water may further carry salts of low solubility like barium sulfate ($BaSO_4$), calcium carbonate ($CaCO_3$) and calcium sulfate ($CaSO_4$), which under certain conditions can precipitate as solid salt. Salt deposition is also called scaling and is a potential problem in pipelines transporting unprocessed well streams containing formation water. Scale precipitation in the reservoir may be seen when seawater is injected into the reservoir with the purpose of enhancing the oil recovery.

Table 5 shows the composition of a gas condensate mixture. Its mutual solubility with pure water and salt water has been measured at high pressure and temperature (Pedersen and Milter 2004). The phase equilibrium data may be seen from Table 6 and the saline water (brine) composition from Table 7. Table 6 gives a good idea of the mutual solubility of water and hydrocarbons at elevated pressure and temperature as well as of the influence of dissolved salts on the gas solubility in water at reservoir conditions. By comparing the gas concentration in the

Table 5. Molar composition of reservoir fluid for which phase equilibrium data mixed with pure water and salt water is shown in Table 6 (Pedersen and Milter 2004). M stands for molecular weight and the C_{10+} density is at 1.01 bar and 15 °C.

Component	mol%	Component	mol%
N_2	0.369	nC_5	0.720
CO_2	4.113	C_6	0.972
C_1	69.243	C_7	2.499
C_2	8.732	C_8	0.732
C_3	4.270	C_9	0.637
iC_4	0.877	C_{10+}	4.571
nC_4	1.641	$C_{10+}\ M$	229.5
iC_5	0.625	C_{10+} Density (g/cm^3)	0.961

pure water with that in the salt water it can be seen that the gas solubility in the water phase is lowered by increasing salt concentration in the water, a thermodynamic effect known as "salting out" (Pitzer 1986).

Figure 8a shows plots of the density of pure water at pressures ranging from 100 to 1000 bar and temperatures between 0 and 200 °C calculated using the model of Keyes et al. (1968) and a computer code published by Hendricks et al. (1973).

The density of a water phase with dissolved salts can be calculated using the correlation suggested by Numbere et al. (1977):

$$\frac{\rho_s}{\rho_w}-1=C_S\left[\begin{array}{c}7.65\times10^{-3}-1.09\times10^{-7}P+C_S\left(2.16\times10^{-5}+1.74\times10^{-9}P\right)\\-\left(1.07\times10^{-5}-3.24\times10^{-10}P\right)T+\left(3.76\times10^{-8}-1.0\times10^{-12}P\right)T^2\end{array}\right] \quad (12)$$

where ρ_s is the salt-water density, ρ_w the density of salt free water at the same T and P, C_s is the salt concentration in weight %, T the temperature in °F and P the pressure in psia. Figure 8b shows plots of the density of a 10 wt% sodium chloride (NaCl) solution in water at pressures ranging from 100 to 1000 bar and temperatures between 0 and 200 °C.

Figure 9 shows the phase envelope of the gas condensate mixture in Table 1 saturated with water at reservoir conditions of 500 bar and 141 °C, where the gas condensate may take up 1.9 mol% water. If a reservoir fluid at reservoir conditions is saturated with water from an underlying water zone, this water must be taken into consideration when producing from the hydrocarbon zone. The water produced with the hydrocarbons may condense as a free water phase in the well or in the pipeline used to transport the well stream to a process facility. This may influence the flow and corrosive properties of the produced well stream and there may also be a risk of gas hydrates (solid ice-like compounds) forming in the pipeline. Formations of solids may potentially lead to plugging of the pipeline and is highly undesirable.

CO$_2$ SEQUESTRATION

For climatic reasons it is desirable to minimize the outlet of carbon dioxide (CO$_2$) to the atmosphere. CO$_2$ is produced when hydrocarbons are burned, for example in power plants. Disposing of CO$_2$ by depositing it in petroleum reservoirs, in aquifers or at the bottom of the sea is called CO$_2$ sequestration (Bennaceur et al. 2004). This may at least at first hand seem to be an attractive way of dealing with the climatic impact of industrial CO$_2$ outlet to the atmosphere. The separation and compression of CO$_2$ from emission streams is however both cumbersome and expensive. The currently most widely used process to capture CO$_2$

Table 6. Measured equilibrium compositions (mol%) for the gas condensate in Table 5 mixed with pure water and mixed with salt water (brine) of the composition in Table 7 (Pedersen and Milter 2004).

	Pure Water		Salt Water	
	Vapor Composition	**Water Composition**	**Vapor Composition**	**Brine Composition**[*]
1000 bar and 35 °C. Feed consists of 0.69 mole gas condensate per mole salt free water.				
H_2O	0.055	99.4331	0.054	99.5845
N_2	0.337	0.0068	0.345	0.0000
CO_2	3.761	0.2130	4.323	0.1798
C_1	69.183	0.3420	68.934	0.2274
C_2	8.751	0.0043	8.580	0.0068
C_3	4.321	0.0002	4.258	0.0008
iC_4	0.898	0.0000	0.883	0.0001
nC_4	1.707	0.0001	1.637	0.0001
iC_5	0.688	0.0003	0.642	0.0000
nC_5	0.787	0.0001	0.740	0.0000
C_6	1.002	0.0000	0.989	0.0000
C_{7+}	8.511	0.0005	8.615	0.0000
1000 bar and 120 °C. Feed consists of 0.74 mole gas condensate per mole salt free water.				
H_2O	0.753	99.3724	0.663	99.5699
N_2	0.343	0.0274	0.357	0.0000
CO_2	4.068	0.1465	4.413	0.0940
C_1	68.583	0.4306	68.328	0.3204
C_2	8.586	0.0164	8.564	0.0127
C_3	4.244	0.0024	4.254	0.0019
iC_4	0.880	0.0001	0.872	0.0001
nC_4	1.666	0.0002	1.604	0.0003
iC_5	0.660	0.0008	0.629	0.0000
nC_5	0.754	0.0002	0.729	0.0000
C_6	0.982	0.0001	0.973	0.0000
C_{7+}	8.481	0.0029	8.614	0.0007
1000 bar and 200 °C. Feed consists of 0.61 mol gas condensate per mole salt free water.				
H_2O	4.683	98.9982	4.235	99.4011
N_2	0.391	0.0063	0.366	0.0000
CO_2	3.816	0.2113	4.188	0.1040
C_1	65.207	0.7455	65.249	0.4652
C_2	8.184	0.0286	8.244	0.0217
C_3	4.048	0.0040	4.102	0.0041
iC_4	0.845	0.0003	0.841	0.0003
nC_4	1.621	0.0005	1.549	0.0006
iC_5	0.663	0.0005	0.635	0.0004
nC_5	0.765	0.0004	0.746	0.0002
C_6	1.010	0.0001	1.029	0.0001
C_{7+}	8.765	0.0045	8.816	0.0024
700 bar and 200 °C. Feed consists of 0.54 mole gas condensate per mole salt free water.				
H_2O	5.725	99.1809	5.094	99.4996
N_2	0.346	0.0056	0.380	0.0000
CO_2	3.795	0.1765	4.168	0.1059
C_1	64.466	0.6011	65.005	0.3750
C_2	8.134	0.0246	8.188	0.0149
C_3	4.040	0.0037	4.038	0.0016
iC_4	0.838	0.0003	0.827	0.0002
nC_4	1.585	0.0005	1.525	0.0004
iC_5	0.641	0.0006	0.609	0.0002
nC_5	0.744	0.0005	0.707	0.0001
C_6	0.980	0.0003	0.972	0.0000
C_{7+}	8.706	0.0056	8.487	0.0021

[*] (excl. salt components)

Table 7. Salt water (brine) composition used in phase equilibrium study reported in Table 6.

Component	mol%
H_2O	97.347
$NaHCO_3$	0.035
$NaCl$	2.404
KCl	0.094
$CaCl_2$	0.075
$MgCl_2$	0.008
$SrCl_2$	0.014
$BaCl_2$	0.024

from flue gas is based on chemical absorption using monoethanolamine (MEA) solvent. Flue gas is bubbled through a solvent in an absorption column and CO_2 is absorbed in the solvent. The CO_2 rich solvent is afterwards led to a distillation unit that strips off the CO_2. This is an expensive process and other methods are being investigated, one being microporous membranes.

Figure 10a shows the vapor pressure curve of CO_2, which compound has a critical point of 31.1 °C and 73.8 bar. The CO_2 density at pressures between 1 and 400 bar and temperatures between 0 and 150 °C may be seen from Figure 10b. The CO_2 density at high pressure is quite high and may exceed that of water. CO_2 may therefore potentially be stored at the bottom of oceans. There are, however, some concerns about making that kind of deposition. CO_2 has a quite high solubility in water and may over time diffuse towards the top of the ocean and be released to the atmosphere (Ennis-King and Paterson 2003). Figure 10c shows the solubility of CO_2 in water as a function of pressure for temperatures of 25 °C and 50 °C (Burd 1968).

Diffusion of CO_2 to the atmosphere may be prevented if CO_2 and water form gas hydrates. These are crystalline ice-like compounds consisting of water lattices with cavities occupied by a gas component (which gas in the actual case will be CO_2). Makogan (1997), Sloan (1998) and Holder and Bisnoi (2000) have given detailed descriptions of gas hydrates. Figure 10d shows the formation conditions of CO_2 hydrates (calculated using model of Munck et al.

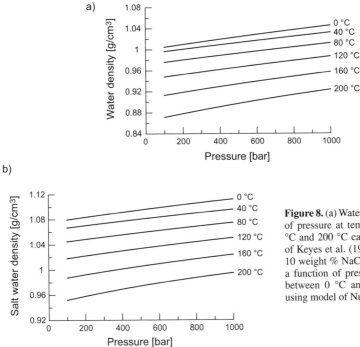

Figure 8. (a) Water density as a function of pressure at temperatures between 0 °C and 200 °C calculated using model of Keyes et al. (1968). (b) Density of a 10 weight % NaCl solution in water as a function of pressure at temperatures between 0 °C and 200 °C calculated using model of Numbere et al. (1977).

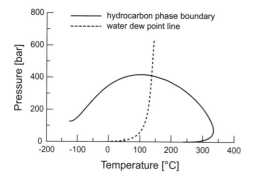

Figure 9. Simulated phase envelope for gas condensate fluid in Table 1 saturated with water at 500 bar and 141 °C.

1988). At 4 °C gas hydrates will be stable at pressures above 30 bar, at which pressures there is little risk of CO_2 bound in gas hydrates to be released to the atmosphere.

Injection of CO_2 into petroleum reservoirs may have a dual advantage, both for CO_2 storage and enhanced hydrocarbon recovery. Exploitation of an oil field normally leads to a pressure drop, because the removed material leaves more space for the remaining reservoir fluid. Gas injection is therefore commonly applied to enhance oil recovery (EOR). The injection gas will help keep the reservoir pressure high. It is desirable that the reservoir fluid remains single-phase, which requires that the injected gas dissolves in the oil phase. In a reservoir with both a gas and an oil phase present the gas will be more mobile than the oil phase and this may possibly lead to a gas break-through. The injection gas may either originate from production from the actual field or it may be produced specifically with the purpose of being used as injection gas. In the latter case the gas will often be carbon dioxide. CO_2 is interesting as an injection gas because it has a high solubility in oil and it is therefore a good candidate as injection gas also from a recovery point of view.

The Norwegian company Statoil is currently undertaking a very interesting CO_2 storage project (Baklid et al. 1996) in the Sleipner gas field, 250 kilometers west of Stavanger in the North Sea. The Sleipner field produces natural gas containing approximately 9% CO_2. To meet the specifications from the Norwegian government, the CO_2 concentration has to be reduced to 2.5%. Statoil captures and separates the CO_2 and has since 1996 injected over 7 million tons of CO_2 into the Utsira formation and plans to continue injection until 2020. The Utsira formation is a saline aquifer at a depth of 800 to 1,000 meters below seabed at the Sleipner field. Geophysical investigations indicate that there is little risk that the stored CO_2 will escape to the atmosphere.

REFERENCES

Baklid A, Korbøl R, Owren G (1996) Sleipner Vest CO_2 Disposal, CO_2 injection into a shallow underground aquifer. SPE 36600, SPE ATCE, Denver, October 6-9

Bennaceur K et al. (2004) CO_2 Capture and Storage – A Solution Within. Oilfield Review 16:44-61

Burd SD Jr. (1968). Phase Equilibria of Partially Miscible Mixtures of Hydrocarbons and Water. Ph.D. Dissertation, The Pennsylvania State University, State College, PA

Dandekar AY (2006) Petroleum Reservoir Rock and Fluid Properties. Taylor and Francis

Ennis-King J, Paterson L (2003) Role of convective mixing in the long-term storage of carbon dioxide in deep saline formations. SPE 84344, SPE ATCE, Denver, October 5-8

Hendricks RC, Peller, IC, Baron, AK (1973) WASP – A Flexible FORTRAN IV Computer Code for Calculating Water and Steam Properties. National Aeronautics and Space Administration, Washington, D.C.

Holder GD, Bisnoi PR (2000) Gas Hydrates - Challenges for the Future. New York Academy of Sciences

Keyes FG, Keenan JH, Hill PG, Moore JG (1968) A Fundamental Equation for Liquid and Vapor Water. Seventh International Conference on the Properties of Steam, Tokyo, Japan

Kobayashi R, Katz DL (1953) Vapor-liquid equilibria for binary hydrocarbon-water systems. Ind Eng Chem 45:440-446

Liebscher A, Heinrich CA (2007) Fluid–fluid interactions in the Earth's lithosphere. Rev Mineral Geochem 65:1-13

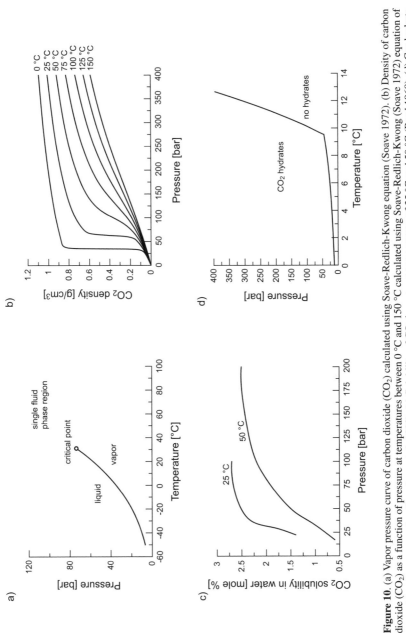

Figure 10. (a) Vapor pressure curve of carbon dioxide (CO_2) calculated using Soave-Redlich-Kwong equation (Soave 1972). (b) Density of carbon dioxide (CO_2) as a function of pressure at temperatures between 0 °C and 150 °C calculated using Soave-Redlich-Kwong (Soave 1972) equation of state with volume correction (Peneloux et al. 1982). (c) Solubility of CO_2 in water at temperature of 25 °C and 50 °C (Burd 1968). (d) Gas hydrate formation curve for carbon dioxide (CO_2) calculated using the model of Munck et al. (1988).

Lindeloff N, Pedersen KS, Rønningsen HP, Milter J (2004) The corresponding states viscosity model applied to heavy oil systems. J Can Petroleum Tech 43:47-53

Makogan YF (1997) Hydrates of Hydrocarbons. PennWell Publishing Company

Montel F, Gouel PL (1985) Prediction of compositional grading in a reservoir fluid column. SPE 14410, SPE ATCE, Las Vegas, NV September 22-25

Munck J, Skjold-Jørgensen S, Rasmussen P (1988) Computations of the formation of gas hydrates. Chem Eng Sci 43:2661-2672

Numbere D, Bringham WE, Standing MB (1977) Correlations for Physical Properties of Petroleum Reservoir Brines. Petroleum Research Institute, Stanford University

Pedersen KS, Fredenslund A, Christensen PL, Thomassen P (1984) Viscosity of crude oils. Chem Eng Sci 39:1011-1016

Pedersen KS, Christensen PL (2006) Phase Behavior of Petroleum Reservoir Fluids. Taylor and Francis

Pedersen KS, Hjermstad HP (2006) Modeling of large hydrocarbon compositional gradient. SPE 101275, Abu Dhabi International Petroleum Exhibition and Conference, Abu Dhabi, U.A.E., November 5-8

Pedersen KS, Lindeloff N (2003) Simulations of compositional gradients in hydrocarbon reservoirs under the influence of a temperature gradient. SPE 84364, SPE ATCE in Denver, Co, October 5-8

Pedersen KS, Milter J (2004) Phase equilibrium between gas condensate and brine at HT/HP conditions. SPE 90309, SPE ATCE, Houston, TX, September 26-29

Pedersen KS, Fredenslund A, Thomassen P (1989) Properties of Oils and Natural Gases. Gulf Publishing Company

Peneloux A, Rauzy E, Fréze R (1982) A consistent correction for Redlich-Kwong-Soave volumes. Fluid Phase Equilibria 8:7-23

Pitzer KS (1955) Volumetric and thermodynamic properties of fluids. I. Theoretical basis and virial coefficients. J Am Chem Soc 77:3427-3433

Pitzer KS (1986) Theoretical considerations of solubility with emphasis on mixed aqueous electrolytes. Pure Appl Chem 58:1599-1610

Reid RC, Prausnitz JM, Poling BE (1987) The Properties of Gases and Liquids. McGraw-Hill

Schulte AM (1980) Compositional variations within a hydrocarbon column due to gravity. SPE 9235, SPE ATCE, Dallas, September 21-24

Sloan ED (1988) Clathrate Hydrates of Natural Gases, 2nd Edition. Marcel Dekker, Inc.

Soave G (1972) Equilibrium constants from a modified Redlich-Kwong equation of state. Chem Eng Sci 27:1197-1203

Reviews in Mineralogy & Geochemistry
Vol. 65, pp. 259-312, 2007
Copyright © Mineralogical Society of America

9

Fluid-Fluid Interactions in Geothermal Systems

Stefán Arnórsson, Andri Stefánsson

Institute of Earth Sciences, University of Iceland
Sturlugata 7, IS-101 Reykjavík, Iceland
stefanar@raunvis.hi.is, as@hi.is

Jón Örn Bjarnason

Iceland GeoSurvey
Grensasvegur 9, IS-108 Reykjavik, Iceland
job@isor.is

INTRODUCTION

The main goal of geothermal geochemistry research is to identify the origin of geothermal fluids and to quantify the processes that govern their compositions and the associated chemical and mineralogical transformations of the rocks with which the fluids interact. The subject has a strong applied component: Geothermal chemistry constitutes an important tool for the exploration of geothermal resources and in assessing the production characteristics of drilled geothermal reservoirs and their response to production. Geothermal fluids are also of interest as analogues to ore-forming fluids. Understanding of chemical processes within active geothermal systems has been advanced by thermodynamic and kinetic experiments and numerical modeling of fluid flow. Deep drillings for geothermal energy have provided important information on the sources and composition of geothermal fluids, their reaction with rock-forming minerals, migration of the fluids, and fluid phase separation and fluid mixing processes.

Based on findings to date, geothermal fluids may be classified as primary or secondary. Primary fluids are those found in the roots of geothermal systems. They may constitute a mixture of two or more fluids, such as water of meteoric origin, seawater and magmatic volatiles. Several processes can lead to the formation of secondary fluids, such as the boiling of a primary fluid that separates it into liquid and vapor and the un-mixing of a very hot brine by its depressurization and cooling. Further, secondary geothermal fluids form by the mixing of deep fluids with shallow ground water or surface water. In this chapter we summarize the geo-hydrological and geochemical features of geothermal systems and delineate the processes that produce the observed chemical composition of the various types of geothermal fluids found in these systems. The main emphasis is, however, on gas chemistry and the assessment of fluid phase separation below hot springs and around discharging wells drilled into liquid-dominated volcanic geothermal systems.

BASIC FEATURES OF GEOTHERMAL SYSTEMS

Geothermal systems consist of a body of hot rock and hot fluid, or hot rock alone, in a particular rock-hydrological situation. The lifetime of individual systems is generally poorly known, as are fluid circulation times. Estimated ages of geothermal systems lie in the range of some 0.1 to 1 million years (Grindley 1965; Browne 1979; Arehart et al. 2002). Fluid circulation times may range from few hundreds of years, or even less, to more than 10,000 years (Arnórsson 1995a; Sveinbjörnsdóttir et al. 2001; Mariner et al. 2006).

1529-6466/07/0065-0009$10.00   DOI: 10.2138/rmg.2007.65.9

The fluids of geothermal systems are essentially meteoric water or seawater by origin or mixtures thereof. Some geothermal fluids contain a significant component of magmatic volatiles. Mixing of these source fluids and their interaction with the rock produces several types of geothermal fluids, as exemplified in Figure 1 for a system associated with an andesitic volcano.

The maximum depth of meteoric or seawater circulation is the brittle/ductile transition depth of rocks (Fournier 1991). This transition has been inferred from the depth of earthquake foci, frequently reaching depths of 5 to 8 km (Klein et al. 1977; Bibby et al. 1995) where temperatures may be 400 °C or higher (e.g., Kissling and Weir 2005). Fluid circulation in geothermal systems is essentially density driven when temperatures at the base of the circulation are above ~150 °C (see also Driesner and Geiger 2007, this volume, for parameters controlling hydrothermal convection).

The ultimate source of heat to geothermal systems is decaying radioactive elements, particularly U, Th and K, and the Earth's gravity field. Rising magma transports heat to the upper crust, and convecting fluids in permeable rocks transport the heat to even shallower depths, creating a geothermal system.

Types of geothermal systems

Several classification schemes have been proposed for geothermal systems. The most important ones are high-temperature and low-temperature systems (Bödvarsson 1961), hot-water (liquid-dominated) and vapor-dominated systems (White et al. 1971) and volcanic and non-volcanic systems (Goff and Janik 2000). High-temperature systems are generally volcanic and low-temperature systems non-volcanic.

Volcanic geothermal systems typically occur in areas of active volcanism where permeability and geothermal gradient are high. The heat source may be a major magma intrusion, a sheeted

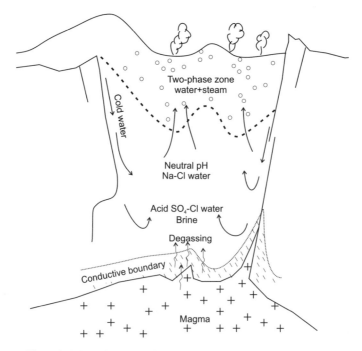

Figure 1. Schematic section of a volcanic geothermal system depicting the origin of different primary fluid types.

dyke complex, or a complex of minor intrusions. In liquid-dominated systems, the fluid is dominantly liquid water by volume (White et al. 1971). In vapor-dominated systems, vapor at ~240 °C fills permeable channels, although liquid water occupies inter-granular pore spaces (Fig. 2). Very hot saline liquid (>400 °C) is known to exist below the vapor-dominated zone (Fournier 1991; Barelli et al. 1995; Gianelli and Ruggieri 2002). Vapor-dominated systems are considered to evolve by the boiling down of earlier liquid-dominated systems, i.e., the supply of heat to the system in relation to fluid circulation rates is sufficient to vaporize a substantial fraction of the circulating fluid.

Geological structure of volcanic geothermal systems

Structurally, no two volcanic geothermal systems are identical. To describe the results of extensive studies of a particular system requires a model, which is unique to that system (e.g., Hochstein and Browne 2000). However, many geothermal systems have some common features, which are the geological structure and the nature of the heat source.

The geological structures of volcanic geothermal systems on diverging and converging plate boundaries differ in some respects. On converging plate boundaries, they are frequently associated with andesitic volcanoes, and the heat source may be magma offshoots from the main magma source or a deeper magma body in the roots of the volcano. On diverging plate boundaries, the heat source is usually a sheeted dyke complex. Volcanic geothermal systems are commonly associated with ring structures or calderas. This structure is the consequence of rapid emptying of a shallow magma body. The magma at converging plate boundaries is characteristically high in volatile components relative to the basaltic magma generated below diverging plate boundaries. This has significant consequences for the composition of the geothermal fluids in these two types of geological settings. The chemical composition and mineralogy of the host rocks also have major influence on the chemical composition of the geothermal fluid.

Temperature and pressure

Measured downhole temperatures in drilled geothermal systems range from ambient to over 400 °C. Depending on P and T the fluids of geothermal systems may be single phase or two-phase. Vapor-dominated systems represent a special type of the latter. Their mobile phase

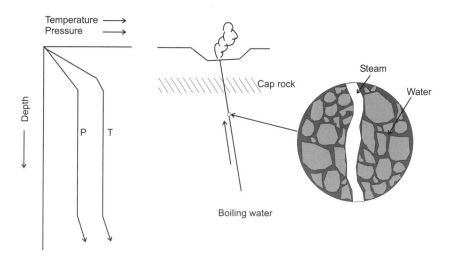

Figure 2. Schematic model of hydrological, fluid, temperature and pressure conditions in a vapor-dominated geothermal system. Based on Fournier (1981).

is vapor and their temperatures are always close to 240 °C, corresponding to the maximum enthalpy of saturated vapor. Geochemical studies indicate that the steam fraction in two-phase systems is typically small, <10 vol% (e.g., Giggenbach 1980; Arnórsson et al. 1990; Gudmundsson and Arnórsson 2002; Karingithi pers. comm.). In two-phase systems, pressure determines temperature at each depth, whereas under single phase conditions, temperature and pressure are independent variables. The fluid temperature at the base of the convection cell is determined by the relative rates of fluid flow and heat conduction in addition to the enthalpy of any magmatic fluid that may be added to the circulating fluid. In systems where the temperature is lower than about 350 to 370 °C, pore fluid pressure is hydrostatic (Fournier 1991). Where higher temperatures have been encountered in drillholes, the pressure is above hydrostatic but probably still below lithostatic (Fournier 1991).

TRANSFER OF HEAT

The mechanism of heat transfer from the magma heat source of volcanic geothermal systems to the circulating fluid is discussed in some detail here, because it is considered to have an important bearing on magma degassing and fluid mixing in the root zones of these systems. The general model for volcanic geothermal systems envisages deep circulation of fluid above and to the sides of a magmatic heat source (Lister 1983; Fournier and Pitt 1985; Fournier 1989), with a thin layer between the magma and the base of fluid circulation, through which heat is transferred conductively (Bödvarsson 1951, Björnsson et al. 1982). Some heat may also be transferred to the geothermal system by convection of saline fluids in a closed loop between the geothermal system and the magma heat source (Truesdell and Fournier 1976; Fournier 1985).

To account for the estimated natural heat loss of some geothermal systems, the conclusion is inescapable that the conductive heat transfer layer between the magma and the base of the fluid circulation must be very thin. Thus, in the subglacial area of Grímsvötn, which has an estimated heat output of 5000 MW (thermal), the estimated average thickness of this layer is 13 m (Björnsson et al. 1982). A recent estimate of the natural heat flow from the relatively small (1 km²) geothermal field at Reykjanes, Iceland, of 130±16 MW (Fridriksson et al. 2006) yields a thickness for the conductive heat transfer layer of about 50 m. In Yellowstone the temperature gradient between magma and the base of the geothermal system has been estimated as 700 to 1000 °C/km and magmatic temperatures may occur at depths as shallow as 3 to 5.5 km (Iyer et al. 1981; Fournier 1989).

Unique data on the mechanism of magma cooling were provided by drillings into the molten lava that formed in 1973 during an eruption on the island of Heimaey just off the south coast of Iceland (Jónsson and Matthíasson 1974). Water was pumped onto the lava to divert its flow from the neighboring village. The drillings into the water-cooled lava revealed a temperature of 100 °C down to a certain depth, followed by a sharp rise to as much as ~1050 °C in a layer that was as thin as 1 m (Fig. 3). The temperature of 100 °C was maintained by rising steam at atmospheric pressure, and the heat forming the steam was transferred by conduction through the 100 to 1050 °C layer from the molten lava to the base of the circulating water. With time, the water circulation caused the conductive layer to migrate downward. When geothermal systems develop over magma intruded into permeable rock, the heat extraction mechanism is likely to be that described above for the lava on Heimaey.

The chemistry of fluids in volcanic geothermal systems should be considered in light of the heat transfer mechanism described above. Close proximity of the circulating fluid to the magmatic heat source also implies closeness of the circulating fluid to gaseous magma components. At magmatic temperatures, these components include H_2O, CO_2, SO_2, H_2, HCl, HF, and many metal-chloride and metal-fluoride species, which upon deposition can form ore

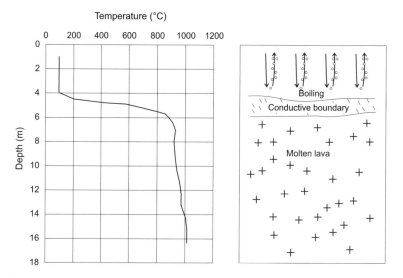

Figure 3. Temperature profile in a hole drilled into molten lava flow on the island of Heimaey off the south coast of Iceland and a heat-transfer model for volcanic geothermal systems. Water was pumped onto the lava. The water percolated to a depth of 4 m where it was all converted into steam. The rising steam kept the lava above 4 m depth at 100 °C. The temperature of the molten lava was ~1000 °C. Heat was transferred conductively through the layer at 4 to 6 m depth to the circulating water. Continued cooling involves downward migration of the conductive layer.

deposits (Buat-Menard and Arnold 1978; Lepel et al. 1978; Le Cloarec et al. 1992; Symonds et al. 1992; Hedenquist and Lowenstern 1994; Hedenquist et al. 1994).

GEOTHERMAL FLUIDS

Primary and secondary geothermal fluids

In this contribution, geothermal fluids at the bottom of the convection cell (base-depth) are termed primary geothermal fluids. They may be a mixture of two or more fluid components such as meteoric and seawater and magmatic volatiles. The main types of primary fluids are Na-Cl waters, acid-sulfate waters and high salinity brines. When primary fluids rise towards the surface, they can undergo fluid phase separation and fluid mixing to form secondary geothermal fluids. The most important processes that lead to the formation of secondary geothermal fluids are:

1) Depressurization boiling to yield boiled water and a steam phase with gas.

2) Phase separation of saline fluids into a hypersaline brine and a more dilute vapor.

3) Vapor condensation in shallow ground water or surface water to produce acid-sulfate, carbon-dioxide or sodium bicarbonate waters.

4) Mixing of CO_2 gas from a deep source with thermal ground water.

5) Mixing of geothermal fluids with shallower and cooler ground water.

Chemical composition of primary fluids

The chemical composition of primary geothermal fluids is determined by the composition of the source fluids and reactions involving both dissolution of primary rock-forming minerals and deposition of secondary minerals, as well as by adsorption and desorption processes. The source fluids are usually meteoric water or seawater or a mixture thereof. Components of connate, magmatic and metamorphic fluids may also be present in geothermal fluids.

Na-Cl waters. The dissolved salt in Na-Cl waters is mainly NaCl. This type of water is the most common in geothermal systems. Chloride concentrations typically range from only a few hundred to a few thousand ppm. They are lowest in waters hosted in basaltic rocks (Sigvaldason and Óskarsson 1976; Arnórsson and Andrésdóttir 1995) but highest in fluids which have interacted with sedimentary rocks containing evaporites (Helgeson 1968; White 1968; Table 1). The salinity of geothermal fluids is determined by the availability of soluble salts. These salts may be leached from the aquifer rock or added to the geothermal fluid by deep magmatic fluids. Alternatively, saline fluids may form through reactions between magmatic HCl and rock-forming minerals.

The concentrations of most major elements in Na-Cl waters are fixed by close approach to local equilibrium with secondary minerals if temperatures are above ~100 to 150°C (e.g., Browne and Ellis 1970; Giggenbach 1980, 1981; Arnórsson et al. 1983; Hedenquist 1990; Gudmundsson and Arnórsson 2005; Karingithi pers. comm.). The only conservative major component in these waters is Cl. The mineral-solution equilibria constrain ion activity ratios and the activities of neutral aqueous species other than Cl-bearing species, including reactive gases like CO_2, H_2S and H_2,which may be largely of magmatic origin. Some systems closely approach redox equilibrium (Seward 1974) while others significantly depart from it (Stefánsson and Arnórsson 2002; Stefánsson et al. 2005).

The concentrations of many trace elements (e.g., Ag, Fe, Cu, Pb, Zn) in Na-Cl geothermal waters are clearly controlled by sulfide mineral deposition (e.g., Simmons and Browne 2000; Reyes et al. 2002). These elements typically form cations in solution. Trace elements that form large simple anions or oxy-anions in solution may have high mobility and even show incompatible behavior (e.g., Br, I, As, Mo, W) (Arnórsson 2003; Arnórsson and Óskarsson 2007).

Acid-sulfate waters. Deep acid-sulfate fluids have been encountered in many volcanic geothermal systems, particularly in association with andesitic volcanoes (Truesdell 1991; Giggenbach and Corrales 1992; Sanchez and Arnórsson 1995; Kiyota et al. 1996; Salonga 1996; Akaku et al. 1997; Hermoso et al. 1998; Gherardi et al. 2002; Matsuda et al. 2005; Tello et al. 2005). The acidity is caused by HCl or HSO_4^- or both, and evidence indicates that it mostly forms by transfer of HCl and SO_2 from the magmatic heat source to the circulating fluid.

When measured at 25 °C, the pH of flashed acid-sulfate water collected at the wellhead may be as low as 2. The pH of the water is near neutral at the high temperature in the aquifer, however. Production of acidity upon cooling is related to the increased acid strength of HSO_4^- with decreasing temperature (Fig. 4).

The most important difference between the Na-Cl and acid-sulfate waters is that the main pH-buffer of the former is CO_2/HCO_3^-, but HSO_4^-/SO_4^{-2} in the latter. Compared to Na-Cl waters, acid SO_4-Cl waters contain higher concentrations of SO_4 and some minor elements, such as Fe and Mg, which are contained in minerals with pH-dependent solubility.

Elevated Cl concentrations (up to 120 ppm by weight) have been measured in superheated vapor at The Geysers and Larderello vapor-dominated fields (D'Amore et al. 1977; Haizlip and Truesdell 1988) and the Krafla liquid-dominated field (Truesdell et al. 1989). The Cl in the vapor is transported as HCl. A high Cl concentration in the vapor is due to evaporation of brine. The Cl concentration of the vapor is, however, also affected by the pH of the brine and the temperature of separation of vapor and brine (Fig. 5).

High salinity waters. Geothermal brines can form in several ways. Brine-forming processes include dissolution of evaporites by water of meteoric origin (Helgeson 1968; White 1968) and reaction between some primary minerals of volcanic rocks and magmatic HCl. Connate hot-

Table 1. Chemical analyses of the major types of geothermal waters. Concentrations are in mg/L.

	1	2	3	4	5	6	7	8	9	10	11	12
h (kJ/kg)[a]	237	1256	1376	1257	1423	1287	995	1315	1839	1653	2306	1850
T °C[b]	0	260	251	248	292	278	248	284	251	262	280	260
Sampl. pressure[c]		25.0									9.8	25.3
pH/°C[d]	4.8/25	5.16/22	6.34/25	8.14/25	8.13/25	3.11/25	7.70/15	8.10/20	9.07/25	8.67/25	7.46/20	8.35/21
SiO_2	550.0	668.8	660.0	361.0	741.0	910.0	557.0	784.0	643.0	855.0	815.0	547.0
B		6.72	159.0	30.0	39.9	21.0	26.2	52.0	6.48	6.80	1.73	2.32
Na	38839	9903.0	5105.0	1764.0	860.0	3117.0	1256.0	910.0	557.0	1283.0	214.4	100.5
K	6250	1314.0	994.0	277.0	175.4	950.0	200.0	155.0	92.0	208.0	43.8	15.6
Mg	34.0	1.05	0.140	0.200	0.190	25.00	0.020	0.010	0.010	0.070	0.049	0.002
Ca	20630	1548.0	285.0	68.4	10.80	82.0	26.7	1.3	0.73	0.66	0.80	0.40
Fe	4.4	0.325	0.180		0.060	282.0	0.005	0.010	0.020	0.020	0.027	0.032
Al		0.092			1.090		0.310		0.660	0.670	1.080	1.646
CO_2[e]	54.8	47.4	14.6	27.7	51.2	0.0	64.1	191.0	74.0	2465.0	200.6	3.5
SO_4	12.0	24.3	20.5	137.6	27.0	508.0	34.2	30.0	28.0	112.0	8.6	20.2
H_2S		3.4	3.6	0.0	1.6	0.0	6.9	6.1	1.02	3.96	53.5	97.8
F	4.0	0.14			3.6				69.0	105.0	1.36	0.80
Cl	103000	20534	9074.0	3026.0	1377.0	6175.0	2183.0	1414.0	764.0	240.0	113.1	28.7

[a] Discharge enthalpy. [b] aquifer temperature. [c] in bar-g. [d] pH/temperature of measurement. [e] Total carbonate carbon as CO_2.

1: Na-Ca-Cl brine, Asal Djibuti (D'Amore et al. 1998)
2: Seawater geothermal system, Reykjanes, well 15, Iceland (Giroud, pers. comm.)
3: Andesitic volcanic geothermal system, Tongonan, well 510, Philippines (Angcoy, pers. comm.)
4: Andesitic volcanic geothermal system, Momotombo, well 2, Nicaragua (Arnórsson 1997)
5: Volcanic system in silicic-andesitic rocks, Zunil, well 3, Guatemala (Arnórsson 1995b)
6: Acidic sulphate-chloride water, Mahanagdong, well 9, Leyte, Philippines (Angcoy, pers. comm.)
7: Volcanic geothermal system in silicic volcanics, Wairakei, well 24, New Zealand (Mahon, pers. comm.)
8: Volcanic system in silicic-andesitic rocks, Broadlands, well 28, New Zealand (Mahon, pers. comm.)
9: Volcanic geothermal systems in basalt to trachyte volcanics, Olkaria, well 2, Kenya (Karingithi pers. comm)
10: Volcanic geothermal systems in basalt to trachyte volcanics, Olkaria, well 301, Kenya (Karingithi pers. comm)
11: Volcanic geothermal system in basalt, Krafla, well 20, Iceland (Gudmundsson and Arnórsson 2002)
12: Volcanic geothermal system in basalt, Námafjall, well 11, Iceland (Gudmundsson and Arnórsson 2002)

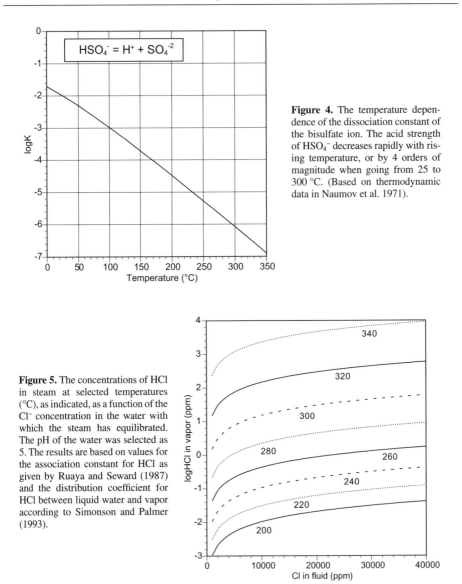

Figure 4. The temperature dependence of the dissociation constant of the bisulfate ion. The acid strength of HSO_4^- decreases rapidly with rising temperature, or by 4 orders of magnitude when going from 25 to 300 °C. (Based on thermodynamic data in Naumov et al. 1971).

Figure 5. The concentrations of HCl in steam at selected temperatures (°C), as indicated, as a function of the Cl^- concentration in the water with which the steam has equilibrated. The pH of the water was selected as 5. The results are based on values for the association constant for HCl as given by Ruaya and Seward (1987) and the distribution coefficient for HCl between liquid water and vapor according to Simonson and Palmer (1993).

water brines have been encountered in sedimentary basins (White 1965). Brines may form by fluid phase separation through cooling and depressurization of moderately saline geothermal fluids in which case they are secondary.

A large region of fluid immiscibility exists in *P-T-X* space for saline geothermal fluids of high temperature (Heinrich et al. 2004; see also Liebscher 2007). The crest of the immiscibility surface (the critical curve) extends from ~25 wt% NaCl at ~700 °C and 1250 bar (~5 km depth for lithostatic pressure) to the critical point of pure water at 374 °C and 221 bar. In volcanic geothermal systems containing saline fluid, the temperature may be too high and the pressure too low to reach the immiscibility surface, in which case a two-phase geothermal fluid exists to the base of the system.

Many metals (Ag, Au, Cu, Mo, Pb, Sn, W, Zn) form complexes with Cl^-, HS^- and OH^- at magmatic temperatures that partition strongly into the magmatic fluid during crystallization (Hedenquist and Lowenstern 1994). As this fluid escapes from the melt into the country rock, these metals together with magmatic gases are transported into the geothermal fluid. Mixing of the magmatic and geothermal fluids and their subsequent interaction with rock-forming minerals leads to brine formation, if the magma is rich in HCl. Cooling and transformation of magmatic SO_2 into H_2S leads to precipitation of metallic sulfides. Porphyry ore-deposits are considered to form in this way (Hedenquist and Lowenstern 1994; Fournier 1999; see also Heinrich 2007a, this volume).

Secondary fluids

Steam-heated acid sulfate waters. In many high-temperature geothermal fields, surface manifestations consist mostly of steam vents (fumaroles), steam-heated surface water and hot intensely altered ground (Fig. 6). Condensation of H_2S-bearing steam by heat loss or mixing with surface water and oxidation of the H_2S leads to the formation of native sulfur, thiosulfate, various polysulfides and ultimately sulfate (Xu et al. 1998, 2000; Druschel et al. 2003). Steam-heated acid-sulfate waters are characterized by low Cl and relatively high sulfate concentrations (Table 2). It is not uncommon that the pH is <1. Due to the low pH, these waters often contain many metals (e.g., Al, Fe, Mn, Cr) in high concentrations. The acid water effectively dissolves the primary minerals of common volcanic rocks leaving a residue rich in amorphous silica, anatase, native sulfur, sulfides, aluminous sulfates and smectite or kaolinite (Sigvaldason 1959; Rodgers et al. 2002).

Carbon-dioxide waters. Thermal and non-thermal waters rich in carbonate carbon (CO_2-waters) are widespread on a global scale (Barnes et al. 1978). They are particularly common in areas of volcanic activity, but are also found in seismically active zones devoid of volcanic activity. Further, CO_2-waters occur at the boundaries of volcanic geothermal systems (e.g., Hedenquist 1990) and around active volcanoes (e.g., Inguaggiato et al. 2005). Carbon-dioxide emissions from active geothermal systems and active volcanoes are largely diffuse and not confined to fumarole and hot spring emissions (Auippa et al. 2004; Chiodini et al. 2004, 2005; Fridriksson et al. 2006). Some CO_2-waters form by mixing of mantle-derived, magmatic or metamorphic CO_2 with ground or surface waters (e.g., White 1957; Arnórsson and Barnes 1983; Chiodini et al. 1995; Cerón et al. 2000; Aka et al. 2001; Marques et al. 2006). In volcanic

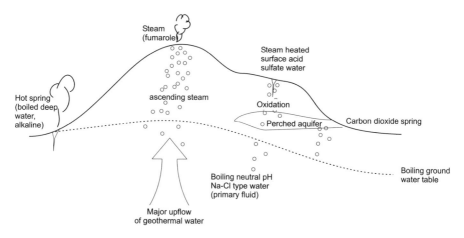

Figure 6. Hypothetical section through the topmost part of a high-temperature geothermal system depicting the formation of steam heated CO_2 and acid sulfate waters.

Table 2. Major and trace element composition of the main types of secondary geothermal waters. Also shown (6-8) is the chemical composition of primary geothermal waters in silicic and basaltic rocks, that are meteoric and seawater by origin, respectively. Concentrations are in mg/L for major components but in µg/L for trace components.

	1	2	3	4	5	6	7	8
Discharge enthalpy[a]							1167	1256
T °C	100	94	60	29.4	65	93.0	238	260
Sampling pressure[b]							18.5	25.0
pH/°C[c]	2.55/22	1.17	6.21/25	6.32	7.98/26	8.61/20	8.80/18	5.16/22
Major components								
SiO_2	226.0	182.0	212.0	67.0	176.2	243.0	500.0	668.8
Na	27.0	535.0	417.7	184.0	209.7	331.0	147.0	9903.0
K	2.44	78.3	30.6	70.0	35.9	9.45	18.0	1314.0
Mg	68.2	15.6	27.2	274.0	1.0	0.001	0.0016	1.05
Ca	94.9	56.1	99.7	727.0	10.0	1.0	2.44	1548.0
CO_2[d]	—	0.0	1145.0	2575.0	300.6	—		
SO_4	1363.0	1598.0	41.3	1078.0	19.9	20.7	290.8	24.3
H_2S	—	<0.01			0.0	1.1		
F	0.06			0.31	4.6	31.6	1.02	0.14
Cl	5.0	1.2	85.8	518.0	20.7	324.0	73.5	20534.0
Trace components								
Al	83300	27500	40.6	108.0	9.4	280	969.0	91.9
As	—	43.4	40.4		0.4	860	105.0	156.0
B	—		518.0	29.0	396.0	3980	1670.0	7860.0
Ba	8.0	117.0	69.5		4.4	<40	0.62	6650.0
Cd	—			26.0		<40	<0.002	0.923
Cr	108	12.5	<0.1		0.07	<90	0.12	0.27
Cs	—	1.0	9.6		0.33		9.66	61.6
Cu	—	4.0	<1.0	58.0	0.09	<140	0.18	1.74
Fe	101000	37600	1546.0	235.0	16.90	23.0	2.6	325.0
Ga	—		0.018		0.12		4.92	2.29
Ge	—		21.3		7.45		24.7	23.0
Hg	—	0.1	0.024		0.004		<0.002	0.068
Li	22.0	4.9	21.3		59.2	3960.0	264.0	3930.0
Mn	2370	313.0	214.0	1959.0	5.1	<120	1.59	1500.0
Mo	—		10.8		47.5		3.87	92.5
Ni				78.0				
P	290		<10		6.1		1.25	15.4
Pb	—	8.1	<0.1	437.0	0.173	<150	<0.01	<0.1
Rb	—		177.0		61.0		206.0	4200.0
Sr	89	45.2	418.0		67.9	8.0	5.16	9120.0
Ti	23		0.05		0.13		0.18	<0.1
Tl	—		0.054		0.001		0.011	17.6
V	376		0.1		1.54		1.73	9.1
W	—		0.19		4.53		6.41	1.0
Zn	127	134.5	0.5	80.0	3.2	<10	0.51	50.9

[a]kJ/kg, [b]bar-g, [c]pH/temperature of measurement. [d]Total carbonate carbon as CO_2
1: Steam-heated surficial acid sulphate water, Krísuvík, Iceland (unpubl. data)
2: Steam-heated surficial acid sulphate water, Mendeleev, Kunashir, Russia (Chelnokov, 2004)
3: Carbon-dioxide water, Lýsuhóll well 6, Iceland (unpubl. data)
4: Carbon-dioxide water, Guadalentín, Spain (Cerón et al., 2000)
5: Mixed high-temperature and cool ground water, Nedridalur, Geysir field, Iceland (unpubl. data)
6: Hot spring in silicic volcanics, Ojo Calinte Spring, Yellowstone, Wyoming, USA (Ball et al. 1998)
7: Volcanic geothermal system in basalt, Krafla, well 21, Iceland (unpubl. data)
8: Seawater geothermal system, Reykjanes, well 15, Iceland (unpubl. data)

geothermal systems, CO_2-waters may form by condensation of CO_2-containing vapor in perched aquifers or by mixing of downward percolating CO_2-rich condensate with the deep primary geothermal fluid (Hedenquist 1990). Finally, CO_2-waters may form by mixing of high-temperature geothermal fluid that has not undergone fluid phase separation with cool ground water (e.g., Arnórsson 1985).

Deuterium (δ^2H) and oxygen-18 ($\delta^{18}O$) data indicate that the CO_2-waters are meteoric by origin. Tritium analyses suggest, at least in some instances, short residence times. Yet, their content of ^{14}C is very low due to extensive dilution by ^{14}C-dead carbon from the deep source (Marques et al. 2006). CO_2-waters are often considerably mineralized (Table 2) because the CO_2 makes the water quite reactive by maintaining relatively low pH, thus increasing the rate of disso-

lution of many common primary rock-forming minerals by enhancing their degree of under-saturation. The low pH may also reduce adsorption of many trace metal cations onto iron-hydroxide or other minerals and in this way increase the mobility of these cations.

Mixed waters. In up-flow zones of geothermal systems ascending boiled or un-boiled water may mix with shallow ground water. Alternatively, the thermal fluid that mixes with the cooler ground water may be two-phase (liquid and vapor). Mixed geothermal waters have been studied with the aim of assessing the temperature of the hot water component in the mixed water, largely for geothermal exploration purposes (e.g., Fournier 1977; Truesdell and Fournier 1977; Arnórsson 1985).

Variably diluted (mixed) geothermal fluids in a particular field can be identified by a negative correlation between temperature and flow rates of springs. A positive correlation between the concentrations of conservative chemical and isotopic components is also typical of mixed waters (Fig. 7). Mixing affects the state of equilibrium between the fluid phase and both primary and hydrothermal minerals and leads to changes in the initial concentrations of reactive components in the mixed water, particularly if the hot fluid component is un-boiled water or two-phase fluid. These changes typically involve an increase in Ca and Mg concentrations and a decrease of Na/K ratios (Arnórsson 1985).

BOILING AND PHASE SEGREGATION

The boiling point or onset of phase separation of an aqueous solution is affected

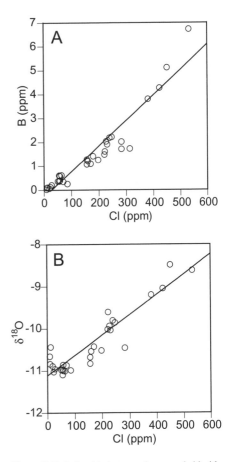

Figure 7. Relationship between boron and chloride (A) and between $\delta^{18}O$ and chloride (B) in geothermal waters from the Landmannalaugar field in Iceland, reflecting mixing between the geothermal fluid and cold ground water. The abscissa intercept in (A) represents the composition of the cold water component in the mixed water. It closely corresponds to that of local precipitation. So does the ordinate intercept in (B).

by its salinity and gas content. Increasing salinity raises the boiling point (Bischoff and Pitzer 1989), whereas dissolved gases lower it. The critical point of pure water is at 374 °C and 221 bars. The corresponding numbers for seawater are 405 °C and 302 bars. Boiling of a rising geothermal liquid starts at a depth where the sum of the water vapor pressure and all dissolved gas partial pressures become equal to the hydrostatic pressure.

The boiling point curve

The solid line in Figure 8 shows how the boiling point of pure water changes with depth. In this figure, pressure has been converted into depth (length of a water column) assuming that water at the surface is at 100 °C. The increase in temperature and pressure with depth corresponds to a column of liquid water that is at the boiling point at all depths. The curve of Björnsson (1966) has been updated here using the IAPWS Industrial Formulation 1997 for the Thermodynamic Properties of Water and Steam. The slim curve in Figure 8 shows how the boiling point for a 3.2 wt% NaCl solution varies with depth. The salt-water curve is based on the experimental data of Bischoff (1984) and Bischoff and Rosenbauer (1984, 1985) on temperature, pressure and specific volume of a 3.2 wt% NaCl solution along the liquid-vapor phase boundary. According to Bischoff and Rosenbauer (1984), a 3.2 wt% NaCl solution is a good proxy for seawater. The boiling point curves for pure water and seawater are quite similar at temperatures below about 300 °C but they diverge considerably at higher temperatures. The critical point of seawater is at about 405 °C and 302 bar. This point occurs at about 5200 m depth in a seawater column that reaches the surface and is at the boiling point at all depths. By comparison, the critical point of pure water would be found at a depth of only 3500 m. The results shown in Figure 8 and those presented by Heinrich et al. (2004) show that the critical temperature and critical pressure (depth) of geothermal fluids vary enormously with salinity.

The following equations describe satisfactorily the boiling point with depth for pure water and a 3.2 wt% NaCl solution (seawater), respectively:

$$D_w = -0.01120276 \cdot T^2 + 7.016900 \cdot 10^{-5} T^3 + 2.395989 \cdot 10^{-23} T^{10} + 21.20911 \cdot \log(T) \quad (1)$$

$$D_s = -0.01091091 \cdot T^2 + 6.614698 \cdot 10^{-5} T^3 + 2.165155 \cdot 10^{-23} T^{10} + 42.04005 \cdot \log(T) \quad (2)$$

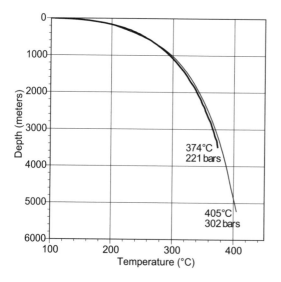

Figure 8. Boiling point curves with depth for pure water (solid line) and 3.2 wt% NaCl solution (thin line). A 3.2 wt% NaCl solution is a good proxy for seawater. The curve for pure water is based on the IAPWS Industrial Formulation 1997 for the thermodynamic properties of water and steam, but that for seawater was retrieved from the experimental data of Bischoff and Rosenbauer 1985) on temperature, pressure and specific volume for a 3.2% NaCl solution along the liquid-vapor boundary.

D_w and D_s denote the depth in m for pure water and a 3.2 wt% NaCl solution, respectively, and T the temperature in °C.

Effect of dissolved gases

The gas content of an aqueous fluid affects its boiling point. A vapor phase begins to form when the sum of the water vapor pressure and the partial pressures of all dissolved gases becomes equal to the hydrostatic pressure (P_h)

$$P_h = P_{H_2O} + \sum_s P_s \tag{3}$$

where P_{H_2O} represents the water vapor pressure and P_s the partial pressure of gas s. The standard molal solubility constant (Henry's Law Constant) of gas s (K_s) in aqueous solution is given by

$$a_s^l = K_s \cdot f_s \tag{4}$$

where K_s is in moles kg^{-1} bar^{-1} and a_s^l designates the activity of the gas in liquid water and f_s its fugacity above the solution. The relationships between the activity (a_s^l) and concentration (m_s^l) of gas s on one hand and the fugacity and partial pressure on the other are given by

$$a_s^l = \gamma_s \cdot m_s^l \tag{5}$$

and

$$f_s = \Gamma_s \cdot P_s \tag{6}$$

Here, γ_s is the activity coefficient of dissolved gas s and Γ_s the fugacity coefficient. The value of Γ_s can be obtained from the dimensionless compressibility factor Z_s and the modified Redlich-Kwong equation (see e.g., Nordstrom and Munoz 1994):

$$\ln \Gamma_s = (Z_s - 1) - \ln \frac{Z_s - b_{R\text{-}K}P}{RT} - \frac{a_{R\text{-}K}}{RT^{1.5}b_{R\text{-}K}} \cdot \ln \left(1 + \frac{b_{R\text{-}K}P}{Z_s RT} \right) \tag{7}$$

Here, Z_s is equal to PV/RT, where V is the molar volume of a real gas, P is pressure, R the gas constant and T absolute temperature. The constants $a_{R\text{-}K}$ and $b_{R\text{-}K}$ are equal to $0.4275R^2T_c^{2.5}/P_c$ and $0.0866RT_c/P_c$, respectively (Nordstrom and Munoz 1994), where T_c and P_c are the critical temperature and pressure, respectively. Nordstrom and Munoz (1994) list values of T_c and P_c for some common gases (Table 3). They also provide data for Z_c, the critical compressibility factor. With T_c, P_c and Z_c, known, Z_s can be obtained from the following equation:

$$\frac{Z_s}{Z_c} = \frac{(P/P_c)(V/V_c)}{T/T_c} \tag{8}$$

The activity coefficient, γ_s in Equation (5), can be obtained experimentally as described by Nordstrom and Munoz (1994) and Anderson (2005). It can also be obtained from the Setchénow equation

$$\log \gamma_s = k_s + m_B \tag{9}$$

where k_s is the Setchénow coefficient and m_B the molar concentration of neutral salt B in solution. As an approximation, m_B in Equation (9) may be taken to be 0.1 (Anderson 2005). The Pitzer equations (see Anderson 2005) provide a more detailed and accurate treatment of uncharged species than Equation (9). For neutral species, the extended Debye-Hückel equation of Helgeson (1969) reduces to

$$\log \gamma_s = B^{\bullet} \cdot I \tag{10}$$

Table 3. Critical constants for selected gases.
From Nordstrom and Munoz (1994).

	T_c (K)	P_c (bar)	V_c (cm³/mol)	Z_c
H_2O	647.3	220.4	56.0	0.229
CO_2	304.2	73.7	94.0	0.274
CH_4	190.6	46.0	99.0	0.288
H_2S	373.2	89.3	98.5	0.284
SO_2	430.8	78.8	122.0	0.268
HF	461.0	65.0	69.0	0.120
HCl	324.6	83.1	81.0	0.249
NH_3	405.6	112.7	72.5	0.242
H_2	33.2	13.0	65.0	0.305
O_2	154.6	50.4	73.4	0.288
N_2	150.8	48.7	74.9	0.291

where I represents the true ionic strength of the solution, i.e., the ionic strength after correction has been made for ion pairing and complexing. Helgeson et al. (1981) updated Equation (10) for neutral species by replacing B^* with revised values.

For dilute solutions, it is a reasonable approximation to take $\gamma_s = 1$, i.e., $a_s^l = m_s^l$. This is, however, not the case for solutions of high ionic strength. For such solutions, Equation (5) needs to be solved using either (9) or (10). At pressures existing in most drilled geothermal systems, Z_s can be taken to be equal to unity for common gases, i.e., the gases can be taken to behave ideally.

Gas pressures in geothermal systems are highly variable, ranging from less than 1 bar, such as at Wairakei in New Zealand and Svartsengi in Iceland to as much as ~100 bar at Broadlands in New Zealand and Olkaria in Kenya. The partial pressures of individual gases in the geothermal systems may be determined by their supply to the geothermal fluid or fixed by specific temperature-dependent mineral-gas or gas-gas equilibria (e.g., Gudmundsson and Arnórsson 2002; Karingithi pers. comm.).

Liquid-vapor separation under natural conditions

Boiling under natural conditions in up-flow zones of geothermal systems is maintained by depressurization of the rising hot fluid. The boiled fluid may emerge at the surface in springs or flow laterally from the up-flow zone underground, depending on the depth of the ground water table of the boiled fluid.

Vapor rising above the ground water table may condense partly or totally by conductive heat loss or by mixing with water in perched aquifers or surface water. In this way, vapor-heated secondary geothermal fluids form. Most of the gases initially present in the deep aquifer fluid will be transferred into the vapor. They may dissolve partly in water of perched aquifers and partly rise to the surface. Dissolution of CO_2 and H_2S in such water may turn it acid and reactive. The H_2S may be oxidized or precipitated as sulfides. Carbon dioxide is transferred into bicarbonate through reaction with the rock and may precipitate to some extent as calcium carbonate.

Vapor-dominated systems

In vapor-dominated geothermal systems, vapor is the continuous phase in fractures, whereas liquid water fills partially or totally intergranular pore spaces (see Fig. 2). Temperature varies insignificantly with depth in these systems and so does pressure due to the low density of the vapor. Both have values close to those corresponding to maximum steam enthalpy (234 °C and 30.1 bar absolute). Vapor-dominated systems are overlain by a cap rock, which may be liquid saturated. Vapor-dominated systems are considered to form by the boiling down of an earlier liquid-dominated system with buoyancy driven segregation of the vapor from the boiling liquid. The extent of vapor formation is probably determined by the balance between the rate of fluid circulation and the rate of heat transfer from the rock to the fluid in the root zones of the system. The best known vapor-dominated fields in the world are Larderello in Italy and The Geysers in California. Part of the Olkaria geothermal system in Kenya had prior to exploitation a steam cap of ~200 m thickness overlying a liquid-dominated system. Pressure drawdown caused by the exploitation of the Wairakei and Svartsengi fields in New Zealand and Iceland, respectively, has led to the formation of steam caps on top of the liquid-dominated systems (Clotworthy 2000; Ármannsson 2003). In both these fields, as well as at Olkaria, the vapor caps have temperatures and pressures close to those of maximum enthalpy for saturated vapor. At Wairakei, the liquid-dominated system below the vapor cap is not much above 240 °C and at Svartsengi, it is close to 240 °C at least down to ~2000 m depth.

Boiling and fluid phase segregation in wells and producing aquifers

Aquifers that are penetrated by wells drilled into high-temperature, liquid-dominated geothermal systems are often at *P-T* conditions insufficient to initiate boiling in the contained geothermal fluid. If the pressure drop caused by discharging such wells is not sufficient to initiate boiling in the aquifer itself, the depth level of first boiling is within the well. Under these conditions, it is a reasonable approximation to treat the aquifer and well as an isolated system, in which case boiling is adiabatic. According to thermodynamics, a system is isolated when no transfer of heat or mass occurs across its boundaries. Hence, considering well and aquifer as an isolated system implies that the discharge enthalpy and chemical composition of the well discharge are the same as those of the aquifer water. By mass conservation, we then have

$$M^{f,t} = M^{d,t} = M^{d,v} + M^{d,l} \tag{11}$$

where $M^{f,t}$, $M^{d,t}$, $M^{d,v}$, and $M^{d,l}$ designate the mass flow rate of aquifer water into the well, mass flow rate of the well discharge, mass flow rate of vapor from the well and mass flow rate of liquid from the well, respectively. The conservation of mass of dissolved components may be expressed by

$$m_i^{f,t} = m_i^{d,t} = m_i^{d,v} X^{d,v} + m_i^{d,l} (1 - X^{d,v}) \tag{12}$$

where $X^{d,v}$ is the vapor mass fraction of the well discharge and therefore $(1 - X^{d,v})$ is the liquid mass fraction. Also, m_i represents the molal concentration of dissolved component i. The superscripts have the same meaning as in Equation (11). For the specific enthalpy (h) we have

$$h^{f,t} = h^{d,t} = h^{d,v} X^{d,v} + h^{d,l} (1 - X^{d,v}) \tag{13}$$

The enthalpy of saturated steam is the sum of the enthalpy of steam-saturated liquid and the latent heat of vaporization (L). Inserting $h^{d,l} + L^d$ for $h^{d,v}$ into (13) and rearranging to isolate $X^{d,v}$ yields

$$X^{d,v} = \frac{h^{d,t} - h^{d,l}}{L^d} \tag{14}$$

Values for $h^{d,l}$ and L^d can be obtained from Steam Tables and a value for $h^{d,t}$ can either be obtained from measurement of the discharge enthalpy or, in the case of sub-boiling aquifer, from evaluation of the aquifer temperature. In a sub-boiling aquifer, the enthalpy of the aquifer fluid ($h^{f,t}$) is simply that of liquid water ($h^{f,l}$) at the aquifer temperature and pressure and can be found in standard Steam Tables. Having obtained $h^{d,t}$, $h^{d,l}$ and L^d, the steam fraction ($X^{d,v}$) can be retrieved from Equation (14). Finally, Equation (12) is used to calculate the concentration of component i in the aquifer water ($m_i^{f,t}$) from analysis of liquid ($m_i^{d,l}$) and vapor ($m_i^{d,v}$) samples collected at the wellhead at vapor pressure P^d.

When a well intersects an aquifer that is two-phase (both liquid and vapor are present), intensive depressurization boiling will start in this aquifer when the well is discharged. In case the pressure drop produced by discharging the well is sufficiently large, boiling can also start in the aquifer even if the aquifer fluid was initially non-boiling. When intensive boiling occurs in producing aquifers, vapor may not only form by depressurization boiling. It can also form by conductive heat transfer from the rock to the fluid: Depressurization boiling lowers the fluid temperature, thus creating a temperature gradient between fluid and aquifer rock and favoring conductive heat transfer from rock to fluid. Addition of heat to the two-phase fluid will, of course, not affect its temperature. It will enhance boiling, i.e., steam formation. Liquid and vapor may segregate (separate partly) in two-phase aquifers, leading to an increase in the vapor to liquid ratio of well discharges. Phase segregation results from the different flow properties of liquid and vapor and the effects of capillary pressure and relative permeability (e.g., Horne et al. 2000; Pruess 2002; Li and Horne 2004). The mass flow rate of each phase is affected by the relative permeability and the pressure gradient, and also its density and viscosity. Adhesive forces between mineral grain surfaces and fluid, which are the cause of capillary pressure, are stronger for liquid than for vapor. In this way, the mobility of liquid is reduced relative to that for vapor. The effect of capillary pressure becomes stronger in rocks of small pores and fractures, i.e., when permeability is low. Typical relative permeability curves are shown in Figure 9. They show the relationship between liquid saturation (volume fraction of total fluid that is liquid) and relative permeability. It can be seen from the figure that the liquid phase is immobile when its volume fraction is below about 0.6. Relative permeability curves, such as those shown in Figure 9, vary with temperature and interface between fluid and rock.

Due to vapor-forming processes and phase segregation in two-phase aquifers, the discharge enthalpy of wells producing from liquid-dominated geothermal reservoirs is often higher than the enthalpy of the initial aquifer fluid, and it is not uncommon that wells drilled into such systems discharge steam only. Wells with discharge enthalpies higher than that of steam-saturated water at the aquifer temperature have been referred to as "excess enthalpy wells." Some of the excess enthalpy may be due to the presence of vapor in the initial aquifer fluid. However, the effects of

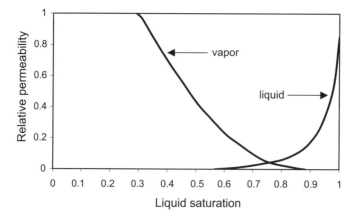

Figure 9. Typical relative permeability functions. They show the relationship between liquid saturation (fraction of liquid water by volume) and relative permeability. It can be seen from this figure that the liquid water phase is immobile when water saturation is below about 0.6. From Pruess (2002).

conductive heat transfer from the aquifer rock to the depressurized fluid, or phase segregation, or both, generally seem to be more important. To calculate the chemical composition of the initial liquid and vapor in the aquifer, the relative contributions of the different processes to the excess discharge enthalpy and the initial vapor fraction of the aquifer fluid need to be evaluated. The calculation procedure, which is rather complex, is detailed in Appendix I.

In general, wet-steam wells and their producing aquifers may be defined as isolated, closed or open systems[1]. Five models are considered in the following, three of which describe open systems. All five are presented in Figure 10. They differ by changes the flowing fluid undergoes with respect to enthalpy and composition between initial aquifer and well, as shown by the arrows in Figure 10. The processes that change the fluid are vapor formation by depressurization boiling, additional vapor formation by conductive heat transfer from the rock to immobile capillary water, enhanced vaporization of the flowing liquid, also by conductive heat transfer from the rock, and fluid phase segregation leading to the retention of some of the boiled liquid in the aquifer by its immobilization. Conductive heat transfer and immobilization of boiled liquid are taken to occur at a constant selected pressure, P^e. This is a simplification. One would expect both conductive heat transfer and immobilization of boiled water to occur over a range of pressures. Since the pressure at which these processes operate cannot be constrained, however, the present choice is reasonable because it is the simplest one. For specific well data, one can study how sensitive the calculated fluid parameters and the composition of the initial aquifer liquid and vapor are to the choice of P^e. It is also advisable to study how sensitive the calculated parameters are to the selection of aquifer temperature.

One of our models assumes that heat transferred conductively from aquifer rock to fluid enhances evaporation of the flowing liquid water (Fig. 10, Model 2). In physical terms, this model is logical when the capillary water has been evaporated to dryness so transfer of heat from rock to fluid involves the flowing fluid.

Figure 10 summarizes the essential features of the five models, and Table 4 lists all their parameters. Some of these are obtained from measurement at the wellhead ($M^{d,t}$, $h^{d,t}$, P^d, $m_i^{d,l}$, $m_i^{d,v}$), others are selected (T^f, P^e) and still others are calculated ($X^{f,v}$, $m_i^{f,l}$, $m_i^{f,v}$, Q^e, $M^{e,v}$, $M^{e,l}$, $M^{f,t}$).

The various models may be specified by setting one or more of the fluid parameters to zero. In Model 1 (see Fig. 10), there is no conductive heat transfer to the fluid ($Q^e = 0$), no retention of liquid by the formation ($M^{e,l} = 0$), and no additional inflow of vapor ($M^{e,v} = 0$). In Model 2, there is neither inflow of additional vapor nor retention of liquid by the formation ($M^{e,v} = 0$ and $M^{e,l} = 0$). There is, however, conductive heat flow to the fluid ($Q^e \neq 0$). Model 3 includes no heat transfer ($Q^e = 0$) and no addition of vapor ($M^{e,v} = 0$), but liquid is retained by the formation ($M^{e,l} \neq 0$). Model 4 involves the addition of vapor to the flowing fluid ($M^{e,v} \neq 0$) and the retention of liquid by the formation ($M^{e,l} \neq 0$), but no conductive heat transfer ($Q^e = 0$). In Model 5, water is retained by the formation ($M^{e,l} \neq 0$) and there is conduction of heat ($Q^e \neq 0$), but no addition of vapor ($M^{e,v} = 0$).

Glover et al. (1981) used the chemistry of well discharges in a simple way to distinguish between excess well enthalpy caused by phase segregation on one hand, and by conductive heat transfer from aquifer rock to fluid on the other. If the concentration of a non-volatile, conservative component like Cl⁻ in the total discharge of a well stays about constant despite variations in excess discharge enthalpy, the cause of the excess enthalpy is conductive heat transfer from the aquifer rock to the flowing fluid (closed system, Fig. 10). If the Cl⁻ concentration stays constant in the total discharge, it will become very high in the liquid phase as the enthalpy approaches that of saturated steam (Fig. 11A). If, on the other hand, the concentration of a conservative

---

[1] In the geothermal industry, the term wet-steam well denotes a well that discharges a mixture of liquid and vapor.

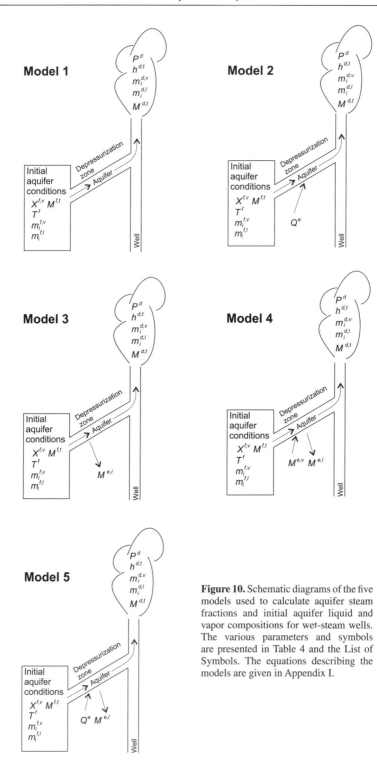

Figure 10. Schematic diagrams of the five models used to calculate aquifer steam fractions and initial aquifer liquid and vapor compositions for wet-steam wells. The various parameters and symbols are presented in Table 4 and the List of Symbols. The equations describing the models are given in Appendix I.

Table 4. Measured, selected and calculated fluid and composition parameters by the five models considered in this contribution. Explanation of parameter symbols are given in Figure 10 and under List of Symbols.

Model	1		2		3		4		5	
Measured parameters	$M^{d,l}$ $h_i^{d,t}$ $m_i^{d,l}$	$M^{d,v}$ P^d $m_i^{d,v}$	$M^{d,l}$ $h^{d,t}$ $m_i^{d,l}$	$M^{d,v}$ P^d $m_i^{d,v}$	$M^{d,l}$ $h^{d,t}$ $m_i^{d,l}$	$M^{d,v}$ P^d $m_i^{d,v}$	$M^{d,l}$ $h^{d,t}$ $m_i^{d,l}$	$M^{d,v}$ P^d $m_i^{d,v}$	$M^{d,l}$ $h^{d,t}$ $m_i^{d,l}$	$M^{d,v}$ P^d $m_i^{d,v}$
Selected parameters	T^f		T^f	T^e	T^f		T^f	T^e	T^f	T^e
Calculated parameters	$X^{f,v}$ $m_i^{f,l}$	$m_i^{f,v}$	$X^{f,v}$ $m_i^{f,l}$	Q^e $m_i^{f,v}$	$X^{f,v}$ $M^{f,t}$ $m_i^{f,l}$	$M^{e,l}$ T^e $m_i^{f,v}$	$X^{f,v}$ $M^{f,t}$ $m_i^{f,l}$	$M^{e,l}$ $M^{e,v}$ $m_i^{f,v}$	$X^{f,v}$ $M^{f,t}$ $m_i^{f,l}$	$M^{e,l}$ Q^e $m_i^{f,v}$

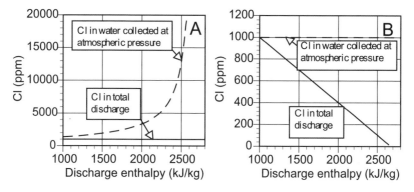

Figure 11. Relation between Cl and discharge enthalpy. A: Excess enthalpy is due to conductive heat transfer from aquifer rock to fluid flowing into well (closed system). B: Excess enthalpy is caused by phase segregation in the producing aquifer (open system).

component in either the liquid (e.g., Cl⁻) or vapor (e.g., Ar) stays constant despite variations in excess enthalpy, the cause is phase segregation. Constant concentration of a conservative component that only occupies the liquid, collected at a particular vapor saturation pressure, implies that its concentration in the total discharge approaches zero as the discharge enthalpy approaches that of saturated steam (Fig. 11B).

For many wet-steam wells, discharge enthalpy may not vary sufficiently with time to make the method of Glover et al. (1981) applicable. In a specific wellfield, one may use many wells with a range of discharge enthalpies to determine whether conductive heat transfer or phase segregation dominates the excess enthalpy. If components like Cl⁻ vary much across the wellfield, they may not be a good choice, but if aquifer temperatures are about constant, SiO_2 is useful because its concentration is determined almost solely by temperature through its equilibrium with quartz. As an example, data from Olkaria in Kenya show how SiO_2 concentrations in the total discharge of wells vary with discharge enthalpy (Fig. 12). They

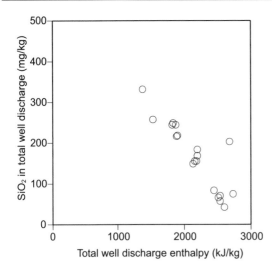

Figure 12. Variation in the silica content in total well discharges of production wells in the Olkaria East production field Kenya. The enthalpy value approaches that of saturated steam when the silica content approaches zero. The observed correlation indicates that excess enthalpy of well discharges is mainly caused by phase segregation in producing aquifers.

show that SiO_2 concentrations in the total discharge approach zero when the discharge enthalpy approaches that of saturated steam, indicating that the excess enthalpy is mainly caused by phase segregation.

INITIAL AND EQUILIBRIUM VAPOR FRACTIONS

The vapor fraction that may be present in an aquifer fluid, prior to induction of extensive boiling by discharging a well, is called the initial vapor fraction. Geochemical methods have been used to evaluate this vapor fraction from the gas content of well discharges, assuming either specific gas-gas or mineral-gas equilibria. When estimated in this way, the initial vapor fraction has appropriately been termed equilibrium vapor fraction.

Giggenbach (1980) estimated equilibrium vapor fractions in the reservoir of three New Zealand geothermal systems based on the assumption of equilibrium for the following reactions:

$$CH_4 + 2H_2O = CO_2 + 4H_2 \tag{15}$$

and

$$2NH_3 = N_2 + 3H_2 \tag{16}$$

The model used by Giggenbach (1980) allows the presence and loss of equilibrium vapor. His model may therefore be considered as a special type of an open model, i.e., one which permits loss or gain of gases from the initial aquifer fluid and a corresponding loss or gain of enthalpy. The retrieved equilibrium vapor fraction values ($X^{f,v}$) ranged from −5% to +1.2% by weight (~−68% and ~+33% by volume), negative values indicating loss of equilibrium vapor from the fluid entering the wells. By assuming $X^{f,v}$ to be zero, Giggenbach (1980) calculated gas-gas equilibrium temperatures for the reactions given above. Gas equilibrium temperatures based on the two reactions considered compare very well. On the other hand, correlation between quartz equilibrium and gas geothermometer temperatures is poor. The model adopted by Giggenbach (1980) did not consider how the excess well enthalpy had developed. This is not important for most of the wells considered in his study, because their discharge enthalpy is close to liquid enthalpy. Two of the wells included in Giggenbach's study, however, discharged dry vapor. The calculated equilibrium vapor fraction for these wells is significantly affected by the mode of formation of the excess vapor (phase segregation or conductive heat transfer from rock to fluid). Conductive heat transfer causes changes in the reaction quotient due to the stoichiometry of the gas-gas reactions considered by decreasing the gas concentrations of the vapor. Phase segregation may also affect the gas-gas reaction quotients, but to a lesser degree, depending on the extent of degassing of the boiling fluid at the pressure at which phase segregation occurs and on the extent of subsequent depressurization boiling. The general conclusions drawn from the study of Giggenbach (1980) are that redox equilibria are closely approached for the reactions shown in Equations (15) and (16), the equilibrium vapor fraction, when present, is small in terms of mass, and for some well discharges gaseous vapor has been lost relative to the initial assumed equilibrium conditions.

D'Amore and Celati (1983) also developed a method to calculate equilibrium vapor fractions in geothermal systems. Like Giggenbach (1980), they assumed gas-gas equilibrium according to the reaction shown by Equation (15) above and also equilibrium according to

$$H_2S = H_2 + 0.5S_2 \qquad (17)$$

They further assumed that re-equilibration between the various species was negligible between reservoir and wellhead. The fundamental assumption was made that the two-phase aquifer fluid is transferred to the wellhead, i.e., their aquifer-well system is isolated. D'Amore and Celati (1983) considered data from several fields, including the vapor-dominated fields at The Geysers, Larderello and Travale. The equilibrium vapor fractions in these fields are in the range of 1 to 60 wt%, 20 to 60 wt% and 10 to 15 wt%, respectively. Corresponding numbers for Cerro Prieto in Mexico are 1 to 6 wt%. A weight fraction of 6% corresponds to 50% by volume at 300 °C. For the fields at Wairakei and Broadlands in New Zealand, no equilibrium vapor fractions could be obtained. D'Amore and Celati (1983) ascribed this to negative equilibrium vapor fraction values, which were not allowed for in their calculations; a result in general agreement with that of Giggenbach (1980).

Unlike the approach of Giggenbach (1980) and D'Amore and Celati (1983), Arnórsson et al. (1990) developed a methodology to estimate the equilibrium vapor fraction in aquifers of discharging wet-steam wells on the assumption of specific mineral-gas equilibria and taking into account the processes of phase segregation and conductive heat transfer from aquifer rock to aquifer fluid. On the basis of hydrological considerations, it is inevitable that both these processes are potentially operative in two-phase aquifers producing into wet-steam wells. The models of Giggenbach (1980) and D'Amore and Celati (1983) may be considered as special cases of the general model of Arnórsson et al. (1990) (see Fig. 10). The differences between the three models lie in the kinds of chemical reactions for which equilibria are assumed for the derivation of $X^{f,v}$. It certainly has many advantages to use gas-gas equilibria rather than mineral-gas equilibria, the latter of which involve the reactive gases CO_2, H_2S and H_2. The main reason is that some of the minerals, which control the reactive gas concentrations, form solids solutions, and therefore their compositions can be expected to vary from one field to another. On the other hand, the gas-gas reactions are always redox reactions, which are slow, so equilibrium may not be attained. Stefánsson and Arnórsson (2002) observed systematic discrepancies between redox couples for >200°C dilute geothermal fluids in Iceland, but good agreement for saline geothermal waters. The discrepancies in the calculated redox potential for dilute geothermal fluids are considered to be due to lack of equilibrium between CO_2, CH_4 and H_2, and between H_2S, SO_4 and H_2. If phase segregation occurs, i.e., if the system aquifer + well is an open system, derivations of equilibrium vapor fractions require information about the concentrations of individual gases in the initial aquifer fluid, whether gas-gas or mineral-gas reactions are used to obtain equilibrium vapor fractions.

Arnórsson et al. (1990) estimated the equilibrium vapor fraction in producing wet-steam wells of the East production field at Olkaria, Kenya by modeling the excess enthalpy according to Figure 10. They used the equations given by Arnórsson and Gunnlaugsson (1985) to represent the concentrations of CO_2, H_2S and H_2 in the initial aquifer water. The equilibrium vapor fractions are in the range of –0.38 to +0.25 wt%, the average being 0.03. Their results indicate that a relatively large part of the vapor discharged from wells is generated from conductive transfer of heat from the aquifer rock to capillary liquid. The results of Arnórsson et al. (1990), described above, are sensitive to the equations selected to describe the equilibrium concentrations of CO_2, H_2S and H_2 in the initial aquifer liquid. According to Karingithi (pers. comm.), equilibrium vapor fractions in the aquifer of selected Olkaria wells are below 0.1 wt%. This estimate should be considered more reliable than earlier estimates of Arnórsson et al. (1990), because the mineral-gas equilibrium constants used by Karingithi are based on analyses of the chemical composition of the minerals in the mineral-gas buffers and on the most recent thermodynamic data bases (Holland and Powell 1998; Robie and Hemingway 1995). Gudmundsson and Arnórsson (2002) estimated the equilibrium vapor fraction for some Krafla wells in Iceland by the method of Arnórsson et al. (1990). They obtained values in the range of –0.07 to +2.22 wt% (corresponding to 0 to 47 % by volume).

The mole fractions of CO_2 in the vapor discharged from wells in the Larderello and The Geysers vapor-dominated fields correspond to CO_2 partial pressures of about 0.7 and 0.1 bar, respectively, according to data presented by Ellis and Mahon (1977). These values correspond well with equilibrium between fluid and the mineral buffer clinozoisite + prehnite + calcite + quartz (see Eqn. 1 in Table 7) with an assumed clinozoisite activity of ~0.1. This indicates that the vapor of these vapor-dominated systems is close to equilibrium with hydrothermal minerals typically found in liquid-dominated systems with temperatures of ~240 °C, i.e., the vapor is equilibrium vapor. The intergranular liquid (see Fig. 2) has apparently closely approached equilibrium with the mineral buffer in question, and the vapor has rapidly equilibrated with this liquid.

GAS CHEMISTRY

Gas chemistry of geothermal fluids is intimately related to boiling of geothermal fluids. Specific gas ratios or gas-gas reactions have been used to model aquifer fluid compositions from data on the gas content of vapor samples collected at the wellhead. Gas concentrations and gas ratios have also been used as geothermometers using both well and fumarole data (e.g., Giggenbach 1981; D'Amore and Celati 1983; D'Amore and Truesdell 1985; Arnórsson et al. 1990). Finally, as discussed in the previous section, gas chemistry has been used to estimate the initial vapor fraction in aquifers producing into wet-steam wells.

Various conventions have been used to present gas compositions of geothermal fluid discharges. They include expressing gas concentration in the vapor phase or as mole fraction of the gas phase, usually as mole percent (volume percent). Various concentration units have also been used (see Henley et al. 1984). These include millimoles/kg of H_2O and millimoles/100 moles of H_2O. A logical selection of concentration units is that used in chemical thermodynamics, i.e., moles (or millimoles) per kg of solvent H_2O, because today gas data interpretation frequently makes use of the principles of chemical thermodynamics. When a separate gas phase is collected, the only choice is to express analytical results in terms of mole percent (volume percent). This is, e.g., the case when samples are collected of gas bubbling through hot spring water or mud pools. It is, however, advantageous to collect samples from fumaroles and well discharges in such a manner that the concentration in the vapor is recorded, because this provides important additional information for data interpretation.

The gas composition of geothermal fluids is highly variable. In volcanic geothermal systems CO_2 is the major gas constituent (Table 5). It may exceed 10% of the volume of the vapor phase but typically it is <1 vol%. Hydrogen sulfide and H_2 are relatively abundant, particularly in systems of low liquid salinity (e.g., samples no. 5 and 9 in Table 5). Methane concentrations

Table 5. Gas concentrations (mmoles/kg) in selected geothermal wells.

	Location[a]	h^f	P^d	T_{qtz}	CO_2	H_2S	H_2	CH_4	NH_3	N_2	Ar
1	Asal, Djibouti, 6				36.1	0.07	0.06	0.033		0.74	0.011
2[b]	Broadlands, NZ, 25	1420	16.2	284	291.1	0.74	2.23	8.02	2.23	3.63	
3	Cerro Prieto, Mex, 5	1284	4.1	289	275.2	15.7	10.1	13.60	5.14	1.74	0.042
4	Kawerau, NZ, 7	1030	7.3	258	47.3	1.93	0.03	0.204	1.68	0.38	
5	Krafla, Iceland, 20	2306	9.8	280	887.0	40.5	24.7	0.320		2.78	0.049
6[c]	Larderello, Italy, 1				726.9	13.3	18.0	8.600		4.10	
7	Mahanagdong, Phil, 9	1287	3.8	278	222.0	13.5	0.05	0.360	0.03	7.03	
8[b]	Momotombo, Nic., 4	1257	0.0	262	297.7	5.96	0.40	0.449		14.91	0.015
9	Námafjall, Icel, 11	1850	25.3	260	142.0	51.4	89.5	0.782		2.47	0.050
10	Olkaria, Kenya, 2	1839	4.8	251	101.0	5.61	2.94	0.722		1.94	
11	Olkaria, Kenya, 301	1653	1.5	262	4260	3.55	1.05	0.833		11.27	
12	Reykjanes, Icel, 15	1256	25.0	260	389.7	10.3	0.78	0.067		7.88	0.122
13	Tongonan, Phil, 510	1376	11.1	251	139.3	10.7	1.08	0.96	0.75	5.90	0.067
14	Zunil, Guatemala, 3	1423	7.3	265	130.8	5.63	0.23	0.017	0.25	1.36	0.006
15	Wairakei, NZ, 72	1295	9.0	246	13.8	0.54	0.04	0.152	0.68	0.22	

h^f: discharge enthalpy, kJ/kg; P^d: sampling pressure, bar-g; T_{qtz}: quartz equilibrium temperature, °C.

[a]The numbers after the local name in column 2 refer to well number. [b]Not the same well from which data are presented in Table 1 but a well from the same area. [c]Dry steam well.

Source of data: 1 (1): D'Amore et al. (1998); 2 (8), 4, 15 (7): Giggenbach (1980); 3: Nehring and D'Amore (1985); 5 (11), 9 (12): Gudmundsson and Arnórsson (2002); 6: D'Amore and Truesdell, 1984; 7 (6), 13 (3): Angcoy (pers. comm.); 8 (4): Arnórsson (1997); 10 (9), 11 (10): Karingithi (pers. comm); 12 (2): Giroud (pers. comm.); 14 (5): Arnórsson (1995b). The numbers in parentheses above refer to the numbers in Table 1 for water sample analysis.

are very variable. They seem to be highest in fluids that have reacted with marine sediments. In some geothermal systems, N_2/Ar ratios are close to those of air-saturated water (e.g., samples no. 3 and 8 in Table 5), but in others they are much higher (e.g., sample no. 9 in Table 5). Elevated N_2/Ar ratios relative to those of atmospheric air may be a reflection of organic source (see, e.g., Mariner et al. 2003; Snyder et al. 2003). In areas of andesitic volcanism on converging plate boundaries, this organic source may ultimately be marine sediments containing organic matter on top of the downward moving lithosphere plate. The andesitic magma may assimilate these sediments, gaining N_2 and other chemical components in the process.

Except for He, noble gas concentrations in geothermal fluids typically are controlled by their concentrations in the source fluid, i.e., air-saturated water (e.g., Mazor and Fournier 1973; Mazor and Truesdell 1984). The concentrations of He, and sometimes also those of Ar, may be elevated due to radiogenic sources. Elevated ^{3}He/^{4}He ratios (5 to over 30 times atmospheric) are typical for volcanic geothermal systems due to the supply of ^{3}He from mantle sources. The highest ratios are found in hot-spot areas such as Hawaii and Iceland (e.g., Welhan et al. 1988; Poreda et al. 1992; Hulston and Lupton 1996).

The gas content of geothermal discharges (fumaroles and wells) has been used to obtain information on the source of the fluid and its temperature. Arnórsson (2000) give a summary of gas geothermometry. Table 6 provides geothermometer equations for the reactions in Table 7. Gas geothermometers may be based on assumptions of specific gas-gas or mineral-gas equilibria or a distribution of isotope ratios between gaseous species. Gas geothermometers include both gas concentrations and gas ratios or a combination of ratios. Early calibrations of gas geothermometers were empirical, i.e., based on drillhole data. Empirical calibration involves correlating specific gas concentrations, gas ratios or a combination of ratios in well discharges with the aquifer temperature of the wells (D'Amore and Panichi 1980; Arnórsson and

Table 6. Temperature equations for gas (steam) geothermometers based on the reactions in Table 7. Arnórsson et al. (2000) provide detailed summary on gas geothermometry equations.

The temperature equations below correspond to the reactions given in Table 7. They are valid in the range 150-350 °C. All minerals appearing in reactions 1 to 10 in Table 7 are taken to be pure (their activity equals 1), except for epidote, prehnite and garnet that form solid solutions. To apply these geothermometers requires knowledge of the acticity of the clinozoisite, Al-prehnite and grossular end members in the respective solid solutions from their analysed compositions.

Reaction[a]

1     $T(°C) = \dfrac{\log(CO_2) + 3.28 + 1.5 \cdot \log(a_{pre}) - \log(a_{czo})}{0.0097}$

2     $T(°C) = \dfrac{\log(CO_2) + 5.072 + 0.6 \cdot \log(a_{gro}) - 0.4 \cdot \log(a_{czo})}{0.01425}$

3     $T(°C) = \dfrac{\log(H_2S) + 6.853 + (2/3) \cdot \log(a_{epi}) - (2/3) \cdot \log(a_{pre})}{0.01343}$

4     $T(°C) = \dfrac{\log(H_2S) + 6.722 + (2/3) \cdot \log(a_{wol}) + (2/3) \cdot \log(a_{epi}) - (2/3) \cdot \log(a_{gro})}{0.01394}$

5     $T(°C) = \dfrac{\log(H_2S) + 6.571 + 2 \cdot \log(a_{epi}) - 2 \cdot \log(a_{gro})}{0.01664}$

6     $T(°C) = \dfrac{\log(H_2S) + 5.537}{0.01197}$

7     $T(°C) = \dfrac{\log(H_2) + 4.686 + (2/3) \cdot \log(a_{epi}) - (2/3) \cdot \log(a_{pre})}{0.007962}$

8     $T(°C) = \dfrac{\log(H_2) + 4.675 + (2/3) \cdot \log(a_{epi}) - (2/3) \cdot \log(a_{gro})}{0.007623}$

9     $T(°C) = \dfrac{\log(H_2) + 7.242 + 6 \cdot \log(a_{epi}) - 6 \cdot \log(a_{gro})}{0.01939}$

10     $T(°C) = \dfrac{\log(H_2) + 4.331}{0.005719}$

[a]See Table 7.

Gunnlaugsson 1985). Calibration has also been based on thermodynamic data for either specific gas-gas or mineral-gas reactions (Giggenbach 1980; Nehring and D'Amore 1984; Arnórsson et al. 1998; Karingithi pers. comm.). Calibrations based on thermodynamic data always rest on an assumption of specific chemical equilibria. This assumption needs to be verified by comparing gas geothermometer results with solute or isotope geothermometers and especially with well temperature data. In the upflow zones of geothermal systems, the vapor may condense partly, either by conductive heat loss or by flowing through water in perched aquifers. Condensation increases the gas concentrations in vapor. To cancel the effect of condensation by conductive heat loss, Giggenbach (1988) calculated the ratios of the reactive gas concentrations (CO_2 and H_2) to the inert gas Ar. Arnórsson (1987) proposed a correction procedure that also involved partial vapor condensation, both by conductive heat loss and mixing with ground water. In the case of mixing, degassing of the liquid was taken into account.

Relatively few attempts have been made to verify or disprove whether specific gas equilibria are closely approached in the aquifers of wet-steam wells. It appears, that equilibrium is frequently closely approached between mineral buffers and CO_2, H_2S and H_2 in the aquifers of wet-steam well discharges (e.g., Arnórsson and Gunnlaugsson 1985; Gudmundsson and Arnórsson 2002; Karingithi pers. comm.). How closely equilibrium is approached, depends on the gas flux from the magma heat source. Thus, in the western part of the Olkaria field in Kenya, CO_2 concentrations are very high in well discharges and apparently not controlled by equilibrium with a mineral buffer. This has been attributed to a high flux of this gas from the magmatic source. In the eastern half of Olkaria, equilibrium between CO_2 and a mineral buffer seems, however, to be closely approached (Fig. 13). One important feature can be observed in Figure 13, which is that different mineral buffers give very similar aqueous CO_2, H_2S and H_2 concentrations at equilibrium when the end-member activities in the solid solution minerals are properly chosen. The same mineral buffer is likely to constrain H_2S and H_2 aquifer liquid concentrations at equilibrium. The studies of Arnórsson and Gunnlaugsson (1985) and Gudmundsson and Arnórsson (2005) indicate that CO_2 in <300 °C aquifers of volcanic geothermal systems are with few exceptions controlled by close approach to equilibrium with the mineral buffer epidote + prehnite + calcite + quartz. Hydrogen sulfide and H_2 concentrations are, on the other hand, controlled by equilibrium with epidote + prehnite + pyrite + pyrrhotite.

In some geothermal systems, equilibrium is attained for the gas-gas reactions shown by Equations (15) and (16) (Giggenbach 1980), but in other systems, data indicate significant departure from equilibrium (Stefánsson and Arnórsson 2002). Whether the concentration of a gas in geothermal fluids, such as CO_2, is controlled by equilibrium with a mineral buffer or by supply from an external source can have profound effects on the depth level of first boiling. Table 7 gives equations that describe how the equilibrium partial pressures of CO_2, H_2S and H_2 vary with temperature for selected mineral buffers. These equations are valid for end-member mineral compositions. More accurate equations can be used for specific areas by taking into account data on the composition of the solid solution minerals (epidote, prehnite, garnet). Figure 14 shows how individual gas partial pressures vary with temperature. Only CO_2 exerts sufficient partial pressures at equilibrium, to significantly affect the boiling point of the liquid when temperatures are above ~300°C. However, H_2 may also contribute at these high temperatures in highly reducing systems. The curves in Figure 14 may be considered to represent minimum values for volcanic geothermal systems associated with an actively degassing magma. At Ohaaki-Broadlands, New Zealand, the CO_2 content of well discharges is up to 6 wt%, leading Grant (1977) to conclude that two-phase fluid existed to at least 2 km depth under natural conditions and that CO_2 partial pressures were as high as ~60 bar at 300 °C. Owing to the two-phase nature of the system, however, rigorous calculation of the CO_2 content of a possible single-phase source fluid and therefore CO_2 partial pressure is impossible (Christenson et al. 2002). In the western part of the Olkaria field in Kenya, CO_2 concentrations of well discharges are very high (see Table 5) causing CO_2 partial pressures to be as high as 80 bars. In well 301 at Olkaria, the temperature of the main producing aquifer is estimated as 262 °C (Karingithi pers. comm.). With this high CO_2 the water will start to boil at a depth of ~1600 m below the water table. The corresponding depth for gas-free water of the same temperature is only 560 m.

Intrusion of fresh magma into the roots of the Krafla geothermal system in Iceland occurred during the period 1975 to 84 (Björnsson et al. 1977). This led to an increase in the gas content of fumarole vapor and well discharges (Ármannsson et al. 1982, 1989). The enhanced gas flux into the Krafla system, which is two-phase, must reflect enhanced boiling. En-

Table 7. Equilibrium constants (as dissolved gas and partial pressure) for mineral–gas reactions that can potentially control aqueous CO_2, H_2S and H_2 concentrations in geothermal systems. They are valid in the range 0–350 °C at P_{sat}. Unit activity was selected for all minerals and liquid water.

	Gas	Reaction	$\log K$ (T)
1	CO_2	$czo + cal + \tfrac{3}{2} qtz + H_2O_l = \tfrac{3}{2} pre + CO_{2,aq}$	$-1.831 - 22843/T^2 - 1344.4/T + 0.00829T - 0.455\log T$
2	CO_2	$\tfrac{2}{5} czo + cal + \tfrac{3}{5} qtz = \tfrac{3}{5} gro + \tfrac{1}{5} H_2O_l + CO_{2,aq}$	$-2.327 - 58215/T^2 - 1829.2/T + 0.01059T - 0.561\log T$
3	H_2S	$\tfrac{1}{3} pyr + \tfrac{1}{3} pyrr + \tfrac{2}{3} pre + \tfrac{2}{3} H_2O_l = \tfrac{2}{3} epi + H_2S_{aq}$	$-2.052 - 215218/T^2 - 1658.7/T + 0.00856T - 0.509\log T$
4	H_2S	$\tfrac{2}{3} gro + \tfrac{1}{3} pyr + \tfrac{1}{3} pyrr + \tfrac{2}{3} qtz + \tfrac{4}{3} H_2O_l = \tfrac{2}{3} epi + \tfrac{2}{3} wol + H_2S_{aq}$	$-1.862 - 253145/T^2 - 1595.2/T + 0.00780T - 0.452\log T$
5	H_2S	$2gro + \tfrac{1}{4} pyr + \tfrac{1}{2} mag + 2qtz + 2H_2O_l = 2epi + 2wol + H_2S_{aq}$	$-1.549 - 383405/T^2 - 1774.0/T + 0.00820T - 0.303\log T$
6	H_2S	$\tfrac{1}{4} pyr + \tfrac{1}{2} pyrr + H_2O_l = \tfrac{1}{4} mag + H_2S_{aq}$	$-2.020 - 188233/T^2 - 1504.4/T + 0.00760T - 0.528\log T$
7	H_2	$\tfrac{4}{3} pyrr + \tfrac{2}{3} pre + \tfrac{2}{3} H_2O_l = \tfrac{2}{3} epi + \tfrac{2}{3} pyr + H_{2,aq}$	$-1.358 - 1420.4/T + 6.777\times10^{-4}T + 5.611\times10^{-6}T^2 - 0.391\log T$
8	H_2	$\tfrac{2}{3} gro + \tfrac{4}{3} pyrr + \tfrac{2}{3} qtz + \tfrac{4}{3} H_2O_l = \tfrac{2}{3} epi + \tfrac{2}{3} wol + \tfrac{2}{3} pyr + H_{2,aq}$	$-1.241 - 1519.9/T + 8.818\times10^{-4}T + 4.693\times10^{-6}T^2 - 0.336\log T$
9	H_2	$6gro + 2mag + 6qtz + 4H_2O_l = 6epi + 6wol + H_{2,aq}$	$1.570 - 5346.3/T + 0.00880T - 6.479\times10^{-6}T^2 + 1.113\log T$
10	H_2	$\tfrac{2}{3} pyrr + H_2O_l = \tfrac{3}{4} pyr + \tfrac{1}{4} mag + H_{2,aq}$	$-1.436 - 1131.3/T - 1.866\times10^{-4}T + 5.377\times10^{-6}T^2 - 0.454\log T$

	Gas	Reaction	$\log P_{gas}$ (bars)
1	CO_2	$czo + cal + \frac{3}{2} qtz + H_2O_l = \frac{3}{2} pre + CO_{2,aq}$	$57.781 - 22843/T^2 - 4792.99/T + 0.00829T + 0.6864\times10^{-6}T^2 - 19.302\log T$
2	CO_2	$\frac{2}{5} czo + cal + \frac{3}{5} qtz = \frac{3}{5} gro + \frac{1}{5} H_2O_l + CO_{2,aq}$	$57.285 - 58215/T^2 - 5277.79/T + 0.01059T + 0.6864\times10^{-6}T^2 - 19.408\log T$
3	H_2S	$\frac{1}{3} pyr + \frac{1}{3} pyrr + \frac{2}{3} pre + \frac{2}{3} H_2O_l = \frac{2}{3} epi + H_2S_{aq}$	$66.723 - 215218/T^2 - 5331.78/T + 0.00856T + 4.07153\times10^{-6}T^2 - 23.070\log T$
4	H_2S	$\frac{2}{3} gro + \frac{1}{3} pyr + \frac{1}{3} pyrr + \frac{2}{3} qtz + \frac{4}{3} H_2O_l = \frac{2}{3} epi + \frac{2}{3} wol + H_2S_{aq}$	$66.913 - 253145/T^2 - 5268.28/T + 0.00780T + 4.07153\times10^{-6}T^2 - 23.013\log T$
5	H_2S	$2gro + \frac{1}{4} pyr + \frac{1}{2} pyrr + \frac{1}{2} mag + 2qtz + 2H_2O_l = 2epi + 2wol + H_2S_{aq}$	$67.226 - 383405/T^2 - 5447.08/T + 0.00820T + 4.07153\times10^{-6}T^2 - 22.864\log T$
6	H_2S	$\frac{1}{4} pyr + \frac{1}{2} pyrr + H_2O_l = \frac{1}{4} mag + H_2S_{aq}$	$66.755 - 188233/T^2 - 5177.48/T + 0.00760T + 4.07153\times10^{-6}T^2 - 23.089\log T$
7	H_2	$\frac{4}{3} pyrr + \frac{2}{3} pre + \frac{2}{3} H_2O_l = \frac{2}{3} epi + \frac{2}{3} pyr + H_{2,aq}$	$23.902 - 2775.68/T + 6.777\times10^{-4}T + 1.49953\times10^{-6}T^2 - 7.357\log T$
8	H_2	$\frac{2}{3} gro + \frac{4}{3} pyrr + \frac{2}{3} qtz + \frac{4}{3} H_2O_l = \frac{2}{3} epi + \frac{2}{3} wol + \frac{2}{3} pyr + H_{2,aq}$	$24.019 - 2875.18/T + 8.818\times10^{-4}T + 0.58153\times10^{-6}T^2 - 7.302\log T$
9	H_2	$6gro + 2mag + 6qtz + 4H_2O_l = 6epi + 6wol + H_{2,aq}$	$26.830 - 6701.58/T + 0.00880T - 10.59047\times10^{-6}T^2 - 5.853\log T$
10	H_2	$\frac{3}{2} pyrr + H_2O_l = \frac{3}{4} pyr + \frac{1}{4} mag + H_{2,aq}$	$23.824 - 2486.28/T - 1.866\times10^{-4}T + 1.26553\times10^{-6}T^2 - 7.420\log T$

The thermodynamic properties of $CO_{2,aq}$, H_2S_{aq} and $H_{2,aq}$ are from Fernandez–Prini et al. (2003). Mineral data are from Holland and Powell (1998) except for pyrite and pyrrhotite that are from Robie and Hemingway (1995). Data on $H_4SiO_4^0$ and quartz solubility are from Gunnarsson and Arnórsson (2000), those on H_2O_l, Ca^{+2}, Fe^{+2} and OH^- from SUPCRT92 program (Johnson et al. 1992) using the slop98.dat data set. The data on $Fe(OH)_4^-$ and $Al(OH)_4^-$ are from Diakonov et al. (1999) and Pokrovskii and Helgeson (1995), respectively. The fluorite solubility equation is from Arnórsson et al. (1982). It is based on Nordstrom and Jenne (1977). Abbreviations for mineral phases are: *cal*: calcite; *czo*: clinozoisite; *epi*: epidote; *flu*: fluorite; *gro*: grossular; *mag*: magnetite; *pre*: prehnite; *pyr*: pyrite; *pyrr*: pyrrhotite; *qtz*: quartz; *wol*: wollastonite.

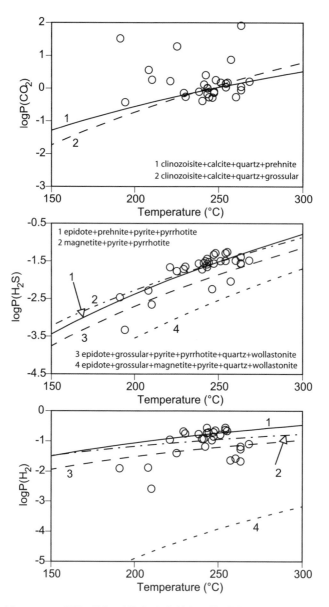

Figure 13. Partial pressures of CO_2, H_2S and H_2 in the initial aquifer fluid producing into wells drilled into the Olkaria geothermal field in Kenya (circles). The curves represent equilibrium constants for selected mineral buffer-gas reactions as shown. In deriving these curves the average analyzed compositions of epidote, prehnite and garnet in the Olkaria system were taken into consideration. Other minerals (calcite, quartz, pyrite, pyrrhotite) in the respective buffers were taken to be pure. The Olkaria data indicate that equilibrium is closely approached in most aquifers between the aquifer water and the mineral buffers epidote + prehnite + calcite + quartz (for CO_2) and epidote + prehnite + pyrite + pyrrhotie (in the case of H_2S and H_2). However, some well discharges (in the west part of the Olkaria field) have CO_2 concentrations higher than those corresponding to equilibrium. It is thought that these high CO_2 concentrations are determined by a high flux of this gas from the magmatic heat source rather than by local mineral buffer equilibrium. Based on data from Karingithi (pers. comm.).

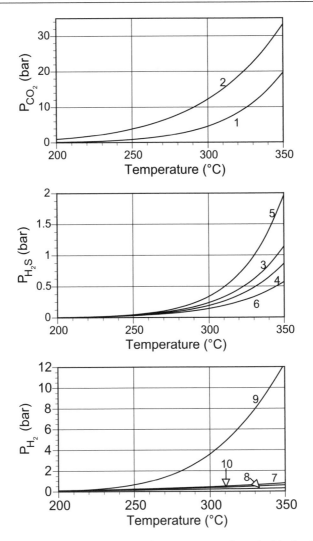

Figure 14. Relationship between temperatures and gas pressures as determined by local equilibria with various mineral buffers. The numbers correspond to the mineral buffer-gas reactions shown in Table 7. If equilibrium is attained, only CO_2 and H_2 will exert a sufficiently high partial pressure above ~300 °C to affect the boiling point of water significantly.

hanced gas flux into a non-boiling geothermal system would shift the depth level of first boiling to greater depths. In both instances the enhanced gas flux will cause some cooling of the fluid.

BOILING AND CHANGES IN MINERAL SATURATION

Changes in fluid composition during boiling and degassing

When geothermal fluids boil extensively, two major changes occur. One involves increased dissolved solids content of the boiled liquid due to vapor formation. The other major change is an increase in pH due to transfer of CO_2 and H_2S ("acid gases") from the liquid into the vapor. These changes and the cooling of the fluid by depressurization boiling cause complicated changes in individual aqueous species activities, as values of association constants for complexes and ion hydrolysis constants are temperature dependent. The pH also plays a major role.

The transfer of dissolved gases from the initial liquid to the vapor is determined by their solubility in liquid and how closely equilibrium distribution is approached between the liquid and vapor. The solubility constant (K_s) of gas species s in aqueous solution is defined by Equation (4) above. The distribution coefficient (D_s) for species s is defined as

$$D_s = \frac{x_s^v}{x_s^l} \tag{18}$$

Here, x_s^v and x_s^l designate the mole fraction of the species in vapor and liquid, respectively. The mole fraction is given by

$$x_s = \frac{N_s}{N_{H_2O} + \sum_i N_i} \tag{19}$$

where N_s stands for the number of moles of volatile species s, and the sum is over all such species except H_2O. If we consider a mass of fluid containing 1 kg of H_2O, then $N_{H_2O} = 55.51$ and

$$x_s = \frac{m_s}{55.51 + \sum_i m_i} \tag{20}$$

where m_s denotes the molal concentration of gas s. For dilute fluids, $\sum_i m_i \ll 55.51$ so

$$x_s \cong \frac{m_s}{55.51} \tag{21}$$

and

$$D_s \cong \frac{m_s^v}{m_s^l} \tag{22}$$

The partial pressure (P_s) of gas s is given by

$$P_s = x_s^v P_{tot} \tag{23}$$

where P_{tot} is the total pressure of all gases, including H_2O. For dilute fluids and moderate pressure, Equation (4) may be replaced by

$$m_s^l \cong K_s P_s \tag{24}$$

since γ_s in Equation (5) and Γ_s in (6) are close to unity in this case. Combining Equations (18), (21), (23) and (24) yields

$$D_s \cong \frac{55.51}{K_s P_{tot}} \tag{25}$$

Due to rapid flow and continuous vapor formation by depressurization boiling of hot fluid rising below hot springs and in wells and their producing aquifers, the mass transfer of gases from liquid to vapor may not be sufficient to establish equilibrium distribution as defined by Equation (25). The liquid will contain more gas than dictated by equilibrium.

The extent of degassing at sampling conditions and in wellhead separators can be established for wet-steam well discharges from analysis of gaseous components in liquid and vapor samples. This is, however, neither possible downhole nor in upflow zones below boiling hot springs. As the extent of degassing during boiling affects the mineral saturation state of the liquid, it is of interest to assess this effect. This is possible with the aid of the WATCH chemical speciation program (Arnórsson et al. 1982). A factor (ç) is used to represent the extent of degassing of the boiled liquid relative to equilibrium degassing. In the program this involves dividing K_s by the selected ç value (between 0.01 and 1). A ç value of 1 corresponds to equilibrium gas distribution between the liquid and vapor. A small value indicates negligible degassing. An equation analogous to Equation (12) permits calculation of the liquid concentration of gas species s at any selected vapor pressure. We have

$$m_s^t = m_s^v X^v + m_s^l (1 - X^v) \tag{26}$$

Combination of (22), (25) and (26) and insertion of ç yields

$$m_s^l = m_s^{t,l} \left[X^v \left(\frac{55.51}{P_{tot} K_s} \varsigma - 1 \right) + 1 \right]^{-1} \tag{27}$$

For Equation (27) to be useful, X^v must be known. For wet-steam wells, a value for X^v can be obtained from the measured discharge enthalpy (h^d) applying Equation (A4). For boiling hot spring fluids, X^v can be obtained from selection of an aquifer temperature by taking the enthalpy of the parent aquifer fluid ($h^{f,t}$) to be equal to that of vapor saturated liquid ($h^{f,l}$) at the aquifer temperature.

Mineral deposition with special reference to calcite

In upflow zones of geothermal systems and in wells where extensive boiling occurs, mineral deposition from the boiling fluid largely occurs in response to its cooling and degassing. Many minerals, such as amorphous silica and metallic sulfides, have prograde solubility (the solubility increases with increasing temperature). Others, including anhydrite, calcite and aragonite, have retrograde solubility with respect to temperature. Cooling causes geothermal waters to become over-saturated with minerals with prograde solubility but under-saturated with those having retrograde solubility. Some minerals, such as calcite, aragonite and metallic sulfides, have pH-dependent solubility. Degassing tends to produce over-saturated water with respect to minerals whose solubility decreases with increasing pH.

The quantity of minerals precipitated from solution is not only determined by the degree of over-saturation but also by the fluid composition and the kinetics of the precipitation reaction. Quartz, adularia and albite, e.g., are abundant in the boiling zone of high-temperature geothermal systems, but do not form in wells due to slow kinetics. In dilute waters, the extent of sulfide mineral deposition is essentially limited by low metal concentrations. In brines, by contrast, sulfide mineral deposition is extensive due to the abundance in solution of cations that form sulfide minerals. Troublesome scale in wells is characterized by phases that readily precipitate from solution, such as amorphous silica and "salts" including calcite, aragonite and anhydrite. Metallic sulfides also precipitate readily from solution.

The solubility of some minerals increases with increasing temperature but decreases with increasing pH; an example being calcite. The combined effects of both processes, together with the rate of the precipitation reaction, determine whether or not the minerals with which the water becomes over-saturated precipitate from solution. Un-boiled geothermal liquids are typically close to being calcite-saturated (e.g., Arnórsson 1989). Extensive degassing by boiling tends to cause an initially calcite-saturated water to become over-saturated. The cooling has the opposite effect due to the retrograde solubility of calcite with respect to temperature. The extent of degassing and cooling determines whether boiling causes an initially calcite-saturated water to become over or under-saturated. Figure 15 shows how the degree of calcite saturation varies during adiabatic boiling for seven selected wells, assuming maximum

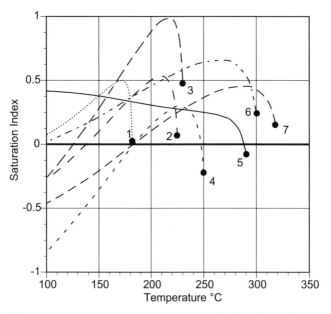

Figure 15. Variation in calcite saturation temperature during adiabatic boiling of fluid discharged from selected wet-steam wells. Equilibrium distribution was assumed for all gases between the liquid water and vapor. Dots indicate aquifer conditions. 1: Hveragerdi, well 2, Iceland; 2: Hveragerdi, well 7, Iceland; 3: Wairakei, well 24, New Zealand; 4: Amatitlan, well 1, Guatemala, 5: Asal, well 1, Djibouti; 6: Krafla, well 6, Iceland; Palinpinon, well 7, Philippines.

(equilibrium) degassing for all analyzed gases (CO_2, H_2S, H_2, CH_4, N_2). The calculated calcite saturation in the aquifer for each well is shown by dots. The departure from saturation is very small for the majority of the wells and is regarded to be within the limit of error when all uncertainties behind the calculations are considered, such as analytical imprecision and selection of aquifer temperature. A similar overall pattern in the variation of the saturation index (SI) with temperature is observed for all the wells. A sharp initial increase in SI is seen, followed by a decline. The initial increase in SI reflects an increase in pH due to CO_2 and H_2S degassing. The pH increase causes a strong increase in the activity of the CO_3^{-2} species. At maximum SI values, the water has been largely degassed, and the subsequent decline in SI is caused by increased calcite solubility with decreasing temperature. The only exception to the general pattern just described is sample 5 in Figure 15 (Nisyros, well 2, Greece). This well discharges a highly saline liquid (~3 times seawater salinity). The slight increase in the SI for this well with decreasing temperature after the initial rise in SI is largely due to dissociation of ion pairs with falling temperature, mostly $CaHCO_3^+$.

Figure 16 shows how the calcite saturation index varies with the extent of degassing during adiabatic boiling of fluid discharged from a well in the Amatitlan field in Guatemala. If degassing is less than 30% of maximum (ç < 0.3), boiling does not produce a calcite over-saturated solution as it does when ç is > 0.3.

MODELING OF AQUIFER FLUID COMPOSITIONS

This section describes modeling of fluid compositions in aquifers that feed boiling hot springs and wells drilled into high-temperature liquid-dominated geothermal systems (wet-steam wells). It is based on analytical data on liquid and vapor samples collected at the surface.

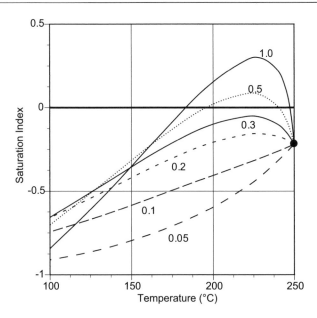

Figure 16. Variation in calcite saturation with temperature during adiabatic boiling of fluid discharged from well 1 at Amatitlan, Guatemala. Degassing was assumed to be variable. The numbers by each curve represent the selected value of the degassing coefficient (ç). A value of 0.05 for ç corresponds to degassing that is 5% of maximum, i.e., equilibrium, degassing and so on.

The modeling assumes that neither mineral precipitation nor dissolution occur between aquifer and surface. Basically, it reconstructs the fluid composition below (beyond) the zone of extensive boiling. Modeling of this kind is essential for studying mineral-fluid equilibria in the aquifer. The model used for hot springs is relatively simple, but it is more involved for the wet-steam wells due to the rather complicated processes in the depressurization zone that forms around discharging wells (see Fig. 10). For hot springs, the boiling between aquifer and surface is taken to be adiabatic.

Boiling hot springs

Boiling, mixing and conductive cooling in upflow zones of geothermal systems are variable and predominantly complex processes that may involve mineral dissolution and precipitation reactions. At the present state of knowledge, satisfactory general boiling-mixing-reaction models cannot be developed to determine initial aquifer fluid compositions from data on hot spring discharges. In the absence of mixing, and when the temperature of the feeding aquifer is not much above 200 °C (for 200 °C water, boiling starts at ~170 m depth), aquifer fluid compositions may be reconstructed reasonably accurately below the zone of boiling, at least when flow rates are sufficiently high to make conductive heat loss of the rising fluid insignificant. Under these conditions, it is logical to assume boiling to be adiabatic.

Many studies have indicated that mineral-solution equilibrium is closely approached for all major components of geothermal fluids except Cl (e.g., Giggenbach 1981; Arnórsson et al. 1983). This greatly constrains the relative activities of aqueous species, and therefore also individual component concentrations, and helps evaluate the validity of the modeling results. For example, cation/proton activity ratios in equilibrated liquids are constant at any specific temperature for a system of specific mineralogy (Fig. 17).

The WATCH chemical speciation program (Arnórsson et al. 1982) version 2.1 (Bjarnason

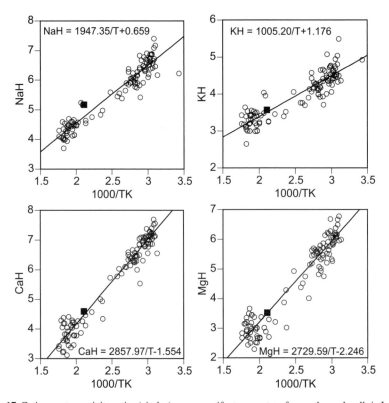

Figure 17. Cation-proton activity ratios (circles) versus aquifer temperature for geothermal wells in Iceland. NaH, KH, CaH and MgH designate the logarithms of the respective cation/proton activity ratios [log(Na+/H+), log(Ca+2/(H+)²), etc]. The solid squares show the calculated cation-proton activity ratios in the aquifer water feeding a boiling hot spring in the Geysir geothermal field, Iceland. The aquifer temperature was taken to be that of last equilibrium with quartz (204 °C). Boiling in the upflow was assumed to be adiabatic. The calculated activity ratios are based on CO_2 degassing that amounts to 3% of maximum degassing. Such degassing gives aquifer water that is just calcite saturated. The equation in the upper or lower part of each diagram represents the best line through the data points.

1994) permits calculation of aquifer water composition from data on the chemical composition of boiling hot-spring waters. Essentially, calculations involve correcting the dissolved solids content of the hot spring water for vapor loss and adding back into the water those gases (CO_2 and H_2S) that were lost with the vapor. To run the program, an aquifer temperature must be selected. The choices include several geothermometer temperatures, including the temperature of last equilibrium with quartz, and an arbitrarily selected temperature value. A value for the degassing coefficient (ç) needs to be chosen as well. One way of doing this, is to study how ç relates to calcite saturation. It is logical to select a ç value that corresponds to calcite saturation in the aquifer at the selected aquifer temperature because, as already pointed out, geothermal reservoir waters are generally close to being calcite-saturated, at least when temperatures are >100 °C. The concentrations of gases other than CO_2 and H_2S in the un-boiled aquifer fluid can be approximated, if data are available on the relative abundance of gases in vapor samples.

To demonstrate the calculation of aquifer fluid composition from analytical data on boiling hot spring water, we have selected one such spring from the Geysir geothermal field in Iceland (Table 8). First the WATCH program was run for different ç values taking the temperature of last equilibrium with quartz (204 °C) to represent the temperature of the un-boiled fluid below the hot spring. For this temperature, the depth level of first boiling is at ~180 m. Figure 18A shows how the calculated calcite saturation in the aquifer liquid varies with the selected degassing factor (ç) values. Also shown in Figure 18 (B and C) are variations with ç in aquifer liquid pH and in vapor CO_2 concentrations. As can be seen from Figure 18A, a value of about 0.03 for ç matches aquifer liquid calcite saturation, indicating that only very limited degassing accompanies boiling. Only 3% of the CO_2 in the initial aquifer liquid is transferred into the vapor relative to the transfer needed to establish equilibrium distribution of CO_2 between liquid and vapor at surface conditions. With this limited degassing, cooling by adiabatic boiling leads to successively greater calcite under-saturation (Fig. 18D).

The calculated cation/proton activity ratios of the aquifer water feed-

Table 8. Chemical composition of boiling hot spring in the Geysir field, Iceland and computed composition of the unboiled aquifer water. Concentrations are in mg/kg.

Component	Hot spring	Aquifer water
pH/°C	8.95/26	7.34/204
SiO_2	358.5	286.8
B	0.835	0.668
Na	228.0	182.4
K	10.25	8.20
Ca	0.85	0.68
Mg	0.005	0.004
CO_2[a]	103.5	156.6
SO_4	147.3	117.8
H_2S	1.43	1.61
Cl	117.0	93.6
F	11.75	9.40

The aquifer temperature was taken to be the last temperature of equilibrium with quartz. Degassing during boiling was taken to be 3% of maximum degassing.

[a]Total carbonate carbon as CO_2.

ing the selected boiling hot spring closely match those for equilibrated geothermal liquids (solid squares in Fig. 17). The deviation from the equilibrium curve in terms of pH is 0.42, 0.07, 0.15 and 0.04 for the respective cation/proton ratios Na/H, K/H, Ca/H and Mg/H. These results are considered to substantiate the calculated pH value and the selected ç value.

Values for ç have been estimated for other boiling hot springs in the Geysir field by the method described above. They are very variable, ranging from almost zero to 1. The area with boiling hot springs in the Geysir field is only a few hundred m across. The results indicate that phase separation within this small area is very variable. Such may also be the case for other areas.

Wet-steam well discharges

For demonstration purposes, the aquifer vapor and liquid compositions have been calculated for one excess enthalpy well in the Olkaria geothermal field, Kenya. The calculations were carried out on the basis of each of the five models described above and delineated in Figure 10. The results are presented in Table 9. Also shown are fluid parameter values which quantify the causes of the excess enthalpy. For Models 1 and 2 (isolated and closed systems, respectively), only one gas controlled by a mineral buffer equilibrium is needed to calculate the equilibrium vapor fraction ($X^{f,v}$). Hydrogen was selected. Its solubility in liquid is low. Therefore, the calculated value for $X^{f,v}$ is sensitive to the concentration of H_2 in the well discharge. The gases used to calculate $X^{f,v}$ by Models 3 to 5 (the open system models) were H_2 and H_2S. These gases were chosen because of their large solubility difference in liquid and because their aqueous concentrations are likely to be controlled by the same mineral buffer. The buffer selected to obtain values for $m_s^{f,l}$ for both these gases was pyrite + pyrrhotite + magnetite. Selection of the buffer epidote + prehnite + pyrite + pyrrhotite would have yielded almost identical results (see Fig. 13).

The results presented in Table 9 are discussed below. We seek to identify which of the five models best describes the actual conditions in the aquifer of the selected Olkaria well.

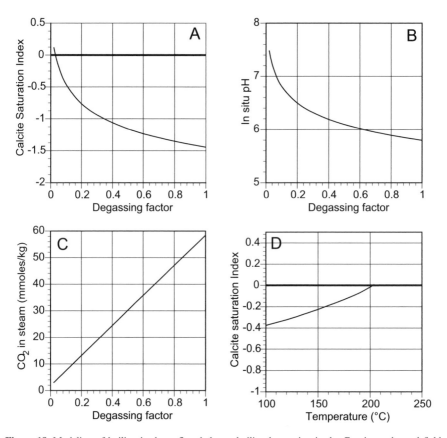

Figure 18. Modeling of boiling in the upflow below a boiling hot spring in the Geysir geothermal field, Iceland. A: Variation in calcite saturation with the extent of CO_2 degassing of the boiling water. B: Variation in pH with CO_2 degassing during boiling. C: CO_2 concentrations in steam versus degassing of the boiling water. D: Calcite saturation in variably boiled water (adiabatic) for CO_2 degassing that is 3% of maximum degassing. Maximum degassing corresponds to equilibrium CO_2 distribution between liquid water and steam.

Models 1 and 2 (isolated and closed systems). By these models, the chemical composition of the initial aquifer fluid is the same as that of the total discharge. The aquifer liquid and vapor compositions, however, differ (Table 9). The calculated silica concentration in the initial aquifer liquid by the isolated system model (Model 1) is 444 ppm, which corresponds to a quartz equilibrium temperature of 245 °C. This compares well with the Na/K geothermometer temperature of 252 °C. By contrast, the silica concentration in the initial aquifer fluid is 314 ppm according to the closed system model (Model 2). This concentration corresponds to a quartz equilibrium temperature of only 214 °C, which is in poor agreement with the Na/K temperature. Measured downhole temperatures in the well, when stabilized thermally, at the depth level of permeable horizons, are 240 to 250 °C, in good agreement with the Na/K geothermometer temperature and the quartz equilibrium temperature found for the isolated system model.

The equilibrium vapor fraction for both the isolated and closed system models, as calculated from the H_2 concentration in the vapor discharged, is close to zero, or 0.0062 by mass. The "physical" vapor fraction (see Appendix I) in the aquifer according to the isolated system model is, on the other hand, very high, or 0.439 by weight, corresponding to a vapor-

Table 9. Calculated chemical composition of the initial liquid and vapor in the aquifer feeding Olkaria well 2 (Kenya) according to the five models considered in this contribution. Concentrations are in mg/kg.

	Weirbox water sample	"Collected" sample[a]	Isolated system 1A[b]	Isolated system 1B[b]	Closed system 2	Open system 3	Open system 4	Open system 5B
Aquifer temp. °C	250							
Discharge enth. kJ/kg	1839							
Sampling pressure, bar-g	5.6							
Steam fraction at sampling	0.555							
Liquid phase								
pH/°C	8.95/20	7.78/163	7.07/250	7.77/250	7.07/250	6.70/250		
SiO_2	576.0	559.6	443.3	443.3	313.8	444.4	480.6	460.9
B	5.90	5.92	4.69	4.69	3.32	4.70	5.08	4.88
Na	526.0	484.7	384.0	384.0	271.8	384.9	416.3	399.2
K	83.0	80.1	63.5	63.5	44.9	63.6	68.8	66.0
Ca	0.56	0.64	0.51	0.51	0.36	0.508	0.55	0.53
Mg	0.010	0.009	0.007	0.007	0.005	0.007	0.008	0.007
Al	0.79	0.574	0.455	0.455	0.32	0.456	0.49	0.47
Fe	0.020	0.017	0.014	0.014	0.010	0.014	0.015	0.014
SO_4	54.0	24.4	19.4	19.4	13.7	19.4	21.0	20.1
Cl	658.0	664.8	526.7	526.7	372.8	528.0	570.9	547.6
F	57.0	60.0	47.5	47.5	33.7	47.7	51.5	49.4
CO_2	92.0	76.6	619.0	63.1	17.6	260.7	260.7	260.7
H_2S	9.86	0.97	62.9	17.6		19.3	19.3	19.3
H_2			0.70		0.70	0.70	0.70	0.70
CH_4			0.80		0.80	0.70	0.70	0.70
N_2			5.9		5.9	7.58	7.58	7.58
Vapor phase								
CO_2		2,421	80,665	3,063	80,665	33,963	33,963	33,963
H_2S		185.4	3,138	213	3,138	922	922	922
H_2		7.16	395.0	9.0	395.0	395	395	395
CH_4		7.12	376.6	9.0	376.6	332	332	332
N_2		79.29	4,673	100	4,673	5,985	5,985	5,985
T^e (°C)						200	220	220
$X^{f,v}$			0.0062	0.439	0.0062	0.00011	0.00011	0.00011
$Q^e/M^{d,t}$					743	0	0	296
$V^{e,v}$				0.383		0	0.159	0
$V^{f,t}$						4.214	4.214	4.214
$V^{e,l}$						3.214	3.373	3.214

[a]Corrected water sample composition for steam loss by boiling from the wellhead steam separator (5.6 bar-g) to the atmospheric silencer but the sample was collected from the weirbox draining the silencer.

[b]1A and 1B refer to gas compositions of "equilibrium" steam and "physical" steam, respectively.

T^e (°C; intermediate zone temperature at which phase segregation, addition of steam and transfer of heat may occur)
$X^{f,v}$ (initial aquifer steam fraction)
$Q^e/M^{d,t}$ (heat added to flowing fluid relative to well discharge)
$V^{e,v}$ ($M^{e,v}/M^{d,t}$; relative mass (to well discharge) of steam added to flowing fluid)
$V^{f,t}$ ($M^{f,t}/M^{d,t}$; relative mass (to well discharge) of initial aquifer fluid that has boiled)
$V^{e,l}$ ($M^{e,l}/M^{d,t}$; relative mass (to well discharge) of boiled water retained in aquifer)

dominated system in terms of volume. Thus, the isolated system model yields a large difference between "equilibrium" and "physical" vapor fractions. The physical vapor fraction gives a very low concentration of CO_2 (and other gases) in the initial liquid (19.3 ppm CO_2) and vapor (10,811 ppm CO_2). The low liquid CO_2 concentration gives rise to anomalously high pH and to strong calcite over-saturation water (saturation index of +0.56).

The discrepancy in the geothermometer results for the closed system model is taken to indicate that this model does not properly account for the excess well discharge enthalpy. The low gas content in the "physical" vapor fraction according to the isolated system model is taken to mean that this model does not explain the excess vapor in the well discharge either. No aquifer fluids are considered to exist with chemical compositions corresponding to those calculated using the isolated and closed system models.

Models 3 to 5 (open system). The equilibrium vapor fraction ($X^{f,v}$) is calculated in the same way for all the open system models and is therefore the same. Calculated gas compositions are also the same. It is possible to use only one gas to calculate $X^{f,v}$ by Model 3. If this is done, the value of T^e must be selected for calculation of the initial aquifer fluid composition. Use of two gases to calculate $X^{f,v}$ according to Model 3, on the other hand, fixes $T^{e\,2}$. There are two versions of Model 5. One assumes that heat (Q^e) is added to the final fluid (Model 5A), but the other that heat is added prior to phase segregation (Model 5B). Models 3 and 5A are very similar when two gases are used to determine $X^{f,v}$. The only difference is that T^e is fixed when $Q^e = 0$ (Model 3) but needs to be chosen when $Q^e \neq 0$ (Model 5A). Both models yield the same initial aquifer liquid and vapor compositions.

The value obtained for the initial vapor fraction is very low, or 0.00011 by mass (0.4% by volume). All models indicate that the "excess" discharge enthalpy is essentially caused by phase segregation. The mass flow rate ($M^{f,t}$) of the initial reservoir fluid is ~4.2 times greater than the mass flow rate ($M^{d,t}$) from the well, so the mass of liquid retained in the formation ($M^{e,l}$) by its immobilization is more than 3 times the mass flow rate of the well discharge. If the value of T^e is selected as 200 °C for Models 4 and 5, both heat flow from the formation (Q^e) in Model 5 and addition of vapor to the flowing fluid ($V^{e,v}$) in Model 4 become zero, but positive at higher values of T^e (see Table 9). The temperature of the aquifer producing into the well is close to 250 °C and the inflow temperature into the well is ~200 °C. The processes that can change the enthalpy of the flowing fluid must therefore be operative in the interval 250 to 200 °C. This narrow interval does not allow large Q^e and $V^{e,v}$ values. Overall, the results presented in Table 9 indicate that phase segregation is largely responsible for the excess well discharge enthalpy. Other studied wells at Olkaria (Karingithi pers. comm.) yield results comparable to that described for well no. 2 above, i.e., low initial aquifer fluid steam fraction and formation of excess discharge enthalpy by phase segregation.

The main obstacle to using the present models to calculate aquifer fluid compositions from wellhead data lies in the assumption of specific mineral-gas equilibria. The mineral buffer selected here (pyrite + pyrrhotite + magnetite) contains minerals whose compositions are not likely to deviate much from pure end-members (mineral activity ~1). The composition of minerals in other potential gas buffers may vary from one field to another, and even show zonation, such as epidote and prehnite. To use such buffers to estimate the initial vapor fraction requires detailed mineralogical studies of the geothermal system in question.

To assess aquifer fluid compositions on the basis of wellhead data, more sophisticated models could be evaluated than those described here. One might, e.g., take into account the gas content of any boiled immobile water, or consider phase segregation and vapor formation by

---

[2]   This can be deduced from (A28) or (A29) and Equation (A23). One of the first two equations fixes $M^{f,t}$ and subsequently (A23) gives $M^{e,l}$. With $M^{f,t}$ and $M^{e,l}$ known, the only unknown variable in Equation (A24) is $h^{e,l}$ that determines T^e.

conductive heat transfer to take place over a range of temperatures. It is, however, questionable whether such elaboration of the present models will improve the results. After all, the models are partly based on the assumption of specific gas-mineral equilibria, and the fluid discharged from the well is likely to be a mixture of many components from sub-systems with different temperatures. Local equilibrium may or may not have been closely approached within all these sub-systems, and the selection of a single aquifer temperature is always a simplification of the real conditions in the geothermal system.

FUTURE DIRECTIONS

The worldwide concern about the consequences of global warming will undoubtedly increase interest in developing geothermal resources, both for power generation and direct use. This is likely to provide much new field data on geothermal systems and spur geochemical studies of geothermal fluids and the processes leading to the formation of ore-deposits.

Hydro-geochemical studies rely on proper sampling techniques, high-quality chemical analyses and data handling. In the case of wet-steam well discharges there is a need to improve sampling techniques and sample treatment (e.g., Arnórsson et al. 2006). The following three points are emphasized. First, liquid and vapor samples should be collected at the same pressure. In many parts of the world it is common practice to collect liquid and vapor samples at different pressures, the liquid from the weirbox at atmospheric pressure and the vapor from the wellhead or a pipeline at an elevated pressure. Samples obtained in this way are not entirely representative of the total well discharge (Arnórsson and Stefánsson 2005). Secondly, the pH of water samples from wet-steam wells should be measured on-site and before onset of silica polymerization. Silica monomers and oligomers are weak acids, and polymerization changes the water pH (Gunnarsson and Arnórsson 2005). To obtain a reliable measurement of pH in amorphous silica over-saturated samples, it is thus necessary to measure it before the onset of silica polymerization. Thirdly, it is desirable to separate sulfur species at collection using resins (Druschel et al. 2003). Upon sample storage, reactions tend to occur among sulfur species. Hydrogen sulfide oxidizes to thiosulfate and other sulfur species, and ultimately to sulfate. Analyzed sulfate concentrations in fluids from wet-steam wells are thus in most cases subject to error, which makes their interpretation with respect to some redox and mineral-solution reactions problematic. Fixing the different species on-site and analyzing them separately solves this problem.

Many trace metals in geothermal fluids form insoluble sulfides. Extensive boiling of geothermal fluids in producing aquifers and wells causes sulfide precipitation. Wellhead samples thus provide inadequate information on sulfide-forming metals in the initial aquifer fluid. There is a need to develop further techniques of collecting downhole samples from wet-steam wells to obtain improved data on the content of many trace elements in the aquifer fluid.

Studies of fossil geothermal systems that have been exposed by erosion are of importance for understanding fluid-rock interaction processes in active geothermal systems and should be pursued. Such studies yield valuable information that cannot be obtained from drillhole data, including the depths of intrusion of magma and the distribution of hydrothermally altered rock in relation to tectonic structures and the magmatic heat source.

Advances in geothermal geochemistry do not only rely on field data. Today, interpretation frequently emphasizes the modeling of observed fluid compositions and alteration mineralogy. Thus, experimental studies of the thermodynamic properties of minerals and aqueous species, and of the kinetics of reactions involving these, are vital, and so are the evaluation and compilation of such data. Thermodynamic data are sorely lacking for numerous aqueous species of significance in geothermal fluids, both for species composed of major elements and those involving trace components. This lack has been addressed by developing algorithms to predict the thermodynamic properties of aqueous species for which experimental data are sparse or lacking (e.g., Helgeson et al. 1981; Shock et al. 1997; Sverjensky et al. 1997). In this context we consider that the density model (Franck 1956, 1961) should be given more attention as a tool for predicting association constants for aqueous species (e.g., Anderson 2005). Nonetheless, predictive methods such as these are not expected to replace experiments anytime soon. In fact, much more experimental work is called for, since computer models of water-rock interaction rely too much on inadequate thermodynamic and kinetic data.

It is generally accepted that hydrothermal mineral-solution equilibria are closely approached with respect to most major components in geothermal systems at temperatures as low as 100 °C, or even lower. Yet, surprisingly few publications seem to concentrate on carefully assessing the state of specific mineral-solution equilibria. On the contrary, it is common that specific equilibria are assumed to prevail, rather than verified. Many hydrothermal minerals form solid solutions. Data on the composition of such hydrothermal minerals are necessary to assess their state of saturation in the fluid. Unfortunately, such data are scarce in the published literature.

With the enormously improved quality of trace element analysis at low cost by ICP-MS during the last decade, a new field has opened up in the study of water-rock interaction in geothermal systems. Recent advances in ion chromatography have also made it possible to determine the concentrations of trace elements in particular oxidation states at sub-ppb levels. Two impeding factors to such work present themselves, however: inadequate thermodynamic data to calculate individual species activities and the relative paucity of representative samples of aquifer fluids in geothermal systems. An alternative to determining the composition of such fluids is to analyze fluid inclusions in hydrothermal minerals using ICP-MS laser ablation.

Mineral-solution equilibria and disequilibria in natural systems are based on the concept of local equilibrium (e.g., Nordstrom and Munoz 1994). All natural systems, such as geothermal systems, are open systems, and samples collected from wells necessarily represent a mixture of many components that have traveled different distances at different velocities from their point of origin to the point of inflow into the well. For data interpretation regarding specific mineral-solution equilibria, a single aquifer temperature is selected. This is an approximation at best. It would be of interest to simulate well discharge compositions by adding together several fluid components in selected proportions, rather than attempting to reconstruct the aquifer fluid composition from wellhead data by assuming a single aquifer temperature.

Deep drillings into the roots of active volcanic geothermal systems are of both scientific and economic interest. Such drillings would hopefully provide information on permeability, the depth of fluid circulation, and the transfer of heat and volatiles from the heat source to the circulating fluid. The economic interest is in the heat stored in the rock and magma of this environment.

LIST OF SYMBOLS

a_{R-K} constant in the modified Redlich-Kwong equation equal to $0.4275R^2T_c^{2.5}/P_c$ where P_c and T_c represent critical pressure and temperature, respectively, and R is the gas constant (see Nordstrom and Munoz 1994)

A_s $m_s^{d,t}/m_s^{f,l}$ (see Eqn. A30)

a_s^l activity of gas s dissolved in liquid water (moles/kg)

B^* constant in the Debye-Hückel Equation (Eqn. 10)

$b_{R\text{-}K}$ constant in the modified Redlich-Kwong equation equal to $0.0866RT_c/P_c$ where P_c and T_c represent critical pressure and temperature, respectively, and R is the gas constant (see Nordstrom and Munoz 1994)

ç   degassing factor (see Eqn. 27)

D_s distribution coefficient for gaseous species s between liquid and vapor; defined by Equation (18)

$h^{d,l}$ specific enthalpy of liquid water in well discharge (kJ/kg)

$h^{d,t}$ specific enthalpy of total well discharge (kJ/kg)

$h^{d,v}$ specific enthalpy of vapor in well discharge (kJ/kg)

$h^{e,l}$ specific enthalpy of steam-saturated water at vapor pressure P^e (kJ/kg)

$h^{e,v}$ specific enthalpy of saturated vapor at pressure P^e (kJ/kg)

$h^{f,l}$ specific enthalpy of initial aquifer water (kJ/kg)

$h^{f,t}$ specific enthalpy of initial aquifer fluid (kJ/kg)

$h^{f,v}$ specific enthalpy of initial aquifer vapor (kJ/kg)

I ionic strength (Eqn. 10)

k_s the Setchénow coefficient defined by Equation (9)

K_s solubility constant (Henry's Law Constant) for gaseous species s (moles kg^{-1} bar^{-1})

L^d latent heat of vaporization at vapor pressure P^d (kJ/kg)

L^e latent heat of vaporization at vapor pressure P^e (kJ/kg)

L^f latent heat of vaporization at vapor pressure P^f (kJ/kg)

m_B molal concentration of salt B in solution in the Setchénow equation (Eqn. 9)

$M^{d,l}$ mass flow rate of water from wet-steam well (kg/s)

$M^{d,t}$ mass flow rate of water and steam (total discharge) from wet-steam well (kg/s)

$M^{d,v}$ mass flow rate of vapor from wet-steam well (kg/s)

$M^{e,l}$ mass flow rate of boiled water retained in aquifer (kg/s)

$M^{e,v}$ mass flow rate of vapor added to flowing fluid at pressure P^e (kg/s)

$M^{f,t}$ mass flow rate of aquifer fluid (kg/s)

$m_i^{d,l}$ concentration of component i in liquid water of well discharge (moles/kg)

$m_i^{d,t}$ concentration of component i in total well discharge (moles/kg)

$m_i^{d,v}$ concentration of component i in vapor of well discharge (moles/kg)

$m_s^{d,t}$ concentration of gas species s in total well discharge (moles/kg)

$m_r^{e,l}$ concentration of component r in liquid water retained in aquifer (moles/kg)

$m_s^{e,v}$ concentration of gas species s in vapor added to flowing fluid at pressure P^e (taken to be zero)

$m_i^{f,l}$ concentration of component i in initial aquifer liquid (moles/kg)

$m_r^{f,l}$ concentration of non-volatile component r initial aquifer liquid (moles/kg)

$m_s^{f,l}$ concentration of gas species s in initial aquifer liquid (moles/kg)

$m_i^{f,t}$	concentration of component i in initial aquifer fluid (moles/kg)
$m_r^{f,t}$	concentration of non-volatile component r initial aquifer fluid (moles/kg)
$m_s^{f,t}$	concentration of gas species s in initial aquifer fluid (moles/kg)
$m_i^{f,v}$	concentration of component i in initial aquifer vapor (moles/kg)
$m_s^{f,v}$	concentration of gas species s in initial aquifer vapor (moles/kg)
P_c	critical pressure (bar)
P_h	hydrostatic pressure (bar)
P_{H_2O}	water vapor pressure (bar)
P_s	partial pressure of gas s (bar)
P^d	discharge vapor pressure (bar)
P^e	vapor pressure at which heat transfer from rock or phase segregation occurs (bar)
P_{tot}	total pressure (bar)
T_c	critical temperature (K)
T^e	intermediate zone temperature (°C)
T^f	aquifer temperature (°C)
Q^e	heat flow from surroundings to aquifer fluid (kJ/kg)
V_c	critical volume
$V^{e,l}$	$(M^{e,l} / M^{d,t})$ relative mass (to well discharge) of boiled (and degassed) water retained in aquifer
$V^{e,v}$	$(M^{e,v} / M^{d,t})$ relative mass (to well discharge) of vapor added to flowing fluid
$V^{f,t}$	$(M^{f,t} / M^{d,t})$ relative mass (to well discharge) of initial aquifer fluid that has boiled
x_s^l	mole fraction of gaseous species s in liquid (defined by Eqn. (19))
x_s^v	mole fraction of gaseous species s in vapor (defined by Eqn. (19))
$X^{d,v}$	vapor fraction in well discharge
$X^{e,v}$	vapor fraction in intermediate zone defined by Equation (A37)
$X^{f,v}$	initial (equilibrium) vapor fraction in aquifer relative to aquifer fluid
Z_c	critical compressibility factor
$Z^{e,v}$	vapor fraction in intermediate zone for Model 5 (Eqn. A49)
Z_s	compressibility factor, equal to PV/RT
γ_s	activity coefficient of gas s
Γ_s	fugacity coefficient of gas s
f_s	fugacity of gas s (bar)

REFERENCES

Aka FT, Kusakabe M, Nagao K, Tanyileke G (2001) Noble gas isotopic compositions and water/gas chemistry of soda springs from the islands of Bioko, São Tomé and Annobon, along with Cameroon Volcanic line, West Africa. Appl Geochem 16:323-338

Akaku K, Kasai K, Nakatsuka K, Uchida T (1997) The source of acidity in water discharged from high temperature geothermal reservoirs in Japan. Proc. 22nd Stanford Workshop on Geothermal Reservoir Engineering 427-434

Anderson G (2005) Thermodynamics of Natural Systems, 2nd ed. Cambridge University Press

Arehart GB, Christenson BW, Wood CP, Foland KA, Browne PRL (2002) Timing of volcanic, plutonic and geothermal activity at Ngatamariki, New Zealand. J Volc Geotherm Res 116:201-214

Ármannsson H (2003) CO$_2$ emission from geothermal plants. International Conference on Multiple Integrated Uses of Geothermal Resources, Reykjavík, Iceland, 56-62

Ármannsson H, Benjamínsson J, Jeffrey AWA (1989) Gas changes in the Krafla geothermal systems, Iceland. Chem Geol 76:175-196

Ármannsson H, Gíslason G, Hauksson T (1982) Magmatic gases in well fluids aid the mapping of flow pattern in a geothermal system. Geochim Cosmochim Acta 46:167-177

Arnórsson S (1985) The use of mixing models and chemical geothermometers for estimating underground temperatures in geothermal systems. J Volc Geotherm Res 23:299-335

Arnórsson S (1987) Gas chemistry of the Krísuvík geothermal field, Iceland, with special reference to evaluation of steam condensation in upflow zones. Jökull 37:31-47

Arnórsson S (1989) Deposition of calcium carbonate minerals from geothermal waters - theoretical considerations. Geothermics 18:33-40

Arnórsson S (1995a) Geothermal systems in Iceland: Structure and conceptual models. I. High-temperature areas. Geothermics 24:561-602

Arnórsson S (1995b) Geothermal Investigations of Geothermal Resources in Guatemala. International Atomic Energy Agency, report GUA/8/009-06

Arnórsson S (1997) Interpretation of chemical and isotopic data on fluids discharged from wells in the Momotombo geothermal field, Nicaragua. International Atomic Energy Agency, report NIC/8/008-04

Arnórsson S (ed) (2000) Isotopic and Geochemical Techniques in Geothermal Exploration, Development and Use: Sampling Methods, Data Handling, Interpretation. International Atomic Energy Agency, Vienna.

Arnórsson S (2003) Arsenic in surface and up to 90 °C ground waters in a basalt area, N-Iceland. Appl Geochem 18:1297-1312

Arnórsson S, Andrésdóttir A (1995) Processes controlling the distribution of B and Cl in natural waters in Iceland. Geochim Cosmochim Acta 59:4125-4146

Arnórsson S, Barnes I (1983) The nature of carbon dioxide waters in Snaefellsnes, western Iceland. Geothermics 12:171-176

Arnórsson S, Gunnlaugsson E (1985) New gas geothermometers for geothermal exploration - calibration and application. Geochim Cosmochim Acta 49:1307-1325

Arnórsson S, Óskarsson N (2007) Molybdenum and tungsten in volcanic rocks and in surface and <100°C ground waters in Iceland. Geochim Cosmochim Acta 71:284-304

Arnórsson S, Stefánsson A (2005) Wet-Steam well discharges: I. Sampling and calculation of total well discharge. World Geothermal Congress 2005, Antalya, Turkey, paper 0870

Arnórsson S, Sigurdsson S, Svavarsson H (1982) The chemistry of geothermal waters in Iceland. I. Calculation of aqueous speciation from 0° to 370°C. Geochim Cosmochim Acta 46:1513-1532

Arnórsson S, Gunnlaugsson E, Svavarsson H (1983) The chemistry of geothermal waters in Iceland. II. Mineral equilibria and independent variables controlling water compositions. Geochim Cosmochim Acta 47:547-566

Arnórsson S, Björnsson S, Muna ZW, Ojiambo SB (1990) The use of gas chemistry to evaluate boiling processes and initial steam fractions in geothermal reservoirs with an example from the Olkaria field, Kenya. Geothermics 19:497-514

Arnórsson S, Fridriksson, T, Gunnarsson I (1998) Gas chemistry of the Krafla geothermal field, Iceland. Intl Symp Water-Rock Interaction, Auckland, New Zealand, 613-616

Arnórsson S. Bjarnason JÖ, Giroud N, Gunnarsson I, Stefánsson A (2006) Sampling and analysis of geothermal fluids. Geofluids 6:1-14

Auippa A, Caleca A, Federico C, Gurrieri S, Valenza M (2004) Diffuse degassing of carbon dioxide at Somma-Vesuvius volcanic complex (Southern Italy) and its relation with regional tectonics. J Volc Geotherm Res 133:55-79

Ball JW, Nordstrom DK, Cunningham KN, Schoonen MAA, Xu Y, Demonge JM (1998) Water chemistry and on site sulfur-speciation data for selected springs in Yellowstone National Park, Wyoming, 1994-1995. US Geol Surv Open-File Report 98-574

Barelli A, Capetti G, Stefani G (1995) Results of deep drilling in the Larderello-Travale/Radicondoli geothermal area. World Geothermal Congress, Florence, Italy:1275-1278

Barnes I, Irwin WP, White DE (1978) Global distribution of carbon dioxide discharges, and major zones of seismicity. US Geol Surv Water-Resources Investigations Open File Report 78-39

Bibby HM, Caldwell TG, Davey FJ, Webb TH (1995) Geophysical evidence on the structure of the Taupo Volcanic zone and its hydrothermal circulation. J Volc Geotherm Res 68:29-58

Bischoff JL (1984) Correction. Earth Planet Sci Lett 69:224

Bischoff JL, Pitzer KS (1989) Liquid-vapor relations for the system NaCl-H$_2$O: Summary of the P-T-x surface from 300° to 500°C. Am J Sci 289:217-248

Bischoff JL, Rosenbauer RJ (1984) The critical point and two-phase boundary of seawater, 200-500°C. Earth Planet Sci Lett 68:172-180

Bischoff JL, Rosenbauer RJ (1985) An empirical equation of state for hydrothermal seawater (3.2 percent NaCl) Am J Sci 285:725-763

Bjarnason JÖ (1994) The speciation program WATCH version 2.1. National Energy Authority, Reykjavík, Iceland

Björnsson A, Saemundsson K, Einarsson P, Tryggvason E, Grönvold K (1977) Current rifting episode in north Iceland. Nature 266:318-323

Björnsson H (1966) Boiling temperatures in wells. National Energy Authority report, Reykjavík

Björnsson H, Björnsson S, Sigurgeirsson T (1982) Penetration of water into hot rock boundaries of magma at Grímsvötn. Nature 295:580-581

Bödvarsson G (1951) Report on the Hengill thermal area. Investigations carried out in the years 1947-1949. J Eng Ass Iceland 1:1-48 [In Icelandic with an English summary]

Bödvarsson G (1961) Physical characteristics of natural heat resources in Iceland. Jökull 11:29-38

Browne PRL (1979) Minimum age of the Kawerau geothermal field, north Island, New Zealand. J Volc Geotherm Res 6:213-215

Browne PRL, Ellis AJ (1970) The Ohaaki-Broadlands hydrothermal area, New Zealand: Mineralogy and related geochemistry. Am J Sci 269:97-131

Buat-Menard P, Arnold M (1978) Heavy-metal chemistry of atmospheric particulate matter emitted by Mount Etna volcano. Geophys Res Lett 5:245-248

Cerón JC, Martín-Vallejo M, García-Rossell L (2000) CO$_2$-rich thermomineral groundwater in the Betic Cordilleras, southeastern Spain: Genesis and tectonic implications. J Hydrogeol 8:209-217

Chelnokov G (2004) Interpretation of geothermal fluid compositions from Mendelev volcano, Kunashir, Russia. The United Nations University Geothermal Training Programme, report 2004:57-82 (ISBN 9979-68-65-9)

Chiodini G, Frondini F, Ponziani F (1995) Deep structures and carbon dioxide degassing in central Italy. Geothermics 24:81-94

Chiodini G, Caliro S, Cardellini C (2004) Origin of the carbon dissolved in the groundwater and derivation of Earth diffuse emission of CO$_2$: the case of the Italian Peninsula. Water-Rock Interaction. Balkema, p 21-29

Chiodini G, Granieri D, Avino R, Caliro S, Costa A, Werner C (2005) Carbon dioxide degassing and estimation of heat release from volcanic and hydrothermal systems. J Geophys Res SE 106 (B8):B08204

Christenson BW, Mroczek EK, Kennedy BM, van Soest MC, Stewart MK, Lyon G (2002) Ohaaki reservoir chemistry: characteristics of an arc-type hydrothermal system in the Taupo Volcanic Zone, New Zealand. J Volc Geotherm Res 115:5382

Clotworthy A (2000) Response of Wairakei geothermal reservoir to 40 years of production. World Geothermal Congress 2000, Kyushu – Tohoku, Japan: 2057-2062

D'Amore F, Celati R (1983) Methodology for calculating steam quality in geothermal reservoirs. Geothermics 12:129-140

D'Amore F, Panichi C (1980) Evaluation of deep temperatures in hydrothermal systems by a new gas geothermometer. Geochim Cosmochim Acta 44:309-332

D'Amore F, Truesdell AH (1985) Calculation of geothermal reservoir temperatures and steam fractions from gas compositions. Trans Geotherm Res Council 9:305-310

D'Amore F, Celati R, Ferrara GC, Panichi C (1977) Secondary changes in chemical and isotopic composition of the geothermal fluids in Larderello. Geothermics 5:153-163

D'Amore F, Daniele G, Abdallah A (1998) Geochemistry of the high-salinity geothermal field of Asal, Republic of Djibouti, Africa. Geothermics 27:197-210

Diakonov II, Schott J, Martin F, Harrichourry JC, Escalier J (1999) Iron (III) solubility and speciation in aqueous solutions. Experimental study and modeling: part 1. Hematite solubility from 60 to 300 °C in NaOH-NaCl solutions and thermodynamic properties of Fe(OH)$_{4(aq)}^-$. Geochim Cosmochim Acta 63:2247-2261

Driesner T, Geiger S (2007) Numerical simulation of multiphase fluid flow in hydrothermal systems. Rev Mineral Geochem 65:187-212

Druschel GK, Schoonen MAA, Nordstrom DK, Ball JW, Xu Y, Cohn CA (2003) Sulfur geochemistry of hydrothermal waters in Yellowstone National Park, Wyoming, USA. III. Ion exchange resin techniques for sampling and preservation of sulfoxyanions in natural waters. Geochem Trans 4:12-19

Ellis AJ, Mahon WAJ (1977) Chemistry and Geothermal Systems. Academic Press, New York

Fernandez-Prini R, Alvarez JL, Harvey AH (2003) Henry's constants and vapor-liquid distribution constants for gaseous solutes in H$_2$O and D$_2$O at high temperatures. J Phys Chem Ref Data 32:903-916

Fournier RO (1977) Chemical geothermometers and mixing models for geothermal systems. Geothermics 5:41-50

Fournier RO (1981) Application of water geochemistry to geothermal exploration and reservoir engineering. Geothermal Systems: Principles and Case Histories, Rybach L, Muffler LJP (eds) Wiley, p 109-143

Fournier RO (1985) Continental scientific drilling to investigate brine evolution and fluid circulation in active hydrothermal systems. Observation of the Continental Crust Through Drilling I. Raleigh CB (ed) Springer-Verlag, p 98-122

Fournier RO (1989) Geochemistry and dynamics of the Yellowstone National Park hydrothermal system. Ann Rev Earth Planet Sci 17:13-53

Fournier RO (1991) The transition from hydrostatic to greater than hydrostatic fluid pressure in presently active continental hydrothermal systems in crystalline rocks. Geophys Res Lett 18:955-958

Fournier RO (1999) Hydrothermal processes related to movement of fluid from plastic into brittle rock in the magmatic-epithermal environment. Econ Geol 94:1193-1211

Fournier RO, Pitt AM (1985) The Yellowstone magmatic-hydrothermal system. Trans Geotherm Council Int. Symp Geotherm Energy, p 319-327

Franck EU (1956) Hochverdichteler Wasserampf II. Ionendissociation von Kcl in H_2O bis 750 °C. Z Phys Chem 8:107-126

Franck EU (1961) Überkritisches Wasser als electrolytisches Lösungsmittel. Angew Chem 73:309-322

Fridriksson T, Kristjánsson BR, Ármannsson H, Margrétardóttir E, Ólafsdóttir S, Chiodini G (2006) CO2 emissions and heat flow through soil, fumaroles, and steam heated mud pools at the Reykjanes geothermal area, SW Iceland. Appl Geochem 21:1551-1569

Gherardi F, Panichi C, Yock FA, Gerardo-Abaja J (2002) Geochemistry of the surface and deep fluids at Miravalles volcano geothermal system (Costa Rica). Geothermics 31:91-128

Gianelli G, Ruggieri G (2002) Evidence of a contact metamorphic aureole with high-temperature metasomatism in the deepest part of the active geothermal field of Larderello, Italy. Geothermics 31:443-474

Giggenbach WF (1980) Geothermal gas equilibria. Geochim Cosmochim Acta 44:2021-2032

Giggenbach WF (1981) Geothermal mineral equilibria. Geochim Cosmochim Acta 45:393-410

Giggenbach WF (1988) Geothermal solute equilibria – derivation of Na-K-Mg-Ca geoindicators. Geochim Cosmochim Acta 52:2749-2765

Giggenbach WF (1991) Chemical techniques in geothermal exploration. In: Application of Geochemistry in Geothermal Reservoir Development. D'Amore F (ed). UNITAR/UNDP Centre on Small Energy Resources, Rome, p 119-144

Giggenbach WF, Corrales SR (1992) Isotopic and chemical composition of water and steam discharges from volcanic-magmatic-hydrothermal systems of the Guanacaste Geothermal Province, Costa Rica. Appl Geochem 7:309-332

Glover RB, Lovelock B, Ruaya JR (1981) A novel way of using gas and enthalpy. New Zealand Geothermal Workshop, Auckland, New Zealand:163-169

Goff F, Janik CJ (2000) Geothermal systems. Encyclopedia of Volcanoes. Sigurdsson H (ed) Pergamon Press, p 817-834

Grant MA (1977) Broadlands – A gas dominated geothermal field. Geothermics 6:9-29

Grindley GW (1965) The Geology, Structure and Exploitation of the Wairakei Geothermal Field, Taupo, New Zealand. New Zeal Geol Surv, DSIR, Bulletin 75

Gudmundsson BT, Arnórsson S (2002) Geochemical monitoring of the Krafla and Námafjall geothermal areas, N-Iceland. Geothermics 31:195-243

Gudmundsson BT, Arnórsson S (2005) Secondary mineral-fluid equilibria in the Krafla and Námafjall geothermal systems, Iceland. Appl Geochem 20:1607-1625

Gunnarsson I, Arnórsson S (2000) Amorphous silica solubility and the thermodynamic properties of aqueous silica in the range 0-350°C. Geochim Cosmochim Acta 64:2295-2307

Gunnarsson I, Arnórsson S (2005) Impact of silica scaling on the efficiency of heat extraction from high-temperature geothermal fluids. Geothermics 34: 320-329

Haizlip JR, Truesdell AH (1988) Hydrogen chloride in superheated steam and chloride in deep brine at The Geysers geothermal field, California. Thirteenth Workshop Geotherm Reserv Eng, Stanford University, California:93-99

Hedenquist JW (1990) The thermal and geochemical structure of the Broadlands-Ohaaki geothermal systems. Geothermics 19:151-185

Hedenquist JW, Lowenstern JB (1994) The role of magmas in the formation of hydrothermal ore deposits. Nature 370:519-527

Hedenquist JW, Aoki M, Shinohara H (1994) Flux of volatiles and ore-forming metals from the magmatic-hydrothermal system of Satsuma Iwojima volcano. Geology 22:585-588

Heinrich CA (2007) Fluid – fluid interactions in magmatic-hydrothermal ore formation. Rev Mineral Geochem 65:363-387

Heinrich CA, Driesner T, Stefánsson A, Seward TM (2004) Magmatic vapor contraction and the transport of gold from the porphyry environment to epithermal ore deposits. Geol Soc Am Bull 32:761-764

Helgeson HC (1968) Geologic and thermodynamic characteristics of the Salton Sea geothermal system. Am J Sci 266:129-166

Helgeson HC (1969) Thermodynamics of hydrothermal systems at elevated temperatures and pressures. Am J Sci 267:729-804

Helgeson HC, Kirkham DH, Flowers GC (1981) Theoretical prediction of the thermodynamic behavior of aqueous electrolytes at high pressures and temperatures. IV. Calculation of activity coefficients, osmotic coefficients, and apparent molal and standard and relative partial molal properties to 600 °C and 5 kb. Am J Sci 281:1249-1516

Henley RW, Truedell AH, Barton PR Jr (1984) Fluid Mineral Equilibria in Hydrothermal Systems. Reviews in Economic Geology 1

Hermoso DZ, Mejorada AV, Rae AJ (1998) Determination of the nature of acidic fluids in the Palinpinon geothermal field, Philippines through the use of $\delta^{34}S$ in sulfates and sulfides. 19th Annual PNOC-EDC Geothermal Conference:65-75

Hochstein MP, Browne PRL (2000) Surface manifestations of geothermal systems with volcanic heat sources. Encyclopedia of Volcanoes, Sigurdsson H (ed) Pergamon Press, p 835-855

Holland TJB, Powell R (1998) An internally consistent thermodynamic data set for phases of petrological interest. J Metamorphic Geol 16:309-343

Horne RN, Satik C, Mahiya G, Li K, Ambusso W, Tovar R, Wang C, Nassori H (2000) Steam-water relative permeability. World Geothermal Congress 2000, Kyushu - Tohoku, Japan, p 2609-2615

Hulston JR, Lupton JE (1996) Helium isotope studies of geothermal fields in the Taupo Volcanic Zone, New Zealand. J. Volc Geotherm Res 74:297-321

Inguaggiato S, Martin-Del Pozzo AL, Aguayo A, Capasso G, Favara R (2005) Isotopic, chemical and dissolved gas constraints on spring water from Popocatepetl volcano (Mexico): evidence of gas-water interaction between magmatic component and shallow fluids. J Volc Geotherm Res 141:91-108

Iyer HM, Evans JR, Zandt G, Stewart RM, Cookley JM, Roloff JN (1981) A deep low-velocity body under the Yellowstone caldera, Wyoming. Delineation using teleseismic P-wave residuals and tectonic interpretations. Geol Soc Am Bull 92:792-798

Johnson JW, Oelkers EH, Helgeson HC (1992) SUPCRT92: A software package for calculating the standard molar properties of minerals, gases, aqueous species and reactions among them from 1 to 5000 bars and 0 to 1000°C. Comp Geosci 18:899-947

Jónsson VK, Matthíasson M (1974) Cooling of the lava at Heimaey - Implementation J Eng Ass Iceland 36:70-82 [In Icelandic]

Kiyota Y, Matsuda K, Shimata K (1996) Characteristics of acid water in the Otake-Hatchubaru geothermal field. 17th Annual PNOC-EDC Geothermal Conference, p 131-135

Kissling WM, Weir GJ (2005) The distribution of the geothermal fields in the Taupo Volcanic Zone, New Zealand. World Geothermal Congress, Antalya, Turkey: paper 1919

Klein FW, Einarsson P, Wyss M (1977) The Reykjanes Peninsula, Iceland, earthquake swarm of September 1972 and its tectonic significance. J Geophys Res 82:865-888

Le Cloarec MF, Allard P, Ardouin B, Giggenbach WF, Sheppard BS (1992) Radioactive isotopes and trace elements in gaseous emissions from White Island, New Zealand. Earth Planet Sci Lett 108:19-28

Lepel EA, Stefansson KM, Zoller WH (1978) Enrichment of volatile elements in the atmosphere by volcanic activity - Augustine volcano. J Geophys Res 83:6213-6220

Li K, Horne RN (2004) Universal capillary pressure and relative permeability model from fractal characterization of rock. Twenty-Ninth Workshop on Geothermal Reservoir Engineering, Stanford University, California: SGP-TR-175

Liebscher A (2007) Experimental studies in model fluid systems. Rev Mineral Geochem 65:15-47

Liebscher A, Heinrich CA (2007) Fluid–fluid interactions in the Earth's lithosphere. Rev Mineral Geochem 65:1-13

Lister CRB (1983) The basic physics of water penetration into hot rock. Hydrothermal Processes at Seafloor Spreading Ridges. Rona PA, Bostrom K, Laubier L, Smith LK Jr (eds) Plenum Press, p 141-168

Mariner RH, Evans WC, Young HW (2006) Comparison of circulation times of thermal waters discharging from the Idaho batholith based on geothermometer temperatures, helium concentrations, and ^{14}C measurements. Geothermics 35:3-25

Mariner RH, Evans WC, Presser TS, White LD (2003) Excess nitrogen in selected thermal and mineral springs of the Cascade Range in northern California, Oregon, and Washington: sedimentary or volcanic origin. J Volc Geotherm Res 121:99-114

Marques JM, Andrade M, Carreira PM, Eggenkamp HGM, Graça RC, Aires-Barros L, Antunes Da Silva M (2006) Chemical and isotopic signatures of $Na/HCO_3/CO_2$-rich geofluids, North Portugal. Geofluids 6:273-287

Matsuda K, Shimada K, Kiyota Y (2005) Isotope techniques for clarifying origin of SO$_4$ type acid geothermal fluid – case studies of geothermal areas in Kyushu, Japan. Use of isotope techniques to trace the origin of acidic fluids in geothermal systems. International Atomic Energy Publication TECDOC-1448:83-95

Mazor E, Fournier, RO (1973) More on noble-gases in Yellowstone National Park hot waters. Geochim Cosmochim Acta 37:515-525

Mazor E, Truesdell AH (1984) Dynamics of a geothermal field traced by noble gases: Cerro Prieto, Mexico. Geothermics 13:91-102

Naumov GB, Ryzhenko BN, Khodakovsky IL (1971) Handbook of thermodynamic data (translation of Russian report) US Geol Surv Report WRD-74-001

Nehring NL, D'Amore F (1984) Gas chemistry and thermometry of the Cerro Prieto, Mexico, geothermal field. Geothermics 13:75-89

Nordstrom DK, Jenne EA (1977) Fluorite solubility equilibria in selected geothermal waters. Geochim Cosmochim Acta 41:175-188

Nordstrom DK, Munoz JL (1994) Geochemical Thermodynamics, 2nd ed. Blackwell Scientific Publications

Pokrovskii VA, Helgeson HC (1995) Thermodynamic properties of aqueous species and solubilities of minerals at high pressures and temperatures: the system Al$_2$O$_3$-H$_2$O-NaCl. Am J Sci 295:1255-1342

Poreda RJ, Craig H, Arnórsson S, Welhan JA (1992) Helium isotopes in Icelandic geothermal systems: 1. ^{3}He, gas chemistry, and ^{13}C relations. Geochim Cosmochim Acta 56:4221-4228

Pruess K (2002) Mathematical modeling of fluid flow and heat transfer in geothermal systems. United Nations University Geothermal Programme, Reykjavík, Iceland. Report to the Earth Science Division, Lawrence Berkeley National Laboratory, University of California

Reyes AG, Trompetter WJ, Britten K, Searle J (2002) Mineral deposits in the Rotokawa geothermal pipelines, New Zealand. J Volc Geotherm Res 119:215-239

Robie RA, Hemingway BS (1995) Thermodynamic properties of minerals and related substances at 298.15 K and 1 bar (10^5 Pascals) pressures and at higher temperatures. US Geol Surv Bull 2131

Rodgers KA, Cook KL, Browne PRL, Campell KA (2002) The mineralogy, texture and significance of silica derived from alteration by steam condensate in three Zealand geothermal fields. Clay Mineral 37:299-322

Ruaya JR, Seward TM (1987) The ion pair constant and other thermodynamic properties of HCl up to 350 °C. Geochim Cosmochim Acta 51:121-130

Salonga ND (1996) Fluid and mineral equilibria in acid NaCl (+SO$_4$) reservoir: The case of Sandawa Collapse, Mt. Apo hydrothermal system. 17th Annual PNOC-EDC Geothermal Conference, p 119-129

Sanchez DR, Arnórsson S (1995) Application of Cl, B tracers and geoindicators to delineate some production characteristics of Mt. Labo geothermal system, Philippines. World Geothermal Congress, Florence, Italy, p 1031-1035

Seward TM (1974) Equilibrium and oxidation potential in geothermal waters at Broadlands, New Zealand. Am J Sci 274:190-192

Shock EL, Sassani DC, Willis M, Sverjensky DA (1997) Inorganic species in geologic fluids: Correlations among standard molal thermodynamic properties of aqueous ions and hydroxide complexes. Geochim Cosmochim Acta 61: 907-950

Sigvaldason GE (1959) Mineralogische Untersuchungen über Gesteinszersetzung durch postvulkanische Aktivität in Island. Beiträge Mineral Petrog 6:405-426

Sigvaldason GE, Óskarsson N (1976) Chlorine in basalts from Iceland. Geochim Cosmochim Acta 40:777-789

Simmons SF, Browne PRL (2000) Hydrothermal minerals and precious metals in the Broadlands-Ohaaki geothermal system: Implications for understanding low-sulfidation epithermal environments. Econ Geol and Bull Soc Econ Geologists 95:971-999

Simonson JM, Palmer DA (1993) Liquid-vapor partitioning of HCl (aq) to 350°C. Geochim Cosmochim Acta 57:1-7

Snyder G, Poreda R, Fehn U, Hunt A (2003) Sources of nitrogen and methane in Central American geothermal setting: Noble gas and I-129 evidence for crustal and magmatic volatile components. Chem Geophys Geosystems 4:Art no 9001, Jan 14, 2003

Stefánsson A, Arnórsson S (2002) Gas pressures and redox reactions in geothermal fluids in Iceland. Chem Geol 190:251-271

Stefánsson A, Arnórsson S, Sveinbjörnsdóttir ÁE (2005) Redox reactions and potentials in natural waters at disequilibrium. Chem Geol 221:289-311

Sveinbjörnsdóttir ÁE, Arnórsson S, Heinemeier J (2001) Isotopic and chemical characteristics of old "ice-age" groundwater, North Iceland. Water-Rock Interaction. Balkema, p 205-208

Sverjensky DA, Shock EL, Helgeson HC (1997) Prediction of the thermodynamic properties of aqueous metal complexes to 1000 degrees C and 5 kb. Geochim Cosmochim Acta 61:1359-1412

Symonds RB, Reed MH, Rose WI (1992) Origin, speciation, and fluxes of trace element gases at Augustine volcano, Alaska – insights into magma degassing and fumarolic processes. Geochim Cosmochim Acta 56:633-657

Tello EH, Tovar RA, Verma MP (2005) Chemical and isotopic study to define the origin of acidity in the Los Humeros geothermal reservoir. Use of isotope techniques to trace the origin of acidic fluids in geothermal systems. International Atomic Energy Publication TECDOC-1448:97-110

Truesdell AH (1991) Origin of acid fluids in geothermal reservoirs. Geothermal Resources Council Trans 15:289-296

Truesdell AH, Fournier RO (1976) Conditions on the deeper parts of the Yellowstone National Park, Wyoming. US Geol Surv Open-File Report 76-428

Truesdell AH, Fournier RO (1977) Procedure for estimating the temperature of a hot water component in a mixed water using a plot of dissolved silica vs enthalpy. US Geol Surv J Res 5:49-52

Truesdell AH, Haizlip JR, Ármannsson H, D'Amore F (1989) Origin and transport of chloride in superheated steam. Geothermics 18:295-304

Xu Y, Schoonen MAA, Nordstrom DK, Cunningham KM, Ball JW (1998) Sulfur geochemistry of hydrothermal waters in Yellowstone National Park. I. The origin of thiosulfate in hot spring waters. Geochim Cosmochim Acta 62:3729-3743

Xu Y, Schoonen MAA, Nordstrom DK, Cunningham KM, Ball JW (2000) Sulfur geochemistry of hydrothermal waters in Yellowstone National Park. II. Formation and decomposition of thiosulfate and polythionate in cinder Pool. J Volc Geotherm Res 97:407-423

Welhan JA, Poreda RJ, Rison W, Craig H (1988) Helium isotopes in geothermal and volcanic gases of the western United States. I. Regional variability and magmatic origin. J Volc Geotherm Res 34:185-199

White DE (1957) Magmatic, connate and metamorphic waters. Geol Soc Am Bull 84:1659-1682

White DE (1965) Saline waters of sedimentary rocks in subsurface environments. Am Assoc Petrol Geol Memoir 4:342-366

White DE (1968) Environments of generation of some base metal ore deposits. Econ Geol 63:301-335

White DE, Muffler LJP, Truesdell AH (1971) Vapor-dominated hydrothermal systems compared with hot-water systems. Econ Geol 66:75-97

APPENDIX 1 –
DERIVATION OF EQUATIONS TO CALCULATE AQUIFER STEAM FRACTIONS AND FLUID COMPOSITIONS FOR WET-STEAM WELLS

In order to calculate the chemical composition of aquifer fluids from analyses of wellhead liquid and vapor samples, one must make some assumptions about the well-aquifer system. The models outlined in Figure 10 constitute five reasonable sets of such assumptions, and they may be represented by rather simple sets of equations. In particular, all five models may be described by the appropriate simplifications of Equations (A1), (A2) and (A3), along with Equations (A4) through (A7).

In what follows, the first superscript indicates the location, or point, in the model. Thus f refers to the feed zone beyond the zone of depressurization (initial aquifer fluid), d to the discharge at the wellhead, and e to the intermediate zone of evaporation (depressurization zone), where heat or vapor may be added and liquid lost. All quantities labeled d are assumed measured, and thus known. The temperature of the aquifer (T^f) is selected at the outset of the calculation. The pressure, or the temperature, of the intermediate two-phase zone is also selected.

The second superscript indicates the fluid phase. Thus, l means liquid, v denotes vapor, and t refers to the total flow, which, in general, is a two-phase mixture. Subscripts indicate components or species; s refers to gases, r to non-volatile components, and i denotes either.

There are three types of conservation equations: for water mass, water enthalpy and mole numbers of chemical components other than water. The enthalpy carried by these components is neglected. Thus we have

$$M^{d,t} = M^{f,t} + M^{e,v} - M^{e,l} \tag{A1}$$

$$h^{d,t} \cdot M^{d,t} = h^{f,t} \cdot M^{f,t} + h^{e,v} \cdot M^{e,v} - h^{e,l} \cdot M^{e,l} + Q^e \tag{A2}$$

$$m_i^{d,t} \cdot M^{d,t} = m_i^{f,t} \cdot M^{f,t} + m_i^{e,v} \cdot M^{e,v} - m_i^{e,l} \cdot M^{e,l} \tag{A3}$$

where M denotes the water mass flow rate, h the specific enthalpy, Q the rate of conductive heat flow from the aquifer rock to the fluid, and m the molal concentration of the chemical component in question.

The overall molal concentration of a component in a two-phase fluid is the sum of the concentrations in the separate phases, weighted by the respective mass fractions of the phases. In the aquifer fluid we thus have

$$m_i^{f,t} = m_i^{f,v} \cdot X^{f,v} + m_i^{f,l} \cdot (1 - X^{f,v})$$ (A4)

where the vapor phase mass fraction is given by

$$X^{f,v} = \frac{h^{f,t} - h^{f,l}}{L^f}$$ (A5)

and L denotes the latent enthalpy of vaporization.

A value for the mass fraction in the aquifer may also be obtained by assuming the concentration of a volatile species s in the liquid phase ($m_s^{f,l}$) to be fixed by equilibrium with a mineral buffer (see Table 7). Substitution of Equations (22) and (25) into (A4) gives

$$m_s^{f,t} = m_s^{f,l} \left[X^{f,v} \left(\frac{55.51}{P_{tot} K_s} - 1 \right) + 1 \right]$$ (A6)

which yields

$$X^{f,v} = \left[\frac{m_s^{f,t}}{m_s^{f,l}} - 1 \right] \cdot \left[\frac{55.51}{P_{tot} K_s} - 1 \right]^{-1}$$ (A7)

when solved for $X^{f,v}$.

We shall refer to the value of the aquifer vapor fraction obtained from Equation (A7) as the "equilibrium" vapor fraction. The vapor fraction value obtained from (A5) may be referred to as the "physical" vapor fraction. The latter can be evaluated independently only for one of the five models, namely Model 1, the isolated system model. Ideally, the two values should be identical. If they differ significantly, then either the model is not valid or the selection of $m_s^{f,l}$ is in error.

There is, of course, an equation equivalent to (A4) for the fluid discharged at the wellhead, namely

$$m_i^{d,t} = m_i^{d,v} \cdot X^{d,v} + m_i^{d,l} \cdot (1 - X^{d,v})$$ (A8)

The concentrations in the separate phases, $m_i^{d,v}$ and $m_i^{d,l}$, are measured, and the surface vapor fraction ($X^{d,v}$) at the known sampling pressure may be found from the measured value of the wellhead fluid enthalpy ($h^{d,t}$), thus

$$X^{d,v} = \frac{h^{d,t} - h^{d,l}}{L^d}$$ (A9)

Model 1: Isolated system

By this model there are no outputs of liquid water from, or inputs of vapor to, the fluid flowing into the well, and no conductive heat flow from the surroundings. Hence $M^{e,v} = 0$, $M^{e,l} = 0$ and $Q^e = 0$ so Equations (A1) through (A3) reduce to

$$M^{d,t} = M^{f,t}$$ (A10)

$$h^{d,t} \cdot M^{d,t} = h^{f,t} \cdot M^{f,t}$$ (A11)

$$m^{d,t} \cdot M^{d,t} = m_i^{f,t} \cdot M^{f,t}$$ (A12)

Thus,

$$h^{d,t} = h^{f,t} \tag{A13}$$

$$m_i^{d,t} = m_i^{f,t} \tag{A14}$$

Now, $X^{f,v}$ may be found from either (A5) or (A7) as explained above. Once $X^{f,v}$ has been found, the concentration ($m_s^{f,v}$) of gas s in the aquifer vapor phase can be computed as follows. For some gases, $m_s^{f,l}$ may be assumed fixed by equilibrium with a mineral buffer (see Table 7) and thus known at the aquifer temperature. For other gases, $m_s^{f,l}$ is found from Equation (A6). In either case, $m_s^{f,v}$ can now be calculated from (A4). The concentration of any non-volatile component r in the initial aquifer liquid may be found from Equation (A4) if we note that $m_r^{f,v} \cong 0$ so

$$m_r^{f,l} = \frac{m_r^{f,t}}{(1 - X^{f,v})} \tag{A15}$$

Model 2: Closed system; conductive heat flow to fluid

In this model, there is conductive heat flow from the surroundings to the fluid, but there is neither an inflow of vapor nor an outflow of liquid. Therefore, $M^{e,v} = 0$ and $M^{e,l} = 0$. Hence

$$M^{d,t} = M^{f,t} \tag{A16}$$

$$h^{d,t} \cdot M^{d,t} = h^{f,t} \cdot M^{f,t} + Q^e \tag{A17}$$

$$m_i^{d,t} \cdot M^{d,t} = m_i^{f,t} \cdot M^{f,t} \tag{A18}$$

so

$$h^{d,t} - h^{f,t} = \frac{Q^e}{M^{d,t}} \tag{A19}$$

$$m_i^{d,t} = m_i^{f,t} \tag{A20}$$

The equilibrium vapor fraction ($X^{f,v}$) can be calculated from Equation (A7). This requires that $m_s^{f,l}$ be known (see Table 7). Subsequently, the value of $h^{f,t}$ can be obtained by use of (A5), since $h^{f,l}$ and L^f are known (from Steam Tables), when the aquifer temperature value has been selected. Finally, since $h^{f,t}$ is now known, Q^e can be obtained from (A19) as $M^{d,t}$ and $h^{d,t}$ are measured quantities.

The right hand side of (A19) represents the heat added per unit mass of well discharge. This heat goes into evaporating water, and consequently the vapor fraction increases at every point downstream over and above that in Model 1. This increase amounts to

$$\Delta X^{e,v} = \frac{Q^e}{M^{d,t} \cdot L^e} \tag{A21}$$

in the intermediate zone, and to

$$\Delta X^{d,v} = \frac{Q^e}{M^{d,t} \cdot L^d} \tag{A22}$$

at the wellhead. The temperature of the intermediate zone (T^e) is selected at the outset of the calculations like T^f, so L^e is known.

The chemical concentrations in the initial aquifer fluid are found as for Model 1. Thus, the values of $m_s^{f,l}$ are calculated from Equation (A6), except for the one gas used to compute the equilibrium vapor fraction. The $m_s^{f,l}$ value for this gas is assumed to be known (Table 7). Values of $m_s^{f,v}$ are then computed from (A4) for all gases. The concentration of non-volatile component r in the initial aquifer liquid is calculated from (A15).

Model 3: Open system; liquid retained in formation

In this case, there is no additional steam flow to the initial aquifer fluid and no conductive heat flow either. Thus, $M^{e,v} = 0$ and $Q^e = 0$. A fraction of the boiled liquid ($M^{e,l}$) is retained in the formation, however, and this constitutes an "outflow" stream. The concentration of volatile components in this outflow is assumed to be negligible, so $m_s^{e,l} = 0$ for all gases s.

Equations (A1) through (A3) now reduce to

$$M^{d,t} = M^{f,t} - M^{e,l} \tag{A23}$$

$$h^{d,t} \cdot M^{d,t} = h^{f,t} \cdot M^{f,t} - h^{e,l} \cdot M^{e,l} \tag{A24}$$

gases:
$$m_s^{d,t} \cdot M^{d,t} = m_s^{f,t} \cdot M^{f,t} \tag{A25}$$

non-volatiles:
$$m_r^{d,t} \cdot M^{d,t} = m_r^{f,t} \cdot M^{f,t} - m_r^{e,l} \cdot M^{e,l} \tag{A26}$$

Inserting Equation (A6) into (A25) gives

$$\frac{M^{d,t}}{M^{f,t}} = \frac{m_s^{f,l}}{m_s^{d,t}} \left[X^{f,v} \left(\frac{55.51}{P_{tot} K_s} - 1 \right) + 1 \right] \tag{A27}$$

Now consider two gases, $s = 1$ and $s = 2$, whose concentrations in the initial aquifer water are fixed by equilibrium with mineral buffers chosen from Table 7. We have one equation of type (A27) for each gas:

$$\frac{M^{d,t}}{M^{f,t}} = \frac{m_1^{f,l}}{m_1^{d,t}} \left[X^{f,v} \left(\frac{55.51}{P_{tot} K_1} - 1 \right) + 1 \right] \tag{A28}$$

and

$$\frac{M^{d,t}}{M^{f,t}} = \frac{m_2^{f,l}}{m_2^{d,t}} \left[X^{f,v} \left(\frac{55.51}{P_{tot} K_2} - 1 \right) + 1 \right] \tag{A29}$$

Equating (A28) and (A29) and solving for $X^{f,v}$ yields

$$X^{f,v} = \frac{A_2 - A_1}{\dfrac{55.51}{P_{tot}} \left[\dfrac{A_1}{K_2} - \dfrac{A_2}{K_1} \right] + A_2 - A_1} \tag{A30}$$

where $A_1 = m_1^{d,t} / m_1^{f,l}$ and $A_2 = m_2^{d,t} / m_2^{f,l}$. Now, $m_1^{d,t}$ and $m_2^{d,t}$ are quantities measured at the wellhead, and $m_1^{f,l}$ and $m_2^{f,l}$ and K_1 and K_2 are known functions of temperature, evaluated at a chosen temperature, T^f, so (A30) gives $X^{f,v}$ directly. Since $M^{d,t}$ is measured, $M^{f,t}$ can now be obtained from either (A28) or (A29). Next, $h^{f,t}$ is computed from (A5). Also, $M^{e,l}$ can be obtained from (A23) using the newly derived value for $M^{f,t}$. Finally $h^{e,l}$ is found from (A25), and this gives T^e, the temperature of the intermediate zone.

Since $M^{f,t}$ is now known, $m_s^{f,t}$ may be calculated from Equation (A25) for gases other than those used to calculate $X^{f,v}$. Subsequently, $m_s^{f,l}$ can be derived from Equation (A6) and finally $m_s^{f,v}$ from (A4). Thus, we have

$$m_s^{f,t} = \frac{M^{d,t}}{M^{f,t}} \cdot m_s^{d,t} \tag{A31}$$

$$m_s^{f,l} = m_s^{f,t} \cdot \left[X^{f,v} \cdot \left(\frac{55.51}{P_{tot} K_s} - 1 \right) + 1 \right]^{-1} \tag{A32}$$

$$m_s^{f,v} = \frac{m_s^{f,t} - m_s^{f,l} \cdot (1 - X^{f,v})}{X^{f,v}} \tag{A33}$$

The concentration of a non-volatile component r in the initial aquifer fluid is related to its concentration in the initial aquifer water and the liquid retained in the aquifer, respectively, by

$$m_r^{f,t} = m_r^{f,l} \cdot (1 - X^{f,v}) \tag{A34}$$

$$m_r^{f,t} = m_r^{e,l} \cdot (1 - X^{e,v}) \tag{A35}$$

Inserting (A34) and (A35) into (A26) and rearranging yields

$$m_r^{f,l} = m_r^{d,t} \cdot M^{d,t} \cdot (1 - X^{f,v})^{-1} \cdot \left[M^{f,t} - \frac{M^{e,l}}{1 - X^{e,v}} \right]^{-1} \tag{A36}$$

where $X^{e,v}$ is the steam fraction in the intermediate zone, immediately after depressurization boiling of the initial aquifer fluid to pressure P^e, but before retention of liquid in the formation. Thus,

$$X^{e,v} = \frac{h^{f,t} - h^{e,l}}{L^e} \tag{A37}$$

Equations (A36) and (A37) yield directly the concentration of non-volatile component r in the initial aquifer liquid.

Model 4: Open system; liquid retained in formation; steam inflow

In this case, some liquid is retained in the formation, as in Model 3. Furthermore, there is assumed to be an addition of vapor to the fluid in the intermediate zone, an "inflow" stream. This steam forms by the boiling of immobile liquid in the formation. The concentration of gas in this vapor is assumed to be negligible, i.e., we take $m_s^{e,v}$ to be zero for all gases. This is a poor approximation at the onset of boiling, but improves as production from the field progresses and the boiling fraction increases. As in Model 3, we also take $m_s^{e,l} = 0$ for all gases. There is no conductive heat flow from the formation to the fluid so $Q^e = 0$. The governing equations for this model are thus

$$M^{d,t} = M^{f,t} + M^{e,v} - M^{e,l} \tag{A38}$$

$$h^{d,t} \cdot M^{d,t} = h^{f,t} \cdot M^{f,t} + h^{e,v} \cdot M^{e,v} - h^{e,l} \cdot M^{e,l} \tag{A39}$$

gases:
$$m_s^{d,t} \cdot M^{d,t} = m_s^{f,t} \cdot M^{f,t} \tag{A40}$$

non-volatiles:
$$m_r^{d,t} \cdot M^{d,t} = m_r^{f,t} \cdot M^{f,t} - m_r^{e,l} \cdot M^{e,l} \tag{A41}$$

Eliminating $M^{e,l}$ between Equations (A38) and (A39) and rearranging yields

$$M^{f,t} \cdot (h^{f,t} - h^{e,l}) = M^{d,t} \cdot (h^{d,t} - h^{e,l}) - M^{e,v} \cdot (h^{e,v} - h^{e,l}) \tag{A42}$$

The temperature of the intermediate zone (T^e), is assumed selected at the outset of the calculations so $h^{e,v}$ and $h^{e,l}$ are known. As for Model 3, we now consider two gases, $s = 1$ and $s = 2$, and obtain $X^{f,v}$ from Equation (A30). The value of $M^{f,t}$ can be obtained from (A28) or (A29) as before, and $h^{f,t}$ from (A5). Next, $M^{e,v}$ is found from Equation (A42), and finally $M^{e,l}$ from (A38).

The chemical component mass balance equations for Models 3 and 4 are the same. Hence Equations (A31) through (A33) give the concentrations of volatile components in the initial aquifer liquid and vapor, other than those used to retrieve $X^{f,v}$.

Addition of vapor at vapor pressure P^e to the fluid in the intermediate zone changes neither the mass flow rate of liquid in the zone nor the concentration of non-volatile component r in the liquid, $m_r^{e,l}$. Since the added vapor is at the vapor pressure of the fluid already in the zone, it merely increases the amount of vapor present there, but in no way affects the liquid phase. Hence, $m_r^{e,l}$ may be obtained from Equations (A36) and (A37) as for Model 3.

Model 5: Open system; liquid retained in formation; conductive heat flow to fluid

In this case, some liquid is assumed to be retained in the formation, as in Models 3 and 4. In contrast to Model 4, there is no addition of vapor to the fluid so $M^{e,v} = 0$. The fluid is, however, assumed to take up heat from the surroundings. As for Models 3 and 4, we take $m_s^{e,l} = 0$ for all gases. As for Models 2 and 4, T^e is selected at the outset of the calculations, so $h^{e,l}$ is known. By Model 5, Equations (A1) through (A3) become

$$M^{d,t} = M^{f,t} - M^{e,l} \tag{A43}$$

$$h^{d,t} \cdot M^{d,t} = h^{f,t} \cdot M^{f,t} - h^{e,l} \cdot M^{e,l} + Q^e \tag{A44}$$

gases:
$$m_s^{d,t} \cdot M^{d,t} = m_s^{f,t} \cdot M^{f,t} \tag{A45}$$

non-volatiles:
$$m_r^{d,t} \cdot M^{d,t} = m_r^{f,t} \cdot M^{f,t} - m_r^{e,l} \cdot M^{e,l} \tag{A46}$$

Again, as for Models 3 and 4, we consider two gases. The equilibrium vapor fraction, $X^{f,v}$, is obtained from Equation (A30) and subsequently $h^{f,t}$ is found from (A5). Since $M^{d,t}$ is a measured quantity, $M^{f,t}$ can be obtained from either (A28) or (A29), using the newly calculated value of $X^{f,v}$. Equation (A43) now yields $M^{e,l}$ directly. Finally Q^e is found from Equation (A44).

The additional vapor fraction, over and above that for Model 1, generated by heat transfer from the formation to the fluid is given by

$$\Delta X^{e,v} = \frac{Q^e + M^{e,l} \cdot (h^{f,t} - h^{e,l})}{M^{d,t} \cdot L^e} \tag{A47}$$

in the intermediate zone and by

$$\Delta X^{d,v} = \frac{Q^e + M^{e,l} \cdot (h^{f,t} - h^{e,l})}{M^{d,t} \cdot L^d} \tag{A48}$$

at the wellhead.

The values for $m_s^{f,l}$ and $m_s^{f,v}$, for gases other than those used to calculate $X^{f,v}$, are obtained from the same equations as for Models 3 and 4, namely (A31) through (A33). The calculation of the concentration ($m_r^{f,l}$) of a non-volatile component in the initial aquifer fluid, however, requires some additional considerations. The reason is that the statement of Model 5 allows for two separate cases, depending on the order in which the two processes in the intermediate zone take place.

On the one hand (case A), liquid may be retained in the aquifer first, followed by the conduction of heat from the formation to the flowing fluid. This case is in essence indistinguishable from Model 3, since the added heat just increases the steam fraction in the wellhead discharge. Since the wellhead mass flow rate ($M^{d,t}$) and chemical concentrations ($m_i^{d,t}$) in the total discharge are assumed measured, the calculation proceeds as for Model 3, and $m_r^{f,l}$ may be found using Equations (A36) and (A37).

On the other hand (case B), conductive heat transfer from the rock to the fluid may precede the retention of liquid in the formation. This is probably the more interesting of the two cases. The vapor fraction of the fluid after the heat transfer, but before phase segregation, is now given by

$$Z^{e,v} = \frac{h^{f,t} + \dfrac{Q^e}{M^{f,t}} - h^{e,l}}{L^e} \tag{A49}$$

and the concentration of a non-volatile component, r, in the initial aquifer liquid may be found from an equation analogous to (A36), namely

$$m_r^{f,l} = m_r^{d,t} \cdot M^{d,t} \cdot (1 - X^{f,v})^{-1} \cdot \left[M^{f,t} - \frac{M^{e,l}}{1 - Z^{e,v}} \right]^{-1} \tag{A50}$$

Reviews in Mineralogy & Geochemistry
Vol. 65, pp. 313-362, 2007
Copyright © Mineralogical Society of America

Fluid Immiscibility in Volcanic Environments

James D. Webster and Charles W. Mandeville

Department of Earth and Planetary Sciences
American Museum of Natural History
Central Park West at 79th Street
New York, New York 10024-5192 U.S.A.
jdw@amnh.org   cmandy@amnh.org

INTRODUCTION

Volatile components enter the atmosphere, oceans, and other surface waters through pre-, syn-, and post-eruptive volcanic processes that involve a variety of fluids. Aluminosilicate-poor fluids include aqueous or carbonic to sulfide-, sulfate-, chloride-, fluoride-, carbonate-, and phosphate-rich compositions in volcanic environments, but other more complex combinations of these constituents may be involved (Giggenbach 1977; Roedder 1984, 1992; Lowenstern 1995). Other, rare gases and dissolved constituents (e.g., H_2, N_2, He, Ar, H_3BO_3, Hg, CH_4 and other hydrocarbon compounds, metals, and metalloids) are also present, but in general they are not sufficiently concentrated to form their own phases or control bulk-fluid composition and volcanic processes. As a result of their diverse compositions, volcanic fluids ranging from vapor to liquid exhibit widely different densities and show strong distinctions in heat capacity, dielectric behavior, viscosity, and other chemical and physical characteristics (Geiger et al. 2006a). The density of a saline liquid in the system $NaCl-H_2O$ at 50 MPa and 800 °C, for example, is more than 7 times that of the coexisting aqueous vapor (e.g., 0.2 gm/cm^3) (Henley and McNabb 1978).

Multiple fluids move, mix, and/or unmix in magma chambers and volcanic conduits, in the root zone of fumaroles, and in the convective hydrothermal systems that are ubiquitous to volcanic environments. Hydrothermal systems are typically centered on magmatic intrusions and may include crater lakes at the top of eruptive conduits. Hydrothermal processes involving two or more fluid phases operate through a variety of geologically relevant, shallow-crustal pressure-temperature-composition conditions. Volcanic fluids occur at supersolidus to subsolidus conditions, meaning that the dense aluminosilicate melt that is common to volcanism may or may not be present. This chapter addresses the evidence for and occurrence of two or more low-density, aluminosilicate-poor fluids in volcanic and related environments. It also describes the role of multiple fluids in the chemical and physical processes that occur in these environments—even though many aspects of volcanic processes are still poorly understood.

BACKGROUND

Volcanoes are surface loci where heat and matter are transferred from the inner Earth and where the temperatures and pressures of magma, rocks, and fluids change rapidly over short distances. Fluid phase equilibria are strong functions of temperature and pressure; so, two-phase volcanic fluids are common given the broad temperature and pressure ranges that characterize volcanoes and their plumbing systems. The involvement of multiple fluids in volcanic processes is important because heat and mass transport in two-fluid systems can be

     DOI: 10.2138/rmg.2007.65.10

significantly different from those in one-fluid systems and because phase separation generates fluids with different compositions, densities, and mobilities as they flow through magma and rock (Duan et al. 2006).

Although two-phase fluids occur in volcanic environments, a set of coexisting fluids may not necessarily represent an immiscible and thermodynamically stable phase assemblage. An *assemblage of immiscible fluids* occurs only in systems containing several homogeneous phases that are in equilibrium (i.e., at the condition of lowest Gibbs free energy). If internal equilibrium applies to each phase, if all phases are in equilibrium with each other, and if all phases are at the same pressure and temperature, then the coexisting phases are considered immiscible (Pichavant et al. 1982; Prausnitz et al. 1999). Coexisting fluids at non-equilibrium conditions (e.g., processes involving fluid mixing, evaporation, boiling, condensation, and others) play a significant role in some volcanic and related hydrothermal processes and in open systems in particular. However, without the essential constraint of thermodynamic equilibrium for the system of interest, our understanding of such processes is limited and less useful.

Evidence for and the consequences of multiple fluids in volcanic systems have been observed and addressed through various methods. These include the sampling of volcanic and hydrothermal vapors and liquids (including crater lake waters); fluid and melt inclusion research; experimental studies; simulations and related theoretical works; and stable isotope investigations. We review the results of these efforts, but for practicality and geological relevance, we restrict this chapter to address fluid immiscibility in terrestrial volcanic and associated environments and we focus primarily on relatively low-density fluids with only minor discussion of aluminosilicate melts. We provide only brief discussions of vapor-liquid phase equilibria and of metal fractionation between vapor and liquid; readers are referred to Driesner and Geiger (2007), Gottschalk (2007), Hack et al. (2007), Heinrich (Fig. 7 of 2007a), Heinrich (2007b), and Liebscher (2007) for additional details.

Before proceeding, we must clarify some terminology. Fluids are physically mobile and commonly dominated by volatile components. We define fluids to include all non-crystalline, polycomponent phases in volcanic and associated magmatic systems. We do not consider, herein, silicate glasses that also are fluids by this definition. Relatively low-density phases are denoted vapor, and higher density phases are referred to as liquid. Low-density phases are commonly dominated by CO_2 and/or H_2O, and higher-density liquids are most commonly enriched in H_2O, electrolytes, and/or molten sulfides as well as carbonates. We refer to electrolyte-bearing fluids as saline; relatively dense fluids that are dominated by aluminosilicate constituents are addressed as melts. Fluid unmixing that involves separation of vapor from liquid at pressures and temperatures less than those of the critical point is referred to as boiling. Alternatively, the separation of liquid from a vapor phase is referred to as condensation.

The fluids of volcanic environments involve a variety of sources, and we discriminate between magmatic, hydrothermal, connate, and fumarolic fluids; seawater; and meteoric waters based primarily on their isotopic compositions and other geochemical characteristics. Magmatic (juvenile) fluids are those sourced exclusively from magma via exsolution from aluminosilicate melt with or without a contribution from the decomposition of volatile component-bearing minerals (e.g., hydrous phases, sulfides, sulfates, and halogen-bearing minerals). Hydrothermal fluids may represent mixtures of purely juvenile magmatic volatile components with meteoric waters (Di Liberto et al. 2002), metamorphic fluids (i.e., fluids derived from decarbonation and/or dehydration processes; Heinrich 2007b), seawater, and/or crater lake waters. The fluids of geothermal environments, developed for their energy potential, are a good example of this. Volcanic fluids may represent mixtures of magmatic, hydrothermal, and meteoric waters with potential contributions from seawater and/or air, and they are emitted from subaerial fumaroles, lava domes, and flows as well as at subaqueous and submarine hydrothermal vents.

OBSERVATIONAL, ANALYTICAL, AND THEORETICAL EVIDENCE FOR MULTIPLE FLUIDS IN VOLCANIC ENVIRONMENTS

Fluid immiscibility in model systems

We review the phase relations of model systems involving H, O, C, S, Cl ± alkali elements because they are the dominant components of most volcanic and hydrothermal fluids. Examples include: H_2O, H_2O-NaCl, H_2O-CO_2, and H_2O-CO_2-NaCl (plus other chloride salts). We begin with the most abundant magmatic volatile component, H_2O, even though its phase equilibria have little relevance to most volcanic fluids. Primary, pure-H_2O fluid inclusions (FI) are rare in crustal rocks because aqueous fluids readily dissolve other components due to the strong solvent properties of H_2O.

The vapor pressure (boiling point) curve for pure H_2O is a univariant line that separates the vapor- and liquid-only fields. The boiling point curve terminates at a critical point at 22.06 MPa and 373.9 °C (Johnson and Norton 1991). At pressure-temperature conditions above the critical point, vapor and liquid are indistinguishable in chemical and physical properties, and both phases merge to form a single supercritical fluid (Nehlig 1991; Diamond 2003).

The model binary system H_2O-NaCl includes a field of partial miscibility involving vapor and saline liquid over a broad range of pressure-temperature conditions relevant to volcanic environments (Sourirajan and Kennedy 1962; Bodnar et al. 1985; Chou 1987; Bodnar 2003; Heinrich 2005). Critical phenomena involving H_2O-NaCl shift to higher pressures and temperatures with addition of NaCl (Fig. 1a), and this system exhibits a continuous critical curve that links the critical pressure and temperature of H_2O with that of NaCl (Sourirajan and Kennedy 1962). The fluids show a strong tendency to unmix at conditions relevant to volcanic environments because the activity coefficients for H_2O and NaCl exceed unity at shallow crustal pressures and magmatic to sub-magmatic temperatures (Anderko and Pitzer 1993; Botcharnikov et al. 2004).

The phase equilibria of H_2O-NaCl can be applied to those of hot seawater. The boiling point for a solution of H_2O and 3.2 wt% $NaCl_{equiv}$ (i.e., the normality-based equivalent NaCl content of seawater) shifts to 29.85 MPa and 407° relative to those of pure H_2O (Bischoff and Pitzer 1985). Fluid boiling can generate quite NaCl-enriched liquids from a single parent fluid that initially contains very little chloride salt. Seawater at 375-425° C and 23-30 MPa, which is equivalent to depths of seafloor volcanism at approximately 2.2 to 2.6 km, separates into vapor containing less than 1 wt% $NaCl_{equiv}$ and saline liquid with as much as 37 wt% $NaCl_{equiv}$ (Bischoff and Pitzer 1985).

Component solubilities and fluid phase equilibria in H_2O-electrolyte systems have also been determined for chloride salts of K, Mg, and Ca. The influence of KCl on equilibrium phase relations in H_2O-NaCl is comparably minor at temperatures up to 900 °C (Chou 1987; Sterner et al. 1988, 1992; Chou et al. 1992). In contrast, critical pressures for $CaCl_2$-H_2O fluids are shifted to slightly greater values than that observed with H_2O-NaCl at equivalent temperature (Bischoff et al. 1996; Duan et al. 2006). The pressure-temperature maxima for vapor plus liquid are at approximately 140 MPa for 600 °C and 180 MPa at 700 °C with ≤ 22 wt% $CaCl_2$ in the system $CaCl_2$-H_2O. The maxima are at approximately 80 MPa for 500 °C and 120 MPa at 600° C with 18 wt% $MgCl_2$ (equivalent to a mole fraction of $MgCl_2$ of ca. 0.04) in the system $MgCl_2$-H_2O. Thus, fluids exsolved from mafic magmas, which contain more Ca and Mg than felsic magmas, are subject to a large pressure-temperature field with vapor plus liquid relative to that of fluids exsolved from felsic magma because of the influence of alkaline earth chlorides on the H_2O-NaCl binary.

Equilibria involving H_2O and alkaline earth chlorides generate HCl. Bischoff et al. (1996) detected as much as 0.1 moles HCl/kg of fluid, due to hydrolytic decomposition of $CaCl_2$:

$$CaCl_2 + 2\,H_2O = Ca(OH)_2 + 2\,HCl \qquad (1)$$

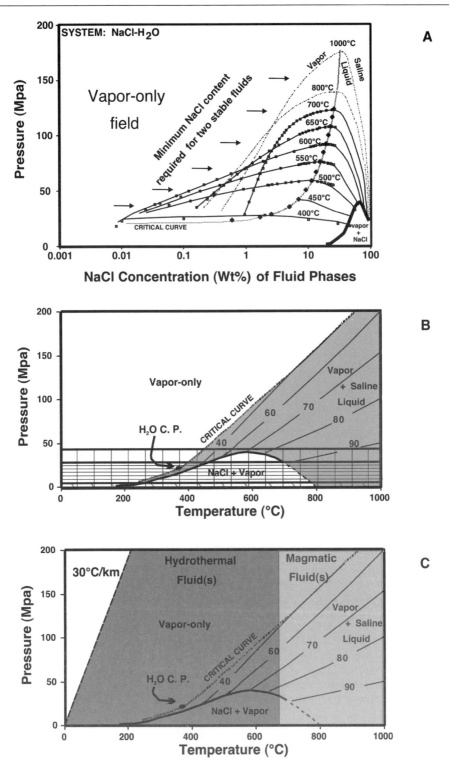

Trace quantities of HCl also form in NaCl-H$_2$O fluids:

$$H_2O + NaCl = NaOH + HCl \tag{2}$$

but hydrolysis is less significant than in the CaCl$_2$-bearing system. Hydrolysis reactions increasingly favor the generation of HCl with decreasing pressure (Fournier 1987); so, the HCl content of aqueous, chloride vapors increases as they ascend toward the surface. These reactions are a primary cause for the acidity of volcanic gases (Shmulovich and Churakov 1998).

Immiscibility involving H$_2$O-CO$_2$ is unlikely at magmatic conditions because critical phenomena are restricted to temperatures ≤ 400° C (Takenouchi and Kennedy 1965; Sterner and Bodnar 1991; Diamond 2003). Thus, H$_2$O-CO$_2$ dominated fluid unmixing may occur only at shallow crustal depths and sub-magmatic temperatures with electrolyte-deficient volcanic fluids (Lowenstern 2001; Diamond 2003). The addition of CH$_4$ to H$_2$O-CO$_2$ increases the size of the two-phase field in pressure-composition space (Duan et al. 1992) marginally.

The addition of NaCl to H$_2$O-CO$_2$, however, greatly expands the vapor-liquid field at magmatic pressures and temperatures (Bowers and Helgeson 1983a,b; Frantz et al. 1992; Joyce and Holloway 1993; Duan et al. 1995, 2003; Shmulovich et al. 1995; Schmidt and Bodnar 2000; Shmulovich and Graham 2004; Anovitz et al. 2004; Heinrich 2007b) because the mutual solubilities of CO$_2$ and NaCl—in the system H$_2$O-NaCl-CO$_2$—are so low that the miscibility gap is large relative to those of any two of the relevant binary systems. Joyce and Holloway (1993) observed that the system H$_2$O-NaCl-CO$_2$ shows evidence of fluid unmixing at 700° C and 500 MPa.

Intensive properties of fluids in volcanic environments

The pressures and temperatures of volcanic fluids vary widely, and the phase relations of fluids depend strongly on these intensive variables. Thus, it is useful to review the intensive properties of fluids in subaerial, subaqueous, and submarine volcanic environments.

The measured temperatures of volcanic vapors as well as thermochemically restored, equilibrium vapor temperatures (i.e., those that have been calculated based on equilibria involving volatile components that are assumed to bear on the system under investigation) can be quite

---

Figure 1 (*on facing page*). Interpretation of the stability of one or two fluids in volcanic and magmatic environments based on phase equilibria of the system NaCl-H$_2$O at varying pressure, temperature, and composition (after Sourirajan and Kennedy 1962; Bodnar et al. 1985). The finely dotted curve designates the NaCl-H$_2$O critical curve, and the bold solid curve shows limits of the NaCl plus vapor stability field (the curve represents coexistence of NaCl, vapor, and saline liquid). Critical point of pure H$_2$O = H$_2$O C.P. (a) Curves (solid and dot-dash) labeled for temperature represent isothermal vapor-liquid solvi. For stable vapor and saline liquid, the NaCl concentration of the vapor must exceed that of the low-NaCl content side of the isothermal solvi (highlighted by horizontal arrows). (b) The rectangular areas express general pressure-temperature ranges for volcanic fluids emitted during degassing at vents of crater lakes (shown with fine diagonal lines), submarine volcanism at ocean floor (as fine vertical lines), and subaerial volcanism (delineated with horizontal hachure marks). Pressures in these environments are either hydrostatic or vaporstatic. Stability fields are also shown for vapor, NaCl plus vapor, and vapor plus saline liquid (light gray field). Labels (40-90) show solubility of NaCl (wt%) in the saline liquid that coexists with vapor. Some volatile components supplied to vent environments are derived from more deeply sourced magmatic fluid(s), and some magmatic fluids undergo vapor-liquid separation well before reaching vents at Earth's surface, in crater lakes, or on the ocean floor. Thus, (c) displays potential pressure-temperature ranges for magmatic vapor with or without saline liquid. The lighter gray area (labeled magmatic fluid(s)) expresses temperatures at which silicate melt and fluid(s) are stable in magmas at ≤ 200 MPa; the temperature range extends from the solidus temperature of water-saturated rhyolitic melt (ca. 680 °C) to basaltic melt temperatures exceeding 1200 °C. The darker gray area expresses fluid(s) stability at submagmatic temperatures. Magmatic fluids may ascend toward surface without the source magma; thus, hydrothermal fluids released from magma can pass through all pressure-temperature space to right of the 30 °C/km geotherm (dashed line) during decompression and cooling. Vapor and saline liquid are stable at temperatures ≥ those of the NaCl-H$_2$O critical curve. See text for discussion.

high (Fig. 1, Table 1). Temperatures of 1185 °C have been recorded, but they can also range down to ambient surface values if significant mixing with air occurs. Temperatures of submarine hydrothermal vent fluids are generally less than 400 °C, but volcanism in ocean basins also involves the extrusion of basaltic lava into seawater so the temperatures of seafloor hydrothermal fluids could, theoretically, exceed 1200 °C. Volcanic lake water temperatures at equilibrium to non-equilibrium steady-state conditions are constrained by the input of heat, the pressure at the base of the lake, and the pressure-temperature boiling curve for a fluid with the composition of the crater lake. Many subaqueous volcanic crater lakes contain liquid sulfur, so their water temperatures are ≥ 119 °C (J. Varekamp, personal communication). Theoretically, however, maximum temperatures could be much higher. Basaltic lava can erupt into crater lakes, so the temperatures of heated lake waters could exceed 1200 °C. The interaction of lava with cold seawater or crater lake water, however, is likely to involve disequilibrium processes because of the huge difference in temperature of the water versus the lava.

The pressure of subaerial degassing is typically near one atmosphere (ca. 0.1 MPa). Interestingly, Giggenbach (1987) and Shmulovich and Churakov (1998) observed that subsurface magmatic degassing and hydrothermal activity can occur at low, near-surface pressures if the magma chambers, magma conduits, or hydrothermal systems are connected to the surface. Stevenson (1993) modeled the flow of magmatic vapor to the surface for a variety of conduits. His calculated pressures for the base of the conduits for Poás volcano and Vulcano, for example, were only a few times atmospheric pressure. Stevenson (1993) suggested that vapor expansion from lithostatic to atmospheric pressure may occur episodically and suddenly as fluids escape from magma. Thus, the pressures of vapor escape from some magmas may be quite low, even for systems that begin magmatic degassing at depths as great as 1.5 km (e.g., at Vulcano in 1983-1987). This is consistent with the observations that high strain rates associated with rapid hydrothermal fluid-release events lead to brittle fracturing of wallrock, and that open fractures cause rapid decompression, from lithostatic to hydrostatic pressure, in some shallow high-temperature volcanic environments (Fournier 1999).

One means of constraining an upper pressure limit for subaerial degassing involves the H_2O concentrations of volcanic matrix glasses. Matrix glasses represent magma that quenched at or

Table 1. Temperatures (°C) of vapors recorded at volcanoes of typical subduction-, hot spot-, and rift-related tectonic environments.

Volcano	Tectonic Setting	Rock Composition	Vapor Temperature	Data Sources
Kilauea, Hawaii	Hot spot	Tholeiitic basalt	1085-1185	[1] [2] [3]
Erta' Ale, Ethiopia	Rift	Alkali basalt	1130	[4]
Galeras, Columbia	Subduction	Basalt	1100	[4]
Poas, Costa Rica	Subduction	Tholeiitic basalt	960-1045	[5]
Etna, Italy	Subduction	Alkali basalt	928	[6]
Kudryavy volcano, Russia	Subduction	Basaltic andesite	920	[7] [8]
Merapi, Indonesia	Subduction	Andesite	900	[9] [10]
Augustine volcano, USA	Subduction	Andesite-dacite	870	[11]
Momotombo, Nicaragua	Subduction	Tholeiitic basalt	844	[12] [4]
Showa-Shinzan, Japan	Subduction	Dacite	800	[4]
Mt. St. Helens, USA	Subduction	Dacite	650	[13]
Vulcano, Italy	Subduction	Basalt-rhyolite	620	[15]

Data Sources: [1] Gerlach (1980); [2] Crowe et al. (1987); [3] Delmelle and Stix (2000); [4] Giggenbach (1996); [5] Barquero (1983); [6] LeGuern (1988); [7] Khorzinskiy et al. (1996); [8] Wahrenberger et al. (2002); [9] Allard (1983); [10] Bernard (1985); [11] Symonds et al. (1990, 1992); [12] Quisefit et al. (1989); [13] Symonds et al. (1994); [14] Gerlach and Casadevall (1986); [15] Wahrenberger (1997)

just below the surface, and such shallowly emplaced bodies of magma may erupt directly to the atmosphere. The maximum H_2O contents of silicate melt and quenched matrix glass are a function of the maximum vapor pressure of H_2O in the magma. Pallister et al. (2005), for example, collected glassy fragments of dacitic lava at Mt. St. Helens, and some of these fragments contain nearly 1.9 wt% H_2O. This H_2O content will cause water saturation in a dacitic melt at ca. 30 MPa. Thus, for the purpose of discussion we use 30 MPa as an upper pressure limit for subaerial degassing. We also note that it is a somewhat arbitrary value because highly explosive eruptions are likely to involve significantly greater pressures of degassing at and near the surface.

The hydrostatic pressures of submarine volcanic environments range from 1-45 MPa (Butterfield 2000). In contrast, degassing in the subaqueous environments of volcanic crater lakes generally involves pressures of 0.1-1 MPa because most lakes are not much deeper than 50-100 m. Hydrostatic pressure is ca. 1 MPa at 100 m depth in freshwater. However, the pressure at the base of the world's deepest volcanic crater lake (Crater Lake, Mt. Mazama, Oregon) is 6 MPa, so we use this estimate for the maximum pressure of subaqueous volcanic degassing.

With increasing depth, subaerial, subaqueous, and submarine volcanic environments grade from shallow to deep hydrothermal systems in and around plutons and their plumbing systems (Figs. 2 and 3). The geochemistry of the plutons is important because they provide

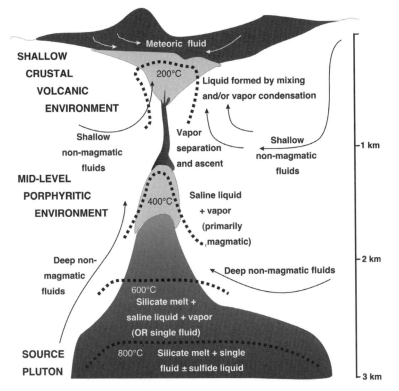

Figure 2. Schematic cross-section diagram (after Heinrich 2005) showing the approximate depths of isotherms (dashed curves) and the types of fluids that are stable in shallow volcanic and underlying plutonic environments. Generalized stability regions of silicate melt + single aqueous fluid ± sulfide liquid, silicate melt plus saline liquid plus vapor, saline liquid plus vapor, vapor, liquid, and non-magmatic fluids. The associated flow paths for these fluids are shown for generalized temperatures ranging from > 800° to < 200 °C. See text for discussion.

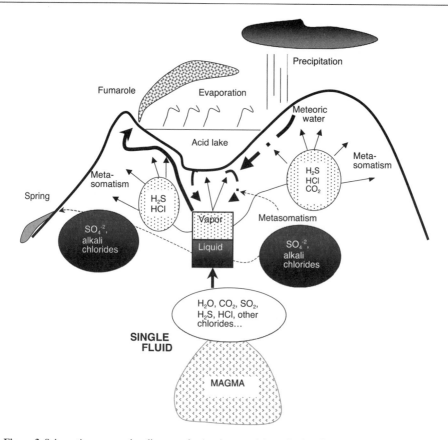

Figure 3. Schematic cross-section diagram of volcanic crater-lake and subsurface magmatic, hydrothermal, and metasomatic environments involving one or two fluids (after Delmelle et al. 2000). The region of immiscible separation of vapor (dotted box) and liquid (gray box) occurs above magma body and beneath crater lake. The flow path for a single magmatic fluid, involving immiscible separation of vapor, feeding fumarolic activity at surface shown as bold solid curve, and the flow paths for meteoric and crater lake waters shown with bold dot-dash curves. Flow paths for magmatic liquid, after vapor-liquid separation, shown with finely dotted curves; liquid contains alkali chloride ions and sulfate ions formed by reaction of SO₂ with H₂O. Liquid may feed hydrothermal springs, metasomatically alter the host rocks, or it may also mix with vapor and/or condense into meteoric water that subsequently enters lake and reacts with lake water. Flow paths for magmatic vapor after fluid separation shown as fine solid curves. The vapor is enriched in acid-forming sulfide, carbonate, and chloride species relative to liquid. Vapor may alter host rocks and/or mix with meteoric water and react with crater lake water. See text for detailed discussion.

volatile components to the fluids of overlying volcanoes. In fact, magmatic volatile component exsolution can begin at pressures as great as 500 MPa (Wallace 2005), so in the discussion that follows we address the consequences of vapor-liquid separation occurring in a deeply sourced, magmatic fluid as it ascends toward the surface.

Magmatic fluids also show broad compositional variability (Table 2). Carbon dioxide and sulfur (largely as H_2S at f_{O_2} values < QFM and SO_2 at higher f_{O_2}) are the two most important components of fumarolic and magmatic fluids after H_2O (Giggenbach 1977, 1997; Gerlach 1980; Stix and Gaonac'h 2000). Additional essential components include HCl and other chloride species (e.g., NaCl, KCl, and others), sulfur species other than H_2S and SO_2, and HF and associated fluoride species (e.g., SiF_4 and others) (Symonds et al. 1992;

Table 2. Chemical compositions (mol%) of volcanic vapors.

Volcano	Augustine Alaska	Augustine Alaska	White-Island New Zealand	Galeras Colombia	Kudryavy Russia	Kilauea Hawaii	Kilauea Hawaii	Erta'Ale Ethiopia	Surtsey Iceland
Tectonic Setting	Subduct	Subduct	Subduct	Subduct	Subduct	Hot Spt	Hot Spt	Rift	Rift
Date	8/28/87	7/6/89	2/7/72	1/6/90	8/7/95	3/25/18	1/14/83	1/23/74	10/15/64
Temp. °C	870	752	620	187	920	1170	1010	1130	1125
H_2O	84.77	96.94	91.1	94.3	95.3	37.09	79.8	77.24	81.13
CO_2	2.27	1.45	4.948	2.223	1.61	48.9	3.15	11.26	9.29
SO_2	6.18	0.25	1.818	NR	1.06	11.8	14.99	8.34	4.12
H_2S	0.68	0.34	0.719	2.303	0.21	0.04	0.622	0.68	0.89
HCl	5.34	0.45	1.157	0.718	0.2	0.08	0.1	0.42	NR
HF	0.086	0.03	0.041	NR	0.096	NR	0.19	0	NR
H_2	0.54	0.54	0.0925	0.0029	0.576	0.49	0.9025	1.39	2.8
Total	99.87	100	99.88	99.55	99.052	98.4	99.755	99.33	98
$\log f_{O_2}$	−12.45	−14.99	NR	NR	NR	−8.38	−10.49	−9.16	−9.8

Data from summaries of Symonds et al. (1994) and Delmalle and Stix (2000).
Tectonic setting: subduct = subduction zone/convergent margin, hot spt = hot spot, rift = spreading center/divergent margin.
NR = not reported.

Oppenheimer 2003). Other, less common gaseous species and dissolved constituents (e.g., H_2, N_2, He, Ar, Rn, H_3BO_3, Hg, CH_4 and associated hydrocarbon compounds, COS, CO, metals, and metalloids) are also present (Giggenbach 1997; Williams-Jones and Rymer 2000; Taran and Giggenbach 2003; Elkins et al. 2006; Oppenheimer et al. 2006).

Magma volumes vary with tectonic setting, and the abundances of volatile components in volcanic environments vary substantially with tectonic setting and magma composition. On average, spreading center volcanism on the seafloor involves ca. 21 km^3 of magma per year (Fisher and Schmincke 1984). This represents about 62% of the magma flux in the crust. Subduction-related volcanism involves ca. 26% and intraplate hot-spot volcanism comprises ca. 12% of the crustal magma flux. The ratio of intrusive to extrusive magma, by volume, is about 9:1 in these settings (Fisher and Schmincke 1984). Although not volumetrically dominant, subduction-related volcanism typically discharges the most volatile component-enriched magmas, so the behavior of fluids in these environments is particularly important. Water comprises > 90 mol% of most volcanic vapors, followed by CO_2 (up to 10 mol%), SO_2 (up to 6 mol%), and HCl (up to 6 mol%) in subduction-related systems (Giggenbach 1987, 1996; Symonds et al. 1994; Delmelle and Stix 2000; Williams-Jones and Heinrich 2005).

This review considers fluid immiscibility in surface and near-surface volcanic environments, and subsequently probes the nature of two fluids at progressively greater depths (Figs. 2 and 3). Although volcanic fluids involve other components, we apply observations from model systems involving NaCl+H_2O±CO_2 to interpret fluid immiscibility. We emphasize these systems primarily because of the strong influence of chloride-bearing salts on vapor-liquid equilibria and critical phenomena because these components are sufficiently abundant to influence volcanic and magmatic processes because they control the solvent behavior of most crustal fluids (Heinrich 2007a), and because more is known about these fluids (Bodnar 2003).

In summary, the low pressures (i.e., less than 50 MPa) and moderate to high temperatures of fluids in subaerial, subaqueous, and submarine vent environments coincide with the vapor and saline liquid field of the NaCl-H_2O binary (Fig. 1b); so, two fluids may occur at volcanic vents at Earth's surface or on the ocean floor or the floors of volcanic lakes. The coexistence of two fluids at magmatic temperatures in these environments may be transitory, involve open-system behavior, and not involve equilibrium, however because it may occur during rapid and violent eruptive behavior. Moreover, the potentially higher pressures and the broad temperature range of plutons and magmatic conduits (Fig. 1a,c) overlap those of the vapor plus saline liquid field for pressures approaching 170 MPa. Thus, fluid immiscibility may begin well before magmatically sourced volcanic fluids reach the surface.

Evidence from volcanic fluids. Information on processes involving immiscible fluids comes from samples of fumarolic vapors and condensed liquids at volcanic vents (Symonds et al. 1994; Giggenbach 1997; Shmulovich and Churakov 1998). Liquid aerosol particles can occur as suspensions in some vapors (Varekamp et al. 1986; Amman et al. 1993), but it is not known if the vapor and liquid were in equilibrium prior to their arrival at the surface. The phases may simply reflect near-surface, liquid saturation of volcanic vapor at disequilibrium conditions.

Saline liquids occur in many volcanic environments and associated magmatic systems (Roedder and Coombs 1967; Solovova et al. 1991; Frezzotti 1992; Roedder 1992; Lowenstern 1994; Webster and Rebbert 1998; Kamenetsky et al. 1999, 2004; Davidson et al. 2005; Kamenetsky 2006; Davidson and Kamenetsky 2007). Chloride- and sulfate-enriched, saline liquids have been observed in the products of deep geothermal wells drilled into active volcanic terrains at Kakkonda, Japan (Kasai et al. 1998); Mofete, Italy (Guglielminetti 1986); and Nisyros, Greece (Chiodini et al. 1993). Evidence of saline liquids has also been observed in FI of some magmatic systems (Cloke and Kessler 1979; Lowenstern 1994). Saline liquids may reflect: (1) exsolution of a chloride- and/or sulfate-bearing fluid from magma, followed by ascent, decompression, and intersection of the fluid with the two-fluid solvus that results in condensation of liquid (Fig. 4) (Roedder 1992; Taran 1992; Hedenquist 1995; Taran et al. 1997), or (2) direct exsolution of relatively saline and dense liquid from magma (Shinohara et al. 1989; Frezzotti 1992; Cline and Bodnar 1994; Lowenstern 1994; Shinohara 1994; Campbell 1995; Candela and Piccoli 1995; Reyf 1997; Veksler and Thomas 2002; Webster and De Vivo 2002; Heinrich et al. 2004; Veksler 2004; Webster 2004). Saline liquids also form by (3) boiling of seawater that is caused by interaction with magma and/or hot volcanic rocks in seafloor environments (Fig. 5) (Fournier 1987; Bischoff 1991; Chiodini et al. 1993), or (4) absorption of HCl- and SO_2-enriched magmatic gases in groundwater followed by neutralization and alkali uptake of these acidic fluids through reactions with wall rocks (Fournier 1987; Giggenbach 1988; Chiodini et al. 1993; Reed 1992a, 1997). Other, less-likely mechanisms include (5) deep circulation of groundwaters causing the dissolution of chloride- and sulfate-dominated sublimates that were generated during previous 'hot' magmatic cycles (Giggenbach 1975; Chiodini et al. 2001), and (6) the interaction of magma with evaporite deposits and resultant dissolution of chloride and/or sulfate by magmatic fluid (Calzia and Hiss 1978; Roedder 1984, 1992).

Extensive boiling of $H_2O \pm CO_2$-dominated fluids followed by efficient physical separation of vapor from liquids in open magmatic and volcanic environments generates strongly anhydrous and hypersaline residual liquids (or salt melts) (Henley and McNabb 1978; Fournier 1987; Shmulovich and Churakov 1998). Progressive fluid ascent leads to continuous vapor-liquid separation because the solvi of aqueous-alkali chloride and/or CO_2-bearing

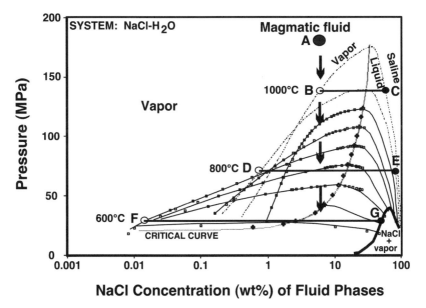

NaCl Concentration (wt%) of Fluid Phases

Figure 4. Figure showing the phase changes that occur during the ascent of a magmatic fluid to a surface vent, by comparing the ascent path with phase equilibria of the binary system NaCl-H$_2$O (curves and fields same as in Fig. 1a). With cooling and decompression, magmatic fluid (A) intersects the 1000 °C solvus at ca. 140 MPa, and the fluid separates into vapor (B) plus saline liquid (C). With continued decompression to ca. 70 MPa and cooling to 800 °C, under closed-system conditions, the NaCl content of the vapor decreases to that at (D) and the salinity of the liquid increases to that at (E). At ca. 30 MPa and 600 °C, the liquid has been converted to crystalline NaCl (G) and vapor. The final salinity of the vapor (F) is ca. 0.01 wt%; whereas, the starting fluid contained > 5 wt% NaCl.

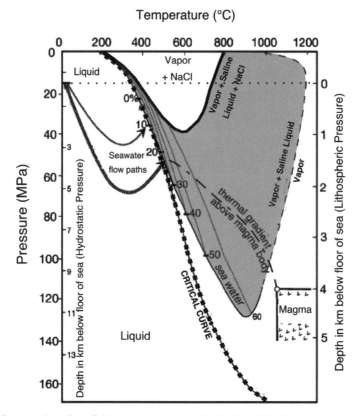

Figure 5. Interpretation of one-fluid versus vapor plus saline liquid stability fields in submarine volcanic environments as a function of pressure and temperature based on modeling of Fournier (1987) and Coombs et al. (2004). For reference, the solid curve with bold squares is the critical curve for NaCl-H₂O. Hachured curves show hypothetical flow paths for heated seawater as it passes through hot basaltic rocks; pressure-temperature field for basaltic magma and thermal gradient above magma shown for reference. The fine solid curve (dashed where inferred) serves as one boundary for the vapor plus saline liquid field, and the other boundary is the bold solid curve designating the stability of vapor, liquid, and crystalline NaCl. The gray area in between both curves is the vapor plus saline liquid field. The labels 0-60 represent solubility of NaCl (wt%) in the saline liquid at various pressure-temperature conditions, and the finely dotted curves express isopleths of NaCl solubility that intersect the seawater phase boundary at labeled black bars. The generation of vapor by boiling of seawater containing 3.2 wt% NaCl$_{equiv}$ at ca. 50 MPa and 500 °C generates saline liquid containing 20 wt% NaCl.

systems open (in composition space) with decreasing pressure, and the liquid-dominated side of the vapor-liquid solvus approaches very high NaCl concentrations at low pressure and high temperature (Fig. 4). Thus, hypersaline liquids may collect as a stably stratified horizon or "hot plate" (Henley and McNabb 1978) located in a transition or "dryout" zone above magma chambers (Hardee 1982). The anhydrous hypersaline liquid may move and ascend through and potentially away from the magma because the density of such liquids is roughly half that of silicate melts. In addition, crystalline salts precipitate (Fig. 4) from this liquid at sub-magmatic temperatures (Sourirajan and Kennedy 1962; Geiger et al. 2005).

These processes have apparently occurred at Kudryavy volcano of the Kuril island arc (Korzhinskii et al. 1996; Shmulovich and Churakov 1998; Wahrenberger et al. 2002). The depth to the Kudryavy vapor source is only 200 meters above the periphery of the magma body, so pressures of degassing are low (Botcharnikov et al. 2003). Moreover, the phases are dominated by H_2O with lesser C and S species and minor HCl and H_2. First-order estimates of the abundances of volatile components initially dissolved in the melt and of the relative quantities of volatile components degassed (Shmulovich and Churakov 1998) suggest that as much as 3×10^6 m^3 of anhydrous hypersaline liquid could have collected and potentially crystallized in 114 years of degassing at Kudryavy. Fluids sampled in the Mt. Vesuvius crater also show strong indications of fluid immiscibility. The fluids included vapor, aqueous liquid, and salt melt; Chiodini et al. (2001) notes that molten salt was observed flowing from fissures in the crater at magmatic temperatures in the first decade following the last eruption of 1944.

Investigations have also addressed the influence of CO_2 in volcanic environments. Carbonate minerals are not common in volcanic sublimates and encrustations associated with aluminosilicate magmas and their fumaroles. The high solubility of CO_2 in vapors combined with the low solubility of CO_2 in most aluminosilicate melts (Holloway 1976; Holloway and Blank 1994) results in efficient fractionation from magma. Carbon dioxide solubility, however, can be quite high in alkalic magma. Thus, CO_2 readily exsolves from most magmas, is largely contained by the magmatic vapor phase (except in environments with CO_2-saturated crater lakes; Bernard and Symonds 1989; Zhang 2000), and it enhances the likelihood of fluid unmixing if electrolytes are present. Interestingly, multiple CO_2-bearing fluids have been observed in rocks of some carbonatitic environments (Roedder 1984; Veksler and Lentz 2006).

The compositions of CO_2-, Cl-, and S-charged vapors and condensed liquids are a function of various processes occurring in volcanic environments (Ellis 1957; Graney and Kesler 1995; Giggenbach 1975, 1987, 1997). Redox processes are important (Giggenbach 1975) because f_{O_2} and sulfur speciation (i.e., the sulfidation state of Einaudi et al. 2003) are so mutually dependent. For instance, aluminosilicate melts exhibit elevated $(S^{6+}/S^{6+}+S^{2-})$ at comparatively high f_{O_2} (i.e., > NNO) (Carroll and Webster 1994). Hydrothermal alteration and dissolution of minerals also play a significant role in modifying volcanic fluid compositions (Giggenbach 1997; Newton and Manning 2005). In addition, vapor and liquid chemistry are influenced by the type of fluid source involved as well as by processes of magma crystallization, fluid exsolution from melt, mixing, contamination, boiling, and condensation.

In summary, the comparison of volcanic vapor compositions with the phase equilibria of simple binary and ternary model systems indicates that a liquid must condense from magmatic and hydrothermal vapors at the low pressures and high temperatures of volcanic vents. In addition, progressive fluid evolution may generate increasingly saline fluids because additional (i.e., secondary, tertiary, etc.) vapor phases may boil off with continued magma ascent and decompression.

Evidence in matrix glasses of subaerially erupted volcanic rocks. The geochemistry of silicate matrix glasses provides evidence of two fluids in eruptive and shallow crustal magmatic environments. Signorelli and Capaccioni (1999) inferred the presence of low-salinity vapor coexisting with saline liquid in the 79 CE magmas erupted at Mt. Somma-Vesuvius, and Signorelli et al. (2001) determined volatile component abundances of samples of the Campanian Ignimbrite of central Italy and called on two coexisting fluids during magmatic degassing. These interpretations were based in part on the comparison of the volatile component contents of matrix glasses with experimentally determined volatile component solubilities (the latter are described in detail below).

Evidence in fluid inclusions and silicate melt inclusions of subaerially erupted volcanic rocks. Assemblages of fluid inclusions containing vapor plus liquid and/or vapor plus saline liquid have been observed in samples of volcanic rocks with the heating/cooling microscope stage at elevated temperatures. Primary fluid inclusions in alkaline volcanic rocks indicate that alkaline magmas coexist with H_2O- ± CO_2-dominated vapors, alkali chloride-rich liquids, and S-bearing liquids (Roedder and Coombs 1967; De Vivo et al. 1993; Vaggelli et al. 1993; De

Vivo and Frezzotti 1994; Lowenstern 1994; De Vivo et al. 1995; Belkin et al. 1996; De Vivo et al. 2006; Kamenetsky 2006). These inclusions occur in phenocrysts of rocks of Ascension Island, Atlantic Ocean, and in Italy at Mt. Somma-Vesuvius, Pantelleria, Campi-Flegrei, Stromboli, and Ponza Island. Roedder (1984) summarized the results of these and other fluid inclusion studies that indicated the formation of alkali chloride-rich liquids and H_2O- ± CO_2-dominated vapors by boiling in terrestrial hydrothermal systems.

Some assemblages of primary fluid inclusions include silicate melt inclusions (Roedder 1984, 1992; Fulignati et al. 2001; Lowenstern 2002; De Vivo et al. 2006; Kamenetsky 2006). The latter are potentially significant because some melt inclusions represent pristine samples of the magma from which they were entrapped (Lowenstern 1995). Melt inclusions may form at depth (described below) or as magma ascends from shallowly emplaced bodies through overlying conduits and dikes. Otherwise, melt inclusions may be trapped at or near the surface.

Vesicles containing immiscible vapor and saline liquid occur in melt inclusions in quartz phenocrysts of volcanic rocks from Pantelleria, Italy (Solovova et al. 1991; Lowenstern 1994). Remarkably, the associated matrix glasses also contain halite crystals that represent the crystallized remnants of a hypersaline liquid (Lowenstern 1994). Inclusions of saline liquids also occur with CO_2-bearing vapor in xenoliths and phenocrysts coexisting with silicate melt inclusions that have been observed in samples from Mt. Somma-Vesuvius, Italy (De Vivo et al. 1993; Kamenetsky et al. 1999, 2004; Kamenetsky 2006). Apparently, the interaction of silicate melt with carbonate-bearing host rocks generated alkali chloride-rich liquids as well as alkali carbonate-dominated and sulfate-bearing liquids that coexisted with silicate melt (Fulignati et al. 2001; Kamenetsky 2006). The interaction of high-temperature saline liquids and/or vapor with carbonate rocks represents one means of generating carbonate liquids in some volcanic-skarn environments.

Melt inclusions in phenocrysts of some high-silica rhyolites show evidence of two magmatic fluids. The volatile component abundances of melt inclusions in quartz and sanidine of the Cerro Toledo rhyolite, Bandelier tuff, New Mexico, and associated modeling of fluid exsolution and evolution processes led Stix and Layne (1996) to conclude that a single fluid exsolved from the felsic magma. However, based on relationships involving Cl, H_2O, and Be in the inclusions, they concluded that this fluid subsequently separated into a Cl-poor vapor and a saline liquid. They also suggested that the initial fluid was present in the magmatic system for at least 160 kyr and that the vapor and saline liquid were stable for < 164 kyr. The presence of vapor and liquid served to buffer the Cl concentrations of the melt (and melt inclusions) at the 0.3 wt% level.

Magmatic sulfide liquids may coexist with aqueous or aqueous-carbonic vapors. Roedder (1979) observed vapor bubbles and sulfide droplets in mafic silicate melts of reheated olivine-hosted melt inclusions in samples of terrestrial and lunar mare basalts. Shallow, submarine MORB and ocean island basaltic magmas also show evidence of the interaction of sulfide liquids with CO_2-enriched vapors (Wallace and Anderson 2000). Moreover, Borisova et al. (2002) observed CO_2-rich fluid inclusions and sulfide melt inclusions coexisting with silicate melt inclusions in olivine phenocrysts of Kerguelen Plateau picritic basalts, but they did not address the consequences of interactions of these fluids on associated magmatic or volcanic processes. Because of their lower temperatures of crystallization, andesitic to rhyolitic magmas are comparatively more likely to contain crystalline pyrrhotite ($Fe_{1-x}S$) than sulfide liquid (Wallace and Anderson 2000). Nevertheless, some of these comparatively evolved magmas are also quite likely to exsolve and contain a second CO_2- and H_2O-rich and S-bearing vapor phase (Whitney 1984), so the silicate melt should coexist with CO_2-charged aqueous vapor and a sulfide liquid at some stages of magma evolution. The 1963 eruption of Gunung Agung, Bali, Indonesia, for example, discharged phenocrysts containing globular pyrrhotite (interpreted to have crystallized from a sulfide liquid) and melt inclusions showing indication of vapor saturation of the same magma, but it is not certain if all three fluids occurred simultaneously (Self and King 1996). Likewise, globules of sulfides have been observed in melt inclusions, trapped at shallow depths, along with CO_2-bearing fluid inclusions in samples from Mt. Etna, Italy (Métrich and Clocchiatti 1989; Métrich et al. 1993). The sulfide globules are considered to represent transitory liquid phases because they are never found in comparatively primitive melt inclusions that were trapped in olivine presumably at greater depths (Burton et al. 2003). These two fluids may not have interacted sufficiently with one another and the melt phase to influence the behavior of other magmatic volatile components and trace elements.

Melt and fluid inclusions in dacite samples of the 1991 eruption of Mt. Pinatubo contain: glass only, vapor-dominated fluid, or aqueous liquids (some of which are saline) with a large bubble and multiple solid phases. The rehomogenization studies of Pasteris et al. (1996) inter- preted the coexisting inclusions of vapor plus liquid to reflect fluid saturation of dacitic magma with derivation of the volatile components from a high-alumina basalt source at greater depth (Hattori 1996). Fluid saturation was followed by separation of the fluid into vapor and liquid, and the fluid unmixing was probably a result of the reduction in pressure associated with ascent of the fluid-saturated magma. Phenocrysts of basalt, andesite, and dacite samples also contain sulfide globules; so, the presence of CO_2-bearing fluid inclusions in plagioclase phenocrysts in- dicates that a sulfide liquid coexisted with the vapor phase as the underlying mafic magma trans- ported volatile components upward into the dacitic magma (Hattori 1996). It is not certain if the sulfide liquid equilibrated with the vapor. Hattori (1996) has suggested that the buoyant ascent of S-rich fluid through magma chambers may have caused the condensation of a sulfide liquid from the fluid and that the transport of SO_2 into the dacitic melt may have also caused oxidation of the reduced sulfur and led to subsequent precipitation of anhydrite. The presence or absence of anhydrite is important because $CaSO_4$ is highly soluble in $NaCl$-H_2O liquids at magmatic temperatures (Newton and Manning 2005); so, the magmatic fluid(s) could have contained ap- preciable sulfate. However, this redox-modifying process has been challenged by more recent work indicating that the oxidation state of the basaltic magma may have actually been relatively high prior to interaction with the more evolved magma fractions (de Hoog et al. 2004).

Some mineralized volcanic systems also exhibit evidence of two-fluid processes. Rock samples from the Cu- and Au-mineralized Bajo de la Alumbrera volcanic complex of north- western Argentina contain primary fluid inclusions, silicate melt inclusions, and sulfide melt inclusions. The inclusion compositions are consistent with the interpretation that the sulfide liquid scavenged > 95% of the Cu and Au in the aluminosilicate melt, and that subsequent interaction of the sulfide liquid with a saline aqueous liquid (with or without coexisting anhy- drite) destabilized the sulfide liquid and released appreciable quantities of the ore metals into the mineralizing aqueous fluid (Halter et al. 2005). The phenocrysts from this complex show abundant textural and chemical evidence for disequilibrium processes, so it is not known if the saline aqueous fluid and sulfide liquid achieved equilibrium or simply interacted by exchanging soluble constituents.

Volcanic crater lakes

Explosive volcanic eruptions in freshwater and seawater environments constitute hazards to life and property. They may involve lahars, floods, tsunamis, and/or destructive base surge eruptive clouds (Mastin and Witter 2000). Of the 7900 volcanic eruptions recorded from 8000 BCE to 2000 CE, roughly 610 occurred in lakes or seawater. These submarine and subaqueous eruptions, which represent only about 8% of known eruptions, were responsible for nearly 20% of the known volcano-related fatalities (Mastin and Witter 2000).

Volcanic crater lakes and the associated hydrothermal and volcanic environments that envelop them display characteristics that reveal the chemical and physical consequences of fluid immiscibility. Subaqueous vent systems at Ruapehu, New Zealand (Christenson 2000), and Kawah Ijen, Indonesia (Delmelle and Bernard 1994; Delmelle et al. 2000), like many others, show evidence of vapor within the crater-lake waters, and this is consistent with the phase relations of the NaCl-H$_2$O binary (Fig. 1). The thermal properties of crater lakes provide a useful diagnostic tool for gauging magma depth beneath volcanoes, studying heat transfer processes, and forecasting eruptive activities (Giggenbach and Glover 1975; Christenson 2000; Varekamp et al. 2000). Crater lake heat budgets are, however, a complex function of numerous processes involving multiple fluid sources (Fig. 3), so their interpretation can be challenging. Nevertheless, the presence of one or more fluids in hydrothermal systems adjacent to these environments is important because the efficiency of heat transfer varies with the number and type of fluids and because vapor-saturated liquid permeable convection is a dominant heat transfer mode (Hardee 1982). Two-fluid convection can be much more efficient at transferring heat than single-fluid convection at subsolidus temperatures (Hardee 1982). Consequently, models of heat flow in crater lakes and their underlying hydrothermal systems must account for the presence of multiple fluids (Geiger et al. 2006a,b; and references cited therein).

The geochemistry of volcanic crater lakes provides information on the abundances of volcanically sourced chlorides, sulfates, and other soluble constituents. Crater lakes range from dilute, meteoric water-dominated compositions to acidic saline liquids, as documented in a systematic geochemical catalog of 373 analyses of volcanic crater-lake waters (Varekamp et al. 2000). Lake compositions reflect volatile component inputs from various sources including: volcanic liquids and vapor, hydrothermal fluids that commonly have reacted with host rocks, and/or meteoric water. Crater lakes act as large condensers of vapors; they sequester reactive magmatic volatile components including HCl, H$_2$S, SO$_2$, HF, CO$_2$, H$_2$CO$_3$, HBO$_2$, and other vapor constituents (Brantley et al. 1987; Poorter et al. 1989; Delmelle and Bernard 1994; Varekamp et al. 2000; Vaselli et al. 2003). The may also entrain high salinity liquids (Fig. 3). Crater lake vapors may or may not be in equilibrium with the aqueous liquid phase. They represent an integrated composition of extended volcanic degassing; so, this geochemical signature is unique relative to those resulting from the standard practice of episodically sampling gases of volcanic fumaroles directly or analyzing them with remote sensing methods. Thus, the study of crater-lake geochemistry and thermal behavior provides an integrated, long-term record of volcanic degassing, albeit modified by the effects of gas exchange with the atmosphere (e.g., oxidation), boiling, evaporation, and water-rock interaction (i.e., typically, extreme hydrolysis involving mineral dissolution and precipitation).

In a broadly framed investigation, Varekamp et al. (2000) described three geochemical types of crater lakes including: (1) mildly acidic to basic, CO$_2$-enriched lakes and (2) H$_2$O-dominated lakes. These lakes are located over volcanoes that are comparatively quiescent, with little if any magmatic input, and contain low abundances of chloride and sulfate ions that typically total less than 3000 ppm. (3) Conversely, acidic to hyperacidic crater lakes, that are associated with active volcanoes, have pH values that may reach (−1) but, more typically, the pH values are near unity. These lakes also have total chloride and sulfate contents approaching 30 wt%, and show an approximate 1:1 correlation between dissolved sulfate and chloride (Varekamp et al. 2000). These pH values are largely derived from acid-forming species in vapors that formed by separation of alkali-enriched saline liquids and acidic vapor from magmatic fluid. Phase separation occurred after initial fluid exsolution from magma but prior to vapor input to the lake (Giggenbach 1974). Molten native S has been observed in many lakes (Giggenbach 1974; Delmelle and Bernard 1994; Delmelle et al. 2000), and it is indicative of reactions at temperatures ranging from ca. 119 °C:

$$2H_2S_{vapor} + SO_{2vapor} = 3S_{elemental} + 2H_2O_{gas/liquid} \qquad (3)$$

(Mizutani and Sugiura 1966) up to 300 °C:

$$3SO_{2\ vapor} + 2H_2O_{gas/liquid} = 2H_2SO_{4\ aqueous} + S_{elemental} \qquad (4)$$

Reaction (4) produces dissolved sulfate ions by disproportionation of SO_2 (Sakai and Matsubaya 1974), and the sulfate ions can subsequently form native sulfur by oxidation of H_2S:

$$H_2SO_4 \text{ aqueous} + 3H_2S_{vapor} = 4S_{elemental} + 4H_2O_{gas/liquid} \qquad (5)$$

or:

$$H_2SO_3 \text{ aqueous} + 2H_2S_{vapor} = 3S_{elemental} + 3H_2O_{gas/liquid} \qquad (6)$$

(Rowe 1994). Crater lake acidity is, thus, also a function of the disproportionation reactions (Giggenbach 1974; Hurst et al. 1991; Christenson 2000). The presence of liquid S in these lakes, however, may not reflect equilibration with the lake water (Reed 1982).

Limnic eruptions are a rare type of subaqueous eruptive behavior involving exsolution of potentially massive quantities of CO_2–rich vapor from the lake. The energetics of degassing can vary dramatically during these eruptions. At one extreme, the thermal energy of the CO_2-saturated aqueous liquid is converted to massive kinetic energy output ($P\Delta V$ work) after rapid gas exsolution, and the energy provides the driving force for explosive expansion (Zhang 2000). At the other end of the energy spectrum, CO_2 may degas comparatively slowly and passively (Zhang 2000), but hereto the consequences can still be deadly. The sudden release of CO_2 vapor from the volcanic crater lake of Lake Nyos, Cameroon, on August 21, 1986, asphyxiated more than 1700 local inhabitants. It is believed that CO_2 vapor, emitted from magma underlying the lake, had accumulated in and saturated the thermally stratified lake waters in CO_2. On August 21st, the vapor-rich layer was disturbed causing it to ascend to depths where the gaseous pressure of CO_2 exceeded the hydrostatic pressure that forced the massive release of CO_2 vapor. The cause of the disturbance that triggered the ascent of the CO_2-rich layer is unknown (Williams-Jones and Rhymer 2000). The CO_2 vapor ponded in topographic lows on land because CO_2 is heavier than air (Barberi et al. 1989; Rice 2000).

Submarine volcanic environments

Investigations of fluids and volcanic rocks in submarine environments provide clear evidence of processes involving vapor plus liquid with or without magma. The presence of saline, seawater-derived liquids is supported by direct analyses of fluids emitted from seafloor hydrothermal vents (Von Damm et al. 1985; Bowers et al. 1988; de Ronde et al. 2003) and through study of FI trapped in minerals of hydrothermally altered seafloor rocks (Kelley and Delaney 1987; Michael and Schilling 1989; Nehlig 1991). Not all saline fluid inclusions are the result of the immiscible separation of vapor from liquid (Kawada et al. 2004), however. Some FI in these environments are the result of vapor and liquid exsolution from magma (Vanko 1986). A detailed description of the role of fluid immiscibility in seafloor hydrothermal systems is addressed in a companion chapter of this volume (Foustoukos and Seyfried 2007).

Evidence in matrix glasses. Seafloor basalt glasses contain elevated Cl abundances that indicate interaction of basaltic magma with saline liquids, and some of the saline liquids may reflect the boiling of seawater. The presence of high Cl concentrations in evolved MORB glasses collected at seafloor spreading centers and hotspot basalts, for example, cannot be explained solely by volatile component enrichment resulting from crystal fractionation (Byers et al. 1984; Michael and Schilling 1989). Basaltic glasses of the Galapagos Islands contain 1000 ppm or more Cl; whereas, primitive basaltic glasses of the Mid-Atlantic ridge and East Pacific rise contain only 10 to 40 ppm (Michael and Schilling 1989). Processes and sources suggested for Cl enrichments of evolved MORB glasses include assimilation of seawater-contaminated and hydrothermally altered rocks, assimilation of halite (Kent et al. 1999b), and reaction of magma and/or quenched matrix glass with saline liquid generated by seawater boiling and concomitant loss of the vapor. It is often concluded that the most likely process responsible for the elevated Cl abundances is the reaction of magma with saline liquid formed by boiling of seawater (Fig. 5). For example, Cl concentrations as high as 1.68 wt% were measured in grains of tholeiitic-basalt matrix glass from the southern submarine flank of Kilauea volcano (Coombs et al. 2004). This led the investigators to conclude that a saline liquid containing 78 wt% H_2O, 13 wt% Cl, 4.4 wt% Na, 2.6 wt% K, 2.6 wt% Ca, and tens to hundreds of parts per million of Pb and alkali and alkaline earth elements exchanged components with basaltic magma and/or hot volcanic rock. The saline liquid was the source of the large Cl concentrations and the associated enrichments in Rb, Ba, Sr, and Pb. Coombs et al. (2004) concluded that the liquid may have formed by heating of seawater at ca. 500 °C and approximately 50-55 MPa as it flowed through hot basaltic rocks beneath the seafloor (Fig. 5).

Seawater readily undergoes phase separation at the depths and equivalent hydrostatic pressures of seafloor volcanic environments (i.e., at generally pervasive low-P conditions) (Henley and MacNabb 1978; Bischoff and Pitzer 1985; Fournier 1987). Boiling of lower-density vapor to generate a saline liquid from heated seawater may well be the rule rather than the exception in submarine hydrothermal systems (Bischoff and Pitzer 1989; Butterfield et al. 1994), but these systems must be sufficiently open to allow the H_2O-enriched vapor to escape. Phase separation is limited, however, to depths less than 11 km of hydrostatic pressure (i.e., at depth below ocean surface) with $\approx$ 3.2 wt% $NaCl_{equiv}$ in an aqueous fluid (Fig. 5) because the vapor-liquid field pinches out completely at pressures > 130 MPa (Fournier 1987; Coombs et al. 2004). In addition, the salinity of the liquid varies with temperature and pressure along the two-phase boundary, so the liquid becomes increasingly saline because of the reduction in pressure that accompanies fluid ascent (Fournier 1987).

Some seafloor glasses indicate a significant role for other volatile components in hydrothermal seafloor environments. The CO_2 contents of some MORB glasses (e.g., 50-400 ppm) exceed experimentally determined CO_2 solubilities of basaltic melts at pressures equivalent to these environments, and some glasses contain small vapor-filled bubbles indicative of vapor saturation of melt (Wallace and Anderson 2000). These glasses may also contain appreciable S (e.g., 800-1400 ppm) and small quantities of quenched immiscible Fe-S-O liquid indicating sulfide liquid saturation of the basaltic melts (Wallace and Anderson 2000). Thus, sulfide liquids and CO_2-enriched vapors are likely to coexist with or without a silicate melt. Significant concentrations of CH_4 also occur in some seafloor-vented, hydrothermal fluids where organic matter in the sediments of these environments interacts with hydrothermal fluids heated by proximal magmatic intrusions and/or hot volcanic rocks (de Ronde et al. 2003). The presence of this non-polar CH_4 species can be significant (Drummond 1981; Lamb et al. 2002) because it: (1) fractionates to the low-density vapor relative to the saline liquid in the ternary system: H_2O-NaCl-CH_4, and (2) modifies f_{O_2}, f_{H_2}, and other variables in two-fluid systems. In addition, processes of fluid unmixing can be complex in some seafloor magmatic systems. Matrix glasses in xenoliths of magnesiohastingsite-rich cumulates of the TUBAF Seamount, near Lihir Island, Papua New Guinea, show evidence of immiscibility involving two silicate melts, a Cu-Fe-S liquid, a Cu-Cl salt-hydrate liquid, and a CO_2-enriched vapor phase (Renno et al. 2004). This study concluded that all five fluids were in equilibrium during magma differentiation.

Evidence in silicate melt inclusions. Chloride enrichments have also been observed in mafic silicate melt inclusions in rocks from ocean islands, and these Cl-enriched inclusions have been interpreted to reflect the interaction of magma with saline liquids that formed by boiling of hot seawater (Kent et al. 1999a). Olivine-hosted melt inclusions in basalts of the Austral Islands, Raivavae and Rapa, for example, contain as much as 2.5 wt% Cl (Lassiter et al. 2002), and, even though other melt inclusions contain far less Cl, the most Cl-enriched inclusions also exhibit elevated (Cl/K_2O) and contain comparatively low abundances of high-field-strength elements. These features led Lassiter et al. (2002) to conclude that the magmas resulted from anatexis of hydrothermally altered and saline liquid-saturated (from modified seawater) ocean crust.

Magma conduits and shallow plutonic magmas that underlie volcanoes

Evidence from hydrothermal systems. Boiling and convecting hydrothermal systems occur in and around magma conduits beneath volcanoes (with or without crater lakes), above and adjacent to magma chambers, and in environments of submarine rifting (Figs. 2 and 3). Hydrothermal fluids (e.g., heated magmatic and metamorphic waters and/or ocean and meteoric waters) alter and cause physical weakening of volcanic rocks and their host rocks, and they modify the permeability of the rocks in these environments. Fluids also transfer mass and heat through porous and fractured solid media there, and hence, exert strong controls on the solubility of trace elements and potential ore metals in the fluids.

Fluid convection drives much of the heat transfer that occurs around bodies of magma. This has been demonstrated in simulations of the convection of fluids through hot rock in volcanic and related magmatic environments. These models can be used to predict the behavior of liquid and vapor phases that are generated by phase separation (Cathles 1977; Henley and McNabb 1978; Faust and Mercer 1979a,b; Burnham and Ohmoto 1980; Carrigan 1986; Graney and Kesler 1995; Fournier 1999; Mastin and Ghiorso 2001; Bai et al. 2003; Kawada et al. 2004; Geiger et al. 2005, 2006a,b). It has been proposed, for instance, that a two-fluid phase envelope develops above bodies of magma and in the environment surrounding the central feeding conduit and/or dikes that transport magma from the chamber to the surface (Giggenbach and Sheppard 1989; Hedenquist 1991; Christenson 2000; Lewis and Lowell 2004). Some environments like these have been modeled as a form of heat pipe (Hardee 1982; Hurst et al. 1991) that may be liquid- or vapor-dominated (McGuinness et al. 1993). In a two-fluid envelope, the upward flow of vapor is accompanied by a downward counterflow of liquid (Hurst et al. 1991), and if overlain by a crater lake, the ascent of hot vapor in a heat pipe plays a key role in maintaining the elevated temperatures of such lakes (Hurst et al. 1991). Interestingly, the temperature profiles of some two-fluid phase systems that envelop magma bodies appear to be constrained to the boiling curves of H_2O-NaCl at the subsolidus conditions relevant to these environments (Hardee 1982; Fournier 1987; Geiger et al. 2006a,b). These environments involve large (fluid/ rock) ratios, and their permeability must be sufficiently high such that the heat capacities and heats of condensation and vaporization of the aqueous electrolyte-bearing vapors and liquids override the heat capacities of the reactions involving the minerals and melt present.

The vapor and liquid flow paths of hydrothermal systems and the phase boundaries and isotherms of systems like those in Figure 2 are transitory because of the efficacy of hydrothermal fluids in transferring heat. In a study of convecting fluids at White Island volcano, New Zealand, Giggenbach and Sheppard (1989) discussed the effects of periodic fluctuations in the input of magmatic vapor into the shallow subvolcanic system. They noted that the fluctuations triggered shifts in the phase boundaries and flow paths of subsurface saline liquids. These shifts also affect the chemical and thermal characteristics of subsurface volcanic-hydrothermal systems (Symonds et al. 1994; Driesner and Geiger 2007) because the densities, viscosities, and other physical properties of vapor vs. liquid vary so significantly (Shinohara 1994).

Evidence from fluid inclusions and silicate melt inclusions of plutonic rocks. Plutons that are located in the same tectonic environments as the volcanoes addressed here also show textural and geochemical evidence for multiple fluid phases in magmas. Some of these plutons may fuel volcanic processes. Blocks of granite at Ascension Island, for example, contain a variety of inclusions in quartz and feldspar phenocrysts: CO_2-enriched fluid inclusions, mixed H_2O-CO_2 fluid inclusions, saline liquid inclusions, and silicate melt inclusions. The granitic blocks were carried to the surface during eruptions of trachyte-composition magma (Roedder and Coombs 1967). The coexistence of multiple inclusion types and the interpretation that some are primary, led the authors to conclude that a magmatic saline, CO_2-bearing aqueous fluid unmixed to form aqueous-carbonic vapor and saline liquid in the presence of silicate melt. These inclusions are comparable to some of those occurring in quartz and feldspar phenocrysts of the Sybille monzosyenite, Laramie anorthosite complex, Wyoming. The latter inclusions indicate that the granitic magma contained immiscible droplets of saline liquid and bubbles of CO_2-enriched vapor (Frost and Touret 1989). Similarly, granites and quartz syenites of the Gardar province, South Greenland, contain CO_2-H_2O vapor-rich inclusions, and the associated nepheline syenites at Gardar contain comparatively reduced CO_2-CH_4 vapor-rich inclusions. Both varieties of these fluid inclusions occur with inclusions of highly saline liquids (Konnerup-Madsen et al. 1985), and indicate the occurrence of multiple magmatic fluids. Moreover, De Vivo et al. (1995) observed similar assemblages of fluid inclusions coexisting with melt inclusions in alkaline xenoliths of Ventotene Island, Italy. Frezzotti (1992) studied melt inclusions and primary fluid inclusions in the Sardinian Mount Genis granite, and determined that a saline melt exsolved directly from the granitic magma. The subsequent evolution of the Mount Genis hydrothermal system occurred at sub-solidus conditions and involved: (1) dilution of the liquid's salinity through interaction with H_2O-bearing host rocks, (2) separation of vapor from saline liquid, and/or (3) subsequent mixing of these fluids with cold, dilute non-magmatic waters.

Mineralized plutons also show evidence of processes involving multiple fluids. High-homogenization temperature, primary CO_2-bearing vapor inclusions that coexist with saline liquid inclusions of Cu- ± Au- ± Mo-mineralized porphyritic and/or epithermal systems imply either direct exsolution of liquid and vapor from magma or the immiscible phase separation of liquid and vapor after exsolution of a single fluid from magma (Henley and McNabb 1978; Cline and Bodnar 1991, 1994; Heinrich et al. 1992; Hedenquist and Lowenstern 1994; Candela and Piccoli 1995; Kamenetsky et al. 1999; Sillitoe and Hedenquist 2003; Webster 2004). The bulk compositions of some systems are analogous to those of the subduction-related volcanic complexes we have discussed previously. Important examples include the systems at Questa, New Mexico (Cline and Vanko 1995); Henderson, Colorado (Kamilli 1978); Santa Rita, New Mexico (Reynolds and Beane 1979); and numerous other examples listed by Roedder (1984). The magmatic vapor phase in two-fluid, plutonic environments like these is generally dominated by H_2O, CO_2, SO_2, H_2S, and HCl, but other gas species can be present. Analyses of primary vapor-rich FI from porphyry Cu deposits and Sn- and W-mineralized granites have identified N_2 (Graney and Kesler 1995), and Kamenetsky et al. (1999) detected N_2 by ion chromatography in primary multiphase fluid inclusions coexisting with melt inclusions in clinopyroxene phenocrysts of the Dinkidi copper-gold porphyry deposit, Philippines. Other fluid inclusions in rocks from the latter deposit are highly enriched in Cl, S, F, and C, and the inclusions are interpreted to represent the separation of immiscible fluids at magmatic temperatures. The saline liquids in systems like these are thought to be instrumental in generating mineralization. The Mariktikan granitic pluton, west Transbaikalia, for example, directly exsolved a liquid that separated into a CO_2-enriched vapor and Mn-, Fe-, and W-enriched saline liquid (Reyf 1997). However, recent research shows that the presence of vapors in mineralized systems can also be quite efficient in the transport of ore metals (Williams-Jones and Heinrich 2005; Heinrich 2007a).

Alkaline and carbonate-rich magmas

Carbonatites and associated kimberlites show evidence for the exsolution and/or separation of multiple fluids, and some of these fluids are involved in mechanisms of carbonatite genesis and evolution. Degassing processes involve alkali-rich and/or calcium-rich carbonate liquids, H_2O-rich fluids, and CO_2-rich fluids of moderate to high density, with or without immiscible saline to hypersaline liquids (Roedder 1984). The compositions of primary melt inclusions in olivine crystals of the Udachnaya-East diamond-bearing kimberlite pipe of Siberia are consistent with the separation of immiscible chloride liquids and carbonate-dominated liquids at temperatures < 600 °C (Kamenetsky et al. 2004). Other, more complex systems also occur. For example, fluorite crystals from the Mushugai-Khuduk calcitic-alkaline carbonatitic complex of southern Mongolia contain a very unusual inclusion assemblage (Andreeva et al. 1998; Veksler and Lentz 2006); the five types of inclusions that are present represent various mixtures of sulfate, phosphate, and halogen-enriched saline hydrothermal liquids (Mitchell 2005). Inclusions in other carbonatites contain sulfate and halogen abundances reaching well into the wt% concentration levels (Veksler and Lentz 2006). These fluids may play a role in carbonatitic magma formation, as noted previously for Mt. Somma-Vesuvius (De Vivo et al. 1995; Fulignati et al. 2001).

Some carbonatite complexes are mineralized with apatite, REE, U, Th, Nb, Ta, Cu, Fe, and other ores. Carbonate-enriched magmas may involve unusual fractionations of volatile components, trace elements, and ore metals between carbonate-, aqueous-, sulfate-, halide-, and phosphate-rich fluids. Further discussion is beyond the scope of this chapter, but Veksler and Lentz (2006) provide a comprehensive review of the relevant literature and describe the great variety of fluid inclusions and melt inclusions in these systems.

Experimental petrology: evidence of multiple fluids coexisting with aluminosilicate melts

The phase equilibria of melt-fluid systems and the solubilities of volatile components in vapor, liquid, and silicate melt are determined by hydrothermal experiments conducted at known pressure-temperature conditions. These experimental data can be combined with the results of thermodynamic modeling to interpret the compositions of volcanic vapors, matrix glasses, fluid inclusions, and melt inclusions. This integrated approach provides insight into processes of fluid mixing and unmixing in volcanic environments. Research of the past two decades has seen development of a method for distinguishing melts saturated in one fluid from melts saturated with two fluids. The method is based largely on the pioneering work of Shinohara et al. (1989) and Shinohara (1994); it only requires accurate knowledge of volatile component solubilities in an aluminosilicate melt as the ratio of (volatile component A/volatile component B) varies. The experiments of the former study and the modeling of the latter determined the phase equilibria and pressure-temperature conditions for vapor-liquid immiscibility, and how vapor-liquid immiscibility controls the volatile component contents of melts and coexisting fluids.

Alkali chloride-bearing aqueous fluids. The phase equilibria of aqueous fluid(s) and silicate melt at fixed pressure and temperature vary with the chloride salt content. At conditions involving melt and one stable, H_2O-bearing fluid phase (at point A, Fig. 6), the pseudo-system H_2O-NaCl-melt is invariant at fixed pressure and temperature, and the melt and aqueous fluid compositions are also invariant. When NaCl is added, the melt composition moves away from point A toward point C, the number of components increases to 3 and the pseudo-system becomes univariant. The NaCl concentration of the melt (C_{NaCl}^{mt}) and the H_2O concentration of the melt ($C_{H_2O}^{mt}$) are free to vary along curve A-C. There is no phase boundary between the H_2O and NaCl apices, so the NaCl and H_2O concentrations of this single fluid (C_{NaCl}^{fl} and $C_{H_2O}^{fl}$, respectively) change from that of the less dense, Cl-deficient vapor to that of the more dense and saline liquid with decreasing (H_2O/NaCl) in the system. This is expressed by the sweep of the tie lines from the vapor to the liquid apices of the pseudo-system. The addition of NaCl increases the C_{NaCl}^{mt} and dilutes the $C_{H_2O}^{fl}$ and the $C_{H_2O}^{mt}$. As the C_{NaCl}^{mt} approaches point C, the $C_{H_2O}^{mt}$ decreases to zero, and at point C, the melt dissolves its maximum, NaCl saturation concentration. This behavior is expressed experimentally by the presence of varying concentrations of H_2O and Cl in the run-product glasses, as the H_2O and NaCl contents of the starting experimental charges are varied from H_2O-rich (point A) to NaCl-rich (point C) conditions.

The solubility behavior at higher temperatures (Fig. 6b) is different than that discussed previously (Fig. 6a) because of the effect of temperature on the phase equilibria of aqueous-electrolyte systems (Fig. 1). At these conditions, the behavior of H_2O and Cl in the melt and fluid are non-Henrian (Webster 1992a,b). With increasing NaCl in the system from point A, the C_{NaCl}^{mt} increases as the $C_{H_2O}^{mt}$ remains constant and equivalent to that at the maximum H_2O saturation value of point A. This behavior continues until the melt dissolves its maximum,

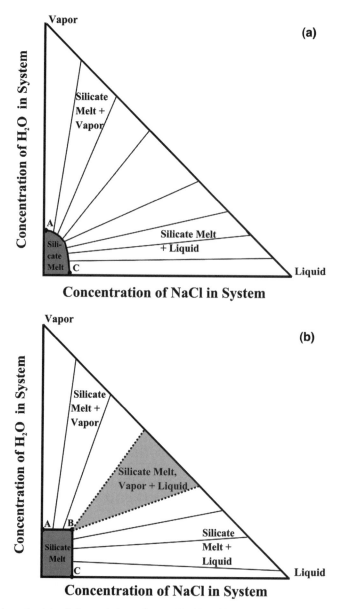

Figure 6. Schematic plots of phase relations of vapor, liquid, and H_2O- and NaCl-bearing silicate melt at similar pressure but different temperatures (after Mathez and Webster 2005). The tie lines (fine solid lines) and relative sizes of stability fields for melt and fluids are interpretive. (a) At comparatively low temperature only one fluid phase is stable throughout the system. The bold curve A-C expresses the H_2O and NaCl concentrations of the fluid-saturated silicate melt (dark gray area). The density and composition of the fluid varies from vapor to saline liquid as the (H_2O/NaCl) of the system increases; there is no phase boundary between vapor and liquid at these conditions. (b) At higher temperature, curve A-B expresses the concentrations of H_2O and NaCl in the vapor-saturated silicate melt (dark gray area) from point A up to but not including point B. The curve C-B expresses those in the liquid-saturated silicate melt from point C up to but not including point B. At point B, the silicate melt intersects the field of vapor plus liquid (light gray area delineated by dotted lines), and the melt is saturated with respect to vapor and liquid. See text for discussion.

NaCl-saturated melt value (i.e., equivalent to that of the C_{NaCl}^{mt} at points B and C). The C_{NaCl}^{fl} (the fluid is a vapor) and the C_{NaCl}^{mt} increase as the (H_2O/NaCl) decreases. At point B, a saline liquid condenses from the vapor and the melt coexists with two fluids. The phase rule requires that the pseudo-system is constrained to an invariant condition with a melt and two stable fluids at point B. The NaCl and H_2O concentrations of all three phases are fixed as long as both fluids coexist with melt (see Shinohara 1994; Candela and Piccoli 1995; Webster 2004). With continued decreasing (H_2O/NaCl) in the system, the $C_{H_2O}^{mt}$, $C_{H_2O}^{vapor}$, $C_{H_2O}^{liquid}$, C_{NaCl}^{vapor}, C_{NaCl}^{liquid}, and C_{NaCl}^{mt} remain constant, while the relative quantities of melt, vapor, and liquid change. This behavior is expressed experimentally as a distinct clustering of H_2O and Cl values of multiple run-product glasses at those of point B, even though the (H_2O/NaCl) ratios of the starting charges of these multiple experiments were distinctly different. When sufficient NaCl is added, all of the vapor is converted to saline liquid and the melt composition moves from B toward C because the system is again univariant and the compositions of the melt and the liquid are free to vary along this curve. From B to C, the C_{NaCl}^{mt} is fixed at that of its maximum, saturation value (implying that the activity of NaCl in the melt remains approximately constant) while the $C_{H_2O}^{mt}$ decreases to zero. The solubility behavior from B to C is expressed experimentally as broadly ranging concentrations of H_2O and relatively invariant NaCl contents of the run-product glasses.

This method establishes the geologically relevant conditions under which a vapor and a saline liquid coexist with metaluminous and subaluminous felsic melts (Shinohara et al. 1989; Malinin et al. 1989; Shinohara 1994; Métrich and Rutherford 1992; Webster 1992a,b; Webster et al. 2006) as well as with melts having intermediate silica contents (Webster et al. 1999). Other melt compositions investigated and shown to coexist with vapor plus saline liquids include alkaline melts (Métrich and Rutherford 1992; Lowenstern 1994; Signorelli and Carroll 2000; Webster and De Vivo 2002; Veksler 2004; Webster et al. 2006) and F-enriched, peralkaline to peraluminous felsic melts (Webster and Rebbert 1998; Veksler and Thomas 2002). These experiments confirm that vapor and saline liquid will exsolve directly from magmas. Similar experiments have also been conducted with mafic melt compositions (Malinin et al. 1989; Webster et al. 1999; Mathez and Webster 2005; Stelling et al. 2006). The results of some of the latter experiments have been equivocal; it is not certain if melt, vapor, and saline liquid or melt plus a single fluid phase were present in all experiments.

The phase relations of granitic melts and alkali chloride-bearing aqueous fluids have also been treated through direct comparison of vapor-liquid solvi for the model system H_2O-NaCl with the minimum melting curve for H_2O-saturated granitic melts (Pitzer and Pabalan 1986). This approach is problematic, however. The H_2O-saturated granitic solidus for the Cl-free system involves an activity of H_2O in the melt ($a_{H_2O}^{melt}$) at or near unity, and hence the solidus is not directly comparable to NaCl-H_2O vapor-liquid solvi because the latter involve a wide range of (NaCl/H_2O) ratios. With increasing NaCl content in the pseudo-system NaCl-H_2O-melt, the $a_{H_2O}^{melt}$ and the $a_{H_2O}^{vapor}$ and $a_{H_2O}^{liquid}$ vary widely as long as the system is not fixed at the invariant pressure-composition-temperature condition (point B in Fig. 6b). In other words, the curve for H_2O solubility in melt and the vapor-liquid solvus curve for NaCl-H_2O are incompatible and should not be compared directly because the $a_{H_2O}^{melt}$ is not always at or near unity over this range in composition.

Hydrothermal experiments also determine the partitioning of volatile components between aluminosilicate melts and fluids, and these data have been integrated with constraints on magmatic volatile component abundances, established from investigations of melt inclusions, to resolve ranges in volatile component abundances of magmatic fluids exsolved at deep to shallow crustal conditions. Partition coefficients (i.e., $D_{Cl} = C_{Cl}^{fl}/C_{Cl}^{mt}$) for felsic melts range from ≤10 to values exceeding 100 (Carroll and Webster 1994), and Cl concentrations in melt inclusions from metaluminous to alkaline felsic systems range from 0.01-1.0 wt% (Webster 2004). It follows from the application of simple mass-balance calculations involving these ranges and an arbitrarily selected initial (fluid/melt) mass ratio of ca. 0.01 that the computed values of the C_{Cl}^{fl} will range from < 0.1 to at least 50 wt%. This range is consistent with estimates of the chloride concentrations of fluids in magmas of intermediate-silica to felsic composition in subduction zone environments ranging from several wt% to > 25 wt% (Kent and Elliot 2002; Wallace 2005). Thus, based on the preceding review of the type system H_2O-NaCl and comparison to these data, it follows that many magmatic fluids have sufficient chloride contents to cause fluid-fluid phase separation during magma ascent and cooling (Fig. 7).

Fluids of andesitic to mafic magmas may contain appreciable transition metal- and alkaline earth-chloride species, in addition to alkali chlorides (Malinin et al. 1989; Webster and DeVivo 2002). It has been described how Ca- and Mg-chloride salts enhance the stability of vapor plus liquid relative to that of NaCl-H_2O fluids, but the detailed fluid-fluid phase equilibria of the former are not well resolved. Moreover, melt inclusions imply that most seafloor rift-related and hot-spot mafic magmatism involves melts containing ≤ 0.2 wt% Cl (Wallace and Anderson 2000; Hauri 2002), and the values of D_{Cl} for such melts are typically ≤ 2 (Webster et al. 1999). The combination of such low Cl abundances in mafic magmas with small values of D_{Cl} implies that most basaltic magmas of seafloor spreading centers and hot spot environments contain too little Cl to generate two coexisting fluids. Conversely, basaltic magmas of subduction zones contain higher Cl concentrations (e.g., 0.2-0.6 wt%) (Wallace and Anderson 2000; Zimmer et al. 2004). With geologically relevant values of D_{Cl} that equal or exceed 2, the NaCl$_{equiv}$ contents of fluid(s) of the latter magmas will exceed 2.4 wt%, and such Cl concentrations cause two-fluid formation at temperatures ≥ 700° C at ≤ 100 MPa (Fig. 7). These comparisons do not account for the capacity of CO_2 to enhance the stability field for vapor and liquid; as discussed in the following, some basaltic magmas contain appreciable CO_2 concentrations.

Alkali chloride-bearing aqueous-carbonic fluids. Only a handful of experiments have addressed equilibria involving melts and saline, aqueous-carbonic fluids, but the preliminary

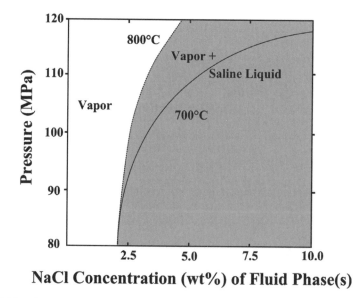

Figure 7. Plot of phase relations for the vapor-dominated side of the vapor-saline liquid solvus (gray area) at 700 °C (solid curve) and 800 °C (dashed curve) and pressures of 80 to 120 MPa for system NaCl-H_2O. The NaCl$_{equiv}$ concentrations of many magmatic fluids, some of which feed volcanoes, are ≥ 2 wt%; thus, NaCl- and H_2O-bearing fluids will boil and generate a saline liquid along with alkali chloride-poor vapor at these conditions. See text for discussion.

results are relevant to vapor-liquid phase relations in volcanic systems. The addition of NaCl and KCl to phonolitic melts coexisting with a H_2O- and CO_2-bearing fluid at 925-1000 °C and 200 MPa, for example, causes strong changes in the mole fraction of H_2O in the melt ($X_{H_2O}^{melt}$) as the $X_{H_2O}^{bulkfluid}$ decreases from 0.7 to 0.3 (Webster 2005). These observations bear on the presence of one vs. two fluids, given that the phase relations of coexisting fluids depend on the activities of volatile components in the system. In a similar study at 200 MPa, Botcharnikov et al. (in press) determined that the addition of CO_2 to the pseudo-system andesite melt-H_2O-Cl at 1050 and 1200 °C caused a simple dilution of the H_2O and Cl in a single fluid at low bulk-system Cl concentrations. At higher Cl concentrations in the bulk system, however, they observed that the stability field of vapor and saline liquid expands with the addition of CO_2. These relationships involving silicate melts are qualitatively consistent with those described previously for the melt-free, type system H_2O-CO_2-NaCl at conditions relevant to volcanism.

The CO_2 in basaltic and andesitic magmas affects fluid phase equilibria at shallow crustal pressures (Bowers and Helgeson 1983a,b; Joyce and Holloway 1993; Duan et al. 1995, 2003; Shmulovich et al. 1995; Shmulovich and Graham 2004). Basaltic magmas can saturate, early and deep, with a CO_2-enriched vapor phase, even at depths ≥ 30 km (Roedder 1984; Gerlach and Taylor 1990; Javoy and Pineau 1991; Hedenquist and Lowenstern 1994; Wallace and Anderson 2000). This led Roedder (1992) to observe that the formation of CO_2-rich fluid inclusions by unmixing of a dense CO_2-bearing fluid from basaltic magmas may be universal. Carbon dioxide abundances can exceed those of H_2O in some basaltic magmas (Giggenbach 1997), and Gerlach et al. (2002) report that the bulk CO_2 content of primary Kilauea magma determined from data on CO_2 emission and magma supply rates is ca. 0.7 wt%. Wallace (2005) estimates that most arc-related basaltic magmas should contain ≥ 0.3 wt% CO_2. He notes that this range far exceeds the CO_2 contents of most arc-derived silicate melt inclusions, and posits that these magmas are likely vapor-saturated at depths of the middle through upper crust. Thus because CO_2 plays a dominant role in the exsolution of magmatic fluid(s) and on phase equilibria of the type system NaCl-H_2O-CO_2, it also exerts a strong control on subsequent fluid unmixing when basaltic magmas ascend toward the surface.

Investigations of melt inclusions show that CO_2 abundances in andesitic to felsic subduction-related magmas range from tens of ppm up to 0.2 wt% (Wallace and Anderson 2000; Newman and Lowenstern 2002; Wallace 2005). However, Wallace (2005) advises against considering the CO_2 contents of evolved, subduction-zone magmas as representing maximum values given that these magmas have likely lost CO_2 to a vapor phase. Nevertheless, CO_2 partitions quite strongly in favor of magmatic vapors (Newman and Lowenstern 2002) because of its low solubility in felsic and andesitic melts at low pressure (Holloway 1976; Holloway and Blank 1994). Thus, these abundances of CO_2 in alkali chloride-bearing magmatic fluids enhance the stability of vapor and liquid during magma ascent and cooling. These observations highlight the need for more experimental data on phase relations in melt-H_2O-CO_2-alkali chloride-transition metal chloride-alkaline earth chloride systems at shallow- to mid-crustal pressures (e.g., ≤ 500 MPa).

Fluids enriched in O-H-C-S-Cl. There are also few published experimental data on fluid-fluid phase relations and volatile component solubility relationships in systems enriched in H_2O, Cl, SO_4^{2+} species, S^{2-} species, and CO_2/CO_3^{2-}. These data are critical given that most magmas are enriched in the volatile components H, C, S, and Cl, and because they affect the phase equilibria of sulfide-, sulfate-, chloride-, carbon dioxide-, and carbonate-bearing vapors and liquids. Recent experimental work involving rhyodacitic melts and fluids containing H_2O, S, and Cl indicates that the addition of S may reduce the tendency for unmixing of H_2O- and Cl-enriched fluids (Botcharnikov et al. 2004). Given the lack of experimental data, some investigators have turned to the analysis of natural assemblages of coexisting fluid inclusions and melt inclusions in phenocrysts to better understand fluid phase equilibria in such systems.

The solubility of S in silicate melts at high $a_{H_2O}^{mt}$ has been determined for pressures ≤ 400 MPa, but solubility data that address a wide a range in (H$_2$O/S) ratios for the pseudo-system silicate melt-H$_2$O-S are virtually nonexistent. At f_{O_2} < NNO, S^{2-} is the dominant S species in silicate melts (Katsura and Nagashima 1974; Scaillet and Macdonald 2006), so basaltic and andesitic magmas exsolve immiscible sulfide (Fe-, S-, and O-bearing) liquids that buffer the f_{S_2} of coexisting melts and fluids at relatively low values (Mathez 1976; Wallace and Carmichael 1992; Carroll and Webster 1994; Scaillet and Pichavant 2003; Moretti and Ottonello 2005). A few experimental studies have determined phase compositions for equilibria involving sulfide liquids and coexisting vapor with or without silicate melt (Haughton et al. 1974; Katsura and Nagashima 1974). Experiments involving basaltic melts coexisting with sulfate and sulfide liquids show that the transition from sulfide to sulfide plus sulfate saturation occurs with increasing S in relatively anhydrous melts (Jugo et al. 2005).

Experimental constraints on magmatic fractionation of S and other constituents between S-bearing vapors and liquids, with or without silicate melt, are few. The immiscible sulfide liquids that are stable at relatively low f_{O_2} buffer the Fe and S contents of melts and any coexisting vapors (Scaillet et al. 1998). The role of vapors in sequestering volatile components and trace elements from a coexisting sulfide liquid, however, is poorly understood. For example, Au and the PGE should fractionate between S-dominated liquids and melts (Jugo et al. 2005). The generation of precious metal-rich silicate melts is favored by either a lack of or the elimination of existing sulfide phases (whether solid or liquid) because the highly siderophile metals Au, Pt, Rh, Ru, Pd, Re, Os, and Ir are readily hosted in sulfides (Mitchell and Keays 1981; Richards 1995; Jugo et al. 2005). The issue of sulfide liquid stability also bears on fluid-fluid processes because sulfide saturation of some silicate melts may retard formation of an aqueous vapor phase (Jugo et al. 1999). In melts characterized by f_{O_2} in the range where the mol% S^{6+} and S^{2-} are approximately equivalent, the likelihood of forming an aqueous vapor phase is less because these systems require higher concentrations of S in the melt for co-saturation with sulfides and anhydrite (Scaillet et al. 1998; Jugo et al. 2005).

Fluorine-bearing fluids. The HF concentrations of some volcanic gases are significant (Symonds et al. 1988), and hydrous as well as anhydrous fluoride phases containing Mg, Al, Na, and/or Ca occur in encrustations on the walls of fumaroles. These minerals reflect the consequences of fluoride alteration at volcanoes including Usu, Erebus, Katmai, and Kilauea (Naboko 1957; summarized by Africano and Bernard 2000). The condensation of F-bearing species from vapor to form HF-enriched liquids represents the main process of generating such fluids (Aoki et al. 1981), so multiple F-bearing fluids should occur at these volcanoes.

The influence of F on the stability and likelihood of two coexisting fluids in magmatic systems is not well resolved. Some experimental studies show that increasing F concentrations increase H$_2$O solubility in felsic aluminosilicate melts (Holtz et al. 1993; Carroll and Webster 1994) and increase the reciprocal solubility of the melt in aqueous fluids (Webster 1990; Carroll and Webster 1994). So, these relationships work to reduce the likelihood of multiple H$_2$O-bearing fluids equilibrating with silicate melts. Conversely, other investigations involving primary FI and coexisting melt inclusions of highly evolved granitoids show evidence of multiple fluids for F- and/or B-enriched felsic systems (Thomas et al. 2005; Thomas and Webster 2006). Thus, the effect of F on fluid-fluid phase relations requires further study.

Constraints from stable isotope geochemistry

Stable isotope geochemistry is used to investigate fluid-vapor-liquid equilibria and define the sources of fluids in volcanic environments. Relatively recent experimental measurements of the fractionation of hydrogen and oxygen isotopes between vapor and saline liquids at temperatures ≤ 600 °C (Shmulovich et al. 1999) allow prediction of the isotopic fractionation that occurs when exsolved magmatic volatile components unmix upon reaching pressure-, temperature-, and composition-conditions favorable for formation of low salinity vapor and saline liquid. Application of isotopic fractionation data to constrain the isotopic composition and origin of the saline liquid component of two-fluid systems will improve as more H_2O-electrolyte systems are investigated experimentally. Most data for the systems Na_2SO_4-H_2O, $CaCl_2$-H_2O, $MgCl_2$-H_2O, $CaSO_4$-H_2O, and $MgSO_4$-H_2O are currently limited to temperatures < 100 °C, so the temperature ranges used in future experiments must increase (Horita et al. 1993). Nevertheless, modeling of hydrogen and oxygen isotope fractionation in low-salinity vapor through use of fractionation factors measured in the pure H_2O liquid-vapor system should yield reasonable results because the salinity of the vapor phase is generally quite low. As a result of the low salinity, the departure of hydrogen and oxygen isotope fractionations in the vapor relative to those measured in pure H_2O vapor and liquid should also be low.

At present, there are few experimental constraints on the fractionation of the stable isotopes of oxygen and hydrogen between liquid and vapor in H_2O-electrolyte systems above the critical temperature of H_2O (e.g., 374 °C) (Berndt et al. 1996; Driesner 1997a,b; Shmulovich et al. 1999; Driesner and Seward 2000). The study of Shmulovich et al. (1999) generated experimental data for temperatures up to 600 °C with up to 33 wt% NaCl in the saline liquid of the H_2O–NaCl system and up to 40 wt% KCl in liquid with the H_2O–KCl system. Their data indicate that deuterium (D) fractionates into the vapor from 350 °C to 600 °C, so liquid is depleted in D relative to the vapor. Oxygen and hydrogen data indicate that the magnitude of isotopic fractionation correlates linearly with the salt concentration of the liquid (Figs. 8a,b) and that fractionation measured in salt-bearing systems is larger than that in pure H_2O vapor and liquid at similar temperatures (Shmulovich et al. 1999).

The dependency of vapor-liquid fractionation on liquid salinity is referred to as the isotopic solution or salt effect (Taube 1954; Horita et al. 1995; Berndt et al. 1996), and it is greater for hydrogen than oxygen. Shmulovich et al. (1999) report that D/H fractionation increases by 0.55‰ per wt% NaCl in liquid, and also showed that the 400 °C data of Driesner (1997a,b), Driesner and Seward (2000), and Berndt et al. (1996) are in good agreement. At constant salt content, D/H fractionation between liquid and vapor decreases with increasing temperature, as expected from the $1/T^2$ dependence of most isotopic fractionations (Fig. 8a). Shmulovich et al. (1999) also report that ^{18}O fractionates preferentially into the liquid over the 350-600° range, and that $^{18}O/^{16}O$ fractionation increases by ~0.05‰ per wt% NaCl in the liquid at constant temperature relative to the oxygen isotope fractionation data from pure H_2O vapor and liquid of Horita et al. (1995) and Driesner (1997a,b; Fig. 8b). The maximum theoretical fractionation of ~28‰ in D/H and 2‰ in oxygen should be possible along the liquid + vapor + halite curve at temperatures of 350-600 °C. The relationship between isotope fractionation and liquid salinity gives rise to a ~28‰ enrichment of δD and a decrease in $\delta^{18}O$ in the vapor by 2‰. These relative changes in δD vs. $\delta^{18}O$ that occur during boiling at halite saturation allow for clear identification of vapors that have undergone this type of fractionation (Fig. 8c), and the significance of these processes is that vapors leaving a hydrothermal system will attain a comparatively heavy D isotopic signature (Fig. 8c). Vapor separation from saline liquids is easily produced by isobaric heating or by isothermal or adiabatic decompression. Moreover, addition of other components (e.g., CO_2 and perhaps SO_2 or H_2S) to a H_2O-, NaCl-, and KCl-dominated system will widen the immiscible vapor-liquid region, thereby initiating isotopic fractionation perhaps earlier during magma evolution (i.e., at higher pressures and/or temperatures).

When modeling isotopic compositions of vapors and liquids in volcanic and hydrothermal systems, the temperature of vapor-liquid separation must be constrained with fluid inclusion homogenization temperatures and/or stable isotope geothermometry. In contrast to the results observed with the boiling effect at halite saturated-conditions described above, processes of single-step boiling of liquids such as the one shown in Fig. 8c (with $\delta D = -33‰$ and $\delta^{18}O = 7.3‰$ at initial temperature of 350 °C, reported from Poas volcano by Rowe 1994), will produce

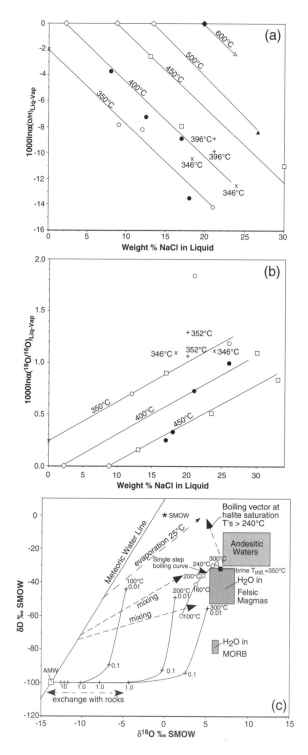

Figure 8. (a) Plot of 1000lnαD/H(Liq-Vap) vs. NaCl concentration in the liquid illustrating fractionation of hydrogen isotopes between liquid and vapor from 350 to 600 °C (modified from Shmulovich et al. 1999). Also shown for reference are 346 °C data (×'s) from Horita et al. (1995) and 396 °C data (+'s) from Driesner and Seward (2000). Critical compositions (where fractionation equals zero) on the x-axis at 400, 450, and 500 °C (open diamonds) are from Bischoff and Pitzer (1989), and at 600 °C (filled diamond) from Shmulovich et al. (1995). y-axis intercept for 350 °C data from pure H_2O vapor-liquid fractionation data of Horita et al. (1995). See text for discussion. (b) Plot of 1000lnα^{18}O/^{16}O(Liq-Vap) vs. NaCl concentration in the liquid illustrating fractionation of oxygen isotopes between liquid and vapor at 350, 400, and 450 °C (modified from Shmulovich et al. 1999). Also shown for reference are 346 °C data (×'s) from Horita et al. (1995) and 352 °C data (+'s) from Driesner and Seward (2000). Critical compositions on x-axis at 400 and 450 °C (open diamonds) are from Bischoff and Pitzer (1989). Fractionation data for pure H_2O at 350 °C (X) are from Horita et al. (1995). (c) Plot of δD vs. δ^{18}O for fluids in volcanic environments. Fields for H_2O in felsic magmas and H_2O in MORB from Taylor (1986), and andesitic waters from Giggenbach (1992). Plotted for reference are data for meteoric water line of Craig (1961). Fluid compositions resulting from water-rock exchange were modeled according to principles of Ohmoto and Rye (1974) and Field and Fifarek (1985) for exchange at 100, 200 and 300 °C with water to rock mass ratios ranging from 0.01 to 10, average meteoric water (AMW) with δD = −100‰ and δ^{18}O = −13‰, and volcanic rock with δD = −70‰ and δ^{18}O = 7.0‰. The trend of isotopic fractionation due to continuous evaporation at 25 °C is from Christenson (2000). Single-step boiling curve for brine with δD = −33‰, δ^{18}O = 7.3‰ and initial temperature of 350 °C with boiling from 300 to 100 °C after Rowe (1994). Standard Mean Ocean Water (SMOW) marked by asterisk. Boiling vector at halite saturation modified after Shmulovich et al. (1999).

fractionation trends for the vapor that are subparallel to the meteoric water line. The depletion of the vapor in δD and $\delta^{18}O$ increases with decreasing temperature of vapor-liquid separation (Fig. 8c). This occurs because D is increasingly concentrated in liquid H_2O with temperature decreases below 221 °C, and ^{18}O is concentrated in the liquid at all temperatures below the critical temperature (Truesdell et al. 1977). Vapors generated by boiling below 221 °C are increasingly depleted in both D and ^{18}O.

Many fluids from volcanic environments that are sampled and analyzed for hydrogen and oxygen isotopes yield δD and $\delta^{18}O$ arrays that span from those plotting within, or near ($\pm$ 1-2‰ in $\delta^{18}O$ and δD), the field of H_2O from felsic magmas (Taylor 1986) or span from the andesitic water field of Giggenbach (1992) to those plotting along trends that project back towards the meteoric water line (Field and Fifarek 1985; Sheppard 1986; Hedenquist and Lowenstern 1994; Taran et al. 1995). Hydrogen and oxygen isotopic data and abundances of dissolved chlorine from low- and high-temperature fumaroles and warm springs at Kudryavy volcano indicate two-component mixing of high-temperature magmatic fluid with meteoric water (Taran et al. 1995). The evidence for mixing includes: a) data from high-temperature (> 700 °C) fumarole samples plotted within, or near the field for andesitic waters (Taran et al. 1989, 1995; Giggenbach 1992) and that these samples had the highest dissolved Cl concentration, and b) the lowest-temperature fumarole samples (< 200 °C) plotted closer to the meteoric water line and these had the lowest dissolved Cl concentrations.

Mixing of magmatic and meteoric water components is not the only process that can account for apparent mixing trends in plots of δD versus $\delta^{18}O$, however. Some of these isotopic trends may arise from water-rock exchange at varying temperatures and water-rock mass ratios (Field and Fifarek 1985; Sheppard 1986). Modeling of hydrogen and oxygen isotope exchange during water-rock interaction places constraints on the likely origin of fluids in hydrothermal systems through application of principles developed by Ohmoto and Rye (1974) and summarized by Field and Fifarek (1985). The final isotopic composition of water (δ_{fw}) after equilibration with rock is given by:

$$\delta_{fw} = \delta_{ir} - \Delta_{r-w} + [(w/r)(\delta_{iw})]/1 + (w/r) \tag{7}$$

where δ_{ir} is the initial isotopic composition of the rock, δ_{iw} the initial (unexchanged) composition of the water, Δ_{r-w} is the fractionation factor between rock and water, and (w/r) is the ratio of exchanged oxygen and hydrogen atoms in the water relative to those in the rock. Unaltered andesites, dacites, and rhyolites typically have $\delta^{18}O$ of 7‰ and δD of −70‰. These rocks also usually contain approximately 50 wt% oxygen and up to 0.11 wt% hydrogen. Therefore, the oxygen and hydrogen isotopic composition of a fluid that has equilibrated with volcanic rock of similar characteristics can be determined from:

$$\delta^{18}O_{wf} = 7 - \Delta_{r-w} + [(1.8R)(\delta^{18}O_{wi})]/(1 + (1.8R)) \tag{8}$$

and

$$\delta D_{wf} = -70 - \Delta_{r-w} + [(100R)(\delta D_{wi})]/(1 + (100R)) \tag{9}$$

In these equations, the coefficients 1.8 and 100 represent the ratios of the wt% oxygen in H_2O (88.8%) relative to that in rock (50%) and the wt% hydrogen in H_2O (11.2%) relative to that in rock (0.112%), and R is the water/rock mass ratio. Examples of water-rock exchange (Field and Fifarek 1985) approximate the fractionation of oxygen isotopes between rock and fluid with plagioclase feldspar (An_{30}) and H_2O (O'Neil and Taylor 1967):

$$1000 \ln\alpha = 2.68(10^6/T^2) - 3.29 \tag{10}$$

The fractionation factor, α, for hydrogen isotopes between fluids and rocks is assumed to be equivalent to that of the chlorite-H_2O fractionation factor (Taylor 1979).

Water-rock exchange trends for D and ^{18}O are illustrated in Figure 8c for constant temperatures of 100°, 200°, and 300 °C with R varying from 10 to 0.01. These data involve an arbitrary value for average meteoric water (AMW) with δD of –100‰ and $\delta^{18}O$ of –13‰. The meteoric waters exchange volatile components with a typical volcanic rock having δD of –70‰ and $\delta^{18}O$ of 7‰. At high water to rock mass ratios ≥ 1.0, the trend lines are horizontal and the D/H signature of meteoric water is retained because the quantity of hydrogen present in H_2O-rich fluid far exceeds that of the rock. At temperatures > 300 °C, the extent of isotopic exchange is greater and the positive shift in $\delta^{18}O$ of the fluid at constant water-rock ratio can be 10‰ (Fig. 8c). However, at lower water to rock ratios (e.g., ≤ 0.1) the hydrogen isotopic composition of the rock begins to affect the δD value of the fluid and shifts in $\delta^{18}O$ of ca. 10‰ and δD of the fluid of up to 20‰ can occur. Similar modeling techniques were utilized by Rowe (1994) to model the origin of fumarole condensate waters at Poas volcano.

Evaporation processes affecting surface waters may also give rise to positive slopes in plots of δD versus $\delta^{18}O$. As described by Varekamp and Kreulen (2000), evaporation from lake surfaces takes place in two steps that operate in adjacent regions: 1) a thin boundary layer with a temperature equivalent to that of the lake water and with a vapor pressure at H_2O saturation forms immediately above the lake surface. The isotopic composition of the H_2O vapor in the boundary layer is related to that of the lake water by the equilibrium fractionation factor, α $(= R_{liq}/R_{vap})$ with R as the ratio of the isotopic species in the two phases. Values of α at varying temperatures are well known (Majoube 1971; Truesdell et al. 1977; Gonfiantini 1986; Horita and Wesolowski 1994). 2) Evaporation also involves kinetic fractionation between H_2O vapor in the open atmosphere and the boundary layer. The H_2O vapor pressure in the open atmosphere is dependent on atmospheric temperature and the relative atmospheric humidity. As a result of differences in the H_2O vapor concentration and isotopic composition between the boundary layer and open atmosphere, H_2O molecules diffuse upwards resulting in kinetic fractionation, and this process enhances the flux of the lighter species. A general equation for kinetic fractionation, $\Delta\varepsilon$, is given by Gat (1996):

$$\Delta\varepsilon = (1 - h)\, n\, \Theta\, CD1000 \qquad (11)$$

where h is the relative atmospheric humidity (defined as the pressure of H_2O vapor in air/ pressure of H_2O vapor in air at saturation and at atmospheric temperature); n is a turbulence parameter with a value of 0.5 (taken for open water bodies under natural conditions); and the weighting factor, Θ, is assumed to be 1. The diffusive fractionation term for ^{18}O, $CD(^{18}O)$ is 28.55‰ and that for deuterium $CD(D)$ is 25.115‰. In crater lakes where the lake temperature is much greater than atmospheric temperature, h is replaced by h^*, the lake temperature-normalized humidity defined by Varekamp and Kreulen (2000) as $h^* =$ (the pressure of H_2O vapor, at atmospheric temperature)/(pressure of H_2O vapor at saturation, at lake temperature). Values of h^* are much smaller than values of h, and the kinetic isotope fractionation is larger for ^{18}O than for D (Gonfiantini 1986; Gat 1996). Thus, plots of δD vs. $\delta^{18}O$ exhibit slopes (e.g., 4-5) that are flatter than that of the meteoric water line (which is a function of h) for evaporation involving warm crater lakes.

The general effect of evaporation is to increase δD and $\delta^{18}O$ of the liquid. In these cases, surface waters that have been subjected to evaporation may descend and absorb magmatic vapors, be heated to the boiling point, or exchange oxygen and hydrogen with host volcanic rocks. Heating results from absorption of magmatic vapor and/or mixing with liquids formed from an exsolved magmatic single-phase fluid; this fluid may also unmix upon decompression, heating, or addition of other volatile components (e.g., CO_2, SO_2, H_2S).

The combination of modeling of magma crystallization, fluid exsolution, and fluid evolution along with fluid inclusion-based constraints and stable isotope data provides insight into late-stage magmatic-hydrothermal processes. For example, Shinohara and Hedenquist (1997) applied δD data of hydrothermal alteration phases and the Cl contents and phase equilibria of fluid inclusions to the exsolution of fluid and the occurrence of two-phase fluid formation for a Cu- and Au-mineralized porphyry of the Philippines. It was determined that open-system separation of vapor (containing HCl and SO_2, with a δD of -20 to $-25\permil$) from a hypersaline liquid was followed by condensation of the vapor into meteoric water which formed an acidic fluid. This fluid generated the advanced argillic alteration of the shallow epithermal deposit that is located over the porphyry mineralization beneath. Simultaneously, the less mobile, coexisting saline liquid (with δD of $-45\permil$ at 550 °C) interacted with other host rocks to produce the characteristic potassium-silicate alteration assemblage of the porphyry environment. Modeling indicates that these processes occurred at about 40-50 MPa, while the system was comparatively hot, and with phenocryst abundances in magma of 30 to 50%. At this stage, modeling further suggests that the vapor (with ≈ 0.7 wt% $NaCl_{equiv}$) comprising approximately 92 wt% of the bulk fluid and saline liquid (≈ 50 wt% $NaCl_{equiv}$) made up the remainder. With continued cooling and crystallization, however, magma convection stalled as the magma became stagnant (at about 50% crystals as predicted by petrographic studies). Consequently, heat loss and the rate of fluid exsolution from the magma chamber were reduced, causing the isotherms to retract progressively deeper into the hydrothermal system. In addition, the vapor-liquid solvus was not intersected by the later stage, cooler magmatic fluids, and only a single fluid was stable. This cooler fluid formed a white mica alteration overprint of the early potassium silicate stage at ~350 °C and 6 wt% NaCl equivalent, similar to the fluid salinity and isotopic composition values predicted by modeling (Shinohara and Hedenquist 1997).

SYNTHESIS AND APPLICATION TO VOLCANIC PROCESSES

Review of two fluids in volcanic environments

This chapter documents a common co-occurrence of two fluids in a variety of subaerial, subaqueous, and submarine volcanic as well as tectonically related-plutonic environments, although much of the evidence of two coexisting volcanic fluids is indirect. Implied constraints come from modeling that supports the presence of two fluids in shallow magmatic and volcanic environments at elevated temperatures and depths equivalent to pressures ≤ 500 MPa (Hedenquist and Lowenstern 1994). The abundances of the predominant volatile components (e.g., H_2O, CO_2, SO_2/H_2S, Cl) emitted from magmas and hot volcanic rocks are well established and allow comparison with binary or ternary model systems in order to predict the likelihood of two fluids in volcanic environments. However, given the lack of experimental data that bear on chemically complex and geologically relevant compositions, equilibria and processes involving fluids in volcanic environments are typically interpreted with regard to the compositionally simple pseudo-system melt-$NaCl$-$H_2O\pm CO_2$. For example, two fluids are stable with as little as 0.1 wt% NaCl (in the bulk integrated fluid) in CO_2-poor systems at ca. 50 MPa and 600 °C and two fluids occur with 1 wt% NaCl (and no CO_2) at ca. 75 MPa and 800 °C (Fig. 1). These conditions apply to many volcanic and magmatic systems as fluids ± magma ascend, decompress, and cool. However, these examples do not account for the influence of CO_2, SO_2, or H_2S on the size of the two-fluid stability field.

In addition to vapor plus aqueous liquid, other potential fluid combinations include vapor plus chloride- and/or sulfate-enriched saline liquids and vapor plus sulfide- or carbonate-rich liquids. However, there are insufficient data on volatile component solubilities, chemical exchanges, and other processes involving sulfide liquids coexisting with aqueous or aqueous-carbonic vapors or liquids to adequately model the latter conditions at present.

Fluids generated by separation of vapor and liquid or vapor and saline liquid may develop chemically reactive characteristics (e.g., strongly acidic properties) that may be quite unlike those present in single-fluid environments because phase separation produces vapor and liquid that are chemically different from each other and from the parent fluid (Butterfield

2000). Acidic vapors generated via fluid boiling and alkaline saline liquids that are formed by condensation processes in volcanic and associated hydrothermal environments, for example, influence a variety of processes as a result of the chemical and physical changes that occur during fluid-fluid exchange reactions. These processes are described below.

Fractionation of volatile components between fluids

Fluid-soluble components fractionate between fluid phases when aqueous or aqueous carbonic vapors form by boiling of coexisting aqueous liquids, saline liquids, and/or sulfide- or carbonate-rich liquids. These processes may involve equilibrium exchange, and in some circumstances they may involve aluminosilicate melts. Knowledge of such fractionation processes in volcanic systems is based on analysis of high-temperature volcanic vapors and condensates as well as experimental partitioning studies (Candela and Piccoli 1995). In addition, recent advances in microanalytical methodologies allow the components of fluid inclusions ± silicate melt inclusions ± sulfide melt inclusions to be analyzed (Ulrich et al. 1999; Bodnar and Student 2006; Halter and Heinrich 2006; Layne 2006; Pettke 2006). If these inclusion assemblages involve multiple fluid types and equilibrium, then one can determine the partition coefficients for volatile components and other constituents.

The solubilities of some magmatic components differ strongly between vapor and aqueous or aqueous-carbonic liquids vs. those in vapor and saline liquids, sulfide liquids or carbonate liquids. The relative solubilities may also change with varying pressure, temperature, and/or bulk system composition (Giggenbach 1980). Aqueous vapors tend to dissolve CO_2, SO_2, H_2S, S_2, HCl, HF, H_2, CO, CH_4, COS, and other volatile components selectively, whereas coexisting saline liquids sequester alkali-enriched salts dominated by chlorides and sulfates but fluorides, phosphates, and carbonates may also occur (Hedenquist and Lowenstern 1994; Symonds et al. 1994; Hedenquist 1995).

It has been observed that "*boiling has two important effects: it directly increases solution concentration, but more importantly, it is a mechanism whereby volatile constituents can be removed, leaving the solution more alkaline and less capable of metal transport*" (Skinner 1997). Thus, processes of fluid-fluid equilibration and/or simple fluid-fluid interaction modify the chemical and thermal characteristics of volcanic and magmatic-hydrothermal systems. Fluid phase separation changes the ionic strength of fluids, and the pH of fluids containing dissociated fluid species at lower temperatures (Cunningham 1978) because acid-forming constituents partition in favor of magmatic vapors whereas other species dissolve preferentially into saline liquids. Note that the tendency for ion-ion association in aqueous fluids decreases with temperature, so the acidity of acid-forming species increases with decreasing temperature in vapor or liquid. These relationships have a fundamentally significant influence over a broad range of fluid compositions for systems undergoing boiling (Reed 1992a,b; Fournier 1999; Einaudi et al. 2003). For example, the loss of CO_2 from a boiling liquid to vapor increases the pH of the residual liquid in most systems. Conversely, the temperature decrease associated with isoenthalpic boiling causes species like HCl and HSO_4^+ that are dissolved in liquids to dissociate to a greater extent. These processes release more H^+ to the liquid and lower its pH (Reed 1992a,b). Fractionation of the reduced species H_2, H_2S, and CH_4 into the vapor may also control Eh and generate a comparatively reduced vapor and an oxidized liquid (Reed 1992b).

These generalizations notwithstanding, acidic liquids also occur in and near some volcanoes (Hedenquist 1995). Giggenbach et al. (2003), for example, describe evidence of acidic liquids at White Island, New Zealand, that are apparently generated by a complex combination of processes. Magmatic fluids interact with seawater that infiltrates the underlying hydrothermal system, and the seawater exchanges components with the hot volcanic rocks, which modifies its chemistry. Other complex interactions between vapors and liquids have been recorded. Magmatic vapors, for example, may condense into meteoric water and generate highly reactive fluids (Einaudi et al. 2003).

The vapors and liquids that form in volcanic and related environments exhibit different physical properties that influence the efficiency of phase separation. Most vapor and liquid phases have large differences in densities (density contrast > 1 gram/cm^3; Anderko and Pitzer 1993), and this facilitates their physical separation in magmas and/or in adjacent, porous volcanic rocks. Physical phase separation occurs in open systems and leads to the ascent of less-dense vapor from relatively dense liquid. Moreover, processes of vapor boiling from a liquid or condensation of liquid from a pre-existing vapor may be episodic or they may occur continuously as the fluid ascends through a volcanic conduit or convects in a hydrothermal system. The continuous ascent of vapor and coexisting saline liquid (in the system H_2O-$NaCl$ at isothermal conditions and pressures < 200 MPa) results in a progressive decrease in the $NaCl$ content of the vapor and an increase in the $NaCl$ concentration of the liquid (Fig. 4). Likewise, decreasing pressure and temperature reduce the $NaCl$ content of the vapor and increase the $NaCl$ concentration of the liquid.

The exchange of volatile components and other species between vapor and liquid modifies the compositions of each fluid phase, and this also influences other processes. Physical separation of vapor from liquid in open systems generates two fluids that are characterized by differing solvi (because of the differences in pressure, temperature, and composition) that subsequently are different from the initial solvus that applied to the first-formed vapor plus liquid. Their physical separation sets each fluid on a separate path of subsequent compositional evolution. With decreasing pressure and temperature, both fluids continue to evolve, and each may exsolve new vapors and/or liquids if these processes occur episodically (Hedenquist 1995; Thomas et al. 2005). Thus, the exchange of components between vapors and liquids and the associated vapor- and/or liquid-exsolution processes can be complex.

Consequences of two fluids in volcanic environments

Dissolution and transport in volcanic fluids. A wide variety of transition metals, heavy metals, and metalloids, including those of potential economic importance, dissolve in, are transported by, and condense (or sublimate) from volcanic vapors. These vapor-soluble components occur at ppm to ppb levels (Williams-Jones and Heinrich 2005).

The speciation of these components is a function of the phase equilibria and of subsequent fractionation, given that C, S, Cl, and F do not partition equally between vapor and liquid. Arsenic, In, Se, and Au (Zreda-Gostynska et al. 1989) and Sb, Ag, Mo, Re, Hg, Cd, Cr, and Ni, for example, occur in vapors from andesitic and basaltic volcanoes (Symonds et al. 1987; Nriagu 1989; Taran et al. 1997; Williams-Jones and Heinrich 2005). Symonds et al. (1992) calculated that Na, K, Zn, Pb, and Cu in vapors are primarily dissolved as chloride species, and that much of the Al and Ca in vapors dissolves as fluoride species. Moune et al. (2006) observed that Y, the REE, Th, and Ba were significantly enriched in the sub-Plinian plume of the 2006 eruption of Hekla volcano, Iceland. Thermodynamically reconstructed compositions of vapors (that account for gas-condensate equilibria and potential mixing processes) emitted from the basaltic-andesite magmas of Kudryavy volcano, Kuril Islands, indicate that most: (1) Ga, In, Tl, Fe, Co, Ni, Na, and K dissolve as chloride species; (2) Ge, Bi, Sn, and Pb dissolve as sulfide and chloride species; (3) Be, Al, and Si as fluoride and oxide species; (4) Ti, Zr, V, Mo, W, and Re as oxyfluoride and oxychloride species; and (5) Hg and Cd are transported as neutral species (Churakov et al. 2000). Vapor-phase geochemistry and vapor-liquid phase equilibria for degassing of Merapi volcano, Indonesia, have also been closely examined. The speciation of more than 25 metals, non-metals, and metalloids in volcanic vapors released from andesitic magma and lava was modeled with SOLVGAS, and Symonds et al. (1987) concluded that most components dissolve as chloride species. Similarly, chloride species dominate the dissolution, transport, and speciation of many components in volcanic vapors of Augustine volcano, Alaska (Symonds et al. 1990, 1992).

Hydrothermal experiments and the compositions of natural fluid inclusions (Audétat and Pettke 2003; Heinrich 2007a) determine volatile component, trace, and ore element partitioning between vapor and liquid with or without a coexisting silicate melt. The focus on ore metals is particularly relevant to processes of alteration and mineralization as vapor separation in open systems allows for selective transport of some metals (Pokrovski et al. 2005). A detailed review is beyond the scope of this chapter and is addressed elsewhere (Heinrich et al. 1992; Figs. 3 and 4 of Heinrich 2007a; Liebscher 2007), so we simply note that experimental and fluid inclusion studies constrain the partitioning of base metals, precious metals, REEs, and other transition elements between alkali chloride- and alkaline earth chloride-dominated saline liquids and vapor phases dominated by H_2O, CO_2, S, Cl, and other volatile components (Williams et al. 1995; see summary of Candela and Piccoli 1995; Shmulovich et al. 2002; Pokrovski et al. 2005; Simon et al. 2005; Williams-Jones and Heinrich 2005).

Hydrothermal alteration. Volcanic crater lakes provide the evidence for extensive leaching of volcanic rocks by hyperacidic fluids (Fig. 3). The hyperacidic lakes in Rincón de la Vieja volcano, Costa Rica, and at Kawah Ijen volcano, Java, Indonesia, for example, contain elevated levels of Cl, S, Ca, Mg, Al, Fe, Mn, Cu, Zn, F, B, and other components (Delmelle et al. 2000; Kempter and Rowe 2000). Thermodynamic modeling of Kawah Putih lake in Patuha volcano, Java, Indonesia, indicates that the lake waters are saturated in silica, barite, pyrite, and a variety of Pb-, Sb-, Cu-, As- and Bi-bearing sulfides (Sriwana et al. 2000).

The fractionation of components between vapor and liquid has fundamental significance to hydrothermal alteration of volcanic and neighboring rock (Reed 1997). As acid-forming species dissolved in mobile vapors pass through and react with rock, the pH of the vapor increases while the vapor leaches components from the rock. The compositional effects of boiling, condensation, mixing, and subsequent fluid-rock interaction have been modeled with the programs CHILLER and SOLVEQ (Reed 1982, 1998). These models can account for the sequestration of CO_2 and H_2S by vapor, followed by acid-neutralization reactions between vapor and rock (forming multiple alteration assemblages) as well as subsequent vapor condensation and fluid mixing processes that modify the dominant fluid species.

Metasomatism modifies permeabilities of altered volcanic rock through processes of dissolution, precipitation, and fracture sealing which can control the extent and timing of the flow of vapor and/or liquid in convecting magmatic-epithermal environments. The role of multiple fluids in processes of hydrothermal alteration of shallow bodies of magma and their crystalline carapaces, as well as the nature of the ductile-brittle transition in hot altered rocks, were recently addressed by Fournier (1999). Vapor and saline liquid exsolved from magma were determined to accumulate in horizontal lenses of hot, plastic rock at lithostatic pressures. This ductile region, involving two fluids, is located adjacent to another region (at comparatively lower temperatures and pressures) that involves the circulation of meteoric waters through brittle rocks at hydrostatic pressures. Both regions are separated by a self-sealed zone of relatively impermeable rock. Rapid, intermittent upsurges of magma and increases in the strain rate and/or increases in fluid pressure cause episodic breaching of the intervening, impermeable zone. These processes may also cause the expulsion of vapor and liquid into the brittle hydrothermal zone followed by cooling as well as heating (Henley and Hughes 2000), decompression, faulting, brecciation, and the deposition of hydrothermal veins (some of which may be mineralized) (Fournier 1999).

Metasomatism involving two fluids in the sealed conduits of some volcanoes may cause hydrothermal explosions. Christenson et al. (2007) recently noted that the lack of seismic or other precursory responses associated with the March 2006 hydrothermal eruption of Raoul island, Kermadec arc, New Zealand, may be an indication of hydrothermal alteration and fracture sealing in and around the volcanic conduit. These processes occurred at the base of a two-phase, vapor-liquid zone. Modeling by Christenson et al. (2007) suggests that the ductile-brittle transition zone which underlies the conduit was critically stressed by development of a vapor-static region and that the failure of the hydrothermal seal resulted from increased pressure in this region due to ascent of magmatic vapor.

The effects of extensive hydrothermal alteration bear on other volcanic processes. The consequences of acidic to hyperacidic vapor phase-driven metasomatism, caused by condensation of the vapor with its complement of acidic species, can be catastrophic given that sector collapse of a hydrothermally altered volcanic edifice can cause destruction via debris flow movement (Lopez and Williams 1993; Zimbelman 1996; Zimbelman et al. 2000, 2005). This issue is of particular concern for Mt. Rainer, Washington (Watters et al. 2000). Moreover, as observed at Mt. St. Helens, tumescence followed by large-scale edifice collapse does trigger subsequent volcanic eruptions (Sillitoe 1994; Elsworth et al. 2004).

Implications of fluid immiscibility on volcanic gas compositions and their use in forecasting eruptive activity. The composition of volcanic fluids (e.g., vapors and their condensates that are collected at surface) yields information on intensive parameters of shallowly emplaced magma (Symonds et al. 1987), and these parameters are extremely useful in the interpretation of magmatic degassing and in forecasting eruptive activity (Giggenbach 1997; Oppenheimer 2003). If magmatic fluids were to ascend directly to the surface without changing along the way, their analysis could provide information on the source magmas and on the potential for and style of impending eruptions. Volcanic fluids are controlled, however, by temperature, pressure, volatile component solubilities in magma, magma mixing, infiltration of other types of fluids, as well as processes of vapor loss and subsequent condensation of liquid. During the latter processes, the dominant magmatic volatile components and their stable isotopes fractionate between vapors and liquids. The dissimilar relative solubilities of these components, in vapor vs. liquid, has important consequences for the observed ratios of species in volcanic fluids, and this is significant because the ratios of some gaseous species have been used to interpret volcanic degassing and eruptive behaviors (Giggenbach 1975; Symonds et al. 1994; Delmelle and Stix 2000). Thus, it is crucial to know if these ratios are sensitive to vapor-liquid equilibration; if they are, then the ratios measured in volcanic vapors will differ from those in the source magmas. The (CO_2/H_2S) ratio of fumarole fluids, for instance, is affected by the condensation of liquid from vapor and by subsequent vapor-liquid exchange (Giggenbach 1995). Not all volatile components exhibit this behavior, however. The solubilities of the noble and more inert gases, N_2 and Ar, are roughly equivalent at elevated temperatures, so the (N_2/Ar) ratio of fluids does not vary significantly as a consequence of fluid boiling or condensation (Giggenbach 1997).

Stable isotope data obtained from fluids in volcanic environments allow for the discrimination of likely fluid sources and are useful for estimating temperatures at which phase separation occurs. For example, the heavy isotope of oxygen always partitions in favor of the liquid thereby leaving the vapor depleted in ^{18}O. These data also help to evaluate likely equilibria among fluids prior to phase separation. Sulfur not only plays a central role in volcano monitoring as SO_2 fluxes, but S isotopes have assisted in the discrimination (Rowe 1994) of two mechanisms of acid sulfate production at Poas volcano: (1) disproportionation of magmatic SO_2, and (2) hydrolysis of native sulfur at the base of the lake. However, the S species that are measured in volcanic gases are a consequence of the fractionation of SO_2, H_2S, and other S compounds between vapor and condensed phases (Scaillet and Pichavant 2003). These fractionation processes can be estimated through collection and interpretation of sulfur isotope data.

The use of chlorine stable isotopes as a geochemical tool for interpreting processes involving two fluids is still at an initial stage (Layne 2006; Liebscher et al. 2006). Thus, the consequences of vapor-liquid exchange and subsequent physical separation of low-density vapor from saline liquid in open systems must be addressed by observation (Giggenbach 1971), theory (Bowers and Helgeson 1983a,b; Shinohara 1994; Duan et al. 1995, 2006), and experiment (Bischoff and Pitzer 1989; Shinohara et al. 1989; Bodnar 2003).

Implications of fluid immiscibility on styles of volcanic eruption. Modeling allows interpretation of the complex interplay between magmatic and hydrothermal processes that takes place prior to and during volcanic eruptions, but most extant models only address two-phase (melt plus phenocryst) or three-phase systems (melt, phenocryst plus one fluid) (Jaupart 2000; Carrigan 2000). The computations of Zhang (2000), however, do address the thermodynamics of gas exsolution from aqueous liquids as applied to limnic and volcanic eruptions.

Observations show that: (a) vapor exsolution and segregation exert fundamental controls on eruption dynamics, (b) magma buoyancy and ascent rate are sensitive to vapor escape, and (c) slight deviations in the extent of vapor loss from magma can make the difference between explosive and effusive eruptive activities (Gerlach et al. 2002; Sparks 2003). The boiling of vapor from an aqueous liquid and/or the condensation of liquid from vapor change the abundance and composition of vapors during fluid ± magma ascent in volcanic systems. The vapor phase has such a strong control on volcanic processes because aqueous or aqueous-carbonic vapors are much more compressible than the liquid, and hence, vapor-bearing magmas are more compressible than liquid-saturated magmas. The compressibility of a pure H_2O liquid plus vapor system, for example, is orders of magnitude greater than that of a pure H_2O liquid (Geiger et al. 2005). Thus, the potential $P\Delta V$ energy (see Burnham and Ohmoto 1980) stored in a given mass of vapor (or vapor-saturated magma) will greatly exceed that in vapor-free environments, so the volume of magmatic vapor is a primary control on the explosive force of volcanic eruptions. In addition, the molar volume of vapor varies not only with pressure, temperature, and the mass of vapor present, but also with the compositions of the gas species that are present because each of the major gaseous species in volcanic vapor is characterized by different molar volumes. Thus, the partitioning of gaseous volcanic species between vapor and liquid must be accounted for when modeling volcanic processes.

Other, related processes involving vapor-liquid equilibria affect the energetics of volatile component degassing because bubble nucleation, coalescence, and growth and formation of magmatic volatile component-charged foams are crucial to passive vs. explosive eruptive behavior (Cashman and Mangan 1994; Sparks 2003; Larsen et al. 2004). Surface tension energy also plays an important role in foam generation, so surface tension and other properties vary between vapor-only, liquid-only, and vapor plus liquid systems.

Boiling fluids and implications for geophysical signals. Volcanologists have detected long-period, volcano-seismic events and determined that these signals can provide information on magma and/or fluid movement and, potentially, on impending eruptive behavior. Long-period signals of tremors permit quantitative estimation of the characteristic physical properties of aluminosilicate melts and/or of lower-density fluids in volcanic systems (Chouet 1996; Kamagai et al. 2001). These events are triggered by pressure fluctuations that may result from highly energetic, non-linear fluid flow in melts and/or fluids as well as from degassing processes (Arciniega et al. 2005).

Interestingly, Rhymer et al. (2000) attributed B-type seismic events to liquid-vapor interactions at the bottom of the Poas crater lake in 1980-1996. Likewise, Vandemeulebrouck et al. (2000) observed hydro-acoustic signals in the crater lake of Kelut volcano, Indonesia, and interpreted that they were due to boiling fluids at the lake bottom.

ACKNOWLEDGMENTS

We sincerely appreciate detailed reviews by Jeff Hedenquist, Axel Liebscher, Peter Nabelek, and Johan Varekamp. Beth Goldoff and Nanette Nicholson provided important assistance with editing and data collation. We appreciate discussions with T. Gerlach, J. Hedenquist, C. Heinrich, B. Scaillet, H. Shinohara, F. Spera, and J. Varekamp, but any errors are our own. Some research cited in this work was supported by NSF award EAR 0308866 to JDW.

REFERENCES

Africano F, Bernard A (2000) Acid alteration in the fumarolic environment of Usu volcano, Hokkaido, Japan. *In:* Crater Lakes. Vol 97. Varekamp JC, Rowe GL (eds) J Volcanol Geotherm, p 475-495

Africano F, Rompaey GV, Bernard A, Le Guern F (2002) Deposition of trace elements from high temperature gases of Satsuma-Iwojima volcano. Earth Planet Space 54:275-286

Aiuppa A, Federico C, Guidice G, Guerrieri S, Paonita A, Valenza M (2004) Plume chemistry provides insights into mechanisms of sulfur and halogen degassing in basaltic volcanoes. Earth Planet Sci Lett 222:469-483

Allard P (1983) The origin of hydrogen, carbon, sulphur, nitrogen and rare gases in volcanic exhalations: Evidence from isotope geochemistry. *In:* Forcasting Volcanic Events. Tazieff H, Sabroux J-C (eds) Elsevier, p 337-386

Amman M, Hauert R, Burtscher H, Siegmann HC (1993) Photoelectric charging of ultrafine volcanic aerosols, detection of Cu(1) as a tracer of chlorides in magmatic gases. J Geophys Res 99:551-556

Anderko A, Pitzer KS (1993) Equation-of-state representation of phase equilibria and volumetric properties of the system NaCl-H_2O above 573 K. Geochim Cosmochim Acta 57:1657-1680

Anderson AT Jr. (2003) Melt (glass ± crystals) inclusions. *In:* Fluid Inclusions: Analysis and Interpretation. Vol 32. Samson I, Anderson A, Marshall D (eds). Mineral Assoc Canada p. 353-364

Andreeva IA, Naumov VB, Kovalenko VI, Kononkova NN (1998) Fluoride-sulfate and chloride-sulfate salt melts in the carbonatite-bearing complex Muchagai-Khuduk, Southern Mongolia. Petrol 6:284-292

Anovitz LM, Labotka TC, Blencoe JG, Horita J (2004) Experimental determination of the activity-composition relations and phase equilibria of H_2O-CO_2-NaCl fluids at 500 °C, 500 bars. Geochim Cosmochim Acta 68:3557-3567

Aoki K, Ishikawa K, Kaniksawa S (1981) Fluorine geochemistry of basaltic rocks from continental and oceanic regions and petrogenetic applications. Contrib Mineral Petrol 76:53-59

Appora I, Eiler JM, Stolper EM (2000) Experimental determination of oxygen-isotope fractionations between CO_2 vapor and Na-melilite melt. Gold Conf (Oxford, UK) 5:149

Arciniega A, Chouet B, Dawson P (2005) Families of long-period seismic signals observed in Popocatepetl volcano, Mexico. EOS Trans, Am Geophys Union 86:G112

Arnórsson S, Stefánsson A, Bjarnason JÖ (2007) Fluid-fluid interactions in geothermal systems. Rev Mineral Geochem 65:259-312

Arthur MA (2000) Volcanic contributions to the carbon and sulfur geochemical cycles and global change. *In:* Encyclopedia of Volcanoes. Sigurdsson H (ed) Academic Press p 1045-1056

Audédat A, Pettke T (2003) The magmatic-hydrothermal evolution of two barren granites: a melt and fluid inclusion study of the Rito del Medio and Canada Pinabete plutons in northern New Mexico (USA). Geochim Cosmochim Acta 67:97-121

Bai W, Xu W, Lowell RP (2003) The dynamics of submarine geothermal heat pipes. Geophys Res Lett 30(3):1108

Barberi F, Chelini W, Marinelli G, Martini M (1989) The gas cloud of Lake Nyos (Cameroon, 1986): results of the Italian technical mission. J Volcanol Geotherm Res 39:125-134

Barquero J (1983) Termometria de la fumarola de Volcan Poas. Bol Vulcanol 13:11-12

Belkin HE, De Vivo B, Lima A, Török K (1996) Magmatic silicate/saline/sulfur-rich/CO_2 immiscibility and zirconium and REE enrichment from alkaline magma chamber margins: evidence from Ponza Island and Pontine archipelago, Italy. Eur J Mineral 8:1401-1420

Bernard A (1985) Les mecanismes de condensation des gaz volcaniques (chimie, mineralogy et equilibres des phases condensees majeures et mineuves). PhD dissertation, University of Bruxelles, Belgium

Bernard A, Symonds RB (1989) The significance of siderite in the sediments of Lake Nyos, Cameroon. J Volcanol Geotherm Res 39:187-194

Bernard A, Symonds RB, Rose WI (1990) Volatile transport and deposition of Mo, W, and Re in high-temperature magmatic fluids. App Geochim 5:317-326

Berndt ME, Seal RR, Shanks WC, Seyfried WE (1996) Hydrogen isotope systematics of phase separation in submarine hydrothermal systems: experimental calibration and theoretical models. Geochim Cosmochim Acta 60:1595-1604

Bischoff JL (1991) Densities of liquid and vapors in boiling NaCl-H_2O solutions: a PVTX summary from 300° to 500 °C. Am J Sci 291:309-338

Bischoff JL, Pitzer KS (1985) Phase relations and adiabats in boiling seafloor geothermal systems. Earth Planet Sci Lett 75:327-338

Bischoff JL, Pitzer KS (1989) Liquid-vapor relations for the system NaCl-H_2O: summary of the P-T-x surface from 300° to 500 °C. Am J Sci 289:217-248

Bischoff JL, Rosenbauer RJ, Fournier RO (1996) The generation of HCl in the system $CaCl_2$-H_2O: liquid-liquid relations from 380-500 °C. Geochim Cosmochim Acta 60:7-16

Bodnar RJ (2003) Introduction to aqueous-electrolyte fluid inclusions. *In:* Fluid Inclusions: Analysis and Interpretation. Samson I, Anderson A, Marshall D (eds) Vol 32, Mineral Assoc Canada, p 81-100

Bodnar RJ, Burnham CW, Sterner SM (1985) Synthetic fluid inclusions in natural quartz. III. Determination of phase equilibrium properties in the system H_2O-NaCl to 1000 °C and 1500 bars. Geochim Cosmochim Acta 49:1861-1873

Bodnar RJ, Cline JS (1991) Fluid inclusion petrology of porphyry copper deposits revisited; Re-interpretation of observed characteristics based on recent experimental and theoretical data. Plinius 5:24-25

Bodnar RJ, De Vivo B (2003) Melt Inclusions in Volcanic Systems. Elsevier

Bodnar RJ, Student JJ (2006) Melt inclusions in plutonic rocks: petrography and microthermometry. *In:* Melt Inclusions in Plutonic Rocks. Vol 36. Webster JD (ed) Mineral Assoc Canada, p 1-25

Borisova AY, Nikogosian IK, Scoates JS, Weis D, Damasceno D, Shimizu N, Touret JLR (2002) Melt, fluid and crystal inclusions in olivine phenocrysts from Kerguelen plume-derived picritic basalts: evidence for interaction with the Kerguelen Plateau lithosphere. Chem Geol 183:195-220

Botcharnikov RE, Holtz F, Behrens H (2007) Volatile solubility in andesites coexisting with H_2O-, Cl-, CO_2-bearing fluids at 200 MPa. Eur J Mineral (in press)

Botcharnikov RE, Behrens H, Holtz F, Koepke J, Sato H (2004) Sulfur and chlorine solubility in Mt. Unzen rhyodacitic melt at 850 °C and 200 MPa. Chem Geol 213:207-225

Botcharnikov RE, Shmulovich KI, Tkachenko SI, Korzhinskii MA, Rybin AV (2003) Hydrogen isotope geochemistry and heat balance of a fumarolic system: Kudriavy volcano, Kuriles. J Volcanol Geotherm Res 124:45-66

Bowers TS, Campbell AC, Measures CI, Spivack AJ, Khadem M, Edmond JM (1988) Chemical controls on the composition of vent fluids at 13°-11°N and 21°N, East Pacific Rise. J Geophys Res 93:4522-4536

Bowers TS, Helgeson HC (1983a) Calculation of the thermodynamic and geochemical consequences of nonideal mixing in the system H_2O-CO_2-NaCl on phase relations in geologic systems: metamorphic equilibria at high pressures and temperatures. Am Mineral 68:1059-1075

Bowers TS, Helgeson HC (1983b) Calculation of the thermodynamic and geochemical consequences of nonideal mixing in the system H_2O-CO_2-NaCl on phase relations in geologic systems: Equation of state for H_2O-CO_2-NaCl fluids at high pressures and temperatures. Geochim Cosmochim Acta 47:1247-1275

Brantley SL, Borgia A, Rowe G, Fernandez JF, Reynolds JR (1987) Poás volcano volcanic lake acts as a condenser for acid metal-rich brine. Nature 330:470-472

Burnham CW (1997) Magmas and hydrothermal fluids. *In:* Geochemistry of Hydrothermal Ore Deposits, 3rd edition. HL Barnes (ed) John Wiley & Sons, p 63-123

Burnham CW, Ohmoto H (1980) Late-stage processes of felsic magmatism. *In:* Granitic Magmatism and Associated Mineralization. Mining Geology Special Issue, No. 8. Ishihara S, Takenouchi S (eds) The Society of Mining Geologists of Japan, p 1-11

Burton M, Allard, P, Muré F, Oppenheimer C (2003) FTIR remote sensing of fractional magma degassing at Mount Etna, Sicily. *In:* Volcanic Degassing. Vol 213. Oppenheimer C, Pyle DM, Barclay J (eds) Geol Soc London Spec Pub, p 281-293

Butterfield DA (2000) Deep ocean hydrothermal vents. *In:* Encyclopedia of Volcanoes. Sigurdsson H (ed) Academic Press, p 857-873

Butterfield DA, McDuff RE, Mottl MJ, Lilley MD, Lupton JE, Massoth GJ (1994) Gradients in the composition of hydrothermal fluids from the Endeavour segment vent field: phase separation and brine loss. J Geophys Res 99:9561-9583

Byers CD, Christie DM, Muenow DW, Sinton JM (1984) Volatile contents and ferric-ferrous ratios of basalt, ferrobasalt, andesite and rhyodacite glasses from the Galapagos 95.5° W propagating rift. Geochim Cosmochim Acta 48:2239-2245

Calzia JP, Hiss WL (1978) Igneous rocks in northern Delaware Basin, New Mexico and Texas. New Mexico Bur Mines Mineral Res Circ 159:39-45

Campbell AR (1995) The evolution of a magmatic afluid: a case history from the Capitan Mountains, New Mexico. *In:* Magmas, Fluids, and Ore Deposits. Vol 23. Thompson JFH (ed) Mineral Assoc Canada, p 291-308

Candela PA, Piccoli PM (1995) Model ore-metal partitioning from melts into vapor and vapor/brine mixtures. *In:* Magmas, Fluids, and Ore Deposits. Vol 23. Thompson JFH (ed) Mineral Assoc Canada , p 101-127

Carrigan CR (1986) A two-phase hydrothermal cooling model for shallow intrusions. J Volcanol Geotherm Res 28:175-192

Carrigan CR (2000) Plumbing systems. *In:* Encyclopedia of Volcanoes. Sigurdsson H (ed) Academic Press, p 219-235

Carroll MR, Rutherford MJ (1988) Sulfur speciation in hydrous experimental glasses of varying oxidation state: results from measured wavelength shifts of sulfur X-rays. Am Mineral 73:845-849

Carroll MR, Webster JD (1994) Solubilities of sulfur, noble gases, nitrogen, chlorine, and fluorine in magmas. Rev Mineral 30: 231-279

Cashman KV, Mangan MT (1994) Physical aspects of magmatic degassing II. constraints on vesiculation processes from textural studies of eruptive products. Rev Mineral 30:447-478

Cathles LM (1977) An analysis of the cooling of intrusives by ground-water convection which includes boiling. Econ Geol 72:804-826

Chacko T, Cole DR, Horita J (2001) Equilibrium oxygen, hydrogen and carbon isotope fractionation factors applicable to geologic systems. Rev Mineral Geochem 43:1-81

Chiba H, Chacko T, Clayton RN, Goldsmith JR (1989) Oxygen isotope fractionations involving diopside, forsterite, magnetite and calcite; application to geothermometry. Geochim Cosmochim Acta 53:2985-2995

Chiodini G, Cioni R, Marini L (1993) Reactions governing the chemistry of crater fumaroles from Vulcano Island, Italy, and implications for volcanic surveillance. Appl Geochem 8:357-371

Chiodini G, Marini L, Russo M (2001) Geochemical evidences of high-temperature hydrothermal brines at Vesuvio Volcano (Italy). Geochim Cosmochim Acta 65:2129-2147

Chou I-M (1987) Phase relations in the system $NaCl-KCl-H_2O$: III. Solubilities of halite in vapor-saturated liquids above 445 °C and redetermination of phase equilibrium properties in the system $NaCl-H_2O$ to 1000 °C and 1500 bars. Geochim Cosmochim Acta 51:1965-1975

Chou I-M, Sterner SM, Pitzer KS (1992) Phase relations in the system $NaCl-KCl-H_2O$: IV. Differential thermal analysis of the sylvite liquidus in the $KCl-H_2O$ binary, the liquidus in the NaCl-KCl-H2O ternary, and the solidus in the NaCl-KCl binary to 2 kb pressure, and a summary of experimental data for thermodynamic-PTX analysis of solid-liquid equilibria at elevated P-T conditions. Geochim Cosmochim Acta 56:2281-2293

Chouet BA (1996) Long-period volcano seismicity: its source and use in eruption forecasting. Nature 380:309-316

Christenson BW (2000) Geochemistry of fluids associated with the 1995-1996 eruption of Mt. Ruapehu, New Zealand: signatures and processes in the magmatic-hydrothermal system. J Volcanol Geotherm Res 97:1-30

Christenson BW, Werner CA, Reyes AG, Sherburn S, Scott BJ, Miller C, Rosenburg MJ, Hurst AW, Britten KA (2007) Hazards from hydrothermally sealed volcanic conduits. EOS Am Geophys Union 88:53-55

Christenson BW, Wood CP (1993) Evolution of a vent-hosted hydrothermal system beneath Ruapehu Crater Lake, New Zealand. Bull Volcanol 55:547-565

Churakov SV, Tkachenko SI, Korzhinskii MA, Bocharnikov RE, Shmulovich KI (2000) Evolution of composition of high-temperature fumarolic gases from Kudryavy volcano, Iturup, Kiril Islands: the thermodynamic modeling. Geochem Int 38:436-451

Clayton RN, Goldsmith JR, Mayeda TK (1989) Oxygen isotopic fractionations in quartz, albite, anorthite and calcite. Geochim Cosmochim Acta 53:725-733

Cline JS, Bodnar RJ (1991) Can economic porphyry copper mineralization be generated by a typical calc-alkaline melt? J Geophys Res 96:8113-8126

Cline JS, Bodnar RJ (1994) Direct evolution of brine from a crystallizing silicic melt at the Questa, New Mexico, molybdenum deposit. Econ Geol 89:1780-1802

Cline JS, Vanko DA (1995) Magmatically generated saline brines related to molybdenum at Questa, New Mexico, USA. *In:* Magmas, Fluids, and Ore Deposits. Vol 23. Thompson JFH (ed) Mineral Assoc Canada, p 153-174

Cloke PL, Kesler SE (1979) Halite trend in hydrothermal solutions. Econ Geol 74:1823-1831

Coombs ML, Sisson TW, Kimura J-I (2004) Ultra-high chlorine in submarine Kilauea glasses: evidence for direct assimilation of brine by magma. Earth Planet Sci Lett 217:297-313

Craig H (1961) Isotopic variations in meteoric waters. Science 133:1702-1703

Craig H, Gordon LI (1965) Deuterium and oxygen[18] variations in the ocean and the marine atmosphere. Stable isotopes in oceanographic studies and paleotemperatures, Spoleto Conferences in Nuclear Geology, CNR, Lab Geologia Nucleare, Pisa, Italy p 1-130

Crowe BM, Finnegan DL, Zoller WH, Boynton WV (1987) Trace element geochemistry of volcanic gases and particles from 1983-1984 eruptive episodes of Kilauea volcano. J Geophys Res 92:13708-13714

Cunningham CG (1978) Pressure gradients and boiling as mechanisms for localizing ore in porphyry systems. J Res US Geol Survey 6:745-754

Davidson P, Kamenetsky V (2007) Primary aqueous fluids in rhyolitic magmas: melt inclusion evidence for pre- and post-trapping exsolution. Chem Geol (in review)

Davidson P, Kamenetsky V, Cooke DR, Frikken P, Hollings P, Ryan C, van Achterbergh E, Mernagh T, Skarmeta J, Serrano L, Vargas R (2005) Magmatic precursors of hydrothermal fluids at the Rio Blanco Cu-Mo deposit, Chile: links to silicate magmas and metal transport. Econ Geol 100:693-978

de Hoog JCM, Hattori KH, Hoblitt RP (2004) Oxidized sulfur-rich mafic magma at Mount Pinatubo, Philippines. Contrib Mineral Petrol 146:750-761 de Ronde CEJ, Massoth GJ, Baker ET, Lupton JE (2003) Submarine hydrothermal venting related to volcanic arcs. *In:* Volcanic, Geothermal, and Ore-forming Fuids: Rulers and Witnesses of Processes within the Earth. Vol 10. Simmons SF, Graham I (eds) Econ Geol Spec Pub, p 91-110

De Vivo B, Danyushevsky LV, Lima A, Kamenetsky VS (2006) Fluid and melt inclusions in sub-volcanic environment from volcanic systems: Examples from the Neapolitan area and Pontine Islands (Italy). *In:* Melt Inclusions in Plutonic Rocks. Vol 36. Webster JD (ed) Mineral Assoc Canada, p 211-237

De Vivo B, Frezzotti ML (1994) Evidence for magmatic immiscibility in Italian subvolcanic systems. *In:* Fluid Inclusions in Minerals: Methods and Applications. Short Course of the WG (IMA) "Inclusions in Minerals," Pontignano (Siena), 1-4 Sept, 1994. De Vivo B, Frezzotti ML (eds) Virginia Polytechnic and State Univ p 346-362

De Vivo B, Frezzotti ML, Lima A (1993) Immiscibility in magmatic differentiation and fluid evolution in granitoid xenoliths at Pantelleria: fluid inclusion evidence. Acta Vulcanol 3:195-202

De Vivo B, Torok K, Ayuso RA, Lima A, Lirer L (1995) Fluid inclusion evidence for magmatic silicate/saline/CO_2 immiscibility and geochemistry of alkaline xenoliths from Ventotene Island, Italy. Geochim Cosmochim Acta 59:2941-2953

Delmelle P, Bernard A (1994) Geochemistry, mineralogy, and chemical modeling of the acid crater lake of Kawah Ijen volcano, Indonesia. Geochim Cosmochim Acta 58:2445-2460

Delmelle P, Bernard A, Kusakabe M, Fischer TP, Takano B (2000) Geochemistry of the magmatic-hydrothermal system of Kawah Ijen volcano, East Java, Indonesia. J Volcanol Geotherm Res 97:31-53

Delmelle P, Stix J (2000) Volcanic gases. *In:* Encyclopedia of Volcanoes. Sigurdsson H (ed) Academic Press, p. 803-815

Di Liberto V, Nuccio PM, Paonita A (2002) Genesis of chlorine and sulphur infumarolic emissions at Vulcano Island (Italy): assessment of pH and redox conditions in the hydrothermal system. J Volcanol Geotherm Res 116:137-150

Diamond LW (2003) Systematics of H_2O inclusions. *In:* Fluid Inclusions: Analysis and Interpretation. Vol 32. Samson I, Anderson A, Marshall D (eds) Mineral Assoc Canada, p 55-78

Dobson PF, Epstein S, Stolper EM (1989) Hydrogen isotope fractionation between coexisting vapor and silcate glasses and melts at low pressure. Geochim Cosmochim Acta 53:2723-2730

Driesner T (1997a) Aspects of Stable Isotope Fractionation in Hydrothermal Solutions. PhD dissertation ETH Zurich, No 11839, Zurich, Switzerland

Driesner T (1997b) The effect of pressure on deuterium-hydrogen fractionation in high-temperature water. Science 277:791-794

Driesner T, Seward TM (2000) Experimental and simulation study of salt effects and pressure/density effects on oxygen and hydrogen stable isotope liquid-vapor fractionation for 4-5 molal aqueous NaCl and KCl solutions to 400 °C. Geochim Cosmochim Acta 64:1773-1784

Driesner T, Geiger S (2007) Numerical simulation of multiphase fluid flow in hydrothermal systems. Rev Mineral Geochem 65:187-212

Drummond SE (1981) Boiling and Mixing of Hydrothermal Fluids: Chemical Effects on Mineral Precipitation. PhD Dissertation, Pennsylvania State University, Pennsylvania

Duan Z, Moller N, Weare JH (1992) An equation of state for the CH_4-CO_2-H_2O system: II. mixtures from 50 to 1000 °C and 0 to 1000 bar. Geochim Cosmochim Acta 56:2619-2631

Duan Z, Moller N, Weare JH (1995) Equation of state for the NaCl-H_2O-CO_2 system: prediction of phase equilibria and volumetric properties. Geochim Cosmochim Acta 59:2869-2882

Duan Z, Moller N, Weare JH (2003) Equations of state for the NaCl-H_2O-CO_2-CH_4 system: phase equilibria and volumetric properties above 573 K. Geochim Cosmochim Acta 67:671-680

Duan Z, Moller N, Weare JH (2006) A high temperature equation of state for the H_2O-$CaCl_2$ and H_2O-$MgCl_2$ systems. Geochim Cosmochim Acta 70:3765-3777

Edmonds M, Pyle D, Oppenheimer C (2001) A model for degassing at the Soufrière Hills volcano, Montserrat, West Indies, based on geochemical data. Earth Planet Sci Lett 186:159-173

Eichelberger JC (1980) Vesiculation of mafic magma during replenishment of silicic magma reservoirs. Nature 288:446-450

Eiler JM (2001) Oxygen isotope variations of basaltic lavas and upper mantle rocks. Rev Mineral Geochem 43:319-364

Einaudi MT, Hedenquist JW, Inan EE (2003) Sulfidation state of fluids in active and extinct hydrothermal systems: transitions from porphyry to epithermal environments. *In:* Volcanic, Geothermal, and Ore-forming Fluids: Rulers and Witnesses of Processes within the Earth. Simmons SF, Graham I (eds) Econ Geol Spec Pub 10:285-313

Elkins LJ, Fischer TP, Hilton DR, Sharp, ZD, McKnight S, Walker J (2006) Tracing nitrogen in volcanic and geothermal volatiles from the Nicaraguan volcanic front. Geochim Cosmochim Acta 70:5215-5235

Ellis AJ (1957) Chemical equilibrium in magmatic gases. Am J Sci 255:416-431

Elsworth D, Voight B, Thompson G, Young SR (2004) Thermal-hydrologic mechanism for rainfall-triggered collapse of lava domes. Geology 32:969-972

Faust CR, Mercer JW (1979a) Geothermal reservoir simulation: 1. mathematical models for liquid- and vapor-dominated hydrothermal systems. Water Resour Res 15:23-30

Faust CR, Mercer JW (1979b) Geothermal reservoir simulation: 2. numerical solution techniques for liquid- and vapor-dominated hydrothermal systems. Water Resour Res 15:31-46

Fazllullin SM, Ushakov SV, Shuvalov RA, Aoki M, Nikolaeva AG, Lupikina EG (2000) The 1996 subaqueous eruption at Academii Nauk volcano, (Kamchatka) and its effects on Karymsky lake. J Volcanol Geotherm Res 97:181-193

Field CW, Fifarik RH (1985) Light stable-isotope systematics in the epithermal environment. *In:* Geology and Geochemistry of Epithermal Systems. Berger BR Bethke PM (eds) Rev Econ Geol 2:99-128

Fischer RV, Schmincke H-U (1984) Pyroclastic Rocks. Springer Verlag

Fournier RO (1987) Conceptual models of brine evolution in magmatic-hydrothermal systems. US Geol Surv Prof Pap 1350:1487-1506

Fournier RO (1999) Hydrothermal processes related to movement of fluid from plastic into brittle rock in the magmatic-epithermal environment. Econ Geol 94:1193-1212

Foustoukos DI, Seyfried Jr. WE (2007) Fluid phase separation processes in submarine hydrothermal systems. Rev Mineral Geochem 65:213-239

Franck EU (1981) Survey of selected non-thermodynamic properties and chemical phenomena of fluids and fluid mixtures. *In:* Chemistry and Geochemistry of Solutions at High Temperatures and Pressures. Vol 1. Rickard DT, Wickman FE (eds) Pergamon Press, p 65-88

Frantz JD, Popp RK, Hoering TC (1992) The compositional limits of fluid immiscibility in the system H_2O-$NaCl$-CO_2 as determined using synthetic fluid inclusion in conjunction with mass spectrometry. Chem Geol 98:237-255

Frezzotti ML (1992) Magmatic immiscibility and fluid phase evolution in the Mount Genis granite (southeastern sardinia, Italy). Geochim Cosmochim Acta 56:21-33

Frezzotti ML (2001) Silicate-melt inclusions in magmatic rocks: applications to petrology. Lithos 55:273-299

Frost BR, Touret JLR (1989) Magmatic CO_2 and saline melts from the Sybille Monzosyenite, Laramie Anorthosite Complex, Wyoming. Contrib Mineral Petrol 103:178-186

Fulignati P, Kamenetsky VS, Marianelli P, Sbrana A, Mernagh TP (2001) Melt inclusion record of immiscibility between silicate, hydrosaline, and carbonate melts: applications to skarn genesis at Mount Vesuvius. Geology 29:1043-1046

Gat JR (1996) Oxygen and hydrogen isotopes in the hydrological cycle. Ann Rev Earth Planet Sci 24:225-262

Geiger S, Driesner T, Heinrich CA, Matthai SK (2005) On the dynamics of $NaCl$-H_2O fluid convection in the Earth's crust. J Geophys Res 110:B07101, 23 p

Geiger S, Driesner T, Heinrich CA, Matthai SK (2006a) Multiphase thermohaline convection in the Earth's crust: I. a new finite element-finite volume solution technique combined with a new equation of state for NaCl-H2O. Trans Por Med 63:399-434

Geiger S, Driesner T, Heinrich CA, Matthai SK (2006b) Multiphase thermohaline convection in the Earth's crust: II. benchmarking and application of a finite element-finite volume solution technique with a NaCl-H2O equation of state. Trans Por Med 63:435-461

Gerlach TM (1980) Evaluation of volcanic gas analysis from Kilauea volcano. J Volcanol Geotherm Res 7:295-317

Gerlach TM, Casadevall TJ (1986) Evaluation of gas data from high-temperature fumaroles at Mount St. Helens. J Volcanol Geotherm Res 28:107-140

Gerlach TM, McGee KA, Elias T, Sutton AJ, Doukas MP (2002) Carbon dioxide emission rate of Kilauea Volcano: implications for primary magma and the summit reservoir. J Geophys Res 107:B9, 2189

Gerlach TM, Taylor BE (1990) Carbon isotope constraints on degassing of carbon dioxide from Kilauea Volcano. Geochim Cosmochim Acta 54:2051-2058

Giggenbach WF (1971) Isotopic composition of waters of the Broadlands geothermal field, New Zealand. New Zeal J Sci 14:959-970

Giggenbach WF (1974) The chemistry of Crater Lake, Mt. Ruapehu (New Zealand) during and after the 1971 active period. New Zeal J Sci 17:33-45

Giggenbach WF (1975) Variations in the carbon, sulfur and chlorine contents of volcanic gas discharges from White Island, New Zeal Bull Volcanol 39:15-27

Giggenbach WF (1977) Chemistry of Indian geothermal discharges. Unpub. UNDP Rpt.

Giggenbach WF (1980) Geothermal gas equilibria. Geochim Cosmochim Acta 44:2021-2032

Giggenbach WF (1987) Redox processes governing the chemistry of fumarolic gas discharges from White Island, New Zealand. Appl Geochem 2:143-161

Giggenbach WF (1988) Geothermal solute equilibria: derivation of Na-K-Mg-Ca geoindicators. Geochim Cosmochim Acta 52:2749-2765

Giggenbach WF (1992) Isotopic shifts in waters from geothermal and volcanic systems along convergent plate boundaries and their origin. Earth Planet Sci Lett 113:495-510

Giggenbach WF (1995) Composition of magmatic components in hydrothermal fluids. *In:* Magmas, Fluids, and Ore Deposits. Vol 23. Thompson JFH (ed) Mineral Assoc Canada , p 247-261

Giggenbach WF (1996) Chemical composition of volcanic gases. *In:* Monitoring and Mitigation of Volcano Hazards. Scarpa R, Tilling R (eds) Springer, p 221–256

Giggenbach WF (1997) The origin and evolution of fluids in magmatic-hydrothermal systems. *In:* Geochemistry of Hydrothermal Ore Deposits. 3rd Edition. Barnes HL (ed) John Wiley & Sons, p 737-796

Giggenbach WF (2003) Magma degassing and mineral deposition in hydrothermal systems along convergent plate boundaries. *In:* Volcanic, Geothermal, and Ore-forming Fluids: Rulers and Witnesses of Processes within the Earth. Simmons SF, Graham I (eds) Econ Geol Spec Pub 10:1-18

Giggenbach WF, Garcia N, Londoño N, Rodriguez A, Rojas L, Calvache M (1990) The chemistry of fumarolic vapor and thermal spring discharges from the Nevado del Ruiz volcanic-magmatic-hydrothermal system. J Volcanol Geotherm Res 42:13-39

Giggenbach WF, Glover RB (1975) The use of chemical indicators in the surveillance of volcanic activity affecting the crater lake on Mt. Ruapehu, New Zeal Bull Volcanol 39:70-81

Giggenbach WF, Sheppard DS (1989) Variation in the temperature and chemistry of White Island fumarole discharges 1972-85. New Zeal Geol Surv Bull 103:119-126

Giggenbach WF, Shinohara H, Kusakabe M, Ohba T (2003) Formation of acid volcanic brines through interaction of magmatic gases, seawater, and rock within the White Island volcanic-hydrothermal system, New Zealand. *In:* Volcanic, Geothermal, and Ore-forming Fluids: Rulers and Witnesses of Processes within the Earth. Simmons SF, Graham I (eds) Econ Geol Spec Pub 10:19-40

Goldfarb MS, Delaney JR (1988) Response of two-phase fluids to fracture configurations within submarine hydrothermal systems. J Geophys Res 93:4585-4594

Gonfiantini R (1986) Environmental isotopes in lake studies. *In:* Handbook of Environmental Isotope Geochemistry. Vol 2. P Fritz JCh Fontes (eds) Elsevier, p 113-168

Gottschalk M (2007) Equations of state for complex fluids. Rev Mineral Geochem 65:49-97

Graham CM, Harmon RS, Sheppard SMF (1984) Experimental hydrogen isotope studies: hydrogen isotope exchange between amphibole and water. Am Mineral 69:128-138

Graney JR, Kesler SE (1995) Gas composition of inclusion fluid in ore deposits: is there a relation to magmas. *In:* Magmas, Fluids, and Ore Deposits. Vol 23. Thompson JFH (ed) Mineral Assoc Canada , p 221-245

Guglielminetti M (1986) Mofete geothermal field. Geotherm 15:781-790

Hack AC, Thompson AB, Aerts M (2007) Phase relations involving hydrous silicate melts, aqueous fluids, and minerals. Rev Mineral Geochem 65:129-185

Halter W, Heinrich CA (2006): Magmatic processes and volatile generation in porphyry-type environments: A laser-ablation ICPMS study of silicate and sulfide melt inclusions. *In:* Melt Inclusions in Plutonic Rocks. Vol 36. Webster JD (ed) Mineral Assoc Canada, p 151-164

Halter W, Heinrich CA Pettke T (2005): Magma evolution and the formation of porphyry Cu-Au ore fluids: evidence from silicate and sulfide melt inclusions. Mineral Depos 39(8):845-863

Hardee HC (1982) Permeable convection above magma bodies. Tectonophys 84:179-195

Harmon RS, Hoefs J (1995) Oxygen isotope heterogeneity of the mantle deduced from global ^{18}O systematics of basalts from different geotectonic settings. Contrib Mineral Petrol 120:95-114

Hattori K (1996) Occurrence and origin of sulfide and sulfate in the 1991 Mount Pinatubo eruption products. *In:* Fire and Mud: Eruptions and Lahars of Mount Pinatubo, Philippines. Newhall CG, Punongbayan RS (eds) Univ Wash Press Seattle, p 807-824

Haughton DR, Roeder PL, Skinner BJ (1974) Solubility of sulfur in mafic magmas. Econ Geol 69:451-467

Hauri E (2002) SIMS analysis of volatiles in silicate glasses, 2: isotopes and abundances in Hawaiian melt inclusions. Chem Geol 183:115-141

Hedenquist JW (1991) Boiling and dilution in the shallow portion of the Waiotapu geothermal system, New Zealand. Geochim Cosmochim Acta 55:2753-2765

Hedenquist JW (1995) The ascent of magmatic fluid: discharge vs. mineralization. In: Magmas, Fluids, and Ore Deposits. Vol 23. Thompson JFH (ed) Mineral Assoc Canada, p 263-289

Hedenquist JW, Lowenstern JB (1994) The role of magmas in the formation of hydrothermal ore deposits. Nature 370:519-527

Heinrich CA (2005) The physical and chemical evolution of low-salinity magmatic fluids at the porphyry to epithermal transition: a thermodynamic study. Mineral Deposita 39:864-889

Heinrich CA (2007a) Fluid – fluid interactions in magmatic-hydrothermal ore formation. Rev Mineral Geochem 65:363-387

Heinrich CA, Ryan CG, Mernagh TP, Eadington PJ (1992) Segregation of ore metals between magmatic brine and vapor: a fluid inclusion study using PIXE microanalysis. Econ Geol 87:1566-1583

Heinrich W (2007b) Fluid immiscibility in metamorphic rocks. Rev Mineral Geochem 65:389-430

Heinrich W, Churakov SS, Gottschalk M (2004) Mineral-fluid equilibria in the system CaO-MgO-SiO$_2$-H$_2$O-CO$_2$-NaCl and the record of reactive fluid flow in contact metamorphic aureoles. Con Mineral Petrol 10.1007/s00410-004-0598-7

Henley RW, Hughes GO (2000) Underground fumaroles: "excess heat" effects in vein formation. Econ Geol 95:453-466

Henley RW, McNabb A (1978) Magmatic vapor plumes and ground-water interaction in porphyry copper emplacement. Econ Geol 73:1-20

Holloway JR (1976) Fluids in the evolution of granitic magmas: Consequences of finite CO_2 solubility. Geol Soc Am Bull 87:1513-1518

Holloway JR, Blank JG (1994) Application of experimental results to C-O-H species in natural melts. Rev Mineral 30:187-230

Holtz F, Dingwell DB, Behrens H (1993) Effect of F, B_2O_3, and P_2O_5 on the solubility of water in haplogranite melts compared to natural silicate melts. Contrib Mineral Petrol 113:492-501

Horita J, Cole DR, Wesolowski DJ (1995) The activity-composition relationship of oxygen and hydrogen isotopes in aqueous salt solutions: III. Vapor-liquid water equilibration of NaCl solutions to 350 °C. Geochim Cosmochim Acta 59:1139-1151

Horita J, Wesolowski DJ, Cole DR (1993) The activity-composition relationship of oxygen and hydrogen isotopes in aqueous salt solutions: I. vapor-liquid water equilibration of single salt solutions from 50 to 100 °C. Geochim Cosmochim Acta 57:2797-2817

Horita J, Wesowlowski DJ (1994) Liquid-vapor fractionation of oxygen and hydrogen isotopes of water from the freezing to the critical temperature. Geochim Cosmochim Acta 58:3425-3437

Huppert HE, Woods AW (2002) The role of volatiles in magma chamber dynamics. Nature 420:493-495

Hurst AW, Bibby HM, Scott BJ, McGuinness MJ (1991) The heat source of Ruapehu Crater Lake; deductions from the energy and mass balances. J Volcanol Geotherm Res 6:1-21

Jambon A (1994) Earth degassing and large-scale geochemical cycling of volatile elements. Rev Mineral 30:479-517

Jaupart C (2000) Magma ascent at shallow levels. *In:* Encyclopedia of Volcanoes. Sigurdsson H (ed) Academic Press, p 237-245

Javoy M, Pineau F (1991) The volatiles record of a "popping" rock from the Mid-Atlantic Ridge at 14°N: chemical and isotopic composition of gas trapped in the vesicles. Earth Planet Sci Lett 107:598-611

Johnson JW, Norton D (1991) Critical phenomena in hydrothermal systems: state, thermodynamic, electrostatic, and transport properties of H_2O in the critical region. Am J Sci 291:541-648

Joyce DR, Holloway JRH (1993) An experimental determination of the thermodynamic properties of H_2O-CO_2-NaCl fluids at high pressures and temperatures. Geochim Cosmochim Acta 57:733-746

Jugo PJ, Candela PA, Piccoli PM (1999) Magmatic sulfides and Au:Cu ratios in porphyry deposits: an experimental study of copper and gold partitioning at 850 °C, 100 MPa in a haplogranitic melt-pyrrhotite-intermediate solid solution-gold metal assemblage, at gas saturation. Lithos 46:573-589

Jugo PJ, Luth RW, Richards JP (2005) An experimental study of the sulfur content in basaltic melts saturated with immiscible sulfide or sulfate liquids at 1300 °C and 1.0 GPa. J Petrol 46:783-798

Kamagai H, Ohmanato T, Nakano M, OOi M, Kubo A, Inoue H, Oikawa J (2001) Very-long-period seismic signals and caldera formation at Miyake Island, Japan. Science 293:687-690

Kamenetsky MB, Sobolev AV, Kamenetsky VS, Maas R, Danyushevsky LV, Thomas R, Pokhilienko NP, Sobolev NV (2004) Kimberlite melts rich in alkali chlorides and carbonates: a potent metasomatic agent in the mantle. Geology 32:845-848

Kamenetsky VS (2006) Melt inclusion record of magmatic immiscibility in mantle and crustal magmas. *In:* Melt Inclusions in Plutonic Rocks, Vol 36, Webster JD (ed) Mineral Assoc Canada, p 81-98

Kamenetsky VS, Danyushevsky LV (2005) Metals in quartz-hosted melt inclusions: Natural facts and experimental artifacts. Am Mineral 90:1674-1678

Kamenetsky VS, Wolfe RC, Eggins SM, Mernagh TP, Bastrakov E (1999) Volatile exsolution at the Dinkidi Cu-Au porphyry deposit, Philippines: a melt inclusion record of the initial ore-forming process. Geology 27:691-694

Kamilli RJ (1978) The genesis of stockwork molybdenite deposits: implications from fluid inclusion studies at the Henderson Mine (abstract). Geol Soc Am Abs Prog 10:431

Kasai K, Sakagawa Y, Komatsu R, Sasaki M, Akuku K, Uchida T (1998) The origin of hypersaline liquid in the Quaternary Kakkonda granite, samples from well WD-1a, Kakkonda geothermal system, Japan. Geotherm 27:631-645

Katsura T, Nagashima S (1974) Solubility of sulfur in some magmas at 1 atmosphere. Geochim Cosmochim Acta 38:517-531

Kawada Y, Yoshida S, Watanabe S-I (2004) Numerical simulations of mid-ocean ridge hydrothermal circulation including the phase separation of seawater. Earth Planet Space 56:193-215

Kelley DS, Delaney JR (1987) Two-phase separation and fracturing in the mid-ocean ridge gabbros at temperatures greater than 700 °C. Earth Planet Sci Lett 83:53-66

Kempter KA, Rowe GL (2000) Leakage of active crater lake brine through the north flank at Rincón de la Vieja volcano, northwest Costa Rica, and implications for crater collapse. J Volcanol Geotherm Res 97:143-159

Kent AJR, Clague DA, Honda H, Stolper EM, Hutcheon ID, Norman MD (1999b) Widespread assimilation of a seawater-derived component at Lo'ihi Seamount, Hawai'i. Geochim Cosmochim Acta 63:2749-2761

Kent AJR, Elliot TR (2002) Melt inclusions from Marianas arc lavas: implications for the composition and formation of island arc magmas. Chem Geol 183:263-286

Kent AJR, Norman MD, Hutcheon ID, Stolper EM (1999a) Assimilation of seawater-derived components in an oceanic volcano: evidence from matrix glasses and glass inclusions from Lo'ihi Seamount, Hawai'i. Chem Geol 156:299-319

Khorzinskiy MA, Tkachenko SI, Bulgarov RF, Shmulovich KI (1996) Condensate composition and native metals in sublimates of high temperature gas streams of Kudriavy volcano, Iturup, Kurile Islands. Geochem Int 34:1057-1064

Kieffer SW (1984) Factors governing the structure of volcanic jets. *In:* Explosive Volcanism: Inception, Evolution, and Hazards. National Academy Press Washington DC, p 143-157

Konnerup-Madsen F, Dubessy J, Rose-Hansen J (1985) Combined Raman microprobe spectrometry and microthermometry of fluid inclusions in minerals from igneous rocks of the Gardar province (south Greenland). Lithos 18:271-280

Korzhinskii MA, Tkachenko SI, Shmulovich KI, Taran YA, Steinberg GS (1994) Discovery of pure rhenium mineral at Kudriavy volcano. Nature 369:51-52

Koster van Groos AF, Wyllie PJ (1969) Melting relationships in the system $NaAlSi_3O_8$-NaCl-H_2O at one kilobar pressure, with petrological applications. J Geol 77:581-605

Lamb WM, Mcshane CJ, Popp RK (2002) Phase relations in the CH_4-H_2O-NaCl system at 2 kbar, 300 to 600 °C as determined using synthetic fluid inclusions. Geochim Cosmochim Acta 66:3971-3986

Larsen JF, Denis M-H, Gardner JE (2004) Experimental study of bubble coalescence in rhyolitic and phonolitic melts. Geochim Cosmochim Acta 68:333-344

Lassiter JC, Hauri, EH, Nikogosian IK, Barsczus HG (2002) Chlorine-potassium variations in melt inclusions from Raivavae and Rapa, Austral Islands: constraints on chlorine recycling in the mantle and evidence for brine-induced melting of oceanic crust. Earth Planet Sci Lett 202:525-540

Layne GD (2006) Application of secondary ion mass spectrometry to the determination of traditional and non-traditional light stable isotopes in silicate melt inclusions. *In:* Melt Inclusions in Plutonic Rocks. Vol 36. Webster JD (ed) Mineral Assoc Canada, p 27-49

Le Guern F (1988) Ecoulements gazeux reactifs a haute temperature: measures et modelisation. Unpub. PhD Dissertation, University of Paris 7

Le Guern F, Gerlach TM, Nohl A (1982) Field gas chromatograph analyses of gases from a glowing dome at Merapi volcano, Java, Indonesia, 1977, 1978, 1979. J Volcanol Geotherm Res 14:223-245

Lewis KC, Lowell RP (2004) Mathematical modeling of phase separation of seawater near an igneous dike. Geofluids 4:197-209

Liebscher A (2007) Experimental studies in model fluid systems. Rev Mineral Geochem 65:15-47

Liebscher A, Barnes J, Sharp Z (2006) Chlorine isotope vapor-liquid fractionation during experimental fluid-phase separation at 400 °C 23 MPa to 450 °C 42 MPa. Chem Geol 234:340-345

Liebscher A, Lüders V, Heinrich W, Schettler G (2006) Br/Cl signature of hydrothermal fluids: liquid-vapour fractionation of bromine revisited. Geofluids 6:113-121

Lima A, De Vivo B, Fedele L, Sintoni MF (2006) Influence of hydrothermal processes on geochemical variations between 79AD and 1944 AD Vesuvius eruptions. *In:* Volcanism in the Campania Plain. Vol 9. De Vivo B (ed), Elsevier, p 235-247

Lopez DL, Williams SN (1993) Catastrophic volcanic collapse: relation to hydrothermal processes. Science 260:1794-1796

Lowenstern JB (1994) Chlorine, fluid immiscibility, and degassing in peralkaline magmas from Pantelleria, Italy. Am Mineral 79:353-369

Lowenstern JB (1995) Applications of silicate-melt inclusions to the study of magmatic volatiles. *In:* Magmas, Fluids, and Ore Deposits. Vol 23. Thompson JFH (ed) Mineral Assoc Canada, p 71-99

Lowenstern JB (2001) Carbon dioxide in magmas and implications for hydrothermal systems. Min Deposita 36:490-502

Lowenstern JB (2002) Bubbles and non-silicate fluids in melt inclusions. *In:* Melt Inclusions: Methods, Applications and Problems. De Vivo B, Bodnar RJ (eds). De Frede Editore-Naples, p 143-148

Lowenstern JB (2003) Melt inclusions come of age. *In:* Melt Inclusions in Volcanic Systems. De Vivo B, Bodnar RJ (eds) Elsevier, p 1-21

Lowenstern JB, Janik CJ (2003) The origins of reservoir liquids and vapors from the Geysers geothermal field, California. *In:* Volcanic, Geothermal, and Ore-forming Fluids: Rulers and Witnesses of Processes within the Earth. Vol 10. Simmons SF, Graham I (eds.) Econ Geol Spec Pub, p 181-195

Majoube M (1971) Fractionnement en oxygene-18 et deuterium entre l'eau et sa vapeur. J Chim Phys 197:1423-1436

Malinin SD, Kravchuk IF, Delbove F (1989) Chloride distribution between phases in hydrated and dry chloride-aluminosilicate melt systems as a function of phase composition. Geochim Int 26:32-38

Mandeville CW, Sasaki A, Saito G, Faure K, King R, Hauri E (1998) Open-system degassing of sulfur from Krakatau 1883 magma. Earth Planet Sci Lett 160:709-722

Mandeville CW, Webster JD, Tappen C, Taylor BE, Timbal A, Sasaki A, Hauri E, Bacon CR (2007) Stable isotope and petrologic evidence for open-system degassing during the climactic and pre-climactic eruptions of Mt. Mazama, Crater Lake, Oregon. (in review)

Mastin LG, Ghiorso MS (2001) Adiabatic temperature changes of magma-gas mixtures during ascent and eruption. Contrib Mineral Petrol 141:307-321

Mastin LG, Witter JB (2000) The hazards of eruptions through lakes and seawater. J Volcanol Geotherm Res 97:195-214

Mathez EA (1976) Sulfur solubility and magmatic sulfides in submarine basalt glass. J Geophys Res 81(23):4269-4276

Mathez EA, Webster JD (2005) Partitioning behavior of chlorine and fluorine in the system apatite-silicate melt-fluid. Geochim Cosmochim Acta 69:1275-1286

Matsuhisa Y (1979) Oxygen isotopic compositions of volcanic rocks from the east Japan island arcs and their bearing on petrogenesis. J Volcanol Geotherm Res 5:271-296

McGonicle AJS, Oppenheimer C (2003) Optical sensing of volcanic gas and aerosol emissions. *In:* Volcanic Degassing. Vol 213. Oppenheimer C, Pyle DM, Barclay J (eds) Geol Soc Lond Spec Pub, p 149-168

McGuinness MJ, Blakeley M, Pruess M, O'Sullivan MJ (1993) Heat pipe stability and upstream differencing. Trans Por Med 11:71-100

Métrich N, Berry A, O'Neill H, Susini J (2005) A XANES study of sulfur speciation in synthetic glasses and melt inclusions. Geochim Cosmochim Acta 69(10S):51

Métrich N, Clocchiatti R (1989) Melt inclusion investigation of the volatile behavior in historic alkaline magmas of Etna. Bull Volcanol 51:185-198

Métrich N, Clocchiatti R (1996) Sulfur abundance and its speciation in oxidized alkaline melts. Geochim Cosmochim Acta 60:4151-4160

Métrich N, Clocchiatti R, Mosbah M, Chaussidon M (1993) The 1989-1990 activity of Etna magma mingling and ascent of H_2O-Cl-S-rich basaltic magma. Evidence from melt inclusions. J Volcanol Geotherm Res 59:131-144

Métrich N, Rutherford MJ (1992) Experimental study of chlorine behavior in hydrous silicic melts. Geochim Cosmochim Acta 56:607-616

Michael PJ, Schilling J-G (1989) Chlorine in mid-ocean ridge magmas: evidence for assimilation of seawater-influenced components. Geochim Cosmochim Acta 53:3131-3143

Mitchell RH (2005) Carbonatites and carbonatites and carbonatites. Can Mineral 43:2049-2068

Mitchell RH, Keays RR (1981) Abundance and distribution of gold, palladium, and iridium in some spinel and garnet lherzolites: implications for the nature and origin of precious metal-rich intergranular components in the upper mantle. Geochim Cosmochim Acta 45:2425-2442

Miyoshi T, Sakai H, Chiba H (1984) Experimental study of sulfur isotope fractionation factors between sulfate and sulfide in high temperature melts. Geochem J 18:75-84

Mizutani Y, Sugiura T (1966) The chemical equilibrium of the $2H_2S + SO_2 = 3S + 2H_2O$ reaction in solfataras of the Nasudake Volcano. Bull Chem Soc Jpn 39:2411-2414

Moretti R, Ottonello G (2005) Solubility and speciation of sulfur in silicate melts: the conjugated Toop-Samis-Flood-Grjotheim (CTSFG) model. Geochim Cosmochim Acta 69:801-823

Moune S, Gauthier P-J, Gislason SR, Sigmarsson O (2006) Trace element degassing and enrichment in the eruptive plume of the 2000 eruption of Hekla volcano, Iceland. Geochim Cosmochim Acta 70:461-479

Mustard R, Kamenetsky VS, Mernagh TP (2003) Carbon dioxide bearing melt inclusions within a gold-mineralized felsic granite. *In:* Mineral Exploration and Sustainable Development, Proceedings of the 7th Biennial SGA Meeting. Vol 1. Eliopoulos DG (eds) Millpress Rotterdam, p 351-354

Mustard R, Ulrich T, Kamenetsky VS, Mernagh TP (2006) Gold and metal enrichment in natural granitic melts during fractional crystallization. Geology 34:85-88

Naboko SI (1957) A case of gaseous fluorine metasomatism at an active volcano. Geochem 23:452-455

Naumov VB, Kovalenko VI (1997) Sulfur concentration in magmatic melts: Evidence from inclusions in minerals. Geochem Internat 35:887-94

Nehlig P (1991) Salinity of oceanic hydrothermal fluids: a fluid inclusion study. Earth Planet Sci Lett 102:310-325

Newman S, Lowenstern JB (2002) VOLATILECALC: a silicate melt-H_2O-CO_2 solution model written in Visual Basic for excel. Comp Geosci 28:597-604

Newton RC, Manning CE (2005) Solubility of anhydrite, $CaSO_4$, in NaCl-H_2O solutions at high pressures and temperatures. J Petrol 46:701-716

Nielsen RL, Sours-Page RE, Harpp KS (2000) Role of a Cl-bearing flux in the origin of depleted ocean-floor magmas. Geochim Geophys Geosyst 1:1999GC000017

Nriagu JO (1989) A global assessment of natural sources of atmospheric trace metals. Nature 338:47-49

O'Neil JR, Taylor HP Jr. (1967) The oxygen isotope and cation exchange chemistry of feldspars. Am Min 52:1414-1437

Obenholzner JH, Schroettner H, Golob P, Delgado H (2003) Particles from the plume of Popocatépetl volcano, Mexico – the FESEM/EDS approach. *In:* Volcanic Degassing. Vol 213. Oppenheimer C, Pyle DM, Barclay J (eds) Geol Soc Lond Spec Pub, p 123-148

Ohmoto H, Rye RO (1974) Hydrogen and oxygen isotopic compositions of fluid inclusions in the Kuroko deposits, Japan. Econ Geol 69:947-953

Ohmoto H, Rye RO (1979) Isotopes of sulfur and carbon. *In:* Geochemistry of Hydrothermal Ore Deposits 2nd ed. Barnes HL (ed) John Wiley & Sons, p 509-567

Oppenheimer C (2003) Volcanic degassing. *In:* The Crust. Treatise in Geochemistry. Vol 3. Rudnick RL (ed) Elsevier p 123-166

Oppenheimer C, Tsanev VI, Braban CF, Cox RA, Adams JW, Aiuppa A, Bobrowski N, Delmelle P, Barclay J, McGonigle AJS (2006) BrO formation in volcanic plumes. Geochim Cosmochim Acta 70:2935-2941

Palin JM, Epstein S, Stolper EM (1996) Oxygen isotope partitioning between rhyolitic glass/melt and CO_2: An experimental study at 550-950 °C and 1 bar. Geochim Cosmochim Acta 60:1963-1973

Pallister JS, Reagan M, Cashman K (2005) A new eruptive cycle at Mount St. Helens? EOS 86(48):499-500

Papale P, Moretti R, Barbato D (2006) The compositional dependence of the saturation surface of $H_2O + CO_2$ fluids in silicate melts. Chem Geol 229:78-95

Pasteris JD, Wopenka B, Wang A, Harris TN (1996) Relative timing of fluid and anhydrite saturation: another consideration in the sulfur budget of the Mount Pinatubo eruption. *In:* Fire and Mud: Eruptions and Lahars of Mount Pinatubo, Philippines. Newhall CG, Punongbayan RS (eds). Univ Wash Press, Seattle, p. 875-891

Pettke T (2006) *In situ* laser-ablation-ICP-MS chemical analysis of melt inclusions and prospects for constraining subduction zone magmatism. *In:* Melt Inclusions in Plutonic Rocks. Vol 36. Webster JD (ed) Mineral Assoc Canada, p 51-80

Pichavant M, Ramboz C, Weisbrod A (1982) Fluid immiscibility in natural processes: use and misuse of fluid inclusion data. Chem Geol 37:1-27

Pineau F, Shilobreyeva S, Kadik A, Javoy M (1998) Water solubility and D/H fractionation in the system basaltic andesite-H_2O at 1250 °C and between 0.5 and 3 kbars. Chem Geol 147:173-184

Pitzer KS, Pabalan RT (1986) Thermodynamics of NaCl in steam. Geochim Cosmochim Acta 50:1445-1454

Pokrovski GS, Roux J, Harrichoury J-C (2005) Fluid density control on vapor-liquid partitioning of metals in hydrothermal systems. Geology 33(8):657-660

Poorter RPE, Varekamp JC, Van Bergen MJ, Kreulen R, Sriwana, T, Vroon PZ, Wirakusumah AD (1989) The Sirung volcanic boiling spring: an extreme chloride-rich, acid brine on Pantar (Lesser Sunda Islands, Indonesia). Chem Geol 76:215-228

Prausnitz JM, Lichtenthaler RN, Gomes de Azevedo E (1999) Molecular Thermodynamics of Fluid-Phase Equilibria. Prentice Hall PTR

Quisefit JP, Toutian JP, Bergametti G, Javoy M, Cheynet B, Person A (1989) Evolution vs. cooling of gaseous volcanic emissions from Momotombo Volcano, Nicaragua: thermochemical model and observations. Geochim Cosmochim Acta 53:2591-2608

Reed MH (1982) Calculation of multicomponent chemical equilibria and reaction processes in systems involving minerals, gases and an aqueous phase. Geochim Cosmochim Acta 46:513-528

Reed MH (1992a) Origin of diverse hydrothermal fluids by reaction of magmatic volatiles with wall rock. Rept Geol Surv Japan 279:135-140

Reed MH (1992b) Computer modeling of chemical processes in geothermal systems: examples of boiling, mixing and water-rock reaction. *In:* Application of Geochemistry in Geothermal Reservoir Development. D'Amore F (ed) Unit Nat Inst Train Res, p 275-298

Reed MH (1997) Hydrothermal alteration and its relationship to ore fluid composition. *In:* Geochemistry of Hydrothermal Ore Deposits. 3rd Edition. Barnes HL (ed) Wiley New York, p 303-366

Reed MH (1998) Calculation of simultaneous chemical equilibria in aqueous-mineral-gas systems and its applications to modeling hydrothermal processes. *In:* Techniques in Hydrothermal Ore Deposits Geology. Vol. Richards J, Larson P (eds) Rev Econ Geo 10:109-124

Reed MH, Spycher NF (1988) Fugacity coefficients for H_2, CO_2, CH_4, H_2O and of H_2O-CO_2-CH_4 mixtures: a virial equation treatment for moderate pressures and temperatures applicable to calculations of hydrothermal boiling. Geochim Cosmochim Acta 52:739-749

Renno AD, Franz L, Witzke T, Herzig PM (2004) The coexistence of melts of hydrous copper chloride, sulfide and silicate compositions in a magnesiohastingsite cumulate, TUBAF Seamount, Papua New Guinea. Can Mineral 42:1-16

Reyf FG (1997) Direct evolution of W-rich brines from crystallizing melt within the Mariktikan granite pluton, west Transbaikalia. Mineral Deposita 32:475-490

Reynolds TJ, Beane RE (1979) The evolution of hydrothermal fluid characteristics through time at the Santa Rita, New Mexico, porphyry copper deposit. Geol Soc Am Abs Prog 11:502

Rhymer H, Cassidy J, Locke CA, Barboza MV, Barquero J, Brenes J, Van der Laat R (2000) Geophysical studies of the recent 15-year eruptive cycle at Poás Volcano, Costa Rica. J Volcanol Geo Res 97:425-442

Rice A (2000) Rollover in volcanic crater lakes: a possible case for Lake Nyos-type disasters. J Volcanol Geotherm Res 97:233-239

Richards JP (1995) Alkali-type epithermal gold deposits: a review. *In:* Magmas, Fluids, and Ore Deposits. Vol 23. Thompson JFH (ed) Mineral Assoc Canada, p 367-400

Richet P, Bottinga Y, Javoy M (1977) A review of hydrogen, carbon, nitrogen, oxygen, sulfur, and chlorine stable isotope fractionation among gaseous molecules. Ann Rev Earth Planet Sci 5:65-110

Roedder E (1979) Origin and significance of magmatic inclusions. Bull Mineral 102:487-510

Roedder E (1984) Fluid Inclusions. Rev Mineral Geochem Vol 12. Mineral Soc America

Roedder E (1992) Fluid inclusion evidence for immiscibility in magmatic differentiation. Geochim Cosmochim Acta 56:5-20

Roedder E, Coombs DS (1967) Immiscibility in granitic melts, indicated by fluid inclusions in ejected granitic blocks from Ascension Island. J Petrol 8:417-451

Rowe GL Jr. (1994) Oxygen, hydrogen, and sulfur isotope systematics of the crater lake system of Poas Volcano, Costa Rica. Geochem J 28:263-287

Rubey HW (1951) Geologic history of seawater. Bull Geol Soc Am 62:1111-1148

Sakai H, Matsubaya O (1974) Isotopic geochemistry of the thermal waters of Japan and its bearing on the Kuroko ore deposits. Econ Geol 69:974-991

Scaillet B, Clemente B, Evans BW, Pichavant M (1998) Redox control of sulfur degassing in silicic magmas. J Geophys Res 103(B10):23,937-23,949

Scaillet B, Macdonald R (2006) Experimental and thermodynamic constraints on the sulphur yield of peralkaline and metaluminous silicic flood eruptions. J Petrol 47:1413-1437

Scaillet B, Pichavant M (2003) Dynamics of magma degassing. *In:* Volcanic Degassing. Vol 213. Oppenheimer C, Pyle DM, Barclay J (eds) Geol Soc Lond Spec Pub, p 23-52

Schmidt C, Bodnar RJ (2000) Synthetic fluid inclusions: XVI. PVTX properties in the system H_2O-NaCl-CO_2 at elevated temperatures, pressures, and salinities. Geochim Cosmochim Acta 64:3853-3869

Self S, King AJ (1996) Petrology and sulfur and chlorine emissions of the 1963 eruption of Gunung Agung, Bali, Indonesia. Bull Volcanol 58:263-285

Sheppard SMF (1986) Characterization and isotopic variations in natural waters. Rev Mineral 16:165-183

Shilobreyeva SN, Devirts AL, Kadik AA, Lagutina YP (1992) Distribution of hydrogen isotopes in basalt liquid-water equilibrium at 3 kbar and 1250 °C. Geochem International 29:130-134

Shinohara H (1994) Exsolution of immiscible vapor and liquid phases from a crystallizing silicate melt: implications for chlorine and metal transport. Geochim Cosmochim Acta 58:5215-5221

Shinohara H, Hedenquist JW (1997) Constraints on magma degassing beneath the Far Southeast porphyry Cu-Au deposit, Philippines. J Petrol 38:1741-1752

Shinohara H, Ilyama JT, Matsuo S (1989) Partition of chlorine compounds between silicate melt and hydrothermal solutions: partition of NaCl-KCl. Geochim Cosmochim Acta 53:2617-2630

Shinohara H, Kazahaya K (1995) Degassing processes related to magma chamber crystallization. *In:* Magmas, Fluids, and Ore Deposits. Vol 23. Thompson JFH (ed) Mineral Assoc Canada, p 47-70

Shinohara H, Kazahaya K, Lowenstern JB (1995) Volatile transport in a convecting magma column: implications for porphyry Mo mineralization. Geology 23:1091-1094

Shmulovich KI, Churakov SV (1998) Natural fluid phases at high temperatures and low pressures. J Geochem Explor 62:183-191

Shmulovich KI, Graham CM (2004) An experimental study of phase equilibria in the systems H_2O-CO_2-$CaCl_2$ and H_2O-CO_2-NaCl at high temperatures and pressures (500-800 °C, 0.5-0.9 GPa) geological and geophysical applications. Contrib Mineral Petrol 146:450-462

Shmulovich KI, Heinrich W, Möller P, Dulski P (2002) Experimental determination of REE fractionation between liquid and vapour in the systems NaCl-H_2O and $CaCl_2$-H_2O up to 450 °C. Contrib Mineral Petrol 144:257-273

Shmulovich KI, Landwehr D, Simon K, Heinrich W (1999) Stable isotope fractionation between liquid and vapour in water-salt systems up to 600° C. Chem Geol 157:343-354

Shmulovich KI, Tkachenko SI, Plyasunova NV (1995) Phase equilibria in fluid systems at high pressures and temperatures. *In:* Fluids in the crust. Shmulovich KI, Yardley BWD, Gonchar GG (eds) Chapman and Hall, p 193-214

Signorelli S, Capaccioni B (1999) Behaviour of chlorine prior and during the 79 A.D. Plinian eruption of Vesuvius (southern Italy) as inferred from the present distribution in glassy mesostases and whole-pumices. Lithos 46:715-730

Signorelli S, Carroll MR (2000) Solubility and fluid-melt partitioning of Cl in hydrous phonolitic melts. Geochim Cosmochim Acta 64:2851-2862

Signorelli S, Carroll MR (2001) Experimental constraints on the origin of chlorine emissions at the Soufriere Hills volcano, Montserrat. Bull Volcanol 62:431-440

Signorelli S, Vaggelli G, Romano C, Carroll MR (2001) Volatile element zonation in Campanian Ignimbrite magmas (Phlegrean Fields, Italy): evidence from the study of glass inclusions and matrix glasses. Contrib Mineral Petrol 140:543-553

Sillitoe RH (1994) Erosion and collapse of volcanoes: causes of telescoping in intrusion-centered ore deposits. Geology 22:945-948

Sillitoe RH, Hedenquist JW (2003) Linkages between volcanotectonic settings, ore-fluid composition, and epithermal precious metal deposits. *In:* Volcanic, Geothermal, and Ore-forming Fluids: Rulers and Witnesses of Processes within the Earth. Simmons SF, Graham I (eds.) Econ Geol Spec Pub 10:315-343

Simon AC, Frank MR, Pettke T, Candela P, Piccoli PM, Heinrich CA (2005) Gold partitioning in melt-vapor-brine systems. Geochim Cosmochim Acta 69:3321-3335

Sisson TW (1995) History and hazards of Mount Rainier, Washington. USGS Open-File Rpt 95-642

Skinner BJ (1997) Hydrothermal mineral deposits: what we do and don't know. *In:* Geochemistry of Hydrothermal Ore Deposits. 3rd Edit. Barnes HL (ed) John Wiley & Sons, p 1-30

Solovova I, Naumov V, Girnis A, Kovalenko V, Guzhova A (1991) High-temperature fluid heterogeneity: evidences from microinclusions in Pantelleria volcanics. Plinius 5:206

Sourirajan S, Kennedy GC (1962) The system H_2O-NaCl at elevated temperatures and pressures. Am J Sci 260:115-141

Sparks RSJ (2003) Dynamics of magma degassing. *In:* Volcanic Degassing. Vol 213. Oppenheimer C, Pyle DM, Barclay J (eds) Geol Soc Lond Spec Pub, p 5-22

Sriwana T, van Bergen MJ, Varekamp JC, Sumarti S, Takano B, van Os BJH, Leng MJ (2000) Geochemistry of acid Kawah Putih lake, Patuha Volcano, West Java, Indonesia. J Volcanol Geotherm Res 97:77-104

Stelling J, Beermann O, Nowak M, Botcharnikov RE (2006) Partitioning of chlorine between hydrous fluid and basaltic melt of Mt. Etna. Berich Deutsch Mineral Ges, Beih Eur J Mineral 18(1):136

Sterner SM, Bodnar RJ (1991) Synthetic fluid inclusions X: experimental determination of P-V-T-X properties in the CO2-H_2O system to 6 kb and 700 °C. Am J Sci 291:1-54

Sterner SM, Chou I-M, Downs RT, Pitzer KS (1992) Phase relations in the system NaCl-KCl-H_2O: V. Thermodynamic –PTX analysis of solid-liquid equilibria at high temperatures and pressures. Geochim Cosmochim Acta 56:2295-2309

Sterner SM, Bodnar RJ (1988) Synthetic fluid inclusions: V: solubility relations in the system NaCl-KCl-H_2O under vapor-saturated conditions. Geochim Cosmochim Acta 52:989-1005

Stevenson DS (1993) Physical models of fumarolic flow. J Volcanol Geotherm Res 57:139-156

Stix J, Gaonac'h H (2000) Gas, plume and thermal monitoring. *In:* Encyclopedia of Volcanoes. Sigurdsson H (ed) Academic Press, p 1141-1163

Stix J, Layne GD (1996) Gas saturation and evolution of volatile and light lithophile elements in the Bandelier magma chamber between two caldera-forming eruptions. J Geophys Res 101:25181-25196

Stoiber RE, Rose WI Jr. (1970) The geochemistry of Central American volcanic gas condensates. Geol Soc Am Bull 81:2891-2911

Symonds RB (1992) Getting the gold from the gas: how recent advances in volcanic-gas research have provided new insight on metal transport in magmatic fluids. Rpt Geol Surv Japan 279:170-175

Symonds RB, Mizutani Y, Briggs PH (1996) Long-term geochemical surveillance of fumaroles at Showa-Shinzan Dome, Usu Volcano, Japan. J Volcanol Geotherm Res 73:177-211

Symonds RB, Reed MH, Rose WI (1992) Origin, speciation, and fluxes of trace-element gases at Augustine volcano, Alaska: Insights into magma degassing and fumarolic processes. Geochim Cosmochim Acta 56:633-657

Symonds RB, Rose WI, Bluth GJS, Gerlach TM (1994) Volcanic-gas studies: methods, results, and applications. Rev Mineral 30:1-66

Symonds RB, Rose WI, Gerlach TM, Briggs PH, Harmon RS (1990) Evaluation of gases, condensates, and SO_2 emissions from Augustine volcano, Alaska: the degassing of a Cl-rich volcanic system. Bull Volcanol 52:355-374

Symonds RB, Rose WI, Reed MH (1988) Contribution of Cl- and F-bearing gases to the atmosphere by volcanoes. Nature 334:415-418

Symonds RB, Rose WI, Reed MH, Lichte FE, Finnegan DL (1987) Volatilization, transport and sublimation of metallic and non-metallic elements in high-temperature gases at Merapi volcano, Indonesia. Geochim Cosmochim Acta 51:2083-2101

Takenouchi S, Kennedy G (1965) The binary system H_2O-CO_2 at high temperatures and pressures. Am J Sci 262:1055-1074

Taran YA (1992) Chemical and isotopic composition of fumarolic gases from Kamchatka and Juril Islands. Rep Geol Surv Jpn 279:183-186

Taran YA, Connor CB, Shapar VN, Ovsyannikov AA, Bilichenko AA (1997) Fumarolic activity of Avachinsky and Koryaksky volcanoes, Kamchatka, from 1993 to 1994. Bull Volcanol 58:441-448

Taran YA, Giggenbach WF (2003) Geochemistry of light hydrocarbons in subduction-related volcanic and hydrothermal fluids. *In:* Volcanic, Geothermal, and Ore-forming Fluid: Rulers and Witnesses of Processes within the Earth. Simmons SF, Graham I (eds.) Econ Geol Spec Pub 10:61-74

Taran YA, Hedenquist JW, Korzhinskii MA, Tkachenko SI, Shmulovich KI (1995) Geochemistry of magmatic gases from Kudryavy volcano, Iturup, Kuril Islands. Geochim Cosmochim Acta 59:1749-1761

Taran YA, Pokrovsky BG, Dubik YM, (1989) Isotopic composition and origin of water from andesitic magmas. Dokl Acad Sci USSR 304:440-443

Taube H (1954) Use of oxygen isotope effects in the study of hydration of ions. J Phys Chem 58:523-528

Taylor BE (1986) Magmatic Volatiles: Isotopic variation of C, H, and S. Rev Mineral 16:185-225

Taylor HP Jr. (1979) Oxygen and hydrogen isotope relationships in hydrothermal mineral deposits. *In:* Geochemistry of Hydrothermal Ore Deposits 2nd Edition. Barnes HL (ed) John Wiley & Sons, p 236-277

Textor C, Sachs PM, Graf H-F, Hansteen TH (2003) The 12900 years BP Laacher See eruption: estimation of volatile yields and simulation of their fate in the plume. *In:* Volcanic Degassing. Vol 213. Oppenheimer C, Pyle DM, Barclay J (eds) Geol Soc Lond Spec Pub, p 307-328

Thomas N, Tait S, Koyaguchi T (1993) Mixing of stratified liquids by the motion of gas bubbles: application to magma mixing. Earth Planet Sci Lett 115:161-175

Thomas R, Förster H-J, Rickers K, Webster JD (2005) Formation of extremely F-rich hydrous melt fractions and hydrothermal fluids during differentiation of highly evolved tin-granite magmas: a complex melt/fluid-inclusion study. Mineral Dep 148:582-601

Thomas R, Webster JD (2006) Understanding pegmatite formation: the melt and fluid inclusion approach. *In:* Melt Inclusions in Plutonic Rocks. Vol 36. Webster JD (ed) Mineral Assoc Canada, p 189-210

Thompson AB, Aerts M, Hack AC (2007) Liquid immiscibility in silicate melts and related systems. Rev Mineral Geochem 65:99-127

Truesdell AH, Nathenson M, Rye RO (1977) The effects of subsurface boiling and dilution on the isotopic compositions of Yellowstone thermal waters. J Geophys Res 82:3694-3704

Ulrich T, Günther D, Heinrich CA (1999) Gold concentrations of magmatic brines and the metal budget of porphyry copper deposits. Nature 399:676-679

Vaggelli G, Belkin HE, Francalanci L (1993) Silicate-melt inclusions in the mineral phases of the Stromboli volcanic rocks: a contribution to the understanding of magmatic processes. Acta Vulcan 3:115-125

Vandemeulebrouck J, Sabroux J-C, Halbwachs M, Surono, Poussielgue N, Grangeon J, Tabbagh J (2000) Hydroacoustic noise precursors of the 1990 eruption of Kelut Volcano, Indonesia. J Volcanol Geotherm Res 97:443-476

Vanko DA (1986) High-chlorine amphiboles from oceanic rocks: product of highly-saline hydrothermal fluids? Am Mineral 71:51-59

Varekamp JC, Kreulen R (2000) The stable isotope geochemistry of volcanic lakes, with examples from Indonesia. J Volcanol Geotherm Res 97:309-327

Varekamp JC, Kruelen R, Poorter RPE, VanBergen MJ (1992) Carbon sources in arc volcanism, with implications for the carbon cycle. Terra Nova 4:363-373

Varekamp JC, Pasternack GB, Rowe GL Jr. (2000) Volcanic lake systematics II. Chemical constraints. J Volcanol Geotherm Res 97:161-179

Varekamp JC, Thomas E, Germani, M, Buseck PR (1986) Particle geochemistry of Etna and Mt. St. Helens. J Geophys Res 91:12233-12248

Vaselli O, Tassi F, Minissale A, Montegrossi G, Duarte E, Fernandez E, Bergamaschi F (2003) *In:* Volcanic Degassing. Vol 213. Oppenheimer C, Pyle DM, Barclay J (eds) Geol Soc Lond Spec Pub, p 247-262

Veklser IV (2004) Liquid immiscibility and its role at the magmatic-hydrothermal transition: a summary of experimental studies. Chem Geol 210:7-31

Veksler IV, Lentz D (2006) Parental magmas of plutonic carbonatites, carbonate-silicate immiscibility and decarbonation reactions: evidence from melt and fluid inclusions. *In:* Melt Inclusions in Plutonic Rocks. Vol 36. Webster JD (ed) Mineral Assoc Canada, p 123-150

Veksler IV, Thomas R (2002) An experimental study of B-, P-, and F-rich synthetic granite pegmatites at 0.1 and 0.2 GPa. Contrib Mineral Petrol 143:673-683

Von Damm KL, Edmond JM, Grant B, Measures BW, Weiss RF (1985) Chemistry of submarine hydrothermal solutions at 21° North, East Pacific Rise. Geochim Cosmochim Acta 49:2197-2220

Wahrenberger CM (1997) Some aspects of the chemistry of volcanic gases. PhD Dissertation, Swiss Federal Institute of Technology, Zurich, Switzerland

Wahrenberger CM, Seward TM, Dietrich V (2002) Volatile trace-element transport in high-temperature gases from Kudriavy volcano (Iturup, Kurile Islands, Russia). *In:* Water-rock Interactions, Ore Deposits, and Environmental Geochemistry. Hellman R, Wood SA (eds) Geochem Soc Spec Pub 7:307-327

Wallace PJ (2005) Volatiles in subduction zone magmas: concentrations and fluxes based on melt inclusion and volcanic gas data. J Volcanol Geotherm Res 140:217-240

Wallace PJ, Anderson AT Jr (2000) Volatiles in magma. *In:* Encyclopedia of Volcanoes. Sigurdsson H (ed) Academic Press, p 149-170

Wallace PJ, Carmichael ISE (1992) Sulfur in basaltic magmas. Geochim Cosmochim Acta 56:1863-1874

Watters RJ, Zimbelman DR, Bowman SD, Crowley JK (2000) Rock mass strength assessment and significance to edifice stability, Mount Rainer and Mount Hood, Cascade Range volcanoes. Pure App Geophys 157:957-976

Webster JD (1990) Partitioning of F between $H_2O \pm CO_2$ fluids and topaz rhyolite melt: implications for mineralizing magmatic-hydrothermal fluids in F-rich granitic systems. Contrib Mineral Petrol 104:424-438

Webster JD (1992a) Fluid-melt interactions in Cl-rich granitic systems: effects of melt composition at 2 kbar and 800 °C. Geochim Cosmochim Actaa 56:659-678

Webster JD (1992b) Fluid-melt interactions involving Cl-rich granites: experimental study from 2 to 8 kbar. Geochim Cosmochim Acta 56:679-687

Webster JD (2004) The exsolution of magmatic hydrosaline melts. Chem Geol 210:33-48

Webster JD (2005) Consequences of exsolution of H_2O-, CO_2-, SO_2-, Cl-bearing volatile phases on the physical and chemical properties of magma. Geochim Cosmochim Acta Spec Supp, A151

Webster JD, DeVivo B (2002) Experimental and modeled solubilities of chlorine in aluminosilicate melts, consequences of magma evolution, and implications for exsolution of hydrous chloride melt at Mt. Somma-Vesuvius. Am Mineral 87:1046-1061

Webster JD, Holloway JR (1988) Experimental constraints on the partitioning of Cl between topaz rhyolite melt and H_2O and $H_2O + CO_2$ fluids: New implications for granitic differentiation and ore deposition. Geochim Cosmochim Acta 52:2091-2105

Webster JD, Kinzler RJ, Mathez EA (1999) Chloride and water solubility in basalt and andesite liquids and implications for magmatic degassing. Geochim Cosmochim Acta 63:729-738

Webster JD, Rebbert CR (1998) Experimental investigation of H_2O and Cl solubilities in F-enriched silicate liquids: implications for volatile saturation of topaz rhyolite magmas. Con Mineral Petrol 132:198-207

Webster JD, Sintoni MF, De Vivo B (2006) The role of sulfur in promoting magmatic degassing and volcanic eruption at Mt. Somma-Vesuvius. *In:* Volcanism in the Campania Plain. Vol 9. De Vivo B (ed), Elsevier, p 219-233

Webster JD, Thomas R (2006) Silicate melt inclusions in felsic plutons: a synthesis and review. *In:* Melt Inclusions in Plutonic Rocks. Vol 36. Webster JD (ed) Mineral Assoc Canada, p 165-188

Weisbrod A (1981) Fluid inclusions in shallow intrusives. *In:* Fluid Inclusions: Applications to Petrology. Vol 6. Hollister LS, Crawford ML (eds) Mineral Assoc Canada, p 241-271

Whitney JA (1984) Fugacities of sulfurous gases in pyrrhotite-bearing magmas. Am Mineral 69:69-78

Whitney JA, Stormer JC (1983) Igneous sulfides in the Fish Canyon Tuff and the role of sulfur in calc-alkaline magmas. Geology 11:99-102

Williams SN, Schaefer SJ, Calavache VML, Lopez D (1992) Global carbon dioxide emission to the atmosphere by volcanoes. Geochim Cosmochim Acta 56:1765-1770

Williams TJ, Candela P, Piccoli P (1995) The partitioning of copper between silicate melts and 2-phase aqueous fluids-an experimental investigation at 1kbar, 800 °C, and 0.5 kbar, 850 °C. Contrib Mineral Petrol 121:388-399

Williams-Jones AE, Heinrich CA (2005) Vapor transport of metals and the formation of magmatic-hydrothermal ore deposits. Econ Geol 100(7):1287-1312

Williams-Jones G, Rymer H (2000) Hazards of volcanic gases. *In:* Encyclopedia of Volcanoes. Sigurdsson H (ed) Academic Press, p 997-1004

Wohletz KH, McQueen RG (1984) Experimental studies of hydromagmatic volcanism. *In:* Explosive Volcanism: Inception, Evolution, and Hazards. National Academy Press Washington DC, p 158-169

Woods AW (1988) The fluid dynamics and thermodynamics of eruption columns. Bull Volcanol 50:169-19

Young SR, Voight B, Duffell HJ (2003) Magma extrusion dynamics revealed by high-frequency gas monitoring at Soufriére Hills volcano, Montserrat. *In:* Volcanic Degassing. Vol 213. Oppenheimer C, Pyle DM, Barclay J (eds) Geol Soc Lond Spec Pub, p 219-230

Zhang Y (2000) Energetics of gas-driven limnic and volcanic eruptions. J Volcanol Geotherm Res 97:215-23

Zimbelman DR (1996) Hydrothermal alteration and its influence on volcanic hazards-Mount Rainier, Washington, a case history. PhD Dissertation, University of Colorado-Boulder

Zimbelman DR, Rye RO, Breit GN (2005) Origin of secondary sulfate minerals on active andesitic stratovolcanoes. Chem Geol 215:37-60

Zimbelman DR, Rye RO, Landis GP (2000) Fumaroles in ice caves on the summit of Mount Rainier-preliminary stable isotope, gas, and geochemical studies. J Volcanol Geotherm Res 97:457-473

Zimmer MM, Plank T, Hauri EH, Nye C, Faust Larsen J, Kelemen PB (2004) Volatile contents in mafic magmas from two Aleutian volcanoes: Augustine and Makushin. EOS Trans, Am Geophys Union 85:V34A-01

Zreda-Gostynska G, Kyle PR, Finnegan D, Prestbo KM (1989) Volcanic gas emissions from Mount Erebus and their impact on the Antarctic environment. J Geophys Res 102(B7):15,039-15,055

Reviews in Mineralogy & Geochemistry
Vol. 65, pp. 363-387, 2007
Copyright © Mineralogical Society of America

Fluid-Fluid Interactions in Magmatic-Hydrothermal Ore Formation

Christoph A. Heinrich

Isotope Geochemistry and Mineral Resources
Department of Earth Sciences
ETH Zürich
8092 Zürich, Switzerland
heinrich@erdw.ethz.ch

INTRODUCTION

Hydrothermal ore deposits have provided some of the first geological evidence for the interaction of multiple fluids in the Earth's interior, including fluid mixing as well as fluid phase separation. Physical evidence comes from veins and breccias documenting the impact of fluids in breaking rocks and creating fracture permeability for ore formation, as a result of the fluid expansion that is commonly associated with phase separation between melts and multiple fluids. Chemical evidence of fluid–fluid interaction includes the precipitation of minerals in veins as a result of fluid boiling or mixing between chemically contrasting fluids. Fluid inclusions record interactions of multiple fluids, especially in magmatic-hydrothermal ore deposits and associated igneous rocks (Fig. 1). In recent years, fluid inclusions have contributed to a more quantitative understanding of the role of fluid–fluid interactions, since new microanalytical techniques allow determination of element concentrations in all types of fluids, from low-density vapor through aqueous and hypersaline liquids to hydrous salt melts (Roedder 1984; Samson et al. 2003).

The present chapter reviews some of the evidence for interactions between multiple fluids in ore-forming hydrothermal systems, focusing particularly on fluid inclusion analyses and their geological context, and building on experimental and theoretical studies that are reviewed in other chapters of this book (Driesner and Geiger 2007; Hack et al. 2007; Liebscher 2007).

Three examples of globally recurring ore-forming environments illustrate the relationships between physical processes of fluid–fluid interaction and their geochemical-mineralogical consequences for the selective enrichment of economically important trace elements. *Orogenic gold* deposits in regional-metamorphic terrains representing the deeper parts of the crust show evidence of mixing and phase separation among aquo-carbonic fluids, but the chemical role of the carbonic vapor phase in these vein systems is still poorly understood. More information exists about fluid compositions and fluid evolution paths in upper-crustal magmatic-hydrothermal ore systems, extending from dioritic to granitic plutons up to the volcanic environment, as introduced by the preceding chapter (Webster and Mandeville 2007). *Granite-related Sn-W veins* provide an example to illustrate the interaction of hot magmatic vapor and hypersaline magmatic liquid with cool surface-derived meteoric water, as an widespread mechanism for ore mineral precipitation by fluid mixing in the upper crust. *Porphyry – epithermal Cu–Mo–Au systems* associated with intermediate-composition calcalkaline magmas demonstrate the role of fluid phase separation in ore-metal fractionation and mineral precipitation, resulting from the large range in density and degree of miscibility of saline fluids between surface and magmatic conditions. These vertically extensive magmatic-hydrothermal systems illustrate how variations in fluid properties in salt–water dominated fluids can give rise to diverse

1529-6466/07/0065-0011$05.00

DOI: 10.2138/rmg.2007.65.11

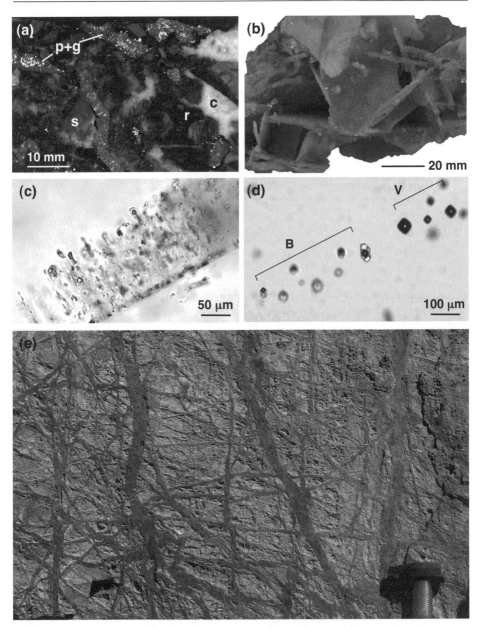

Figure 1. Geological evidence for fluid phase separation in ore-forming magmatic-hydrothermal systems. (a) 'Bonanza grade' gold (~1 wt%) that was precipitated together with fine-grained pyrite in a fluidized breccia resulting from instantaneous boiling and expansion of magmatic fluids ascending from an underlying andesite porphyry intrusion. The breccia consists of adularia–sericite altered clasts of sedimentary wall rock (s) and nodules of (originally colloidal?) pyrite with bright specks of gold (p+g), both cemented by roscoelite (r; ~ vanadium-rich muscovite) and late calcite (c) (Zone VII ore body, Porgera, Papua New Guinea; Ronacher et al. 2004). (b) Bladed manganocalcite, the characteristic habit of calcite precipitated from aqueous epithermal fluids as a result of the pH change following loss of CO_2 to a separating vapor phase (Madan, Bulgaria; Kostova et al. 2004). (c) Growth band in a euhedral quartz crystal marked by

(*caption continued on facing page*)

fluid evolution paths, which in turn control metal enrichment in distinct porphyry-style and epithermal ore deposits.

This chapter concentrates on ore-forming hydrothermal systems involving interaction among H_2O–CO_2–NaCl dominated fluids at temperatures above 150 °C. Hydrocarbon bearing sedimentary basins are an other crustal environment, in which fluid–fluid interactions are important for base-metal ore formation. This can include sulfide deposition by chemical redox reactions between metals and liquid hydrocarbons, or the physical consequences of natural gas formation that may lead to rapid expulsion of hot basin brines. Both processes may be essential for the formation of Mississippi Valley type and other low-temperature sediment-hosted base-metal deposits (e.g., Cathles and Adams 2005). These processes are not further explored in this paper, but are touched upon in the chapter of Pedersen and Christensen (2007).

CHEMICAL CONSEQUENCES OF FLUID PHASE SEPARATION

Principles of hydrothermal ore formation

Hydrothermal ore deposits form by enrichment of rare elements through processes of selective dissolution and re-precipitation of minerals (Kesler 2005). Chemical gradients in mineral solubility and in the stability of metal-transporting complexes are therefore a key to hydrothermal ore formation. Temperature gradients in crustal fluid flow systems contribute importantly to these essential chemical gradients (e.g., Yardley 2005), but the following review will show that steep and highly non-linear variations in the physical properties of variably miscible fluids are equally important for ore formation. It is no coincidence that most major ore deposits form in the upper crust, where advection of magma and hot fluids into cooler rocks establishes steep temperature gradients, and where brittle deformation of rocks favors the formation of vertically extensive vein networks and temporarily creates sharp gradients in pressure, density and miscibility of crustal fluids.

Variations in density and degree of miscibility of crustal fluids are dominated by their composition in terms of the three major components, H_2O–CO_2–NaCl (see Driesner and Geiger 2007; Liebscher 2007; Gottschalk 2007). The physical and chemical properties of these fluids in turn affect the chemical stability of dissolved species, including ore forming components. These are the economically valuable metals themselves, as well as diverse species of sulfur (SO_2, SO_4^{2-}, H_2S, HS^-) that are essential as complex-forming ligands to stabilize certain metals in solution (e.g., aqueous $Au(HS)_2^-$; Stefánsson and Seward 2005), or provide the essential anion for ore mineral precipitation (sulfides).

Figure 1 (*caption continued*).

numerous primary fluid inclusions, showing variable phase ratios of aqueous liquid and gas bubbles; these reflect the entrapment of random proportions of liquid and vapor that coexisted in a boiling epithermal Pb-Zn-Cu sulfide vein (Baia Sprie, Romania: Grancea et al. 2002). (d) Healed micro-fracture that trapped coexisting hypersaline liquid (oversaturated with halite and sylvite after cooling to ambient temperature; inclusions B lower left) and a low-salinity vapor with a density of ~ 0.1 g cm^{-1} (V, large gas bubble with barely visible layer of liquid, upper right part of former crack); Yankee Lode quartz–cassiterite veins, Mole Granite, Australia (Audétat et al. 1998, 2000a, see Fig. 6, below); also note cover photo of this book for a similar "boiling trail" from a porphyry in Arizona). (e) Outcrop photograph of the characteristic "stockwork" veining of porphyry-type ore deposits. Vein opening results from forceful hydraulic fracturing of the porphyry intrusion, due to the positive volume change upon crystallization of hydrous magma to less-hydrous rock plus magmatic-hydrothermal fluid (Burnham and Ohmoto 1980). Magma crystallization, fracturing, pressure release, and separation of the exsolved magmatic fluid into voluminous Cu-rich magmatic vapor plus minor hypersaline liquid occur almost simultaneously, and are immediately followed by the precipitation of quartz + Cu-Fe-sulfides + gold ± molybdenite (Rosia Poieni, Romania; Damman et al. 1996, and own unpublished data).

Fluid phase separation, in particular, has profound geochemical effects on the formation of metallic mineral deposits (Roedder 1971; Ellis 1979; Drummond and Ohmoto 1985; Heinrich et al. 1999). Two principal effects can be distinguished. First, the loss of gaseous components during low-pressure boiling in epithermal veins and active geothermal systems can dramatically affect mineral solubility, even when metals and anions are entirely transported in the aqueous liquid phase. Second, fractionation of ore metals themselves can occur by separation of dense vapor from a hypersaline liquid, particularly in magmatic-hydrothermal environments at P-T conditions above the critical point of pure water. Due to the different transport properties of the two phases, this can lead to large-scale segregation of ore forming components and the generation of distinct fluids resulting in particular types of ore deposits. These two fundamental effects are first illustrated separately in the following two subsections, but later will be linked as parts of a continuous fluid evolution in vertically extensive porphyry – epithermal ore systems.

Mineral precipitation by low-pressure boiling

Aqueous liquid is the dominant agent for mass transport and mineral precipitation in epithermal ore deposits and active geothermal systems, such as those in Iceland or the Taupo Volcanic Zone of New Zealand (Arnórsson and Stefánsson 2007). Depending on the actual temperature gradient along its ascent path, this liquid may boil by separating bubbles of a low-density vapor phase. This vapor consists of common gas species like CO_2, CH_4, H_2S and H_2O, but the concentration of ionic solutes including salts (chloride, carbonate, sulfide, sulfate and metals, with possible exception of Hg, see Barnes and Seward 1997) in the vapor phase is negligible. Boiling nevertheless affects the ionic composition of the remaining aqueous liquid, because volatile elements like S contribute important metal-complexing ligands in aqueous solution, and because CO_2 (or H_2CO_3) and H_2S are weak acids in aqueous solution. Selective loss of CO_2 and H_2S to the vapor phase shifts the acid–base balance in the remaining liquid towards more alkaline conditions, e.g., by reducing the CO_2/HCO_3^- ratio according to equilibria such as $HCO_3^- + H^+ = H_2O + CO_2$ (Ellis 1979; Berger and Henley 1989; Arnórsson and Stefánsson 2007). The redox state of hydrothermal fluids is also affected by the shifting of liquid–vapor equilibria resulting from vapor loss, notably by reactions among sulfur species (H_2S, HS^-, SO_2 and SO_4^{2-}), which are a strong function of temperature and water fugacity (Drummond and Ohmoto 1985; Carrol and Webster 1994; Webster and Mandeville 2007).

The aqueous solubility of ore and gangue minerals including gold, base-metal sulfides and carbonates depends on the concentrations of complexing ligands as well as on pH. As a result, vapor separation affects competing fluid/mineral reactions which, depending on initial fluid composition, can lead to either dissolution or precipitation of minerals (Drummond and Ohmoto 1985). Multicomponent thermodynamic modeling shows that gold-rich 'bonanza' ore zones can form within more extensive epithermal vein systems by selective removal of gaseous H_2S to the vapor phase, where boiling starts at a certain depth below the local paleo-watertable (e.g., Spycher and Reed 1989; Cooke and McPhail 2001; Ronacher et al. 2004; Figs. 1a, 2). The loss of H_2S to the vapor phase leads to de-complexation of gold bisulfide complexes (e.g., $Au(HS)_2^-$ in solution; Stefánsson and Seward 2004), resulting in oversaturation and efficient precipitation of gold in the veins. Removal of H_2S and CO_2 also makes the residual fluid more alkaline by increasing the HCO_3^-/CO_2 and HS^-/H_2S ratio, which can lead to co-precipitation of base-metal sulfides like galena, sphalerite or pyrite. The resulting pH increase also leads to oversaturation of carbonates, or the crystallization of hydrothermal adularia from an initially mildly acid fluid that previously was in equilibrium with muscovite as the dominant wall-rock alteration mineral (Berger and Henley 1989). Bladed calcite and vein-hosted adularia are therefore useful 'boiling zone indicators' for epithermal gold exploration (Fig.1b; Simmons et al. 2005).

Ore-metal fractionation between vapor and hypersaline liquids

As the density of water vapor increases with rising pressure along the liquid–vapor equilibrium curve (Fig. 2c of Liebscher 2007), its ability to hydrate ions and to complex metal

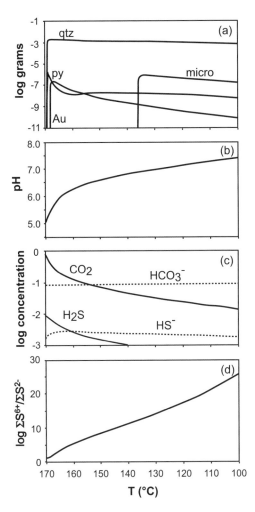

Figure 2. Mass-transfer calculation illustrating the coupled thermal and chemical consequences of adiabatic boiling of an aqueous liquid at conditions proposed for epithermal gold ore deposition at Porgera (modified from Ronacher et al. 2004 with data from M.H. Reed, pers. comm. 2007). Phase separation causes cooling (horizontal axis) and mineral precipitation (a, grams of minerals precipitated). Rapid deposition of gold + pyrite (cf. Fig. 1a) results from a combination of decomplexation of gold bisulfide complexes and an increase in solution pH (b), as the aqueous solution is progressively depleted in volatile CO_2 and H_2S, whereas quartz deposition is a more gradual effect of simple fluid cooling. The gas species concentration in the solution decreases by selective loss to the vapor phase (c) while the aqueous HCO_3^- and HS^- ions remain in the liquid. The resulting pH increase finally leads to saturation in K-feldspar (micro; a) from an initially muscovite-saturated solution, and commonly also of carbonates (see Figs. 1a,b; not shown in this computed example due to the low initial Ca concentration assumed in the starting solution).

species increases dramatically. Recent experiments demonstrate that the solubility of many ore minerals in vapor increases with increasing water vapor density (e.g., Williams-Jones et al. 2002). This increase is contrary to the expectations if vapor transport of metals were dominated solely by their dry volatility (e.g., Krauskopf 1964; Eastoe 1982), and demonstrates that increasing hydration of aqueous volatile species is a key chemical factor determining vapor transport of metals and other solute compounds (Williams-Jones and Heinrich 2005).

In saline fluid systems, two-phase conditions of coexisting vapor and liquid extend far beyond the critical point of water. Increasing temperature and increasing vapor density both enhance the solubility of most minerals in the vapor. Experiments designed to separately sample coexisting saline liquid and lower-salinity vapor phases in large-volume autoclaves (e.g., Pokrovski et al. 2005, 2006a,b; Fig. 3) or by entrapment in synthetic fluid inclusions (e.g., Simon et al. 2005, 2006) have confirmed the high metal concentrations in dense vapor and are starting to provide quantitative data about the degree of metal partitioning between the two coexisting fluids.

Recent vapor solubility experiments were motivated by geological observations of significant metal transport in volcanic fumaroles (e.g., Hedenquist et al. 1993; Symonds et al. 1994; Taran et al. 1995; Webster and Mandeville 2007) and by observations of even higher ore-metal concentrations in vapor inclusions from magmatic-hydrothermal ore deposits. The first recognition of major vapor transport of copper was based on the observation of consistently large chalcopyrite daughter crystals in vapor inclusions of porphyry-copper deposits (Eastoe 1978, Sawkins and Scherkenbach 1981). With the advent of microanalytical techniques for single fluid

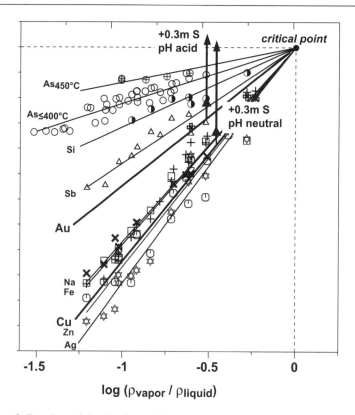

Figure 3. Experimental data for the partitioning of a range of elements between NaCl-H$_2$O-dominated vapor and hypersaline liquid, plotted as a function of the density ratio of the two phases coexisting at variable pressures (modified from Pokrovski et al. 2005; see also Liebscher 2007, Figs. 13, 14). As required by theory, the fractionation constant of all elements approaches 1 as the two phases become identical at the critical point for all conditions and bulk fluid compositions. Chloride-complexed elements, including Na, Fe, Zn but also Cu and Ag are enriched to similar degrees in the saline liquid, according to these experiments in S-free fluid systems. Hydroxy-complexed elements including As, Si, Sb, Sb and Au reach relatively higher concentrations in the vapor phase, but never exceed their concentration in the liquid ($m_{vapor}/m_{liquid} <$ 1). Preliminary data by Pokrovski et al. (2006a,b) and Nagaseki and Hayashi (2006) show that the addition of sulfur as an additional complexing ligand increases the concentration ratios for Cu and Au in favor of the vapor (arrows); in near-neutral pH systems (short arrows) the increase is minor, but in acid and sulfur-rich fluids (long arrows) the fractionation constant reaches ~ 1 or more, explaining the fractionation of Cu and Au into the vapor phase as observed in natural fluid inclusions (cf. Fig. 4 below).

inclusions, the metal-transporting capability of vapor can now be quantified (Lowenstern et al. 1991; Heinrich et al. 1992, 1999; Bodnar 1995; Damman et al. 1996; Vanko et al. 2001). Of particular importance to ore formation is the recognition that the fractionation between coexisting vapor and hypersaline liquid phases is highly element-specific, with the tendency of some ore-forming elements including Cu, Au, As and B to partition into the low-density vapor phase while other ore metals including Fe, Zn and Pb preferentially enter the hypersaline liquid (Audétat et al. 1998; Heinrich et al. 1999; Ulrich et al. 1999; Baker et al. 2004; Fig. 4).

These observations showed that, contrary to previous belief emphasizing the role of liquids as the only ore fluids of importance, the vapor phase can be a highly effective ore-forming fluid. In fact, vapor can contain higher concentrations of ore metals (relative to H$_2$O) than most other geological fluids known before (e.g., Cu > 1 wt%, Au > 1 ppm; Ulrich et al.

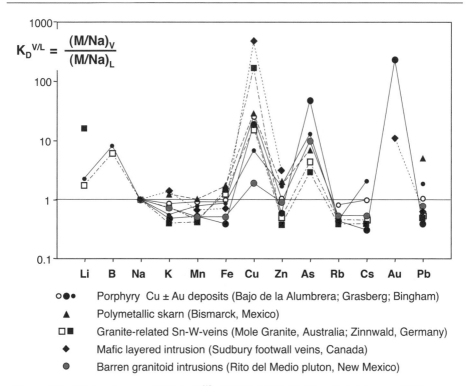

$$K_D^{V/L} = \frac{(M/Na)_V}{(M/Na)_L}$$

○●●   Porphyry Cu ± Au deposits (Bajo de la Alumbrera; Grasberg; Bingham)

▲   Polymetallic skarn (Bismarck, Mexico)

□■   Granite-related Sn-W-veins (Mole Granite, Australia; Zinnwald, Germany)

◆   Mafic layered intrusion (Sudbury footwall veins, Canada)

●   Barren granitoid intrusions (Rito del Medio pluton, New Mexico)

Figure 4. Empirical exchange coefficients $K_D^{V/L} = (M_V/Na_V) / (M_L/Na_L)$ for selected elements (M) between coexisting vapor (V) and hypersaline liquid (L), based on microanalysis of natural inclusion assemblages such as the one shown Figure 1d. The spidergram shows the degree of fractionation of certain elements (B, Cu, As, Au) into the vapor phase (V), whereas typically chloride-complexed metals like K, Mn, Fe, Zn, Rb, Cs and Pb fractionate into the liquid phase (L) to a similar degree as Na. The large variations in exchange coefficients for Cu and Au probably reflect variable degrees of sulfur complexation in the vapor phase (see Fig. 3; simplified from Williams-Jones and Heinrich 2005, including data from Heinrich et al. 1999, Audétat and Pettke 2003, Baker et al. 2004 and other sources).

1999). The data also demonstrate that phase separation between vapor and liquid can be an important step in the segregation of elements to form certain types of ore deposits.

The physico-chemical reason for the specific partitioning behavior of elements between coexisting hypersaline liquid and vapor is still subject to speculation and somewhat conflicting evidence, as discussed by Mavrogenes et al. (2002), Williams-Jones and Heinrich (2005), Simon et al. (2006) and Philippot and Cauzid et al. (2006). Concentration differences in vapor and liquid generally are the greater, the more different the densities of the two fluids are. The concentrations of all components must become identical in the two phases, where pressure and temperature approach the critical conditions of the respective fluid system so that the distinction between vapor and liquid vanishes (Fig. 3; cf. Palmer et al. 2004). In saline two-phase systems, salt components like alkalis and alkaline earth elements, and many ore-forming metals including Fe, Mn, Pb, and Zn fractionate to a similar degree in favor of the liquid (Figs. 3, 4), probably due to chloride complexation in the more saline phase. The relative preference of As and B for the vapor phase can be explained by the salting-out effect of oxyanions or neutral hydroxy complexes such as $As(OH)_3$ in highly saline liquids, similar to that of Si in $Si(OH)_4$ (Newton and Manning 2000). Experiments by Pokrovski et al. (2006a) and Simon et al. (2005, 2006) indicate that Cu and Au fractionate in favor of the liquid phase in chloride–water systems (Fig. 3).

Cu and Au concentrations in the vapor phase increase slightly with the addition of reduced sulfur to the system, if pH conditions are near-neutral, and preliminary experiments with added elemental sulfur yielding acid and very sulfur-rich experimental conditions (Nagaseki and Hayashi 2006; Pokrovski et al. 2006b) reproduce the observations from natural fluid inclusions of significant fractionation of Cu into the vapor phase (Fig. 3, arrows). These results seem to confirm the interpretation that complexation of Cu and Au with sulfur-bearing ligands (but not necessarily HS^-; Pokrovski et al. 2006b) is responsible for variable degrees of preferential partitioning of these metals into low-salinity magmatic vapor. Gold-solubility experiments in S-rich single-phase fluids of low salinity are also consistent with gold concentrations of several ppm (Stefánsson and Seward 2004), as observed in natural vapor inclusions (Ulrich et al. 1999; Heinrich et al. 2004).

OROGENIC GOLD DEPOSITS: CRUSTAL-SCALE ORE SYSTEMS

Orogenic or mesothermal lode gold deposits span an apparent continuum in regional-metamorphic terrains of Archaean to Tertiary age. Unlike all other types of ore-forming hydrothermal systems, gold deposition is not restricted to a characteristic depth level in the crust. It extends from lower-crustal granulite-facies conditions (e.g., Ridley and Hagemann 1999) to upper-crustal environments, although most large deposits form between 5 and 15 km depth (Groves et al. 2003; Goldfarb et al. 2005). Orogenic gold deposits are confined to the vicinity of crustal fault and shear zones acting as fluid-focusing structures. There is ongoing debate about the ultimate origin of the ore fluids, which could include metamorphic devolatilization, exsolution from lower-crustal magmas, or ascent of a deep component of mantle-derived CO_2. There is growing consensus, however, that the fluids originate from the Earth's interior (e.g., Pettke et al. 2000), rather than by deep convection of meteoric water (cf. Nesbitt et al. 1986).

The ore fluids are invariably $CO_2 \pm CH_4$ bearing and of relatively low salinity (typically 0-10 wt% NaCl eq.), but have a greatly variable content of water ranging from ~90% in widespread aquo-carbonic to almost 0% water in purely carbonic fluids. In some of the best-preserved and most intensely-studied examples in young orogenic belts, the incoming ore fluid is a water-rich aquo-carbonic fluid with 5-15% CO_2 and a few percent of salts (e.g., Brusson, Western Alps: Diamond 1990; New Zealand: Craw 1992). The same type of fluid, typically recorded as 3-phase inclusions with a gas bubble + liquid CO_2 + a predominant aqueous liquid, is widespread in most of the older deposits (Robert and Kelley 1987; Ridley and Diamond 2000; Mernagh 2001; Polito et al. 2001).

In high-grade gold veins of the Western Alps, there is unambiguous petrogaphic evidence that the aquo-carbonic fluid was trapped early in the vein crystallization history, and then underwent phase separation to an essentially pure carbonic vapor and a CO_2-depleted residual liquid recorded in later quartz generations and secondary fluid inclusions (Diamond 1990; Pettke et al. 2000). The early fluids are known to be salt-poor but H_2S-rich (Yardley et al. 1993). Phase separation leads to loss of sulfur to the vapor phase, resulting in gold precipitation from the remaining liquid, as a widespread mechanism for the deposition of coarse vein-hosted gold in this and in many other orogenic gold–quartz vein deposits (Diamond 1990; also Walsh et al. 1988; Ridley and Diamond 2000; Hagemann and Lüders 2003). Gold mineralization by phase separation can extend to several kilobars of pressure, because the two-phase immiscibility region expands when the fluids are CO_2-rich as well as moderately saline (Liebscher 2007). Gold deposition thus occurs by a process similar to fluid boiling in shallow epithermal systems (Fig. 2), but can extend to much greater crustal depth because of the greater CO_2 content of the fluids. Phase separation is probably triggered by local pressure drops to sub-lithostatic conditions, by faulting and transient opening of vein space in tectonically active regimes that alternate between ductile and brittle rock deformation (e.g., the fault-valve model; Robert and Kelly 1987; Sibson et al. 1988; Cox et al. 1995).

Methane (CH_4) is present in many orogenic vein deposits, ranging from a minor component besides CO_2 in aquo-carbonic fluids to CH_4-dominated carbonic fluids (Ridley and Diamond 2000). Its origin and role in mesothermal gold mineralization is not entirely clear. Methane is less soluble in aqueous liquid than CO_2, so that the presence of CH_4 expands the field of two-phase coexistence towards higher pressures (Liebscher 2007). Admixture of methane derived from carbonaceous metasedimentary country rocks (Naden and Shepherd 1989) or fluid-rock reaction in the periphery of gold-mineralized alteration systems in reduced and Fe-rich magmatic rocks (Polito et al. 2001) can therefore lead to phase separation and result in gold deposition by gold bisulfide decomplexation. However, methane is also a strong reductant, and the chemical effects of fluid mixing are expected to result in gold oversaturation by reduction of any Au(+I) complex to Au(0), irrespective of possibly attending phase separation.

Some Precambrian orogenic gold deposits and entire provinces contain ubiquitous carbonic fluid inclusions composed entirely of dense $CO_2 \pm CH_4$, but practically no H_2O-bearing fluids (Nigeria: Garba and Akande 1992; Ashanti, Ghana: Mumm et al. 1997; Campbell–Red Lake, Canada: Chi et al. 2006). The observations by Mumm et al. (1997) are somewhat ambiguous and were disputed (Klemd 1998) because of the widespread hydrous alteration in the deposits and the known possibility of fluid inclusions to selectively lose water as a result of post-entrapment strain of the host quartz (Hollister 1990; Bakker and Jansen 1994). Indeed, minor strain is ubiquitous in all mesothermal quartz samples, except for the youngest and shallowest deposits like Brusson in the Western Alps. The recent study of Chi et al. (2006) of the Campbell–Red Lake deposit shows systematic variations in composition and density of petrographically distinct carbonic inclusion assemblages, convincingly arguing against an origin of these carbonic inclusions by post-entrapment water loss. These authors therefore propose gold ore formation by a single-phase carbonic fluid. However, neither analyses of well-preserved natural fluid inclusions nor experimental data currently exist to assess whether essentially water-free carbonic fluids are capable of transporting gold, quartz, and alteration components, to form gold deposits that have essentially similar mineralogical and geochemical characteristics as other orogenic gold deposits, which clearly were formed by water-rich aquo-carbonic fluids undergoing local phase separation.

FLUID MIXING AND ORE DEPOSITION: GRANITE-RELATED Sn-W VEINS

Our main resources of copper, molybdenum, tin and tungsten, as well as important deposits of gold, silver, lead, zinc and other metals are formed by hydrothermal processes operating at 1-10 km depth beneath the Earths surface, where saline fluids exsolve from hydrous magmatic intrusions, ascend and cool, and locally interact and mix with water from the hydrosphere. The recurrent association of mineral deposit types (e.g., tin-tungsten veins, porphyry-copper deposits, epithermal gold-silver deposits) with certain fluid inclusion characteristics indicates that fluid–fluid interaction is an essential requirement for selective redistribution of trace-elements and the formation of economic ore bodies.

Mixing of magmatic and external fluids in the roof and contact zones of granitic intrusions is widespread in contact metamorphism (W. Heinrich 2007, this volume) and commonly contributes to the formation of economic mineral deposits of Sn, W, U, base and precious metals as well as skarns (e.g., Williamson et al. 1997; Vallance et al. 2001; Somarin et al. 2004; Bettencourt et al. 2005; Černý et al. 2005). This section illustrates the role of fluid mixing in ore deposition, summarizing the results of geological, isotopic, fluid-analytical and thermodynamic modeling studies of the Mole Granite magmatic-hydrothermal system in Eastern Australia as example (Herbert and Smith 1978; Eadington 1983; Sun and Eadington 1987; Plimer et al. 1991; Rankin et al. 1992; Heinrich et al. 1992; Heinrich and Ryan 1992; Henley et al. 1999; Audétat et al. 1998, 2000a,b; Mavrogenes et al. 2002; Pettke et al. 2005;

Schaltegger et al. 2005). Results were obtained over several decades of research, as analytical and theoretical modeling approaches were successively improved. Recent inclusion analyses confirm earlier interpretations and model calculations, demonstrating that the deposition of rich cassiterite lodes as well as regionally zoned minor deposits of Sn, W, Mo, Cu, As, Pb, Zn and topaz were controlled by mixing of a hot hypersaline liquid of magmatic origin with low-salinity sediment-equilibrated meteoric water.

The Mole Granite in northern New South Wales is a large laccolith of homogeneous alkali feldspar granite of Early Triassic age (Blevin et al. 1996; Schaltegger et al. 2005). Its roof is exposed over an area of about 20×40 km, with low-angle intrusive contacts dipping outward beneath partly eroded turbiditic metasediments. Hundreds of mines and occurrences, exploited in the early decades of the 20th century as the then leading source of tin worldwide, show a zonation from Sn and W veins in the granite itself, grading out through polymetallic deposits with Cu, Sn and As to distal Zn-Pb-Ag veins (Henley et al. 1999; Audétat et al. 2000a). This classic zonation extends over a horizontal distance of several kilometers and correlates with decreasing fluid salinity and ore-deposition temperature from >600 °C and >60 wt% NaCl eq. hypersaline liquids in W-mineralized pegmatites to nearly salt- free fluids in base metal sulfide veins (Eadington 1983). The T-X_{NaCl} variation in turn correlates with $\delta^{18}O$ / δD fluid isotope compositions decreasing form $+10$ / $-50‰$ to -15 / $-120‰$, consistent with mixing of a magmatic fluid with isotopically light meteoric water in a cool continental climate prevailing in southern Australia during Permo-Triassic times (Sun and Eadington 1987). Sulfur isotope data from the Mole Granite and other granite-related ore systems in the district (Herbert and Smith 1978) also follow the same mineralogical correlation. High-temperature molybdenite has the highest $\delta^{34}S \sim +2 \pm 2‰$, and sulfur isotope ratios gradually decrease through the zoning sequence of cassiterite-associated arsenopyrite, chalcopyrite, Fe-sulfides and sphalerite to $-10‰$ in distal galena and tetrahedrite, on a trend towards even more negative $\delta^{34}S$ values ($< 20‰$) in diagenetic pyrrhotite in unaltered metasediments, as indicated by isotopic and geological data (Heinrich and Ryan 1992). These correlations suggest that regionally zoned ore formation involved fluid mixing, with the addition of a component of isotopically light sulfur mobilized from the country rocks by meteoric water heated to about 280 °C.

First analyses of Sn, Cu, Fe, Zn and Pb concentrations in pre-ore high-temperature hypersaline fluids from quartz crystals in the Yankee Lode tin deposit, obtained by micro-PIXE (Heinrich et al. 1992), were used as input for a multicomponent thermodynamic reaction model, simulating addition of increasing proportions of a cooler fresh water that had previously been equilibrated with pyrite + pyrrhotite bearing semipelitic metasediments (Heinrich and Ryan 1992; Fig. 5). The results predict a mineral precipitation sequence of cassiterite $\rightarrow$ arsenopyrite $\rightarrow$ Cu-Fe-sulfides $\rightarrow$ sphalerite and galena that matches the regional mineral zonation, the paragenetic sequence of mineral precipitation in polymetallic veins, and the correlated ore-deposition temperatures, fluid salinities and isotopic constraints available at that time.

More recently, fluid evolution and mineral precipitation in the Mole Granite were studied in detail by Audétat et al. (1998, 2000a,b), as a first quantitative application of Laser-Ablation ICP Mass Spectrometry (LA-ICPMS) of fluid and melt inclusions. Individual quartz crystals from quartz–cassiterite–tourmaline veins from Yankee Lode, located within the quenched roof zone of the granite, record a complex succession of up to 29 fluid entrapment events, of hypersaline liquid as well as clearly coexisting lower-salinity vapor inclusions (Fig. 6). Liquid/vapor fractionation in one assemblage of clearly coexisting vapor and liquid inclusions is represented by open square symbols in Figure 4, showing that the vapor phase is highly enriched in Cu and As. However, neither of these elements precipitated anywhere in this entirely sulfide-free deposit. Microthermometric and metal-concentration data for the liquid inclusions show some scatter in Cu, but the ratios of all other major and trace elements in successively trapped inclusion assemblages display a very systematic variation.

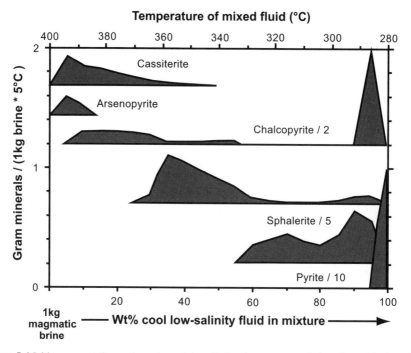

Figure 5. Multicomponent thermodynamic model predicting the sequence of mineral precipitation when a hot hypersaline magmatic liquid (brine, analyzed for initial ore metal concentrations by micro-PIXE; Heinrich et al. 1992) is mixed with increasing proportions of a cooler meteoric fluid of low salinity that had previously interacted with the reduced quartzofeldspathic metasediments surrounding the mineralizing Mole Granite (Australia). Vertical axis shows grams of minerals precipitated from 1 kg of input brine over a cooling interval of 5 °C, resulting from variable additions of cool fluid. Numbers next to the mineral names indicate scaling factors for convenent plotting, e.g., sphalerite/5 reflecting the more than 5 times larger Zn concentration in the original brine compared to Sn in cassiterite. Modified from Heinrich and Ryan (1992) using thermodynamic data from Heinrich (1990).

This variation in successive inclusion events records a process of selective mineral precipitation by fluid mixing. Concentration ratios of K, Mn, Zn, Cs, Rb and several other elements maintain a constant concentration ratio, but decrease rapidly from inclusion generation 21 onward, parallel with the NaCl concentration determined from microthermometry. The results indicate progressive dilution of the metal-rich magmatic fluid by essentially pure water, up to a mixing ratio of 1:300. In the very first steps of fluid mixing between steps 21 and 23, the inclusions record a small drop in non-reactive element concentrations indicating a magmatic: meteoric mixing ratio of ~1:3, but a very steep drop in Sn concentration to a few percent of its initial value. Fluid generation 21 corresponds to the particular growth zone of the quartz crystals at which the deposition of cassiterite crystals begins. A little later in the growth history of the quartz crystal, a steep drop in B concentration follows, and zone 24 marks the beginning of tourmaline precipitation.

Cassiterite and tourmaline deposition thus did not result from phase separation (which occurred continuously throughout the growth history of the vein; cf. Fig. 4), nor as a result of magmatic fluid cooling alone (the temperature at stages 18-19 was lower than that of stages 21-22; Fig. 6). Instead, cassiterite and tourmaline precipitation resulted from chemical reactions driven by the mixing of magmatic and meteoric fluids, as shown by the much steeper drop in the respective element concentrations compared with the dilution monitored by the non-

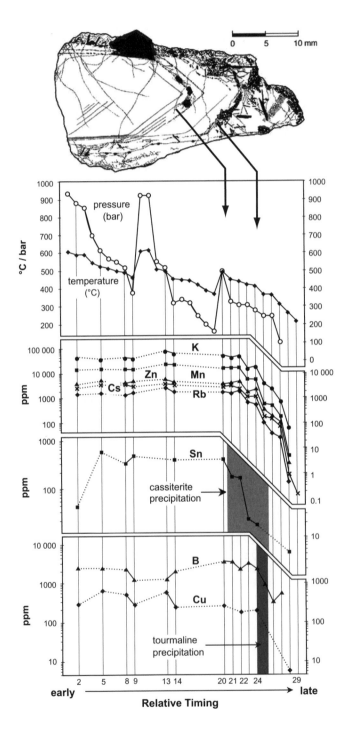

Figure 6. Results from a detailed fluid inclusion study of a quartz–cassiterite–tourmaline vein in a small but high-grade tin deposit in the Mole Granite (Australia; modified from Audétat et al. 1998). Petrographic timing of 29 fluid entrapment events in this zoned crystal (horizontal axis) was the basis for quantitative analysis of several inclusions in each assemblage by LA-ICPMS. The results reveal that cassiterite precipitation was caused by a small amount of essentially salt-free meteoric water to a hypersaline magmatic liquid (brine), consistent with earlier thermodynamic prediction (Fig. 5). Besides Sn, the magmatic hypersaline liquid initially contained high concentrations of K, Mn, Zn, Cs and Rb, which were not chemically reacting by vein mineral deposition, and thus serve as a quantitative monitor for progressive dilution of the granite-derived brine. Likewise, Cu did not precipitate, probably due to loss to an escaping magmatic vapor phase and a deficiency of sulfide in the variably mixed aqueous liquids (cf. Fig. 5).

reactive components. Cassiterite-depositing reactions were probably driven by a combination of detabilization of Sn-Cl-complexes and the pH-neutralizing and oxidizing contributions by the wall-rock-equilibrated groundwater, as predicted by the earlier thermodynamic reaction modeling shown in Figure 5.

LIQUID – VAPOR EVOLUTION IN PORPHYRY – EPITHERMAL SYSTEMS

The spectrum of Cu-Mo-Au mineralization and wall-rock alteration in porphyry-type and epithermal ore deposits can be linked by a process chain, in which fluid phase separation and segregation of vapor from hypersaline liquid play a key role in selective ore metal enrichment. By reviving the classic "vapor plume model" proposed by Henley and McNabb (1978), it is possible to link the diverse processes evolving in space and time, as hydrous magmas are emplaced in the upper crust and variably dense and saline fluids exsolve from the magma, ascend, cool, and interact with each other and with fluids of the hydrosphere. Geological observation and experimental data together show that selective element transfer and the formation of mineralogically and geochemically distinct ore deposit types are controlled, to a large degree, by the physical properties of fluids in the salt–water system, allowing a range of distinct fluid evolution paths illustrated schematically in Figures 7 to 10. This range of evolution paths is consistent with recent physical flow modeling, summarized in the chapter of Driesner and Geiger (2007).

Geological observations

Porphyry-Cu±Mo±Au deposits and epithermal Au-Ag±As±Pb±Zn veins can be considered as a family of ore deposits forming at characteristic but partly overlapping depth and temperature regions in the top 10 km of magmatic-hydrothermal systems. They are usually linked with upper-crustal calcalkaline magmatism related to active subduction (Hedenquist and Lowenstern 1994; Sillitoe and Hedenquist 2003; Seedorff et al. 2005). Porphyry deposits are, as indicated by their name, hosted by porphyritic stocks, rising above the roof of larger dioritic to granodioritic plutons (Dilles and Einaudi 1992), and commonly intruding the base of andesitic volcanoes (e.g., Sillitoe 1973; Hedenquist and Lowenstern 1994; Ulrich et al. 2001; Proffett 2003). Copper-iron sulfides (chalcopyrite and bornite) associated with variable amounts of molybdenite and gold, are precipitated in numerous quartz veinlets centered on the porphyry stock and extending into adjacent wall rocks (Fig. 1e). Host and wall rocks are pervasively altered to secondary biotite and K-feldspar replacing ferromagnesian silicates and plagioclase (potassic alteration).

Epithermal deposits occupy open space in large veins, in breccia columns, or in zones of secondary porosity generated by intense leaching of wall-rock silicates (Simmons et al. 2005). Epithermal deposits of various subtypes commonly occur above or adjacent to porphyry-style ore bodies. Systems exposed by rotation of fault blocks and partial erosion indicate a continuum of fluid processes extending over a vertical depth interval in excess of 6 km (Dilles and Einaudi 1992). Where the two ore deposit types intersect in the field, epithermal veins generally overprint and postdate porphyry-style mineralization, although the two mineralization styles can be almost coeval within the uncertainty of isotopic age measurements (Arribas et al. 1995). Epithermal gold veins commonly grade downward into quartz–pyrite veins associated with sericite alteration. The underlying porphyry-type deposits are crosscut by similar quartz–pyrite veins, overprinting the earlier potassic alteration and porphyry-copper mineralization (Gustafson and Hunt 1975). These field observations, combined with fluid inclusion and stable isotope data (Hedenquist et al. 1998; Watanabe and Hedenquist 2001) indicate that the quartz–pyrite veins are the linking channelways for low-salinity magmatic fluids ascending through the cooling porphyry towards the epithermal veins, where they deposit precious metals closer to the surface (Sillitoe and Hedenquist 2003; Heinrich 2005).

Petrologic, stable-isotope and fluid inclusion data indicate that potassic alteration in porphyry-style ore deposits is produced by single-phase magmatic fluids (barren core at Bingham: Redmond et al. 2004; porphyry-Cu-Mo ore at Butte, Montana: Rusk et al. 2004), or occurs more commonly in the stability field of coexisting hypersaline liquid + vapor, cooling together from temperatures above 600 °C to about 350 °C (e.g., Alumbrera: Ulrich et al. 2001; Bingham: Roedder 1971; Redmond et al. 2004). The main stage of bornite and chalcopyrite precipitation usually occurs near the end of this cooling interval between 450 and 320 °C, as indicated by a rapid decrease in Cu concentration in the cooling fluids. This compositional evolution was documented by combined petrography, microthermometry and LA-ICPMS analysis of fluid inclusions (Ulrich et al. 2001; Landtwing et al. 2004; Redmond et al. 2004; Klemm et al. 2007b). At the Bingham (Utah) and El Teniente deposits (Chile), chloride/water mass balance associated with the separation of a single-phase input fluid into two coexisting fluids implies that the vapor phase is greatly in excess relative to the small amount of hypersaline liquid (Williams-Jones and Heinrich 2005; Klemm et al. 2007b). Predominance of the vapor is also indicated by textural evidence for quartz dissolution (Redmond et al 2004; Landtwing et al. 2005), which can only be explained by a dominantly low-salinity fluid cooling through the small P-T-X window of retrograde quartz solubility (Fournier 1999).

Quartz dissolution, fluid inclusion analysis and salt–water mass balance data consistently indicate that a low- to intermediate-salinity vapor with an initially very high Cu concentration (Cu ~ Na ~ K ~ Fe ~ 1 wt%) was the dominant phase causing porphyry-Cu-Au mineralization at Butte, Bingham and El Teniente. By contrast, Mo in the practically Cu-free porphyry-molybdenum deposit of Questa (New Mexico) was precipitated dominantly from a residual hypersaline liquid, after the originally abundant Cu in the input fluid had partitioned into a vapor phase that left the region of the ore body (Klemm et al. 2007a). Such element separation and selective molybdenum precipitation is analogous to the formation of the Yankee Lode tin veins discussed above (Audétat et al. 1998, Fig. 6).

Epithermal precious-metal deposits of the high-sulfidation type (Hedenquist and Lowenstern 1994) are characterized by early leaching and kaolinite–alunite alteration, caused by extremely acid fluids generated by the partitioning of volatile chloride and sulfur species into magmatic vapor and the subsequent disproportionation of SO_2 into H_2S and sulfuric acid, H_2SO_4. The same type of alteration characterizes fumaroles of hot magmatic vapor emanating form active volcanoes (Webster and Mandeville 2007). The precipitation of pyrite, As-Sb-sulfosalts, base-metal sulfides and eventually gold in epithermal ore deposits occurs within the top ~1 km below the land surface, and typically after the acid kaolinite–alunite alteration, mostly in the temperature interval of 250 to 150 °C. Ore-depositing fluids are aqueous liquids of low to intermediate salinity (2-10% NaCl eq.; data review in Heinrich 2005), but still partly or even dominantly of magmatic origin based on stable isotope compositions (Muntean et al. 1990; Vennemann et al. 1993; Alderton and Fallick 2000; Wallier et al. 2006). Hot springs of liquid water are typical surface expressions of shallow epithermal gold deposits such as McLaughlin (Nevada; Sherlock 1996), or the still steaming and perhaps actively forming Ladolam deposit (Lihir, Papua New Guinea; Carman 2003; Simmons and Brown 2006). The gold-depositing solutions are weakly acid, with a pH consistent with observations of stable muscovite or illite at the economic mineralization stage. This matches the interpretation that gold-bearing epithermal fluids ascends through subjacent quartz–pyrite veins, along which the original magmatic acidity is neutralized by feldspar-destructive wall-rock alteration reactions (Gammons and Williams-Jones 1997; Heinrich et al. 2004; Heinrich 2005).

Fluid evolution paths in Cu-Au mineralizing systems

Figures 7 to 10 schematically illustrate a range of fluid evolution paths expected to occur in vertically extensive magmatic-hydrothermal systems (see Fig. 3 in Webster and Mandeville 2007), interpreted from geological and fluid inclusion observations, experimental data summarized in Williams-Jones and Heinrich (2005) and hydrodynamic modeling using the NaCl–H$_2$O model system by Geiger et al. (2005, 2006) and Driesner and Geiger (2007).

The evolution in space and time, of hydrous magmatic intrusions, characteristic alteration assemblages, porphyry-style and epithermal ore bodies, and fumaroles or hot springs at the paleo-surface, can be explained by fluids exsolving from magmas with essentially similar H$_2$O, chloride and sulfur contents. Varying fluid properties and mineral compositions are controlled, to a first order, by the depth of magmatic fluid exsolution and the temperature distribution along the flow path of the fluids from the magma to the Earth's surface. Time-dependent distribution of pressure and temperature in turn depend on the rate of fluid production, the permeability structure of the rocks above the fluid-generating magma chamber, and the rate of heat loss by conduction to the wall rocks and by thermal interaction of the magmatic fluids with convecting external fluids (Driesner and Geiger 2007).

Figures 7 to 10 represent four fluid evolution paths in hydrothermal systems above parent intrusions emplaced at variable depths (see Liebscher 2007, Fig. 2, for introduction of the NaCl-H$_2$O phase diagram shown in simplified form on the left side of each figure). Four sets of diagrams, each set illustrating a characteristic fluid evolution path (left-side of each figure, big arrows in perspective *P-T-X*$_{NaCl}$ phase diagram from Driesner and Heinrich 2007; see Fig. 2 in Liebscher 2007). Each fluid evolution path results from the emplacement of a hydrous magma at a characteristic depth, as indicated by the cartoons (right-side of figure); these give a schematic illustration of fluid exsolution, a fluid ascent channel and the physical behavior of the fluid phase(s) within it. In the cartoon and along the big arrows in the phase diagram, gray-scale shading qualitatively indicates fluid density from black (= liquid-like, ~ 1 g cm^{-1}) to white (= low-density vapor at atmospheric pressure), to illustrate the processes of expansion, contraction, condensation (separation of droplets) and boiling (bubble formation). The shape of the fluid channelway in the cartoon indicates the degree of hydraulic conductivity to the surface, corresponding to an inferred pressure–depth diagram (center right). Pressures range from supra-lithostatic (active hydraulic fracturing due to the volume excess produced by crystallization of hydrous magma to less-hydrous rock + fluid; Burnham and Ohmoto 1980; Fig. 1e) to below the cold hydrostatic gradient (equivalent to the load of a 1 g cm^{-1} liquid water column) in periods and regions approaching vapor-static conditions.

Figure 7 schematically shows a fluid evolution path that is likely to result when a hydrous magma is emplaced close to the surface, for example in a feeder to a stratovolcano. Unless sudden fluid exsolution over a large depth interval terminates in a catastrophic explosion, the

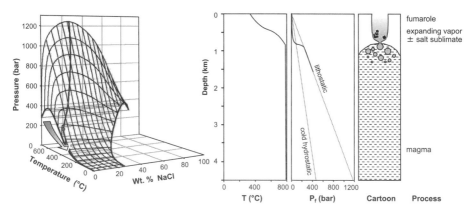

Figures 7. The case of shallow magma emplacement and direct expansion of magmatic fluid to low-density vapor and fumarolic salt sublimates. See text for details.

crystallization of the shallow magma column leads to expulsion of a high-temperature but low-density magmatic vapor, manifested by minor eruptions intermittent with high-temperature fumarolic activity. The vapor is rich in acid volatiles (SO_2, H_2S, HCl) but the concentration of salts and ore metals in the high-temperature vapor—although significant (e.g., White Island: Hedenquist et al. 1993; Kudryavy: Taran et al. 1995)—is relatively low compared with metal concentration in ore-forming fluids recorded by fluid inclusions. Metal transport is limited because the density of the fumarole vapor is too low for significant hydration of ions and aqueous complexes (Williams-Jones et al. 2002; Williams-Jones and Heinrich 2005). On cooling, the magmatic fluid will intersect the vapor+halite stability field, as illustrated by the big arrow entering the salt-water model phase diagram in Figure 7 (left) from its back plane. Initially at near-surface lithostatic pressure, the magmatic vapor will rapidly expand on ascent, and precipitate any salts and metals as sublimates at high temperature (e.g., >600 °C at Kudryavy), as illustrated by the cartoon on the right side of Figure 7. Acid leaching (vuggy quartz) and kaolinite–alunite alteration of wall rocks occur where fumarolic vapor cools to ~100 °C and mixes with ambient groundwater. Some gold deposits are interpreted to have formed in such an environment (e.g., Pascua: Chouinard et al. 2005), but in most epithermal gold deposits acid leaching leads to barren alteration, which predates but prepares the ground for later economic mineralization by aqueous liquids, as discussed in the context of Figure 10 (below).

Figure 8 illustrates an evolution that may result if magmatic fluid is produced from an intrusive front at >1 km depth. Depending on depth and the initial OH/Cl ratio of the magma, the exsolving fluid may be a single-phase low-salinity fluid of vapor-like to intermediate density (e.g., at Bingham; Redmond et al. 2004), or the silicate melt may directly exsolve two coexisting fluid phases, a vapor and a hypersaline liquid (Bodnar et al. 1995; Audétat et al. 2001). If an initially single-phase fluid expands on ascent, as a result of relatively high temperatures and/or increasing hydraulic connectivity to the surface above a zone of hydraulic fracturing, its evolution path will intersect the two-phase surface on the vapor side of the critical curve, leading to the condensation of a small quantity of hypersaline liquid. This is indicated in Figure 8 by the branching arrow in the phase diagram, and by the droplets in the cartoon. The physical hydrology of this phase separation is being studied by numeric modeling (Driesner and Geiger 2007). It is intuitive that the denser, more viscous and preferentially wetting, but volumetrically subordinate hypersaline liquid will rise more slowly than the

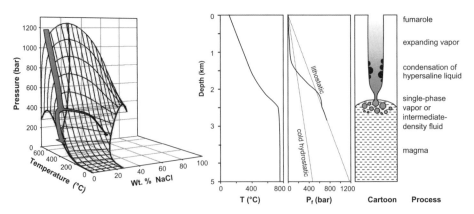

Figure 8. Fluid evolution resulting from magma emplacement at a few kilometers depth, with intense hydraulic fracturing leading to a pressure drop, instantaneus condensation of minor hypersaline brine and expansion of the dominant vapor phase. This results in wholesale precipitation of dissolved metals, as observed in some shallow-level porphyry-Cu-Au deposits including Bingham (cf. Driesner and Geiger 2007, Figs. 11A,B and 12A,B). See text for details.

buoyant low-viscosity vapor. Continued pressure drop causes the predominating vapor phase to further expand, to lose more salt to the liquid phase, and eventually to enter the halite saturation region. Due to its higher initial density, the magmatic fluid is capable of transporting high concentrations of ore metals out of the magma, in particular Cu and Au, but progressive expansion of the vapor will lead to wholesale deposition of all vapor-transported metals in a vertically confined interval. Fluid inclusion data from Bingham (Redmond et al. 2004; Landtwing et al. 2005; C. Furrer, ETH Zürich, unpublished data) indicate that such a fluid evolution is optimal for producing a high-grade "ore shell," which is characteristic for the relatively shallow—and economically particularly attractive—gold-rich porphyry-Cu-Au-(-Mo) deposits (Silltoe 1997). Molybdenite at Bingham dominantly occurs in separate, structurally late veins in and below the Cu-Au ore shell. These veins can be interpreted as products of the residual hypersaline liquid, similar to the process proposed for the formation of the Questa porphyry-Mo deposit (Klemm et al. 2007a) and the Sn veins of Yankee Lode (Audétat et al. 1998; Fig. 6).

Figure 9 illustrates a third type of fluid evolution, which is likely to separate copper from some of the gold, and may lead to distinct but spatially associated porphyry-copper-(gold) and high-grade epithermal gold-(silver) deposits. As in the previous case (Fig. 8), the vapor initially condenses out some hypersaline liquid after intersecting the two-phase surface on the vapor side of the critical curve (bifurcation of the arrow in Fig. 9, left). It then ascends and cools at slightly higher pressures within the single-phase stability field. As a key characteristic of the salt–water phase diagram, its critical curve (the crest of the tunnel-shaped two-phase surface) swings from low to increasingly higher salinity with increasing pressure and temperature. As a consequence, the low-to-moderate-salinity vapor that initially coexisted with a hypersaline liquid at high temperature and pressure, can cool and contract (increase its density without any heterogeneous phase change) to a single-phase aqueous liquid of the same salinity. If this liquid again intersects the two-phase curve at any temperature below the critical temperature of pure water (e.g., at the tip of the large arrow in Fig. 9), it will boil by separation of bubbles of low-density vapor. The physical process of brine condensation, vapor to liquid contraction and finally low-pressure boiling is depicted schematically in the cartoon in Figure 9, and probably has important chemical consequences for ore formation. Early Cu-Fe-sulfide precipitation and condensation of hypersaline liquid in the temperature range of porphyry-copper ore formation

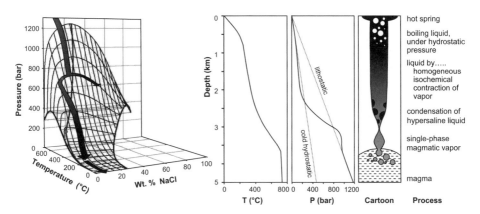

Figure 9. Fluid evolution resulting from deeper magma crystallization, allowing initial condensation of hypersaline liquid and subsequent contraction of the ascending vapur phase to an aqueous liquid, if the vapor cools along any pressure path above the two-phase surface; subsequent near-surface boiling of this low-salinity magmatic liquid optimizes gold transport and deposition in epithermal deposits (cf. Driesner and Geiger 2007, Figs. 11E,F and 12C-F). See text for details.

will selectively remove Fe and some of the Cu from the original magmatic fluid, leaving the buoyant vapor relatively enriched in sulfur, arsenic, gold and the remainder of the copper. On cooling and contraction of this vapor to a low-salinity liquid, it can retain its excess of sulfur relative to chalcophile metals. Sulfur is essential as a ligand for complexation and effective transport of gold at temperatures below 300 °C, and thermodynamic modeling indicates that the sulfur-rich (but relatively Fe and base-metal poor) vapor-derived fluid can readily transport 1-10 ppm gold into the near-surface epithermal environment (Heinrich et al. 2004; Heinrich 2005). Here, low-pressure boiling may lead to rapid oversaturation and high-grade gold ore deposition as discussed above. However, starting from a 100-1000 times more gold-rich fluid than observed in meteoric-dominated geothermal systems (which is taken as the basis for the modeling shown in Fig. 2), correspondingly higher gold grades are likely to result.

Figure 10 shows the final case of a single-phase fluid of low to intermediate density that exsolves and cools at sufficient pressure to never intersect the two-phase surface. As a typical example, the porphyry-Cu-Mo ore at Butte (Montana) was formed from extremely Cu-rich single-phase fluids of this type, probably at even higher temperatures and pressures than those depicted in the diagrams of Figure 10 (Rusk et al. 2004). Hedenquist et al. (1998) proposed that low-salinity aqueous liquids forming the Lepanto epithermal Cu-As-Au deposit of Lepanto (Philippines) was also derived by cooling and contraction of a single-phase magmatic fluid of intermediate density. This fluid was probably exsolved from the deeper part of a hidden pluton that had generated the intimately associated Far South East deposit some 0.3 Ma earlier (Arribas et al. 1995). This high-grade porphyry-Cu-Au deposit formed at only ~ 2 km paleodepth or less, based on geological reconstruction. Early fluid inclusions associated with pre-ore potassic alteration record a fluid stage documenting coexistence of hypersaline liquid and low-density vapor, more similar to the hotter and shallower fluid regime illustrated in Figure 8. The overlying Lepanto deposit in turn shows extensive vuggy quartz alteration consistent with a near-surface fumarole stage (~ Figs. 7, 8), followed by the introduction of pyrite, Cu-As sulfides and gold by a low-salinity aqueous liquid along a fluid evolution path similar to the one shown in Figures 9 or 10 (Arribas et al. 1995; Hedenquist et al. 1998).

Consistent field relationships between contrasting mineralization and alteration stages, documented from many magmatic-hydrothermal ore districts worldwide, indicate that the four fluid evolution paths indicated in Figures 7 to 10 are commonly superimposed on each other in a characteristic time sequence (Sillitoe and Hedenquist 2003). As illustrated by the example of

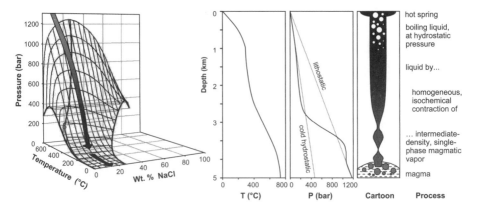

Figure 10. Fluid evolution of a low- to intermediate-salinity magmatic fluid exsolving as a single phase and cooling at high pressure, so that it never intersects the two-phase surface during its ascent to the surface. See text for details.

Far South East – Lepanto (Hedenquist et al. 1998; see also Heinrich 2005), the instantaneous distribution of fluid states merely represents a snapshot in a characteristic, larger-scale evolution of magmatic-hydrothermal systems. The characteristic succession of fluid evolution paths can result from the progressive retraction of the fluid-producing magma front to greater depth, as a hydrous intrusion cools, crystallizes and expels magmatic-hydrothermal fluids into cooler rocks (Driesner and Geiger 2007).

SUMMARY AND CONCLUSIONS

The enrichment of chemical elements in hydrothermal ore deposits results from dissolution, transport and reprecipitation operating on the scale of orogens, batholiths and large sedimentary basins. Fluid–fluid interactions of two types—mixing of chemically contrasting fluids, and separation of immiscible fluids—commonly provide the chemical driving force required for selective element enrichment. Clear evidence for interaction between different fluids is recorded by fluid inclusions, which can now be quantitatively analysed by LA-ICPMS and other techniques, and placed into relative time sequence by careful geology and petrographic observations.

Instantaneous oversaturation of certain minerals to form major vein deposits of sulfide and oxide ore minerals can occur as a result of mixing between hot magmatic fluids with metamorphic or meteoric fluids, which are not only cooler but also have different chemical compositions. Effective and highly selective precipitation of other ore metals, for example gold, results from phase separation, due to the partitioning of ligand-forming, and acid-base or redox controlling species from the metal transporting liquid to an escaping vapor phase.

Separation of vapor and liquid in saline fluid systems, at pressures and temperatures well above the critical point of pure water, is found to be a major fractionation process for the ore metals themselves. Thus, low-salinity vapor (also containing volatile CO_2 and certain sulfur species) preferentially transports Cu, Au as well as B, As and Sb. By contrast, coexisting hypersaline liquids are enriched in chloride-complexed metals including Na, K, Fe, Mn, Zn, Pb and many other minor salts. If the two fluids physically segregate as a result of their different density and viscosity, distinct evolution paths through pressure–temperature space leads to the formation of different ore deposit types. High-salinity liquid solutions (magmatic-hydrothermal brines) preferentially generate deposits of Sn, Mo, or base-metals including Pb and Zn. Low-salinity fluids, either directly derived from a single-phase magmatic fluid or generated by separation of a vapor phase from a residual hypersaline liquid, can be extremely Cu and Au rich and generate porphyry-type and epithermal Cu ± Au ± As deposits. Gold enrichment in low-temperature epithermal deposits is optimized if the vapor is first depleted in iron relative to sulfur (which is required for gold complexation), followed by cooling of the vapor at elevated pressure. Such a fluid evolution path leads to homogeneous contraction of the vapor to a low-salinity aqueous liquid, which can again boil closer to the surface and precipitate its gold content in rich epithermal veins. Similar processes may operate in deeper parts of regional-metamorphic terrains, where the presence of CO_2 extends the possibility of two-phase fluid stability, leading to the formation of mesothermal or orogenic gold deposits.

Geological and fluid-inclusion evidence shows that fluid–fluid interactions are systematic and lead to globally recurrent deposit types, worldwide and throughout Earth's history. Comparison of fluid properties recorded by natural fluid inclusions with experimental fluid phase relationships in the H_2O – NaCl ± CO_2 model systems indicates that the physical properties of fluids, rather than detailed variations of local geology, exert the prime control on the distinct processes leading to ore formation. This implies that understanding the physics and chemistry of these fluids has predictive potential, which can be applied in resource exploration and extraction.

ACKNOWLEDGMENTS

Extensive discussion with colleagues and graduate students at ETH Zürich and elsewhere have greatly contributed to this review. I would like to thank Andreas Audétat, Steve Kesler and Axel Liebscher for careful and constructive reviews of the paper. Mark Reed and Elisabeth Ronacher kindly provided raw data for replotting their published diagram shown in Figure 2. Nicole Hurtig and Barbara Kuhn assisted with drafting of figures. Continuing financial support by ETH Zürich and the Swiss National Science Foundation are gratefully acknowledged.

REFERENCES

Alderton DHM, Fallick AE (2000) The nature and genesis of gold-silver-tellurium mineralization in the Metaliferi Mountains of western Romania. Econ Geol 95:495-515

Arnórsson S, Stefánsson A, Bjarnason JÖ (2007) Fluid-fluid interactions in geothermal systems. Rev Mineral Geochem 65:259-312

Arribas A, Hedenquist JW, Itaya T, Okada T, Concepcion RA, Garcia JS (1995) Contemporaneous formation of adjacent porphyry and epithermal Cu-Au deposits over 300 ka in Northern Luzon, Philippines. Geology 23:337-340

Audétat A, Günther D, Heinrich CA (1998) Formation of a magmatic-hydrothermal ore deposit: Insights with LA-ICP-MS analysis of fluid inclusions. Science 279:2091-2094

Audétat A, Günther D, Heinrich CA (2000a) Causes for large-scale metal zonation around mineralized plutons: Fluid inclusion LA-ICP-MS evidence from the Mole Granite, Australia. Econ Geol 95:1563-1581

Audétat A, Günther D, Heinrich CA (2000b) Magmatic-hydrothermal evolution in a fractionating granite: A microchemical study of the Sn-W-F-mineralized Mole Granite (Australia). Geochim Cosmochim Acta 64, 3373-3393.

Audétat A, Pettke T, Thomas R, Bodnar RJ (2001) Characterization of the magmatic-hydrothermal transition in barren vs. mineralized granites. 11th Goldschmidt Conference, Hot Springs VA, May 2001. Abstract # 3152 (CD-ROM)

Audétat A, Pettke T (2003) The magmatic-hydrothermal evolution of two barren granites: A melt and fluid inclusion study of the Rito del Medio and Canada Pinabete plutons in northern New Mexico (USA). Geochim Cosmochim Acta 67:97-121

Baker T, van Achterberg E, Ryan CG, Lang JR (2004) Composition and evolution of ore fluids in a magmatic-hydrothermal skarn deposit. Geology 32:117-120

Bakker RJ, Jansen JB (1994) A mechanism for preferential H_2O leakage from fluid Inclusions in quartz, based on TEM observations. Contrib Mineral Petrol 116:7-20

Barnes HL, Seward TM (1997) Geothermal systems and mercury deposits. *In* Geochemistry of Hydrothermal Ore Deposits. Barnes HL (ed.), Wiley, p 699-736

Berger BR, Henley RW (1989) Advances in the understanding of epithermal gold-silver deposits, with special reference to the Western United States. *In*: The Geology of Gold Deposits: The Perspective in 1988. Keays RR, Ramsay WHR, Groves DI (eds) Econ Geol Monograph Series 6:405-423

Bettencourt JS, Leite WB, Goraieb CL, Sparrenberger I, Bello RMS, Payolla BL (2005) Sn-polymetallic greisen-type deposits associated with late-stage rapakivi granites, Brazil: fluid inclusion and stable isotope characteristics. Lithos 80:363-386

Blevin PL, Chappell BW, Allen CM (1996) Intrusive metallogenic provinces in eastern Australia based on granite source and composition. Trans Royal Soc Edinburgh - Earth Sci 87:281-290

Bodnar RJ (1995) Fluid-inclusion evidence for a magmatic source for metals in porphyry copper deposits. *In*: Magmas, Fluids and Ore Deposits. Thompson EJFH (ed) Mineralogical Association of Canada Short Course 23:139-152

Bodnar RJ, Burnham CW, Sterner SM (1985) Synthetic fluid inclusions in natural quartz. III. Determination of phase equilibrium properties in the system H_2O-NaCl to 1000 °C and 1500 bars. Geochim Cosmochim Acta 49:1861-1873

Burnham CW, Ohmoto H (1980) Late-stage processes of felsic magmatism. Mining Geology Special Issue 8:1-11

Carman GD (2003) Geology, mineralization, and hydrothermal evolution of the Ladolam gold deposit, Lihir Island, Papua New Guinea. Soc Econ Geol Special Pub 10:247-284

Carroll MR, Webster JD (1994) Solubilities of sulfur, noble-gases, nitrogen, chlorine, and fluorine in magmas, volatiles in magmas. Rev Mineral 30:231-279

Cathles LM, Adams JJ (2005) Fluid flow and petroleum and mineral resources in the upper (<20-km) continental crust. Econ Geol 100th Anniversary Volume, p 77-110

Cerný P, Blevin PL, Cuney M, London D (2005) Granite-Related Ore Deposits. Econ Geol 100th Anniversary Volume, p 337-370

Chi GX, Dube B, Williamson K, Williams-Jones AE (2006) Formation of the Campbell-Red Lake gold deposit by H_2O-poor, CO_2-dominated fluids. Mineral Deposita 40:726-741

Chouinard, A, Williams-Jones AE, Leonardson RW, Hodgson CJ, Silva P, Téllez C, Vega J, Rojas F (2005) Geology and genesis of the multistage high-sulfidation epithermal Pascua Au-Ag-Cu deposit, Chile and Argentina. Econ Geol 100:463-490

Cooke DR, McPhail DC (2001) Epithermal Au-Ag-Te mineralization, Acupan, Baguio district, Philippines: Numerical simulations of mineral deposition. Econ Geol 96:109-131

Cox SF, Sun SS, Etheridge MA, Wall VJ, Potter TF (1995) Structural and geochemical controls on the development of turbidite-hosted gold quartz vein deposits, Wattle Gully mine, central Victoria, Australia. Econ Geol 90:1722-1746

Craw D (1992) Fluid evolution, fluid immiscibility and gold deposition during cretaceous recent tectonics and uplift of the Otago and Alpine Schist, New-Zealand. Chem Geol 98:221-236

Damman AH, Kars SM, Touret JLR, Rieffe EC, Kramer J, Vis RD, Pintea I (1996) PIXE and SEM analyses of fluid inclusions in quartz crystals from the K-alteration zone of the Rosia Poieni porphyry-Cu deposit, Apuseni mountains, Rumania. Eur J Mineral 8:1081-1096

Diamond LW (1990) Fluid inclusion evidence for P-V-T-X evolution of hydrothermal solutions In late-alpine gold-quartz veins at Brusson, Val d'Ayas, Northwest Italian Alps. Am J Sci 290:912-958

Dilles JH, Einaudi MT (1992) Wall-rock alteration and hydrothermal flow paths about the Ann-Mason porphyry copper deposit, Nevada - a 6 km vertical reconstruction. Econ Geol 87:1963-2001

Driesner T, Geiger S (2007) Numerical simulation of multiphase fluid flow in hydrothermal systems. Rev Mineral Geochem 65:187-212

Driesner T, Heinrich CA (2007) The System H_2O-NaCl. I. Correlation formulae for phase relations in temperature-pressure-composition space from 0 to 1000 °C, 0 to 5000 bar, and 0 to 1 X_{NaCl}. Geochim Cosmochim Acta (in press)

Drummond SE, Ohmoto H (1985) Chemical evolution and mineral deposition in boiling hydrothermal systems. Econ Geol 80(1):126-147

Eadington PJ (1983) A fluid inclusion investigation of ore formation in a tin mineralized granite, New England, New South Wales. Econ Geol 78:1204-1221

Eastoe CJ (1978) Fluid inclusion study of Panguna-porphyry-copper-deposit, Bougainville, Papua-New-Guinea. Econ Geol 73(5):721-748

Eastoe CJ (1982) Physics and chemistry of the hydrothermal system at the Panguna porphyry copper-deposit, Bougainville, Papua New-Guinea. Econ Geol 77:127-153

Ellis JA (1979) Explored geothermal systems. *In:* Geochemistry of hydrothermal ore deposits. 2nd edition. Barnes HL (ed), Wiley and Sons, p 632-683

Fournier RO (1999) Hydrothermal processes related to movement of fluid from plastic into brittle rock in the magmatic-epithermal environment. Econ Geol 94:1193-1211

Gammons CH, Williams-Jones AE (1997) Chemical mobility of gold in the porphyry-epithermal environment. Econ Geol 92:45-59

Garba I, Akande SO (1992) The origin and significance of nonaqueous CO_2 fluid inclusions in the auriferous veins of Bin Yauri, Northwestern Nigeria. Mineral Deposita 27:249-255

Geiger S, Driesner T, Heinrich CA, Matthäi SK (2006) Multiphase thermohaline convection in the earth's crust: II. Benchmarking and application of a finite element - Finite volume solution technique with a NaCl- H_2O equation of state. Transport Porous Media 63:435-461

Geiger S, Driesner T, Heinrich CA, Matthäi SK (2005) On the dynamics of NaCl-H_2O fluid convection in the Earth's crust: J Geophys Res - Solid Earth 110:B07101

Gottschalk M (2007) Equations of state for complex fluids. Rev Mineral Geochem 65:49-97

Goldfarb RJ, Baker T, Dubé B, Groves DI, Hart CJR, Gosselin P (2005) Distribution, character, and genesis of gold deposits in metamorphic terranes. Econ Geol 100th Anniversary Volume, p 407-450

Grancea L, Bailly L, Leroy J (2002) Fluid evolution in the Baia Mare epithermal gold/polymetallic district, Inner Carpathians, Romania. Mineral Deposita 37:630-647

Groves DI, Goldfarb RJ, Robert F, Hart CJR (2003) Gold deposits in metamorphic belts: Overview of current understanding, outstanding problems, future research, and exploration significance. Econ Geol 98:1-29

Gustafson LB, Hunt JP (1975) Porphyry copper-deposit at El-Salvador, Chile. Econ Geol 70(5):857-912

Hack AC, Thompson AB, Aerts M (2007) Phase relations involving hydrous silicate melts, aqueous fluids, and minerals. Rev Mineral Geochem 65:129-185

Hagemann SG, Lüders V (2003) P-T-X conditions of hydrothermal fluids and precipitation mechanism of stibnite-gold mineralization at the Wiluna lode-gold deposits, Western Australia: conventional and infrared microthermometric constraints. Mineral Deposita 38(8):936-952

Hedenquist JW, Lowenstern JB (1994) The role of magmas in the formation of hydrothermal ore- deposits. Nature 370(6490):519-527

Hedenquist JW, Arribas A, Reynolds TJ (1998) Evolution of an intrusion-centered hydrothermal system: Far Southeast-Lepanto porphyry and epithermal Cu-Au deposits, Philippines. Econ Geol 93(4):373-404

Hedenquist JW, Simmons SF, Giggenbach WF, Eldridge CS (1993) White-Island, New-Zealand, volcanic-hydrothermal system represents the geochemical environment of high-sulfidation Cu and Au ore deposition. Geology 21:731-734

Heinrich CA (2005) The physical and chemical evolution of low to medium-salinity magmatic fluids at the porphyry to epithermal transition: a thermodynamic study. Mineral Deposita 39:864-889

Heinrich CA (1990) The chemistry of tin (-tungsten) ore deposition. Econ Geol 85:529-550

Heinrich CA, Ryan CG, Mernagh TP, Eadington PJ (1992) Segregation of ore metals between magmatic brine and vapor - a fluid inclusion study using pixe microanalysis. Econ Geol 87:1566-1583

Heinrich CA, Ryan CG (1992) Mineral paragenesis and regional zonation of granite-related Sn-As-Cu-Pb-Zn deposits: a chemical model for the Mole Granite district (Australia) based on PIXE fluid inclusion analyses. *In:* Fluid–Rock Interaction. Kharaka YK, AS Maest (eds) Balkema, p 1583-1587

Heinrich CA, Driesner T, Stefánsson A, Seward TM (2004) Magmatic vapor contraction and the transport of gold from porphyry to epithermal ore deposits. Geology 32:761-764

Heinrich CA, Günther D, Audétat A, Ulrich T, Frischknecht R (1999) Metal fractionation between magmatic brine and vapor, determined by microanalysis of fluid inclusions. Geology 27(8):755-758

Heinrich W (2007) Fluid immiscibility in metamorphic rocks. Rev Mineral Geochem 65:389-430

Henley HF, Brown RE, Stroud WJ (1999) The Mole Granite – extent of mineralization and exploration potential. *In:* Regional Geology, Tectonics and Metallogenesis of the New England Orogen. Flood PG (ed) University of New England, Armidale, Australia, p 385-392

Henley RW, McNabb A (1978) Magmatic vapor plumes and ground-water interaction in porphyry copper emplacement. Econ Geol 73:1-20

Herbert HK, Smith JW (1978) Sulfur isotopes and origin of some sulfide deposits, New England, Australia. Mineral Deposita 13:51-63

Hollister LS (1990) Enrichment of CO_2 in fluid inclusions in quartz by removal of H_2O during crystal-plastic deformation. J Struct Geology 12:895-901

Kesler SE (2005) Ore-forming fluids. Elements 1:13-18

Klemd R (1998) Comment on the paper by Schmidt, Mumm et al.: High CO_2 content of fluid inclusions in gold mineralization's in the Ashanti belt, Ghana: a new category of ore forming fluids? (Mineralium Deposita 32:107-118, 1997). Mineral Deposita 33:317-319

Klemm L, Pettke T, Heinrich CA (2007a) Fluid and source magma evolution of the Questa pophyry Mo deposit, New Mexico, USA. Mineralium Deposita (in press)

Klemm L, Pettke T, Heinrich CA, Campos E (2007b) Hydrothermal evolution of the El Teniente deposit (Chile): porphyry Cu-Mo ore deposition from low-salinity magmatic fluids. Econ Geol (in press)

Kostova B, Pettke T, Driesner T, Petrov P, Heinrich CA (2004) LA ICP-MS study of fluid inclusions in quartz from the Yuzhna Petrovitsa deposit, Madan ore field, Bulgaria. Schweiz Mineral Petrograph Mitt 84:25-36

Krauskopf KB (1964) The possible role of volatile metal compounds in ore genesis. Econ Geol 59:22-45

Landtwing MR, Pettke T, Halter WE, Heinrich CA, Redmond PB, Einaudi MT (2005) Causes for Cu-Fe-sulfide deposition in the Bingham porphyry Cu-Au-Mo deposit, Utah: combined SEM-cathodoluminescence petrography and LA-ICPMS analysis of fluid inclusions. Earth Planet Sci Lett 235:229-243

Liebscher A (2007) Experimental studies in model fluid systems. Rev Mineral Geochem 65:15-47

Lowenstern JB, Mahood GA, Rivers ML, Sutton SR (1991) Evidence for extreme partitioning of copper into a magmatic vapor-phase. Science 252:1405-1409

Mavrogenes JA, Berry AJ, Newville M, Sutton SR (2002) Copper speciation in vapor-phase fluid inclusions from the Mole Granite, Australia. Am Mineral 87:1360-1364

Mernagh TP (2001) A fluid inclusion study of the Fosterville Mine: a turbidite-hosted gold field in the Western Lachlan Fold Belt, Victoria, Australia. Chem Geol 173(1-3):91-106

Mumm AS, Oberthür T, Vetter U, Blenkinsop TG (1997) High CO_2 content of fluid inclusions in gold mineralizations in the Ashanti Belt, Ghana: A new category of ore forming fluids? Mineral Deposita 32(2):107-118

Muntean JL, Kesler SE, Russell N, Polanco J (1990) Evolution of the Monte Negro acid sulfate Au-Ag deposit, Pueblo Viejo, Dominican-Republic - important factors in grade development. Econ Geol 85(8):1738-1758

Naden J, Shepherd TJ (1989) Role of methane and carbon-dioxide in gold deposition. Nature 342(6251):793-795

Nagaseki H, Hayashi K (2006) The effect of sulfur on the vapor-liquid distribution of Cu and Zn in boiling hydrothermal fluid by SRXRF microanalysis of fluid inclusions. Int Mineral Assoc Conf, Kobe, Abstract: 99

Nesbitt BE, Murowchick JB, Muehlenbachs K (1986) Dual origins of lode gold deposits in the Canadian Cordillera. Geology 14:506-509

Newton RC Manning CE (2000) Quartz solubility in H_2O-NaCl and H_2O-CO_2 solutions at deep crust-upper mantle pressures and temperatures: 2-15 kbar and 500-900 degrees C. Geochim Cosmochim Acta 64:2993-3005

Palmer DA, Simonson JM, Hensen JP (2004) Partitioning of electrolytes to steam and their solubilities in steam. *In:* Aqueous Systems at Elevated Temperatures and Pressures. Palmer DA, Fernández-Prini RJ, Harvey AH (eds) Elsevier Academic Press, p 409-439

Pedersen KS, Christensen P (2007) Fluids in hydrocarbon basins. Rev Mineral Geochem 65:241-258

Pettke T, Diamond LW, Kramers JD (2000) Mesothermal gold lodes in the north-western Alps: A review of genetic constraints from radiogenic isotopes. Eur J Mineral 12:213-230

Pettke T, Audétat A, Schaltegger U, Heinrich CA (2005) Magmatic-to-hydrothermal crystallization in the W-Sn mineralized Mole Granite (NSW, Australia) - Part II: Evolving zircon and thorite trace element chemistry. Chem Geol 220:191-213

Philippot P, Cauzid J (2006) Contrasting Cu-complexing behaviour in vapour and liquid fluid inclusions from the Yankee Lode deposit, Mole Granite, Australia. Geochim Cosmochim Acta 70:A489-A489

Plimer IR, Lu J, Kleeman JD (1991) Trace and Rare-Earth Elements in cassiterite - sources of components for the time deposits of the Mole Granite, Australia. Mineral Deposita 26: 267-274

Pokrovski GS, Roux J, Harrichoury JC (2005) Fluid density control on vapor-liquid partitioning of metals in hydrothermal systems. Geology 33(8):657-660

Pokrovski GS, Borisova Y, Harrichoury JC (2006a) The effect of sulfur on vapor-liquid partitioning of metals in hydrothermal systems: An experimental batch-reactor study. Abstract. Goldschmidt Conference, Geochimica Cosmochimca Acta 33:657-660

Pokrovski GS, Tagirov BR, Schott J (2006b) Speciation of gold and associated metals at hydrothermal conditions: In-situ experimental approaches and physical-chemical modelling. Colloque TRANSMET Abstracts, Nancy: 91-94.

Polito PA, Bone Y, Clarke JDA, Mernagh TP (2001) Compositional zoning of fluid inclusions in the Archaean Junction gold deposit, Western Australia: a process of fluid-wall-rock interaction? Aust J Earth Sci 48:833-855

Proffett JM (2003) Geology of the Bajo de la Alumbrera porphyry copper-gold deposit, Argentina. Econ Geol 98:1535-1574

Rankin AH, Ramsey MH, Coles B, Vanlangevelde F, Thomas CR (1992) The composition of hypersaline, iron-rich granitic fluids based on Laser-ICP and Synchrotron-XRF Microprobe analysis of individual fluid inclusions in topaz, Mole Granite, Eastern Australia. Geochim Cosmochim Acta 56: 67-79.

Redmond PB, Einaudi MT, Inan EE, Landtwing MR, Heinrich CA (2004) Copper deposition by fluid cooling in intrusion-centered systems: new insights from the Bingham porphyry ore deposit, Utah. Geology 32:217-220

Roedder E (1971) Fluid inclusion studies on the porphyry-type ore deposits at Bingham, Utah, Butte, Montana, and Climax, Colorado. Econ Geol 66:98-118

Roedder E (ed) (1984) Fluid Inclusions. Rev Mineralogy, Volume 12. Mineralogical Society of America

Ronacher E, Richards JP, Johnston MD (2000) Evidence for fluid phase separation in high-grade ore zones at the Porgera gold deposit, Papua New Guinea. Mineral Deposita 35:683-688

Ronacher E, Richards JP, Reed MH, Bray CJ, Spooner ETC, Adams PD (2004) Characteristics and evolution of the hydrothermal fluid in the North zone high-grade area, Porgera gold deposit, Papua New Guinea. Econ Geol 99:843-867

Ridley J, Hagemann SG (1999) Interpretation of post-entrapment fluid-inclusion re-equilibration at the Three Mile Hill, Marvel Loch and Griffins Find high-temperature lode-gold deposits, Yilgarn Craton, Western Australia. Chem Geol 154:257-278

Ridley JR, Diamond LW (2000) Fluid chemistry of orogenic lode gold deposits and implications for genetic models. *In:* Gold in 2000. Hagemann SG, Brown PE (eds), Rev Econ Geol 13:146-162

Robert F, Kelly WC (1987) Ore-forming fluids in Archean gold-bearing quartz veins at the Sigma mine, Abitibi Greenstone Belt, Quebec, Canada. Econ Geol 82:1464-1482

Rusk B, Reed MH, Dilles JH, Klemm L (2004) Compositions of magmatic-hydrothermal fluids determined by LA-ICPMS of fluid inclusions from the porphyry copper-molybdenum deposit at Butte, Montana. Chemical Geology 210:173-199

Samson I, Anderson A, Marshall D (eds) (2003) Fluid Inclusions – Analysis and Interpretation. Min Assoc Can Sh Course, Volume 32

Sawkins FJ, Scherkenbach DA (1981) High copper content of fluid inclusions in quartz from northern Sonora: implications for ore genesis theory. Geology 9:37-40

Seedorff E, Dilles JH, Proffett JM, Einaudi M, Zurcher L, Stavast WJA, Johnson DA, Barton MD (2005) Porphyry deposits: characteristics and origin of hypogene features. Econ Geol 100th Anniversary Volume, p 251-298

Schaltegger U, Pettke T, Audétat A, Reusser E, Heinrich CA (2005) Magmatic-to-hydrothermal crystallization in the W-Sn mineralized Mole Granite (NSW, Australia) - Part I: Crystallization of zircon and REE-phosphates over three million years - a geochemical and U-Pb geochronological study. Chem Geol 220:215-235

Sibson RH, Robert F, Poulsen KH (1988) High-angle reverse faults, fluid-pressure cycling, and mesothermal gold-quartz deposits. Geology 16:551-555

Simmons SF, White NC, John DA (2005) Geological characteristics of epithermal precious and base metal deposits. Econ Geol 100th Anniversary Volume, p 455-522

Sherlock RL (1996) Hydrothermal alteration of volcanic rocks at the McLaughlin gold deposit, northern California. Can J Earth Sci 33:493-508

Sillitoe RH (1973) The tops and bottoms of porphyry copper deposits. Econ Geol 68:799-815

Sillitoe RH (1997) Characteristics and controls of the largest porphyry copper-gold and epithermal gold deposits in the circum-Pacific region. Aust J Earth Sci 44:373-388

Sillitoe RH, Hedenquist JW (2003) Linkages between volcano tectonic settings, ore-fluid compositions and epithermal precious metal deposits. *In*: Volcanic, Geothermal and Ore-forming Fluids: Rulers and Witnesses of Processes within the Earth. Simmons SF, Graham I (eds) Econ Geol Special Pub 10: 315-343

Simmons SF, Brown KL (2006) Gold in magmatic hydrothermal solutions and the rapid formation of a giant ore deposit. Science 314:288-291

Simon AC, Frank MR, Pettke T, Candela PA, Piccoli PM, Heinrich CA (2005) Gold partitioning in melt-vapor-brine systems. Geochim Cosmochim Acta 69:3321-3335

Simon AC, Pettke T, Candela PA, Piccolli PM, Heinrich CA (2006) Copper partitioning in a melt-vapor-brine-magnetite-pyrrhotite assemblage. Geochim Cosmochim Acta 70:5583-5600

Somarin AK, Ashley P (2004) Hydrothermal alteration and mineralisation of the Glen Eden Mo-W-Sn deposit: A leucogranite-related hydrothermal system, southern New England Orogen, NSW, Australia. Mineral Deposita 39: 282-300

Spycher NF, Reed MH (1989) Evolution of a broadlands-type epithermal ore fluid along alternative P-T paths - implications for the transport and deposition of base, precious and volatile metals. Econ Geol 84:328-359

Stefánsson A, Seward TM (2004) Gold (I) complexing in aqueous sulphide solutions to 500 °C at 500 bar. Geochim Cosmochim Acta 68:4121-4143

Symonds RB, Rose WI, Bluth GJS, Gerlach TM (1994) Volcanic gas studies - methods, results and applications. Rev Mineral 30:1-66

Sun S, Eadington PJ (1987) Oxygen isotope evidence for the mixing of magmatic and meteoric waters during tin mineralization in the Mole Granite, New South Wales, Aust Econ Geol 82:43-52

Taran YA, Hedenquist JW, Korzhinsky MA, Tkachenko SI, Shmulovich KI (1995) Geochemistry of magmatic gases from Kudryavy Volcano, Iturup, Kuril Islands. Geochim Cosmochim Acta 59:1749-1761

Ulrich T, Günther D, Heinrich CA (1999) Gold concentrations of magmatic brines and the metal budget of porphyry copper deposits. Nature 399:676-679

Ulrich T, Günther D, Heinrich CA (2001) Evolution of a porphyry Cu-Au deposit, based on LA-ICP-MS analysis of fluid inclusions: Bajo de la Alumbrera, Argentina. Econ Geol 96:1743, correctly reprinted in 2002 vol 97:1888-1920

Vallance J, Cathelineau M, Marignac C, Boiron MC, Fourcade S, Martineau F, Fabre C (2001) Microfracturing and fluid mixing in granites: W-(Sn) ore deposition at Vaulry (NW French Massif Central). Tectonophysics 336:43-61

Vanko DA, Bonnin-Mosbah M, Philippot P, Roedder E, Sutton SR (2001) Fluid inclusions in quartz from oceanic hydrothermal specimens and the Bingham, Utah porphyry-Cu deposit: a study with PIXE and SXRF. Chem Geol 173:227-238

Vennemann TW, Muntean JL, Kesler SE, Oneil JR, Valley JW, Russell N (1993) Stable isotope evidence for magmatic fluids in the Pueblo-Viejo epithermal acid sulfate Au-Ag deposit, Dominican-Republic. Econ Geol 88:55-71

Wallier S, Rey R, Kouzmanov K, Pettke T, Heinrich CA, Leary S, O'Connor G, Tamas CG, Vennemann T, Ulrich T (2006) Magmatic fluids in the Breccia-hosted epithermal Au-Ag deposit of Rosia Montana, Romania. Econ Geol 101:923-954

Walsh JF, Kesler SE, Duff D, Cloke PL (1988) Fluid inclusion geochemistry of high-grade, vein-hosted gold ore at the Pamour Mine, Porcupine-Camp, Ontario. Econ Geol 83:1347-1367

Watanabe Y, Hedenquist J (2001) Mineralogic and stable isotope zonation at the surface over the El Salvador porphyry copper deposit, Chile. Econ Geol 96:1775-1797

Webster J, Mandeville C (2007) Fluid immiscibility in volcanic environments. Rev Mineral Geochem 65:313-362

Williams-Jones AE, Heinrich CA (2005) 100th Anniversary special paper: Vapor transport of metals and the formation of magmatic-hydrothermal ore deposits. Econ Geol 100:1287-1312

Williams-Jones AE, Migdisov AA, Archibald SM, Xiao ZF (2002) Vapor-transport of ore metals. *In*: Water-rock Interaction: a Tribute to David A. Crerar. Hellmann R, Wood SA (eds) The Geochemical Society, Special Publication, p 279-305

Williamson BJ, Stanley CJ, Wilkinson JJ (1997) Implications from inclusions in topaz for greisenisation and mineralisation in the Hensbarrow topaz granite, Cornwall, England. Contrib Mineral Petrol 127: 119-128

Yardley BWD (2005) Metal concentrations in crustal fluids and their relationship to ore formation. Econ Geol 100:613-632

Yardley BWD, Banks DA, Bottrell SH, Diamond LW (1993) Post-metamorphic gold quartz veins from NW Italy - the composition and origin of the ore fluid. Mineral Mag 57:407-422

Reviews in Mineralogy & Geochemistry
Vol. 65, pp. 389-430, 2007
Copyright © Mineralogical Society of America

12

Fluid Immiscibility in Metamorphic Rocks

Wilhelm Heinrich

Department 4, Chemistry of the Earth
GeoForschungsZentrum Potsdam
Telegrafenberg
D 14473 Potsdam, Germany
whsati@gfz-potsdam.de

INTRODUCTION

Evidence of fluid immiscibility in metamorphic rocks comes, in the best of all worlds, from fluid inclusion observations. In fact, fluid immiscibility should be anticipated over the full range of metamorphic conditions provided that appropriate bulk fluid compositions are present. In a review paper entitled *"Metamorphic Fluids: the Evidence from Fluid Inclusions,"* Crawford and Hollister (1986) summarized: "The role of fluid immiscibility in metamorphism was only recognized after serious study of fluid inclusions in metamorphic rocks was underway in the late 1970s and has yet to attract the attention of more than a handful of petrologists studying metamorphic rocks." Since then, tremendous progress has been made. A huge amount of fluid inclusion studies have convincingly shown that immiscible fluids can be present in rocks of all bulk compositions and at all *P-T* conditions. Thermodynamic databases of minerals, solid solutions and fluid mixtures along with the development of Gibbs free energy minimization programs now allow for detailed prediction of the evolution of phase assemblages and mineral compositions in *P-T* space. Significant progress has also been made in understanding reaction kinetics, diffusional and convective mass transport and the interpretation of fluid-rock interaction during metamorphism. Given the vast fluid inclusion evidence of immiscibility it would appear, however, that phase petrology and mass transport in any specific rock where fluids are involved are not correctly interpreted if possible or proven immiscibility is not considered. Relatively few studies directly link mineral reactions with immiscible fluids. Therefore, this review aims at working out the effects of fluid immiscibility on the progress of metamorphic reactions and fluid transport over a wide range of conditions.

Questions arise directly from the fluid inclusion record in metamorphic rocks. This chapter, however, is organized the other way round. At first it summarizes phase relations of fluid systems pertinent to metamorphic rocks obtained by laboratory experiments and thermodynamic calculations. It then addresses how calculated mineral reactions are affected by immiscible fluid systems, and how immiscible fluids are produced and evolved when metamorphic mineral reactions proceed. The impure limestone plus H_2O-CO_2-salt system is theoretically explored, for which by far the most information is available and which may stand for other fluid-rock systems. A section on the physical behavior of immiscible fluids follows, again mainly based on experimental results. The second part of the paper reviews numerous field examples of contact, regional, and subduction related metamorphic rocks where fluid immiscibility played a significant role during petrogenesis and mass transport. Late events in metamorphic rocks such as formation of hydrothermal systems, late vein formation, ore deposits, etc. where fluid immiscibility and fluid segregation is of paramount importance are not explicitly addressed here. Instead, the review focuses on the link between mineral reactions and immiscible fluids.

1529-6466/07/0065-0012$05.00   DOI: 10.2138/rmg.2007.65.12

FLUID PHASE RELATIONS

Phase relations of simple and complex fluid systems are reviewed in detail by Gottschalk (2007) and Liebscher (2007). Some simple binary systems are very briefly summarized here. The basic features of a few ternary model systems relevant for fluid evolution in metamorphic rocks are recapitulated, the most important of which is H_2O-CO_2-$NaCl$, and their topologies in *P-T* space are shown. How they are related to specific mineral reactions is outlined below.

Binary fluid systems

Several binary metamorphic fluids are immiscible over a range of metamorphic conditions including non-electrolyte H_2O-CO_2, H_2O-CH_4, and H_2O-N_2 mixtures (Fig. 1). The consolute point in the pure H_2O-CO_2 system lies at 275 °C, which is below most temperatures of metamorphic interest. The miscibility gaps of H_2O-CH_4 and H_2O-N_2 expand to somewhat higher temperatures and fluids of these binaries may be immiscible up to about 400 °C (Crawford and Hollister 1986; Diamond 2003). The miscibility gap of H_2O-CO_2-CH_4 lies, dependent on *P* and *T,* somewhere between the two binaries (Zhang and Franz 1992). Generally, H_2O-CO_2, H_2O-CH_4, and H_2O-N_2 mixtures behave rather similar, and so do probably ternary and quaternary mixtures between them.

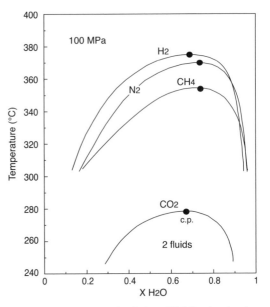

Figure 1. *T-X* plot (mol fraction) at 100 MPa showing the solvi of the H_2O-CO_2, H_2O-CH_4, H_2O-H_2, and H_2O-N_2 binaries. Adapted from Diamond (2003).

Binary H_2O-salt systems are summarized in Figure 2. In the metamorphic context immiscibility is mainly an issue of pressure, and above 200 MPa all H_2O-metal chloride mixtures are completely miscible at the relevant metamorphic temperatures. From NaCl over KCl and $MgCl_2$ to $CaCl_2$ the solvus opens more widely. Critical points at 700 °C lie at about 120 MPa for the NaCl and KCl, 150 MPa for $MgCl_2$, and 190 MPa for $CaCl_2$-H_2O binaries. Saline brines coexist with low density, low salinity vapor. The diagrams highlight the importance of immiscibility at the low-pressure, high temperature regime of contact metamorphic environments and associated hydrothermal systems, even if no additional fluid components are present.

Addition of non-polar components such as CO_2, CH_4 and N_2 to H_2O-electrolyte mixtures dramatically enlarges the two fluid solvus, giving rise to fluid immiscibility over the full range of metamorphism. Brines of varying salinity are ubiquitous in most sedimentary and metamorphic environments and supply of significant amounts of a CO_2 (N_2, CH_4) component to saline water by any process may drive the fluid system into immiscibility. The system H_2O-CO_2-$NaCl$ is a close approximation to real metamorphic fluids, where NaCl may stand for other metal chloride components, and CO_2 for N_2 and CH_4.

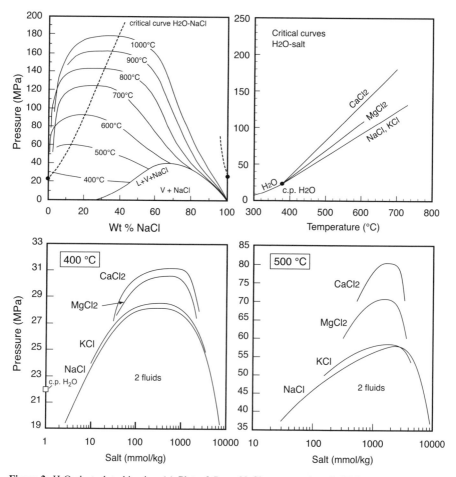

Figure 2. H$_2$O-electrolyte binaries. (a) Plot of P vs. NaCl concentration. Solid lines show the solvus boundaries at the respective temperatures. Dashed line is the critical curve, the three phase curve L+V+NaCl separates the L+V from the V+NaCl field. (b) P-T plot showing critical curves of H$_2$O-salt binaries. c.p. is the critical point of H$_2$O. (c) P vs. salt concentration showing the solvus boundaries for NaCl, KCl, MgCl$_2$ and CaCl$_2$-H$_2$O binaries at 400 °C. (d) As before, at 500 °C. Data for NaCl-H$_2$O are from Sourirajan and Kennedy (1962) and Bodnar et al. (1985), all other data from Shmulovich et al. (1995).

Phase relations in the system H$_2$O-CO$_2$-NaCl

A large body of experimental work on phase relations of the H$_2$O-CO$_2$-NaCl ternary at varying P and T conditions is available (Takenouchi and Kennedy 1965; Naumov et al. 1974; Gehrig 1980; Gehrig et al. 1986; Gibert et al. 1998; Kotel'nikov and Kotel'nikova 1990; Johnson 1991; Frantz et al. 1992; Shmulovich and Plyasunova 1993; Schmidt et al. 1995; Shmulovich and Graham 1999, 2004; Schmidt and Bodnar 2000). Based on the data of Gehrig (1980), Bowers and Helgeson (1983a,b) developed a modified Redlich-Kwong (MRK) equation of state (EOS) for that ternary. This MRK-EOS is, however, unable to accurately predict fugacities of fluid components near the solvus boundaries and at moderate and high salinities. An EOS more convenient for rigorous thermodynamic modeling of H$_2$O-CO$_2$-NaCl fluids was presented by Duan et al. (1995), which can be applied to a T range of 300-1000 °C and P up to 600 MPa. It calculates activities of fluid components for compositions of up to 30 wt% NaCl (relative to H_2O-NaCl) in good agreement to experiments and up to 50 wt% NaCl with somewhat less accuracy.

Following the delineation of Trommsdorff and Skippen (1986), Skippen and Trommsdorff (1986), and Heinrich et al. (2004) the main features of the phase topology of ternary H_2O-salt-polar gas systems are exemplified for the H_2O-CO_2-NaCl ternary (Fig. 3). Isothermal-isobaric sections are set into a *P-T* plot ranging to 1000 °C and 600 MPa along with phase relations of the binaries H_2O-CO_2, H_2O-NaCl, CO_2-NaCl and the melting curve of halite. The essential points are as follows: (1) The critical curve of the H_2O-CO_2 binary, indicated as L+V = single phase fluid (F), lies always below 350 °C except at conditions very close to the critical point of pure water (T_c). A binary H_2O-CO_2 mixture is always homogeneous at the relevant *P-T* conditions. (2) The topology of the CO_2-NaCl binary is controlled by the melting curve of halite (Hl) at high *T* (Hl=L). The critical curve is not well known but it is clear that $NaCl^{liquid}$ and CO_2 are extremely immiscible, even at very high *T*. Fluids of the CO_2-NaCl binary are likely two-phase for any geological process in the crust. At *T* below the melting curve of halite, a field of halite-liquid equilibrium occupies the NaCl-rich part of the CO_2-NaCl binary (II and V in Fig. 3). Below 800 °C, halite coexists with almost pure CO_2 because of the very low solubility of halite in CO_2 vapor (I and IV). (3) In the H_2O-NaCl binary, halite coexists with a single-phase brine

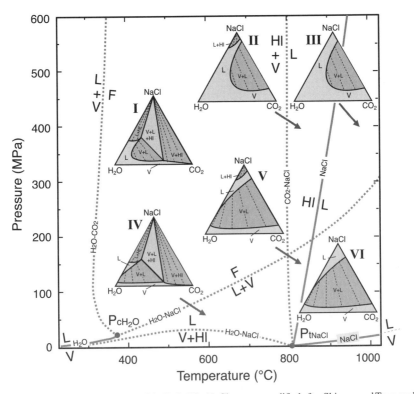

Figure 3. Pressure-temperature plot of the H_2O-CO_2-NaCl ternary, modified after Skippen and Trommsdorff (1986) and Heinrich et al. (2004). Components are given in mol%. Binary curves and distinct isothermal-isobarical sections through the ternary are calculated with the EOS of Duan et al. (1995). Data for halite and $NaCl_{liquid}$ are from the Glushko (1976-1982) database. Thin broken lines within the two-phase fields indicate calculated compositions of coexisting liquid and vapor. Pc_{H_2O} is the critical point of water, Pt_{NaCl} the triple point of NaCl.

at P above the critical curve (indicated as L+V = F). At lower pressure, a V+L field appears (IV in Fig. 3). At very low P a V+Hl field exists which is bound by the V+Hl = L curve. Above the halite melting curve, at high P a single brine exists (III), and at low P NaCl-rich brine coexists with NaCl-poor vapor (VI). (4) In the H_2O-CO_2-NaCl ternary, phase relations of type I and IV are characteristic for the major part of the relevant P-T space. The peculiar property of these topologies is that large fields of V+L and V+L+H equilibrium occur. In the water-rich part, NaCl-rich, CO_2-poor brine coexists with NaCl-poor vapor, which is richer in CO_2. The V+L+H field shrinks with increasing T in favor of the V+L field because of increasing solubility of NaCl in the brine. (5) The extension of the V+L field is strongly dependent on P, particularly near the H_2O apex. It is widely open at low P (IV and V) and shrinks in favor of the one phase field as P increases (I, II, and III). (6) Two main differences exist between the phase relations previously shown by Trommsdorff and Skippen (1986), which are based on the Bowers and Helgeson MRK-EOS, and that shown in Figure 3, based on the Duan et al. EOS: (i) The calculated V-L tie lines in sections I and IV are significantly steeper than those estimated by Trommsdorff and Skippen (1986). This is because a large V+H field exists at low and medium T, where the X_{CO_2} of vapor coexisting with halite-saturated brine cannot exceed 0.6 at most conditions. (ii) The NaCl concentrations in the H_2O-CO_2 fluid coexisting with brine at low and medium P are much lower than estimated by Trommsdorff and Skippen (1986) and are even exaggerated in sections I, IV, V, and VI. This is particularly important for low-P contact-metamorphism where very small amounts of NaCl are sufficient to create a solvus between brine and H_2O-CO_2-bearing vapor. (7) Thus, at most conditions from contact to granulite facies metamorphism a three component H_2O-CO_2-NaCl fluid system may consist of at least two phases.

H_2O-CO_2-$CaCl_2$

The shape of the immiscibility gap of the H_2O-CO_2-$CaCl_2$ ternary has been experimentally determined by Zhang and Frantz (1989), Shmulovich and Plyasunova (1993), and Shmulovich and Graham (2004). Data cover the range from 0.1 to 0.9 GPa and 500 to 800 °C (Fig. 4). Similar to H_2O-CO_2-NaCl, $CaCl_2$ is strongly partitioned into the water-rich liquid and likewise, small amounts of $CaCl_2$ cause immiscibility to exist over a wide P-T range. Comparing the two systems Figure 5 shows significantly wider solvi for $CaCl_2$ compared to NaCl at the same conditions and their shrinking towards higher P and T (Shmulovich and Graham 2004). The slopes

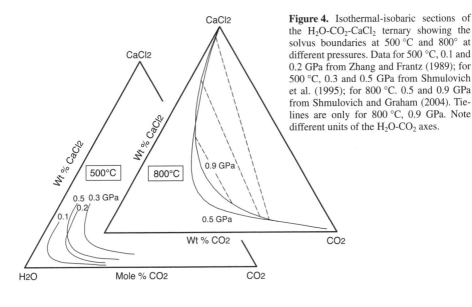

Figure 4. Isothermal-isobaric sections of the H_2O-CO_2-$CaCl_2$ ternary showing the solvus boundaries at 500 °C and 800° at different pressures. Data for 500 °C, 0.1 and 0.2 GPa from Zhang and Frantz (1989); for 500 °C, 0.3 and 0.5 GPa from Shmulovich et al. (1995); for 800 °C. 0.5 and 0.9 GPa from Shmulovich and Graham (2004). Tie-lines are only for 800 °C, 0.9 GPa. Note different units of the H_2O-CO_2 axes.

of the tie-lines between brine and vapor are flatter in the $CaCl_2$-bearing system. So far, there is no EOS available for this system.

H_2O-CH_4-NaCl

Experimental data are available up to 600° and 250 MPa and a wide range of salinities (Krader and Franck 1987; Lamb et al. 1996, 2002). Isothermal isobaric sections are shown in Figure 6. Duan et al. (2003) presented an EOS for this system that is in good agreement with the experimental data and that reliably predicts parameters up to about 1000 °C and 500 MPa. The system shows many similarities to H_2O-CO_2-NaCl over large compositional ranges at metamorphic conditions. Duan et al. (2003) also developed an EOS for the quaternary system H_2O-CO_2-CH_4-NaCl. There are no experimental data available for this and other complex systems.

PHASE RELATIONS AND FLUID EVOLUTION IN THE CaO-MgO-SiO_2-H_2O-CO_2-NaCl MODEL SYSTEM

H_2O and CO_2 are involved in many reactions during progressive metamorphism of siliceous limestones and carbonate-bearing argillites. Bowers and Helgeson (1983a,b) were the first who quantitatively calculated the effect of fluid immiscibility on mineral reactions of the system CaO-MgO-SiO_2-H_2O-CO_2-NaCl. Using their modified MRK-EOS of the fluid ternary in combination with the thermodynamic database of Helgeson et al. (1978), they calculated the positions of the reaction curves in *P-T-X* space relative to the fluid solvus. They were presented, for example, in conventional *T-X* plots at constant *P* (Figs. 7-9). Though their MRK-EOS

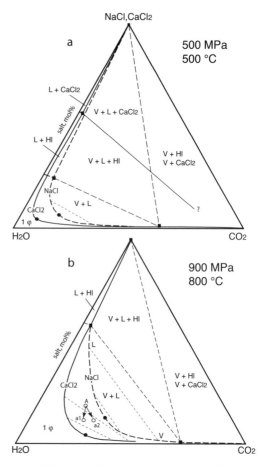

Figure 5. Phase equilibria in the systems H_2O-CO_2-NaCl (thick dashed lines) and H_2O-CO_2-$CaCl_2$ (solid lines) at (a) 500 MPa, 500 °C and (b) 900 MPa, 800 °C derived from synthetic fluid inclusion experiments (Shmulovich and Graham 1999, 2004), including data from Joyce and Holloway (1993), all in mol%. Thin dashed lines depict V-L tie-lines within the solvus. Black dots show the consolute points, black squares salt saturation of the respective systems. At 900 MPa, 800 °C the salt saturation of H_2O-$CaCl_2$ is estimated at 80-85 mol% of $CaCl_2$ (not drawn in b). Note the significantly wider solvi for $CaCl_2$ compared to NaCl and the shrinking of both solvi towards higher *P,T* conditions. Note also the much steeper V-L tie-lines within the solvi and much lower CO_2 concentrations in the vapor compared to earlier assumptions of the NaCl-bearing system behavior (Bowers and Helgeson 1983a,b; Skippen and Trommsdorff 1986). See text for discussion of points A, a1 and a2. (Simplified from Shmulovich and Graham 2004).

as well as the thermodynamic database may have had some shortcomings the essential points are perfectly addressed: (1) Many isobarically univariant reactions in siliceous limestones are cut by the fluid solvus if NaCl is present. (2) Isobarically univariant equilibria become invariant

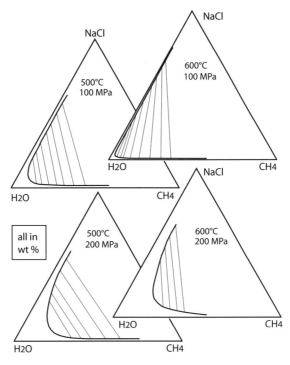

Figure 6. Isothermal-isobaric sections of the H₂O-CH₄-NaCl ternary showing the solvus boundaries and tie-lines at 500 °C and 600 °C, 100 and 200 MPa. Data from Lamb et al. (1996, 2002).

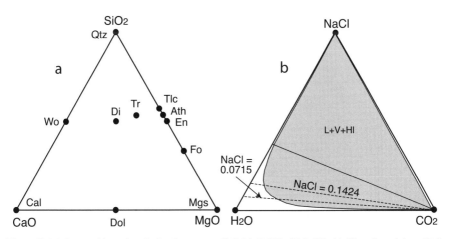

Figure 7. (a) Composition triangle for the system CaO-MgO-SiO₂-H₂O-CO₂-NaCl projected from H₂O, CO₂ and NaCl showing positions of minerals considered by Bowers and Helgeson (1983b) for stability calculations in impure limestones. Abbreviations of minerals after Kretz (1983). (b) Bulk compositions in the H₂O-CO₂-NaCl ternary corresponding to pseudobinaries for which X_{NaCl} = 0.0715 and 0.1423 (see Fig. 8 and 9). Fluid solvus is schematic.

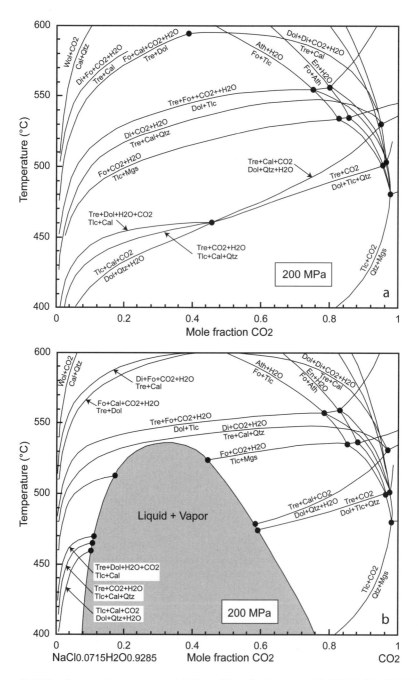

Figure 8. T-X_{CO_2}-diagram depicting mineral fluid-equilibria in the system CaO-MgO-SiO$_2$-H$_2$O-CO$_2$-NaCl at 200 MPa. (a) NaCl-free; (b) at $X_{NaCl} = 0.0715$, i.e., along lower dashed line in Figure 7b. Dots represent invariant points. Thermodynamic database for minerals is from Helgeson et al. (1978); EOS of ternary fluids is that of Bowers and Helgeson (1983a). Redrawn from Bowers and Helgeson (1983b). Abbreviations of minerals after Kretz (1983).

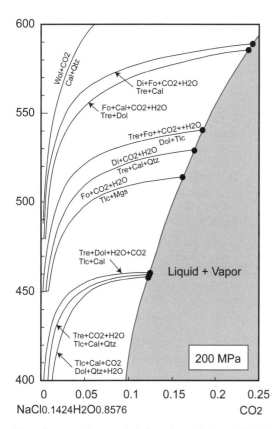

Figure 9. T-X_{CO_2}-diagram depicting mineral-fluid equilibria in the system CaO-MgO-SiO_2-H_2O-CO_2-$NaCl$ at 200 MPa and $X_{NaCl} = 0.1473$, i.e., along upper dashed line in Figure 7b. Only equilibria at the water-rich side of the solvus are shown. Modified after Bowers and Helgeson (1983b).

at the solvus boundaries. Invariant points existing with a single fluid may disappear (Fig. 8a,b). (3) If one of the conjugate fluids is able to segregate from the system, for example by boiling off a lower density CO_2-rich vapor that is produced by decarbonation, extensive reaction may occur at the solvus boundary without significantly increasing the amount of CO_2 in the water-rich fluid. Such process would increase the salinity of the water-rich fluid, giving rise to wider opening of the solvus. Thus, highly saline solutions may be generated from low salinity fluids by immiscibility and fluid segregation. (4) Similarly, direct production of a conjugate fluid pair within the solvus by a single reaction would also allow for extensive reaction. (5) NaCl not only dramatically enlarges the H_2O-CO_2 solvus and raises the consolute point, i.e., the crest of the solvus in a T-X-plot, above 600 °C at 200 MPa and $NaCl_{0.1424}$ $H_2O_{0.8576}$ (Fig. 9) but also increases the nonideality of the H_2O-CO_2 fluid mixture. Therefore, univariant reactions at the water-rich side of the solvus are shifted to higher T with increasing salinity of the brine. Also, mineral assemblages that would otherwise be stable only at low temperatures in the presence of an H_2O-rich fluid occur at much higher T in presence of coexisting liquid and vapor phases. (6) Mineral assemblages that are stable in the presence of a single phase H_2O-CO_2 fluid over a narrow fluid composition may coexist with fluids exhibiting a wide range in the H_2O-CO_2-$NaCl$ ternary. (7) Greenwood's (1975) simple concept of buffering of pore fluids by mineral reactions does not apply for two immiscible fluid systems. Mineral assemblages cannot be used to infer fluid compositions at conditions in the same way as in pure H_2O-CO_2 systems. (8) Failure to account for the possible effects of fluid immiscibility on equilibrium mineral assemblages may lead to serious errors in interpretation of phase relations in metamorphic rocks. In particular, reaction isograds may be misinterpreted if immiscibility is overlooked. Bowers and Helgeson (1983b) give further examples of the immiscibility effect on simple dehydration reactions such as $Ms + Qtz \leftrightarrow Kfsp + And + H_2O$ and on the activity ratios of ionic species characteristic of mineral equilibria in the system CaO-MgO-SiO_2-H_2O-CO_2-$NaCl$.

The T-X plots given in Figures 8 and 9 are simplified because they represent pseudobinary sections, which do not allow for reading the effect of changing salinities when a reaction occurs at the solvus boundaries. Skippen and Trommsdorff (1986) used a more appropriate approach

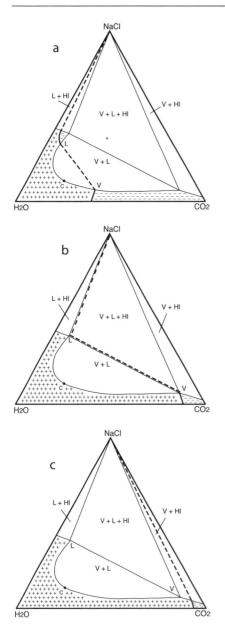

by plotting reaction curves onto the H$_2$O-CO$_2$-NaCl ternaries at fixed P and T. Figure 10 is a showcase of the calculated position of a single decarbonation reaction, Dol + Qtz + H$_2$O ↔ Tr + Cal + CO$_2$, at different temperatures relative to the fluid composition(s). The equilibrium curve crosses the two phase regions L+V and L-Hl. At higher T a three phase assemblage L+V+Hl in addition to the mineral assemblage is possible. At even higher T the equilibrium curve leaves the two fluid solvus and a very CO$_2$-rich vapor plus halite is present. The illustration is schematic insofar as other reactions may interfere with the reaction sequence at distinct *P-T-X* conditions (see below, Fig. 18b).

Decarbonation in presence of brine not only produces an immiscible CO$_2$-rich vapor but also induces phase separation and boiling off where, in this case, the lower density vapor may leave the system (e.g., Trommsdorff et al. 1985; Trommsdorff and Skippen 1986; Yardley and Bottrell 1988; Heinrich 1993; Heinrich and Gottschalk 1994; Heinrich et al. 2004). If so, water is continuously transferred from brine to vapor, which boils off, thus increasing the salinity of the brine and increasing the CO$_2$ content of the vapor as reaction proceeds. Figure 11 shows the fluid evolution for a schematic decarbonation reaction when boiling operates. As equilibrium temperatures are overstepped the reaction may start in the one fluid field. Due to CO$_2$ production, the fluid composition meets the solvus and exsolves CO$_2$-rich vapor. Increasing T, advancing reaction and boiling drives both fluids towards salt saturation. Further reaction along A'B' in Figure 11 finally transfers all water from brine to CO$_2$-rich vapor (see Fig. 10c). With respect to brine the reaction runs dry, and salt crystals are found in the rock (Trommsdorff et al. 1985; Gómez-Pugnaire et al. 1994).

Figure 10. Isothermal-isobaric sections of the H$_2$O-CO$_2$-NaCl ternary (mol%) depicting progress of the decarbonation reaction dolomite + quartz + H$_2$O ↔ tremolite + calcite + CO$_2$ at amphibolite facies conditions. Temperature increases from (a) to (c). Thick solid lines indicate equilibrium conditions at which reaction progress is zero. + indicate regions where the reaction proceeds to the right, - to the left. (a) Equilibrium crosses the two phase regions V+L and L+Hl. (b) Equilibrium includes the three-phase triangle V+L+Hl, indicating stability of all the three phases with the mineral assemblage at appropriate bulk compositions. (c) Equilibrium crosses the phase boundary V+Hl. Modified after Trommsdorff and Skippen (1986). Note that the solvus boundaries are based on extrapolations of Gehrig's (1980) experimental data. Recent data suggest much lower salt concentrations in CO$_2$ rich fluids, a larger V+Hl field, and steeper tie-lines between salt-rich and CO$_2$-rich fluids (see text).

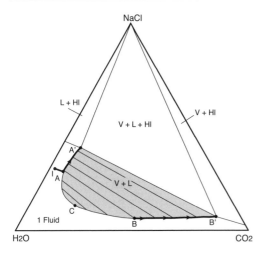

Figure 11. Phase relations in the H_2O-CO_2-NaCl ternary (mol%) for a schematic decarbonation reaction under amphibolite facies conditions. Point I indicates the initial composition of an evolving fluid that hits the two fluid solvus at A. Further decarbonation drives liquids along AA′, and coexisting vapors along BB′. Preferential loss of vapor evolves residual liquids to salt saturation and halite precipitation in A′ and B′. C is the consolute point of the solvus at distinct P,T conditions where liquid and vapor become indistinguishable. Modified after Trommsdorff and Skippen (1986). Solvus boundaries are based on extrapolations of Gehrig's (1980) experimental data. Recent data suggest much lower salt concentrations in CO_2 rich fluids, a larger V+Hl field, and steeper tie-lines between salt-rich and CO_2-rich fluids (see text).

In a more general context Trommsdorff and Skippen (1987) considered interaction of aqueous brine with dehydration, decarbonation, hydration-decarbonation and dehydration-decarbonation equilibria at low, medium, and high grade conditions (Fig. 12). Depending of *P-T* conditions, shape of the fluid solvus, reaction type, stoichiometric coefficients and initial brine composition, fluids generated by the proceeding reactions may hit or miss the solvus. If decarbonation is involved at low and medium grade conditions fluids almost inevitably meet it. Compositions of conjugate fluids may evolve in very different directions. At granulite facies conditions the solvus significantly shrinks towards the CO_2-NaCl side. Nevertheless, hydration-decarbonation reactions will also produce immiscibility at these conditions. For any particular reaction mass balance must be considered. The relative abundances of hydrous and carbonate minerals and the stoichiometric coefficients of the phases that are consumed and produced control the compositional path of the evolving fluid(s). Moreover, different porosities and rates of porosity collapse will have a significant effect. Hydration-decarbonation reactions are likely to lead to immiscibility under almost all conditions, whereas dehydration reactions within rocks with very low porosity and low salinity are unlikely to produce immiscibility even if they also generate CO_2, because the proportion of salt in the system is small. If water is produced this could rapidly lower the salinity in the initial pore fluid, particularly if porosity is low. Hydration, on the other hand, increases salinity because most hydrous silicates strongly prefer OH over Cl during fluid-solid exchange. Immiscibility can be produced in siliceous metamorphic rocks even without involvement of carbonate-bearing phases simply by lowering the H_2O-concentration of any pre-existing homogeneous H_2O-CO_2-NaCl fluid, for example by hydration processes during retrogression.

Combining the approaches of Bowers and Helgeson (1983b) and Skippen and Trommsdorff (1986), Heinrich et al. (2004) presented calculations of phase relations in the system CaO-MgO-SiO_2-H_2O-CO_2-NaCl at 300° to 1000 °C and up to 600 MPa. Thermodynamic properties of minerals are from the internally consistent dataset of Gottschalk (1997), fugacities of fluid components are calculated using the EOS of Duan et al. (1995). Calculations are based on a Gibbs free energy minimization program that allows for input of any bulk composition in the CMS subsystem and also for non-ideal solid solutions. Twenty-four solid phases are considered (Fig. 13) and results are shown by a series of isothermal-isobaric sections, an example of which is presented in Figure 14 for bulk composition A (see Fig. 13). Shown are stability regions of isothermal-isobarically divariant solid assemblages at 50 MPa and 200 MPa, which are separated by univariant reaction curves (black lines). Reaction curves are invariant (dotted

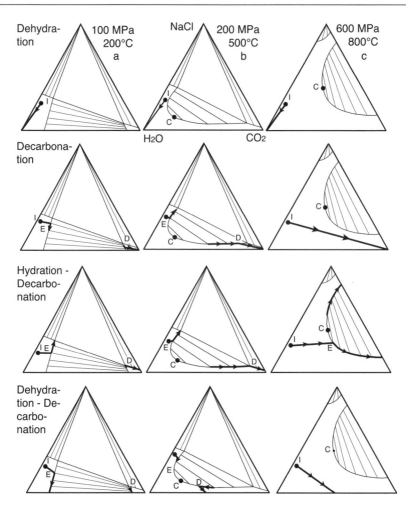

Figure 12. Schematic sections in the H_2O-CO_2-NaCl ternary along with fluid evolution paths for metamorphic dehydration, decarbonation, hydration/decarbonation and dehydration/decarbonation reactions. (a) low T/low P (~200 °C, 100 MPa); (b) 500 °C, 200 MPa; (c) 800 °C, 600 MPa. (I) indicates composition of initial fluid, arrows indicate direction of fluid evolution with reaction progress, (E) indicates entry of bulk fluid compositions into the fluid solvus, (D) departure of the bulk fluid from the solvus, (C) is the consolute point. Dehydration drives fluid towards the H_2O apex, decarbonation towards the CO_2 apex, hydration/decarbonation towards the NaCl-CO_2 join, and dehydration/decarbonation towards the H_2O-CO_2 join. Modified after Skippen and Trommsdorff (1986).

lines) where a second fluid or halite additionally appears. Isothermal-isobaric invariant points in the one fluid field, where two reactions intersect do not become visible because they lie outside of the applied plotting range in steps of 50 °C. For both pressures, the well-known sequence of the prograde index minerals talc, tremolite, diopside, wollastonite, monticellite, and akermanite develops with increasing T. Because they are formed by decarbonation reactions, including hydration and dehydration for reactions involving talc and tremolite, they first appear at the water-rich side of the ternary diagrams. As T increases, reaction curves are generally shifted towards the CO_2 apex eventually intersecting the two fluid solvus and, in some cases, arriving again in a one fluid field where the fluid phase is CO_2-rich, almost NaCl-free, and coexists

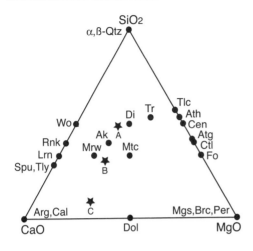

Figure 13. Composition triangle for the system CaO-MgO-SiO₂-H₂O-CO₂-NaCl projected from H₂O, CO₂ and NaCl showing positions of minerals considered by Heinrich et al. (2004) for stability calculations in impure limestones and siliceous dolomites. A, B, and C (stars) denote different bulk compositions used in calculations of stable phase assemblages by Gibbs free energy minimization methods, using the database of Gottschalk (1997) for minerals and fluid activities extracted from Duan et al. (1995).

with halite. This behavior is seen, for example, in sections at 450 °C and 500 °C, 200 MPa. At 450 °C, two assemblages appear, Qtz + Cal + Dol and Qtz + Tr + Cal, which are separated by the reaction Dol + Qtz + H₂O ↔ Tr + Cal + CO₂. This reaction curve lies in the one fluid field, very close to the solvus boundary. At a few degrees higher, the reaction shifts into the solvus, where two immiscible fluids are produced. At 500 °C, the reaction curve has already moved into the vapor + halite field (dotted line) where a single CO₂-rich, NaCl-free fluid along with halite is present. At the water-rich side, two new assemblages appear in the one-fluid field, Qtz + Di + Cal and Wo + Di + Cal, due to the reactions Cal + Qtz + Tr ↔ Di + CO₂ + H₂O, and Cal + Qtz ↔ Wo + CO₂, respectively, the latter evolving in a comparable way from 500 °C up to 700 °C (right column). Similar diagrams were also presented for dolomite-dominated bulk compositions (C in Fig. 13) and also for *P* up to 500 MPa (Heinrich et al. 2004).

The diagrams highlight the relative positions of mineral assemblages and reaction curves with respect to the two-fluid solvus, the two fluid + halite, and the vapor + halite stability field. At 50 MPa and above 450 °C, minute amounts of NaCl in the fluid system lead most reactions into the two fluid solvus or in the vapor + halite field, implying that at the low *P*, high *T* regime of contact metamorphism immiscibility plays the dominant role. As *P* increases at identical *T* (*see* respective diagrams at 200 MPa), mineral stability fields and reaction curves shift towards the H₂O apex, whereas the solvus shrinks in the opposite direction. It is obvious that the position of any reaction relative to the two fluid solvus may be very sensitive to small *P* and *T* variations. This is particularly true for regional metamorphic conditions.

The system CaO-MgO-SiO₂-H₂O-CO₂-NaCl is reviewed in some detail here because it probably provides a good approximation for understanding fluid immiscibility and immiscibility producing processes in metamorphic rocks in general. Water-electrolyte-CH₄ or N₂–bearing systems with or without CO₂ would show similar behavior at most metamorphic conditions. So far, no attempt has been made to link fugacities of fluid components at solvus boundaries in other systems (e.g., available for H₂O-CH₄-NaCl; Duan et al. 2003) with possible mineral reactions at whichever conditions or bulk compositions. CO₂ (CH₄, N₂) producing mineral reactions in presence of brine is probably the most effective way to generate large amounts of immiscible fluids during metamorphism. Other processes, which may occur in any rock, include simple hydration reactions (not in Fig. 12) where a homogeneous CO₂ (CH₄, N₂) bearing brine is driven into the solvus by decreasing H₂O content and increasing salinity. Hydration preferentially occurs during retrogression and may produce salt saturation and halite precipitation without meeting the solvus if CO₂ is initially low (Markl and Bucher

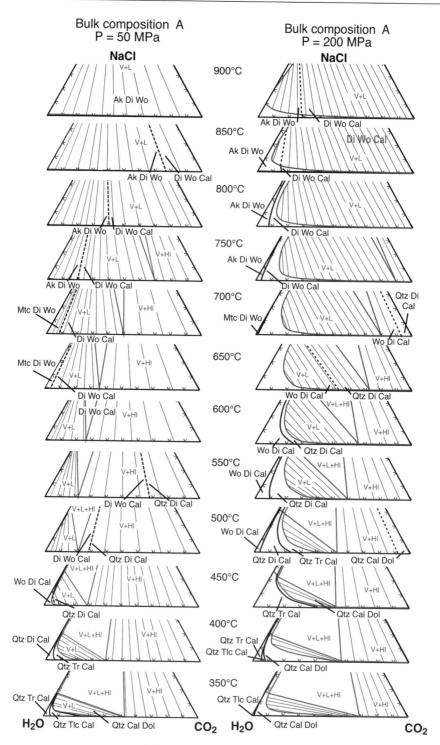

Figure 14. Caption on facing page.

1998). Mineral reactions are not a requisite. Any pre-existing free fluid may enter the solvus along a distinctive *P-T* trajectory. This is most probable at decompression-heating or isothermal decompression paths because solvi, in general, expand drastically with decreasing *P* and moderately with increasing *T*.

Shmulovich and Graham (2004) explored another important immiscibility producing process that may occur at lithological boundaries in the lower crust. They considered a boundary between a carbonate or basic rock and a quarzo-feldspatic or pelitic rock both in presence of a ternary H_2O-CO_2-brine. In a rock-dominated fluid-rock system, the cation compositions of the fluids depend on the mineralogy of the respective host rocks. For constant Cl concentrations, the saturated fluid in the quarzo-feldspatic rock will have dominantly Na-K cations, and the carbonate or basic rock dominantly Ca cations. Cation exchange occurs if transport pathways by grain edge-flow or hydrofracture are available at or across the boundary and if solid-fluid equilibration is rapid. Because the $CaCl_2$ solvus is wider than that of NaCl (Fig. 5; Liebscher 2007) immiscibility occurs. Irrespective of the transport mechanism, the change in cation composition along a diffusion profile or advection pathway results in unmixing of the fluid near the lithological boundary. Depending on whether Na is incorporated into feldspar or fluid during exchange, the reaction can be represented either as

$$2\,NaCl_{aq} + CaAl_2Si_2O_8 + 4\,SiO_2 \leftrightarrow CaCl_{2aq} + 2\,NaAlSi_3O_8 \tag{1}$$

or

$$2\,NaCl_{aq} + CaCO_3 + H_2O \leftrightarrow CaCl_{2aq} + 2\,NaOH + CO_2 \tag{2}$$

In reaction (1) the fluid bulk composition is shifted from A (in Fig. 5b) in a direction away from the NaCl apex to point a1, and in reaction (2) towards the CO_2 apex, to point a2. Irrespective whether the starting fluid is homogeneous or on the two-phase surface, every fluid evolves by the same process, because Ca or CO_2 is additionally produced. Shmulovich and Graham (2004) pointed out that the change of phase state during cation exchange between fluid and buffering host rocks may take place over a wide interval of *P-T-X_{fluid}* conditions, but that the compositional limits for the phenomenon depend on *P* and *T*. For example, from Figure 5, the limiting conditions for unmixing events on boundaries between Na-rich and Ca-rich lithologies at 800 °C and 0.9 GPa may be estimated as $0.4 < X_{H_2O} < 0.7$ and $X_{NaCleq} > 0.04$. At 500 °C and 0.5 GPa the equivalent limits are $\sim 0.5\,X_{H_2O} < 0.9$ and $X_{NaCleq} > 0.2$. These compositions are geologically realistic, although composition and amount of free fluids in the lower crust are still a matter of debate (e.g., Yardley and Valley 1997; Shmulovich and Graham 2004).

PHYSICAL BEHAVIOR OF IMMISCIBLE FLUIDS

The record of fluid inclusions in metamorphic rocks: problems with selective entrapment and post-entrapment modifications

Fluid immiscibility between saline aqueous fluids and $CO_2 \pm N_2 \pm CH_4$ rich vapors can be anticipated over the whole range of metamorphic conditions. Unequivocal evidence is given by fluid inclusion investigations if the two fluids are shown having been trapped along defined or estimated solvus boundaries both in terms of composition and density, and

Figure 14. (*on facing page*) Isothermal-isobaric sections of the system H_2O-CO_2-NaCl showing the stability regions for mineral assemblages in siliceous dolomites (composition A; Fig. 13) at various temperatures and 50 and 200 MPa. *Grey:* fluid phase relations. In the one fluid phase field, isobaric isothermal divariant three solid assemblages are separated by univariant reaction curves (*solid black lines*), within the solvus the two reaction points (*connected by dotted tie-lines*) are isothermal-isobarically invariant. Invariant points in the one-fluid field, where two reactions intersect, do not appear in these sections but are present at distinct *P-T-X* conditions (modified from Heinrich et al. 2004).

also texturally. In regional metamorphic and subduction related rocks this is, unfortunately, only rarely given because (1) fluid inclusions inevitably change compositions and densities during deformation and uplift, and/or (2) fluids are likely to be selectively trapped as they are produced or transported through the rock matrix. The verification of fluid immiscibility at peak metamorphism indicated by mineral assemblages is a very difficult task as fluid inclusions may pretend or mask immiscibility, and the interpretation of metamorphic phase assemblages with regard to activities of components in presumably immiscible fluids appears rather impossible.

The presence of nearly pure $CO_2\pm N_2\pm CH_4$ fluid inclusions in amphibolite and granulite facies rocks has intrigued petrologists since decades (e.g., Touret 1971, 1977, 1981; Crawford et al. 1979; Sisson et al. 1981; Touret and Olsen 1985; Crawford and Hollister 1986; Hollister 1988, 1990; Lamb 1988; Olsen 1988; Van Reenen and Hollister 1988; Bakker and Jansen 1990, 1991; Whitney 1992; Johnson and Hollister 1995; and many more). Near peak *P-T* high salinity brines accompanying high-density carbonic fluid inclusions in granulites are much less obvious and widespread than carbonic inclusions (Touret 1985, 1995; Herms and Schenk 1992, 1998; Touret and Huizenga 1999; van den Berg and Huizenga 2001) and in many cases it remains unclear if immiscible brine ever coexisted. However, the absence of brine inclusions does not necessarily imply that brines were not there, particularly if one considers that trapping mechanisms are highly selective. Moreover, the presence of metasomatic K-feldspar microveins in granulites have been interpreted as remnants of high salinity aqueous fluids (Perchuk and Gerya 1993; Hansen et al. 1995; Franz and Harlov 1998: Harlov et al. 1998). The source of these brines may have been connate water inherited from the sediments or introduced from mafic intrusions.

Formation of nearly pure CO_2 inclusions in medium and high-grade metamorphic rocks were interpreted by two mechanisms (1) post-entrapment extraction of H_2O from H_2O-CO_2 fluids by selective removal during crystal-plastic deformation (Fig. 15a; Crawford and Hollister 1986; Hollister 1988, 1990; Johnson and Hollister 1995) or by preferential H_2O leakage by means of mobile dislocations (Bakker and Jansen 1990, 1991) and (2) preferential trapping of the immobile CO_2 from an immiscible brine-CO_2 system (Fig. 15b,c; Hollister 1990; Holness 1992; Gibert et al. 1998). If (1) operates, immiscibility is not necessarily involved. Case (2) relies on the fact that brine and CO_2-rich fluid have different wetting properties in a quartz matrix at metamorphic conditions (Watson and Brenan 1987). Brines form an interconnected network along quartz edges whereas CO_2 forms large isolated bubbles at the interstitials (Fig. 16; Gibert et al. 1998). Brine can escape by porous flow leaving CO_2 behind that eventually is trapped as nearly pure CO_2 inclusions. The different dihedral angles (Figs. 15 and 17) would allow for

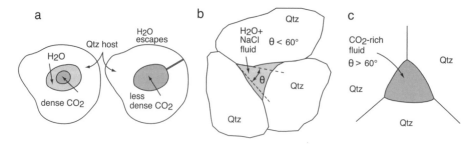

Figure 15. (a) Sketch illustrating the process of "wicking" (Crawford and Hollister 1986) where a two phase H_2O-CO_2 inclusion is transformed to a pure CO_2 inclusion of lower density by preferential water loss during crystal-plastic deformation of the quartz host. (b, c) Sketches illustrating different wetting angles (θ) for intergranular fluids (Watson and Brenan 1987). In a quartz matrix, wetting angles for water and brines are <60° allowing for a connected fluid network. For CO_2 wetting angles are >60 °C and CO_2-rich fluids form isolated pockets (modified after Hollister 1990).

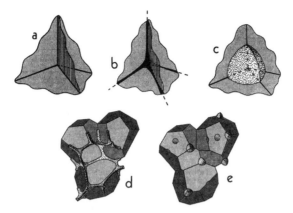

Figure 16. Sketch showing the equilibrium shape of a small volume of fluid at a grain corner. (a) in a polycrystalline aggregate of quartz with (b) $\theta < 60°$, (c) $\theta > 60°$. (d, e) same effect at the scale of a few grains. [Used by permission of E. Schweizerbart'sche Verlagsbuchhandlung, from Gibert et al. (1998) Eur J Mineral, Vol. 10, Fig. 6, p. 1117.]

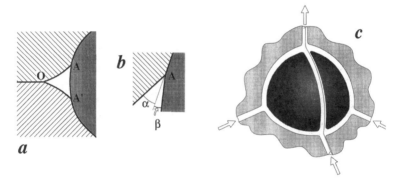

Figure 17. Sketch showing (a) textural relationships between two grains of quartz and immiscible brine (white) and CO_2 (gray). (b) detail of the three phase junction. (c) two fluid geometry at a grain corner (schematic). The CO_2-rich fluid forms a rounded pore surrounded by channels filled with brine, arrows indicate brine flow in a non-reactive system. [Used by permission of E. Schweizerbart'sche Verlagsbuchhandlung, from Gibert et al. (1998) Eur J Mineral, Vol. 10, Fig. 8, p. 1120.]

pervasive flow of brines out of the rock but not for CO_2-rich fluids. Gibert et al. (1998) argued that this mechanism accounts for the widespread CO_2 and often absent brine inclusions in granulites. Further support for selective entrapment is given by the fact that many granulites show strong mobilities of HFSE and REE (Stähle et al. 1987; Newton 1992; Harlov et al. 2006) which is difficult to ascribe to a CO_2-rich fluid only because of its poor ability to dissolve and transport such elements (Meen et al. 1989; Ayers and Watson 1991). Holness and Graham (1991) experimentally showed the opposite behavior for a carbonate matrix. In an H_2O-CO_2-NaCl-calcite system very saline brines and H_2O-CO_2 fluids with X_{CO_2} close to 0.5 have wetting angles smaller than 60°, whereas those of low salinity H_2O and CO_2-rich fluids are larger. The observed bimodal distribution of dihedral angles suggests that fluids trapped as inclusions in calcite grains in marbles may be predominantly H_2O-rich or very CO_2-rich, and of low salinity (Holness and Graham 1991). The highly contrasting wetting abilities of coexisting immiscible fluids enable selective entrapment of the more immobile fluid phase. The "other fluid" (Touret 1995) is squeezed out and not recorded in fluid inclusions. If so, lack of conjugate fluid inclusions does not disapprove immiscibility (Sisson et al. 1981). On the contrary, selective entrapment may be considered as a consequence of fluid immiscibility behavior provided that the mechanism acts effectively. However, the argument that migration of different fluids depends on equilibrium wetting angles presupposes that the rock has an equilibrium texture at the time of reaction. For carbonate-dominated rocks undergoing progressive continuous reaction this may be the case, but some reactions in carbonates appear to occur rapidly, giving rise to non-equilibrium textures and transient secondary porosity (Yardley and Lloyd 1989). It is unclear if wetting angles could play any significant role for fluid transport in such transient systems.

Aside from the different trapping mechanisms, rock's exhumation path can place existing fluid inclusion populations under over or under-pressure conditions that lead to changes of compositions and densities of fluid inclusions. In the regional metamorphic context, most inclusions are prone to collapse under internal pressure along regional retrograde metamorphism and various degrees of deformation and partial or complete reequilibration wipes out the information that originally may have been stored by fluid inclusions. Such processes may even pretend entrapment of conjugate fluids. Hurai and Horn (1992) described boundary layer-induced immiscibility in naturally reequilibrated quartz hosted H_2O-CO_2-NaCl inclusions. Overpressure of 60 to 150 MPa led to decrepitation of initially homogeneous inclusions. Very thin fractures and cracks formed which have not been connected with quartz surfaces. The fluid was internally recycled and structural and electrostatic forces in very thin fractures resulted in separation of CO_2-rich and saline H_2O-rich inclusions, the latter having been partly refilled by reprecipitated quartz. The compositional and density differences between aqueous and CO_2 inclusions could not be interpreted in terms of classical immiscibility under equilibrium conditions. Hurai and Horn (1992) argued that surface forces strongly attack fluids entering re-equilibration related cracks, and that these forces can cause compositional and density differences between crack and bulk fluids in polycomponent systems in general (Janàk et al. 1996, 1999; Hurai et al. 2000) Considering that selective trapping and various mechanisms of post-entrapment modifications may act in concert during the metamorphic cycle, fluid inclusionist's quandary in correctly interpreting immiscibility processes related to mineral reactions is obvious.

Fluid phase separation and two fluid flow in metamorphic rocks

Yardley and Bottrell (1988) considered the implications of two-phase flow for the reaction history in metamorphic rocks by applying simple semi-empirical transport principles developed for water-oil or water-gas mixtures. If metamorphic fluids behave analogous, it would mean that where one fluid is being produced by reaction it will flow out of the intergranular pore space or along microcracks while the other remains immobile, except insofar as it is soluble in the more abundant fluid. This is valid not only for the generation of vapor from liquid but also for the opposite. Because the "second fluid" is removed, any conventional fluid-rock buffer is out of work and extensive reaction at either side of the solvus may take place. Yardley and Bottrell (1988) argued that this effect not only accounts for production of salt-saturated fluids in marbles, but also for extensive consumption of carbonates due to formation of immiscible CO_2, and removal of graphite from metapelites through generation of immiscible CH_4. The same may hold for generation of immiscible N_2 due to breakdown of ammonium-bearing K-minerals. That reaction-produced immiscible fluids separate effectively and move independently is also supported by the fluid inclusion record in metamorphic rocks. Direct textural evidence of trapping of two immiscible fluid pairs is rare, and so is indication of heterogeneous trapping where two conjugate fluids were trapped in different amounts within the same inclusion (Yardley and Bottrell 1988), the latter being typical in porphyry systems (C. Heinrich 2007). It is probably this behavior that prevents reaction-induced fluid immiscibility to be easily recognized via fluid inclusion observations. This is apart from selective trapping at the grain scale and post-entrapment modifications of fluid inclusions during the metamorphic cycle.

Geophysical consequences of fluid unmixing

Shmulovich and Graham (2004) suggested that fluids in the critical state of unmixing, i.e., close to the consolute point may amplify seismic reflections that result from different fluid densities. Velocities of P- and S-waves strongly depend on the compressibility of the pore fluid when the pore fluid pressure is close to the minimum value of the differential stress (Zatsepin and Crampin 1997). Fluid systems at the critical state have in fact very high compressibilities that may easily account for the effect. Shmulovich and Graham (2004) proposed a scenario using transport-induced immiscibility at lithological boundaries (see above, Fig. 5). If a single phase fluid drives into the solvus, some supersaturation is required for nucleation of the "second" fluid. The degree and rate of supersaturation depends on various factors notably on the interface surface tension. The assumed diffusion process is continuous until equilibrium concentrations of Na and Ca are attained, whereas the limit of supersaturation depends on the fluid properties. During the process the "second" fluid will nucleate and separate and this will occur repeatedly. Acoustic energy can initiate the nucleation of the "second" phase and be absorbed during the process. As long as the fluid is interconnected nucleation and separation will continue, resulting in a very high impedance contrast when the fluid system is in the critical state. This mechanism may possibly explain numerous deep crustal reflections in many geological provinces. Critical unmixing phenomena at lithological boundaries may lead to dispersion of acoustic energy and to amplification of seismic reflection even when rock layers have comparable densities. Shmulovich and Graham (2004) argued that, in principle, similar effects could be achieved even in homogeneous rocks if fluid unmixing takes place after crossing the fluid solvus along a distinctive *P-T* gradient, or as a consequence of mineral reactions releasing the immiscible component, or if melts are involved.

Warner (2004) argued against the hypothesis of seismic reflectivity produced by free water in the lower crust. He showed that porosities of several percent or thin sheet-like pores with high crack density are required to produce broad seismic reflections. Rocks in textural equilibrium do not contain thin pores and require porosities in excess of 5 % to explain the reflection coefficients. He argued that the lower crust will compact, water will be expelled, and sharp boundaries in porosity and pore pressure will diffuse in geologically short times. He concluded that free water cannot explain seismic reflections and that there were good geophysical and petrological reasons for supposing that the lower crust in the stable continental interior typically would contain little or no free water. It is not clear, however, if Warner's (2004) arguments would also hold true for immiscible fluids.

FIELD STUDIES

Evidence of fluid immiscibility in metamorphic rocks relies on fluid inclusion investigation. Where coexisting immiscible fluids were not trapped during metamorphic processes, confirmation of immiscibility is difficult or impossible. Fluid inclusions may be modified after entrapment, particularly in regional metamorphic rocks. Fluids may be selectively trapped so that only one fluid is recorded in mineral hosts. In this case, one may still argue for immiscibility if composition and density of that fluid falls onto a calculated solvus boundary at distinct *P* and *T* derived from mineral equilibria. In fact, many speculations on fluid immiscibility in regional metamorphic rocks are based on this argument. Still, field study's best case is if coexisting fluids are trapped, their compositions and densities fit with experimentally or calculated solvus boundaries at *P* and *T*, and if these comply with the respective mineral equilibria at that conditions. Relatively few field studies meet this criterion. These are highlighted in this section.

The seminal studies: marbles from Campolungo, Lepontine Alps

Siliceous carbonate rocks at Campolungo have been metamorphosed and metasomatized to the lower amphibolite grade as indicated by development of tremolite (Trommsdorff et al. 1985; Skippen and Trommsdorff 1986; Trommsdorff and Skippen 1986). Tremolite and calcite are produced by Dol + Qtz + H_2O ↔ Tr + Cal + CO_2 and form an armoring texture separating quartz veins and dolomite nodules. Mainly two types of fluid inclusions are present in quartz hosts (Mercolli 1982; Walther 1983) and also in calcite and tremolite (Trommsdorff et al. 1985): a very CO_2-rich fluid and a (Na,K)Cl brine with up to 50 wt% dissolved salt. Halite and sylvite solid inclusions are present in quartz. Estimations of P,T conditions from fluid inclusion compositions and densities are about 200 MPa, 500 °C (Mercolli 1982) and 300 MPa, 500 °C (Walther 1983). This agrees with estimated T from the Mg content in calcite coexisting with other solids. There is clear evidence that the Tr + Cal - producing reaction proceeded in the two-fluid solvus up to salt saturation at this conditions [reaction (1) in Fig. 18] and that fluid inclusions recorded the fluid evolution. In principle, this holds irrespective of which EOS and solvus boundaries are chosen. In Figure 18, the reaction

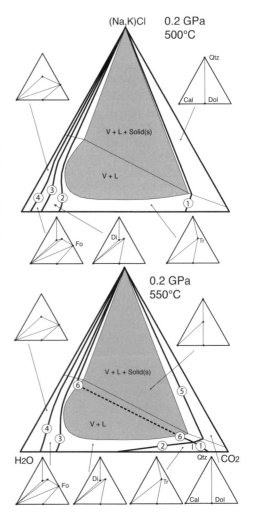

Figure 18. Relationships at 500 °C and 550 °C, 0.2 GPa between mineral equilibria in marbles and fluids in the system H_2O-CO_2-NaCl-KCl (mol%) at Campolungo, Switzerland. Phase relations for carbonate-silicate equilibria are shown on triangular diagrams that represent projections onto the CaO-MgO-SiO_2 base of the system CaO-MgO-SiO_2-CO_2-H_2O. Only assemblages including calcite have been considered. Mineral abbreviations after Kretz (1983). Reaction labels: (1) Do + Qz ↔ Tr + Cc; (2) Tr + Cc + Qz ↔ Di; (3) Tr + Do ↔ Fo + Cc; (4) Tr + Cc ↔ Di + Fo; (5) Do + Qz ↔ Di; (6) Tr + Cc ↔ Di + Do. Modified after Skippen and Trommsdorff (1986). Solvus boundaries are based on extrapolations of Gehrig's (1980) experimental data. Recent data suggest much lower salt concentrations in CO_2 rich fluids, a larger V+Hl field, and steeper tie-lines between salt-rich and CO_2-rich fluids. Thus, invariant point I at 550 °C/0.2 GPa does probably not exist because it falls within the fluid solvus.

attains salt saturation at 500 °C. Figure 14, calculated with the EOS of Duan et al. (1995; right column, 3rd and 4th diagram from below), suggests slightly lower T at 200 MPa for this process because the reaction enters the solvus at about 450 °C, and has already proceeded far into the L+Hl field at 500 °C. Diopside is absent in these rocks, indicating that no low salinity brine was present at these conditions. The regional metamorphism in the Central Alps near Campolungo attained P of 0.6 to 0.7 GPa (Frey et al. 1980). The much lower pressure indicates that metasomatism occurred either at less than lithostatic conditions or was a late event during uplift. An interesting feature is the development of large acicular tremolite in radiating growth at Campolungo. Trommsdorff and Skippen (1986) argued that such textures are typical for growth in presence of immiscible fluids. Similar textures of tremolite, diopside, and wollastonite are readily found in many metacarbonates and may be taken as a hint to fluid immiscibility processes during metasomatism.

Contact metamorphism

Contact aureoles differ from geothermal systems in that conductive heat transfer, steep temperature gradients and near lithostatic fluid pressures prevail. Most if not all contact aureoles are exposed to infiltration to various degrees by magma-derived fluids. Rock compartments in aureoles are therefore not isochemical. However, the nature of reactive fluid flow in more than 30 carbonate-bearing contact aureoles worldwide has been subject to substantial research summarized in reviews by Ferry and Gerdes (1998), Baumgartner and Valley (2001) and Ferry et al. (2002). Actual or possible immiscibility during reactive fluid flow is therefore reviewed in some detail here.

Patapedia thermal zone, Quebec. The zone is located at the intersection of fault zones and was intruded by a swarm of felsic porphyritic dykes (Williams-Jones 1982; Williams-Jones and Ferreira 1989). The formation of the thermal zone was the result of heat transfer from a porphyry-style hydrothermal system dominated by fluids of low to moderate salinity. Calcsilicate bodies cut by the dykes have outwardly telescoping prograde metamorphic isograds formed at a lithostatic pressure of 70-100 MPa. Fluid inclusions in the calcsilicates have been trapped during prograde metamorphism at 450°-500 °C and include (1) low to moderate salinity fluids, (2) a high salinity brine with up to 63 wt% $NaCl_{eq}$, and (3) a CO_2-rich vapor. Low to moderate salinity fluids were interpreted as primary magma-derived and CO_2 vapor as produced by decarbonation reactions in the calcsilicate bodies. Most metamorphic reactions took place at immiscibility and large volumes of CO_2 have been produced. Williams-Jones and Ferreira (1989) calculated the mineral reactions relevant for the development of distinctive isograds for a bulk fluid of 12 wt% NaCl (Fig. 19).

Layered carbonate-shale sequences in the Oslo Rift. This is a classical location where Goldschmidt (1911) developed the interpretation of phase assemblages by equilibrium thermodynamics. Contact-metamorphism by shallow felsic Permian plutons into layered carbonate-shale sequences have been revisited by Jamtveit et al. (1992a,b) and Jamtveit and Andersen (1993). Infiltration of magma-derived fluids partially mixed with meteoric fluids produced three stages of skarn formation mainly along bedding and lithological boundaries. Stages are (1) at 350-400 °C induced by moderate saline fluids with significant CO_2 contents, (2) at 300-400 °C where heterogeneous fluids existed namely high salinity brine, low salinity liquid, and CO_2-rich vapor, (3) at T down to 200 °C with very low salinity liquid present. Typical calc-silicate assemblages developed, which were calculated in P-T-X space assuming a homogeneous H_2O-CO_2 fluid. Detailed fluid inclusion investigation in the intrusion rocks, but not in the skarns, and stable isotope analysis of the skarn minerals revealed a complex degassing history of the pluton that affected the country rocks to various degrees. Even though fluid immiscibility caused by mineral reactions was not explicitly considered, Jamtveit and Andersen (1993) concede the case for decarbonation reactions within the H_2O-CO_2-salt solvus with local generation of CO_2-rich fluids. More

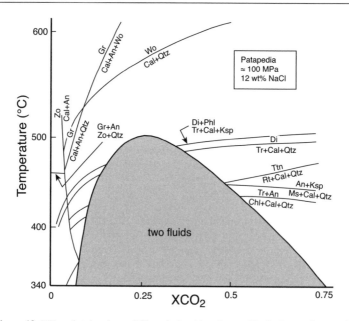

Figure 19. T-X_{CO_2} plot showing stability relationships observed in the Patapedia aureole, Quebec. Solvus is for 12 wt% NaCl as observed in fluid inclusions. Redrawn from Williams-Jones and Ferreira (1989).

important, they provide a fluid evolution model around the cooling pluton, which may be valid for shallow intrusions in general (Fig. 19; see also C. Heinrich 2007). At the early stage fluid released from the magma directly separates into $V_1 + L_1$. Vapor is more easily transported upwards and may subsequently exsolve brine at upper levels when P decreases (left column) resulting in $V_2 + L_2$. The right column of Figure 20 is a snapshot of the evolution after some time elapsed. A fracture zone penetrates the brittle part of the intrusion along which blow-out of aqueous fluids and melt from the roof of the intrusion occurred. The hydrosaline liquid L_1 becomes salt-saturated during pressure release and cooling, precipitates salt and forms a lesser saline liquid L_3 at the later metamorphic stage. Again, vapor exsolves fluid to $V_2 + L_2$, the latter being much less saline because of the flatter tie-lines at 400 °C, 50 MPa. The model is based on fluid inclusions trapped in the magmatic rocks and illustrates that vapors and brines may attain a large variety of compositions and densities, irrespective of reaction with the carbonate-bearing wallrocks. On the other hand, it demonstrates the large salinity range and density ranges of fluids that may be transported into country rocks around cooling plutons.

Notch peak aureole, Utah. A large amount of papers haven been published on fluid-rock interaction in this aureole during the last two decades (e.g., Labotka et al. 1988; Nabelek and Labotka 1993; Cui et al. 2001, 2003; Nabelek 2002; and references therein). The pluton is a composite of granite and quartz-monzonite. Calcareous argillite layers interbedded with siliceous limestones acted as aquifers for magmatic fluids, for example giving rise to typical wollastonite and diopside isogrades in the Big Horse Limestone Member of the Orr Formation. Modal amounts of phases and phase equilibria indicate that throughout the wollastonite zone the reacting system was fluid-dominated, whereas in the diopside zone, beyond the wollastonite isograd, it was rock-buffered. The wollastonite isograd located at 464 °C was interpreted as an "infiltration isograd." There is no indication of fluid immiscibility by fluid inclusion observations. Labotka et al. (1988) tentatively calculated the position of the tremolite, diopside, and wollastonite forming reactions (Fig. 21) relative to the H_2O-CO_2-

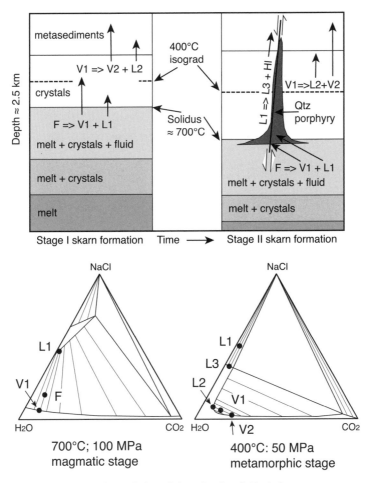

Figure 20. Geologic model for the evolution of skarn-forming fluids during contact-metamorphism of layered shale-carbonate sequences in the Oslo rift. The columns represent the situation during stage I and stage II skarn formation along with topologies of the ternary system H_2O-CO_2-NaCl at magmatic (*lower left*) and metamorphic (*lower right*) conditions. Modified from Jamtveit and Andersen (1993).

NaCl ternary using the Bowers and Helgeson (1983b) data at 400 °C and 500 °C. It results that Cal + Qtz $\leftrightarrow$ Wo + CO_2 and Cal + Qtz + Tr $\leftrightarrow$ Di + CO_2 + H_2O would have occurred in the one-fluid field, irrespective of salinity. If the infiltrating fluid was saline, however, Dol + Qtz + H_2O $\leftrightarrow$ Tr + Cal + CO_2 would have proceeded within the solvus at prograde conditions, and talc-forming reactions would have been suppressed. In the Weeks Formation of the same aureole, Nabelek and Labotka (1993) and Nabelek (2002) located the wollastonite isograd at slightly below 550 °C at 150 MPa. Again, the isograd represents a sharp front separating zones of no reaction from that of complete reaction. Fluid immiscibility was not considered.

Merida contact-metamorphic deposit, Spain. Granitic and dioritic melts intruded a sequence of limestones with interbedded metachert layers (Fernandez-Caliani et al. 1996). Contact metamorphism produced typical wollastonite, diopside, and tremolite zones with increasing distance from the contact. Extensive masses of very pure wollastonite developed between marble and metachert where individual wollastonite crystals are up to 5 cm in size.

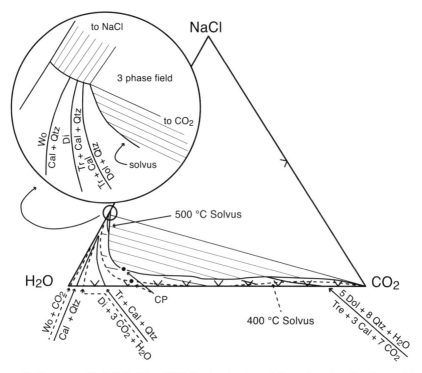

Figure 21. The system NaCl-H$_2$O-CO$_2$ at 200 MPa showing the stability regions for mineral assemblages in siliceous dolomites at 500 °C (solid lines) and 400 °C (dashed lines) adapted from Labotka et al. (1988) and Labotka (1991). Solvus boundaries and fugacity coefficients for calculation of mineral equilibria are from Bowers and Helgeson (1983b).

Conditions were estimated at 550°-700 °C at 60-70 MPa. Wollastonite contains two types of fluid inclusions, both of which are high temperature, high salinity brines. Total salinities are 52 wt% for type I, and up to 69 wt% for type II. The dominant salt is CaCl$_2$ (39 and 45 wt% respectively). X_{CO_2} in both brines is close to 0.05. Vapor-rich inclusions were not observed and fluid immiscibility not considered. The authors, however, suggest that wollastonite was formed by the reaction of aqueous fluids of moderate salinity with calcite at the marble-granite contact or along marble-quartzite bedding planes, and that the high salinity brine is an exhaust fluid or reacted solution formed during the wollastonite-producing reaction. Given the low *P*, large amounts of CO$_2$ produced, high salinity of the trapped liquids, and CO$_2$ contents of the saline liquids that are about at CO$_2$ saturation, it is highly probable that fluid immiscibility occurred, even though the CO$_2$-rich vapor was not trapped.

Bufa del Diente contact aureole, NE-Mexico. The effects of fluid immiscibility on mineral reactions and isotopic resetting at the Bufa del Diente (BD) aureole have been explored by a series of papers (Heinrich 1993; Heinrich and Gottschalk 1994, 1995; Heinrich et al. 1995, 2004; Romer and Heinrich 1998; Romer et al. 2005). Limestones with subhorizontal interbedded chert and metaargillite layers were intruded by a tertiary alkali syenite magma. Massive marbles exhibit mineralogical and oxygen isotope evidence for very limited fluid infiltration and developed typical tremolite, diopside and wollastonite zones at internally buffered fluid compositions. The wollastonite isograd is at about 580 ± 20 °C and 75 ± 25 MPa. Around metachert layers, however, up to 4 cm thick wollastonite rims formed by the reaction Cal + Qtz ↔ Wo + CO$_2$ reach far into the tremolite zone (Fig. 22a,b). These layers acted as

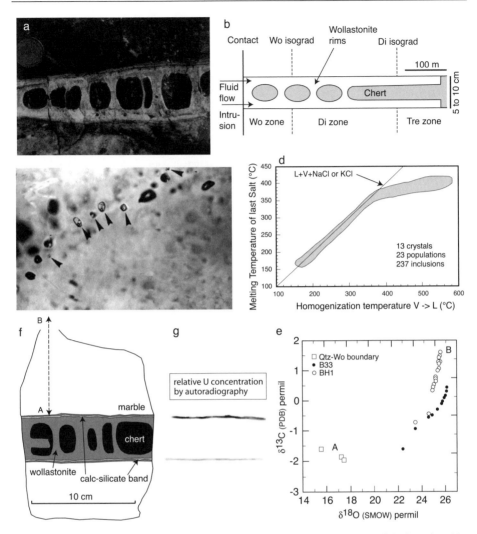

Figure 22. Features from the Bufa del Diente aureole, NE-Mexico, quarry north of the intrusion. (a) Wollastonite-rimmed metachert layers in contact-metamorphic marbles. (b) Sketch depicting infiltration of magmatic fluids along metachert layers. Infiltrating fluids drive the reaction Cal+Qtz ↔ Wo + CO$_2$ far into the diopside and tremolite zones as mapped in impermeable marbles. (c) High salinity brines (*arrows*) and CO$_2$-rich fluid inclusions trapped along a microcrack in quartz host. (d) T_h vs T_m (salt) of fluid inclusions in quartz hosts of a specimen from the wollastonite isograd, indicating continuous influx and trapping of magmatic fluids during the cooling history. (e) $\delta^{18}O - \delta^{13}C$ covariation trends of calcite in overlying marble along profile A-B in (f). (f) Sketch of a hand specimen from the wollastonite isograd. Note thin calc-silicate bands at the border between wollastonite and marble, which represent small metaargillite remnants of limestone after calcite removal. (g) Relative U concentration by autoradiography. The upper metaargillite band contains 40 times more U than the lower one. From Heinrich (1993), Heinrich et al. (1995) and Romer and Heinrich (1998).

channelways for magma-derived brines, allowing for substantial wollastonite formation down to about 470 °C in the infiltrated layers. Fluid inclusions in metachert quartz hosts from the wollastonite isograd show brines with salinities of up to 73 wt% (NaCl+KCl) that are nearly salt saturated at trapping conditions along with CO$_2$-rich vapor (Fig. 22c). In some of the metachert inclusions the vapor is not CO_2-rich but CH_4-rich and CO_2-free, formed by reaction with graphite initially present in the metachert. Compositions and densities of brine and vapors indicate trapping conditions of about 560 °C at 75 MPa and suggest simultaneous trapping on the solvus boundaries. Brines were continuously trapped throughout the cooling history of the aureole (Fig. 22d). The low-density CO_2-rich vapor was produced when decarbonation proceeded in the two-fluid solvus. $\delta^{18}O$ values of wollastonite of as low as 11 ‰ reflect the magmatic signature of brines. Gradients in $\delta^{18}O$ profiles across the wollastonite rim show very reduced values at the Qtz-Wo boundary indicating that the main influx occurred there. $\delta^{13}C$ and $\delta^{18}O$ of calcite are significantly altered in overlying marbles whereas at the footwall they are not. This suggests an advection process caused by fluid that moved upwards. $\delta^{18}O - \delta^{13}C$ covariation trends into the hanging wall marble across the alteration front shows that this fluid was CO_2-rich (Fig. 22e), thus representing the vapor phase produced at the solvus boundary. Its isotopic composition at the inlet of the chromatographic column may result from various proportions of decarbonation, exchange with magmatic-signatured fluid, and equilibration with isotopically light graphite occurring in chert. Mass balance considerations suggest that all of the produced CO_2 was lost as an H_2O-CO_2 fluid into hanging wall marble. The BD-rocks provide a convincing field example for the model of Yardley and Bottrell (1988) who argued that where two immiscible fluids coexist, the one produced by the reaction tends to move out of the rock and leave the other, i.e., the brine, behind.

Generation, segregation and transport of the CO_2-rich fluid can also be traced by trace element fractionation behavior. On both sides along wollastonite-marble boundaries, minute, <1 mm wide veneers of a fine-grained calcsilicate assemblage including phlogopite, diopside, vesuvianite, titanite, and many others developed. These formed by reaction with the accumulated, very small argillite content of the limestones during calcite consumption (Fig. 22f). Radioautography (Fig. 22g) shows U contents in the upper veneer that are 40× higher than in the lower one, and no significant traces of U are present in other portions of the rock. This implies that U was transported upwards from the aquifer through the developing wollastonite rims to the veneers where it was scavenged by minerals that easily accommodate U, for example vesuvianite or titanite. The upwards-moving immiscible H_2O-CO_2 fluid was responsible for that, implying that U transport by a CO_2-rich fluid is possible. U-Pb isotope data and distinctly lowered Th/U ratios demonstrate that U mainly stems from the intrusion and was transported with brine along the aquifer to the reaction site where at least part, if not all, of it had fractionated into the CO_2-rich fluid as reaction proceeded. Only the example of U is given here. Likewise, Romer and Heinrich (1998) and Romer et al. (2005) used the isotope systematics of Li, B, Sr, Nd, and Pb to explore their behavior during reactive fluid flow, fluid immiscibility and wall-rock alterations in the BD aureole.

The BD rocks provide yet another simple and illustrative example of reaction-induced fluid segregation at isothermal and isobaric conditions (Heinrich and Gottschalk 1994; Heinrich et al. 2004). At a deeper level of the intrusion, where P,T conditions close to the contact are estimated at about 680 °C and 120 MPa, subhorizontal metaargillite bands were infiltrated by magmatic fluids (Fig. 23a). Again, thick-bedded impure marbles were impervious to magmatic fluids and the fluid evolution in marble was internally buffered. Typical "wollastonite-zone" assemblages consisting of Cal + Di + Wo + Kfs + Tnt developed. By fluid infiltration the calcareous argillite bands are completely decarbonated resulting in Mel + Wo + Phl + Vs + Prv - bearing assemblages. The evolution of the reaction Di + Cal $\leftrightarrow$ Ak + CO_2 from marble towards the boundary is traced in Figure 23. The profile illustrates progressive replacement of previously formed prograde diopside (Fig. 23c) by melilite along a distance of 10 cm towards the aquifer (Fig. 23c-f). Large fluid inclusions, both high salinity brines as well as CO_2-bearing vapor-rich inclusions are abundant (Fig. 23b). Further into the marble, diopside is intact and fluid inclusions are absent. The reaction textures along with presence of inclusions with conjugate fluids indicate that the reaction was controlled by brine

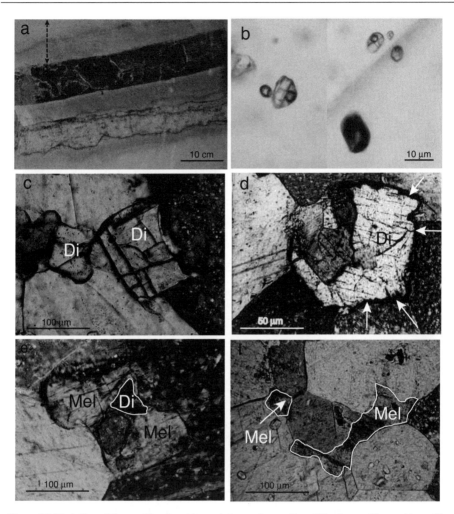

Figure 23. Evolution of the reaction diopside + calcite ↔ akermanite + CO$_2$ along a 12 cm wide profile (dashed line) in (a) from metaargillite layer into overlying marble from the Bufa del Diente aureole, quarry south-east of the intrusion. (a) Metargillite (above) and wollastonite (below) bands in marbles. Note large vesuvianite idioblasts within the metargillite band. (b) Fluid inclusions in marble calcite crystals 1 cm from the boundary showing high salinity brines with NaCl and KCl daughter crystals coexisting with H$_2$O-CO$_2$-rich inclusions. (c) Intact diopsides 12 cm from the boundary. Here and further into overlying marble, melilite is absent as are fluid inclusions. (d) Thin rims of melilite around diopside 9 cm from the boundary. (e) Diopside 7 cm from the boundary is almost completely replaced by melilite. (f) Interstitial melilite within calcite 4.5 cm from the boundary. There are no diopside remnants here. Modified after Heinrich and Gottschalk (1994) and Heinrich et al. (2004).

infiltration from aquifer into marble, fluid immiscibility and subsequent fluid segregation at constant *P,T*-conditions, and that infiltration occurred not earlier than at peak temperatures. The progress of the reaction depends on small variations in pressure and temperature (Fig. 24). Three isothermal isobaric sections are calculated for 660 °C, 680 °C, and 700 °C at 120 MPa, a_{Di} =1, a_{Ak} = 0.5, and the simplified H$_2$O-CO$_2$-NaCl fluid system. If CO$_2$ free brine infiltrated the marble at 660 °C, the reaction would have occurred within the one-fluid field, independent of salinity (Fig. 24a). The reaction progress would have been very small, particularly at initial

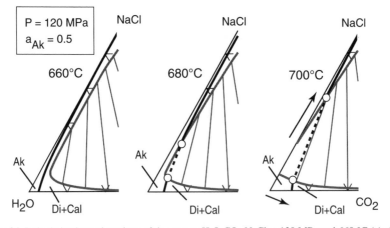

Figure 24. Isobaric isothermal sections of the system H_2O-CO_2-NaCl at 120 MPa and 660 °C (a), 680 °C (b) and 700 °C (c) with the position of the reaction $Cc + Di \leftrightarrow Ak + CO_2$ calculated for $a_{Ak} = 0.5$ (Heinrich and Gottschalk 1994). Components are given in mol%. Dashed line connects two coexisting fluids stable with $Cc + Di + Ak$. At 700 °C, circles indicate the intersections of the reaction curve with the two-fluid solvus. The brine has about 58 wt% NaCl, the vapor 6 wt% NaCl at $X_{CO_2} = 0.06$. Arrows indicate the compositional evolution of the two fluids along which melilite formation may have occurred (see text). Adapted from Heinrich et al. (2004).

brine salinities of >10 mol% NaCl, and the reaction would have stopped when the brine had dissolved about 1 mol% of the generated CO_2. Only minute amounts of melilite would have been produced. If the infiltrating brine were already CO_2 saturated at about 3 to 5 mol% CO_2 (corresponding to 10 to 60 mol% NaCl) no reaction would have taken place. At 680 °C, there is a very small field where the reaction curve intersects the solvus. Fluid immiscibility may occur for infiltrating brines having NaCl concentrations between 4 and 14 mol% (Fig. 24b). The assemblage $Cc + Di + Mel$ cannot coexist with brine having salinities of >14 mol%, however. Given maximum salinities of about 65 wt% detected in fluid inclusions, it is most likely that the reaction occurred at 700 °C, provided that pressure is correctly estimated at 120 MPa. At these conditions, any NaCl-H_2O fluid at the inlet is just single-phase, irrespective of salinity (Fig. 2a). As the reaction started, minute amounts of produced CO_2 drove the fluid into the solvus. Water-rich CO_2-bearing fluid exsolved and the salinity of the brine increased with proceeding reaction along the profile (*arrows in* Fig. 24c). The reaction terminated when the brine arrived at 30 mol% NaCl corresponding to 58 wt% NaCl, broadly in line with the observed salinity of fluid inclusions. Having attained their respective equilibrium compositions (circles in Fig. 24c) both fluids could move without inducing further reaction, as long as no pressure or temperature gradients existed. The brine did not reach salt saturation because the gap between the reaction curve and the solvus boundary increases towards the NaCl apex. A striking point is that large fluid inclusions are only abundant along the 10 cm wide reaction profile. Obviously, significant entrapment occurred only where the two conjugate fluids were actually produced by boiling. The same process is also recorded for the reactions $Cal + Kfs + Di + H_2O \leftrightarrow Phl + Wo + CO_2$ and $Cal + Tnt \leftrightarrow Prv + Wo + CO_2$ which initiate at about 10 cm and are completed at about 2 cm from the boundary. Calculated equilibria are in a similar range than that of melilite formation (Heinrich and Gottschalk 2004).

This simple small-scale example is reviewed in some detail because it illustrates a common process in the inner portions of contact metamorphic aureoles: decarbonation induced by brine infiltration, fluid immiscibility and fluid segregation. Sharp mineral zone boundaries occurring on a cm-scale and often interpreted as isograds probably result from this process. Progress of calc-silicate reactions driven by reactive fluids and resulting fluid flows are delicately dependent on each particular reaction in *P-T* space relative to the solvus of the H_2O-CO_2-salt system, which in turn drastically expands with increasing temperature and decreasing pressure. It is clear that examination of this process on an aureole-scale, where large temperature gradients exist and pressure gradients are temporarily built up, becomes a very difficult task.

Modeling reactive fluid flow in contact aureoles. It has been shown for many contact aureoles that fluid infiltration controls the mineralogical and stable isotope compositions of rocks during metamorphism. Often, significant shifts in oxygen isotope ratios indicate substantial flow of magma-derived fluids into the host rocks. The progress of fluid-producing mineral reactions has been interpreted in terms of time-integrated fluid fluxes by combining fluid advection/dispersion models with the spatial arrangement of mineral reactions and isotopic resetting. Ferry and Gerdes (1998), Baumgartner and Valley (2001) and Ferry et al. (2002) provided global reviews of contact metamorphic fluid flow in metacarbonate host rocks worldwide. Models based on treatment of continuum mechanics in porous media that combine homogeneous fluid flow with mineral reactions and oxygen isotopic exchange have attempted to show that the spatial disposal of such fronts as well as the geometry of the isotopic front itself can be used to determine, or to speculate on, important parameters including: composition and sources of fluids, migration pathways and direction of fluid flow, time-integrated fluxes, fluid transport mechanisms in terms of advection/diffusion/dispersion, validity and degree of local mineralogical and isotopic equilibrium, isotope exchange kinetics, and time-spans of fluid-rock interaction (e.g., Labotka et al. 1988; Jamtveit et al. 1992a,b; Nabelek et al. 1992; Bowman et al. 1994; Gerdes et al. 1995; Cartwright and Buick 1996; Dipple and Ferry 1996; Cartwright 1997; Cook et al. 1997; Ferry et al. 1998; and many others summarized by Baumgartner and Valley 2001 and Ferry et al. 2002).

However, all of the presented models generally assume that the magma-derived infiltrating fluid is pure water and that decarbonation/dehydration reactions occurred in presence of a single phase H_2O-CO_2 fluid. Based on the simple fact that alkaline and calc-alkaline magmas do not exsolve pure water but rather brines or brine + H_2O-NaCl vapor (Shinohara 1989; Shinohara et al. 1989; C. Heinrich 2007; Webster and Mandeville 2007), Heinrich et al. (2004) challenged the validity of such models, arguing that fluid segregation inevitably takes place at low pressures of contact-metamorphism within the inner, hotter parts of aureoles. Since existing models do not allow for the effect that fluid immiscibility has on flow patterns, fluid flow in contact-metamorphic carbonates is probably not adequately described. If a two-fluid flow regime is established in the inner part of the aureole fluids would move separately. With regard to oxygen isotopes, the emergence of a second fluid would affect their distribution behavior. For example, in the pure H_2O-NaCl system up to 600 °C and 50 MPa, $^{18}O/^{16}O_{(L-V)}$ equilibrium fractionation is a nearly linear function of salinity and is about 2‰ at all temperatures near salt saturation (Shmulovich et al. 1999). Simple boiling in the H_2O-NaCl system produces isotopically light vapor and heavier brine. Simple decarbonation such as Cal + Qtz ↔ Wo + CO_2 produces CO_2 enriched in ^{18}O. The process becomes substantially complex and one may expect that salt-poor vapors with varying H_2O/CO_2 ratios in equilibrium with brine + solids would develop a wide range in $^{18}O/^{16}O$ isotope ratios as brine changes salinity and oxygen isotopic composition along its pathway.

Fluid immiscibility across aureole-scale T-gradients. Specific decarbonation reactions may proceed in the inner aureole within and in outer portions outside the solvus (Heinrich et al. 2004) illustrated by the model reaction Cal + Qtz ↔ Wo + CO_2 at 150 MPa (Fig. 25). Below 560 °C the reaction occurs always with a one-phase water-rich fluid, irrespective of salinity. The reaction curve hits the solvus at 560 °C at moderate salinities. Towards higher temperatures, the reaction proceeds within the solvus. Above 640 °C it occurs again with a single phase fluid

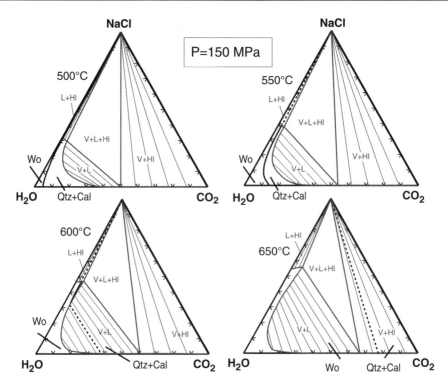

Figure 25. Isobaric isothermal sections of the system H_2O-CO_2-NaCl at 150 MPa and 500 °C, 550 °C, 600 °C, and 650 °C denoting the position of the reaction Cc + Qz $\leftrightarrow$ Wo + CO_2. Components are given in mol%. At 500 °C and 550 °C, Cc + Qz + Wo always coexist with one fluid, and at very high salinities, together with single fluid + halite. At 600 °C, Cc + Qz + Wo may coexist with brine (about 60 wt% NaCl) plus H_2O+CO_2-fluid (X_{CO_2} = 0.38), or only with brine having more than 60 wt% NaCl. At 650 °C, Cc + Qz + Wo coexist with an H_2O-CO_2-fluid of X_{CO_2} = 0.78 plus halite, and brine is not stable with this assemblage. Modified after Heinrich et al. (2004).

which now is CO_2-rich and NaCl-free, and in presence of halite instead of brine. Field studies have shown that the spatial distribution of wollastonite records the geometry of infiltration-driven fluid flow in rocks with quartz and calcite as main constituents (e.g., Heinrich 1993; Ferry et al. 2001; Nabelek 2002). Characteristically, sharp reaction fronts developed, separating the upstream part, where decarbonation has gone to completion, from the downstream part, where no reaction occurred. These reaction fronts are conventionally recognized as infiltration isograds. At 150 MPa, a distinct temperature window of 560 °C to 640 °C exists during which reaction induced fluid immiscibility is possible. The prograde reaction initially proceeds in presence of one fluid and changes to two-fluid behavior above 560 °C. As fluid flow advances further into the aureole, reaction induced fluid immiscibility and segregation may take place down to a point where the peak temperature is about 560 °C. Beyond that, only brine is stable with the assemblage. Similarly, infiltration of late retrograde fluids would shift the system back to fluid miscibility along the whole aureole during cooling.

Infiltration-driven wollastonite isograds have been mapped in calcite-bearing contact-metamorphic sandstones from the Mt. Morrison pendant, California (Ferry et al. 2001) and in calc-silicate layers and interfaces between calc-silicate and carbonate layers of the Weeks formation, Notch Peak aureole, Utah (e.g., Nabelek and Labotka 1993; Nabelek 2002; Cui et al. 2003). Intrusions are granitoids and pressure was estimated close to 150 MPa in both cases.

Peak temperature at the wollastonite isograd was 560 °C at Mt. Morrison based on mineral equilibria (Ferry et al. 2001), and somewhat below 550 °C at Notch Peak based on heat flow estimates (Nabelek and Labotka 1993; Nabelek 2002). Fluid immiscibility was not considered by these authors. However, if one follows the arguments of Heinrich et al. (2004; see also Shinohara 1989) that magma-derived fluid is always saline it is striking that the position of the wollastonite isograd in both aureoles coincides with that temperature at which reaction-induced fluid segregation caused by brine influx would halt. The same holds for the BD aureole ($P \sim 75$ MPa), where the Wo-isograd in infiltrated layers is at about 470 °C, coinciding with the solvus boundaries at that conditions. Heinrich et al. (2004) speculated that this coincidence is probably not accidental. If so, it implies that the position of infiltration-driven wollastonite isograds, i.e., the outer bounds of the main down-temperature flow, were perhaps determined by the P-T-X conditions up to which fluid immiscibility occurred. Wollastonite isograds have been interpreted as boundaries between high aqueous fluid-flux region on its higher-grade side and up-temperature, low fluid flux on its lower-grade side (e.g., Nabelek 2002; Cui et al. 2003) with different geometries of the flow system observed in the inner and outer portions of an aureole (Ferry et al. 2001; Cui et al. 2003). Fluid immiscibility in the inner portions would easily explain the different flow regimes.

By conventional interpretation preferential flow of homogeneous H_2O-CO_2 fluids within wollastonite zones is triggered by volume loss of the solids and increased pore pressure due to CO_2 production. Permeability is transiently increased (Buick and Cartwright 2002; Milsch et al. 2003) and buoyancy drives fluid flow upward and down-temperature near intrusion wall-rock contacts (e.g., Nabelek 2002). With advancing down-temperature reactive flow the produced CO_2 must be removed from the reaction sites in order to maintain the reaction. This is somewhat difficult to interpret if homogeneous H_2O-CO_2 fluid is assumed. Several possibilities have been suggested, such as diffusional transport in a stagnant fluid away from the reaction front (Labotka et al. 1988), diffusion of water against the back flow created by decarbonation and CO_2 production at the front (Balashov and Yardley 1998), strong local fluid expulsion at the front followed by dilution of CO_2-rich fluids by further infiltration of water, or by dilution with CO_2-poor sedimentary fluid, which may be present beyond the reaction front in the cooler portions of an aureole (Cui et al. 2003). In contrast, reaction-induced fluid immiscibility and segregation, "boiling," provides a simple and effective mechanism for removal of low-density, salt-poor H_2O-CO_2 vapor away from the front and also for strongly enhanced solid-fluid reactions (Yardley and Bottrell 1988). That the main fluid flow was confined to portions of the three aureoles where reaction-induced immiscibility may have occurred possibly results from this behavior. Finally, one may even speculate that the entire patterns of magmatic fluid flow into aureoles were ultimately controlled by reaction-induced fluid immiscibility and fluid segregation, apart from pre-metamorphic structural controls such as bedding, lithologic contacts, and faults.

Regional and subduction-related metamorphism

Under regional metamorphic conditions immiscibility requires a fluid system composed of H_2O + salt + non-polar components such as CO_2, N_2 and CH_4. Salinity might be inherited from sedimentary pore waters or evaporite material, and initial salt concentrations may be modified by a variety of processes (Yardley and Graham 2002). Saline fluids can also be introduced into rocks by crystallizing magmas in the lower crust. CO_2 is provided by decarbonation, or likewise, by magma degassing. Another process is oxidation of graphite due to H_2O + C $\leftrightarrow$ CH_4 + CO_2 which, in combination with mineral redox reactions, produces CO_2 or CH_4-rich fluid component (Connolly and Cesare 1993; Pattison 2006). Oxidation of the NH_4-bearing component in K-bearing minerals generates N_2, e.g., by $NH_4AlSi_3O_8 + O_2 \leftrightarrow Al_2SiO_5 + SiO_2 + H_2O + N_2$ or similar reactions involving NH_4-bearing amphiboles and micas (Andersen et al. 1993; Pöter et al. 2004). Since reactions act in concert, brines of varying salinity as well as complex water- CO_2-CH_4-N_2 fluids may evolve. Changes in *P-T* conditions, particularly during retrogression, may drive initially homogenous fluid to immiscibility, as does influx of compositionally different fluids during various stages of a rock's evolution.

Unlike in contact aureoles the source and evolution of immiscible fluids is often unclear or speculative in regional metamorphic rocks. Fluids frequently lose the compositional and isotopic memory of their sources when transported through rocks during various metamorphic cycles. Numerous fluid inclusion studies use composition and density of inclusion populations mainly to derive information on the retrograde *P-T* paths of the rock, considering fluid immiscibility processes rather as a by-product of the rock's exhumation history. Examples that connect compositions of immiscible fluids to specific mineral compositions or mineral reactions are rare. A number of field examples for which immiscibility was reported is reviewed here, the compilation is, however, not comprehensive.

Eclogites. Selverstone et al. (1992) calculated fluid activity ratios via mineral equilibria between mm to cm-wide layers in banded mafic eclogites from the Tauern Window, Austria, metamorphosed at 625 °C and 2 GPa. Strong variations in calculated a_{H_2O} existed between the layers whereas calculated a_{CO_2} was nearly constant. Banding consists of alternations of omphacite-, garnet-, clinozoisite-, zoisite-, dolomite-, and phengite-rich assemblages. Calculations show that the results are consistent with the existence of saline brines and carbonic fluids, the latter sometimes with substantial amounts of N_2, and also with the coexistence of immiscible pairs of both in some of the layers. Fluid inclusions in a variety of mineral hosts including quartz revealed the presence of different brines with salinities of up to 39 wt%. X_{CO_2} varies from 0 to 0.18 in the brines, to 0.2 in the N_2-CO_2 fluids, and up to 1.0 in a few pure CO_2 inclusions in quartz. Besides NaCl and KCl, $CaCl_2$ and $MgCl_2$ are also present. The combination of mineral equilibria, calculated fluid activity ratios, and fluid inclusion evidence strongly supports that immiscible fluids occurred in the banded eclogites, even if there is no sound information about the solvus boundaries of the H_2O-$CO_2(N_2)$-salt system at high pressure. The densities of the inclusions were somewhat modified due to recrystallization and decrepitation during uplift. The record of fluid variability implies that fluid transport was very limited to local flow along layers and that no large-scale mixing during devolatilization occurred at depths of 60-70 km. Selverstone et al. (1992) interpreted the lack of mixing of aqueous fluids in terms of the different wetting behavior of fluids of different compositions (Fig. 26). Nonwetting fluids (water-rich or carbonic) would be essentially immobile, whereas wetting fluids (saline brines) could migrate more easily along an interconnected network.

Similarly, Klemd et al. (1992) described small-scale local fluid gradients between immiscible fluids in interlayered eclogite facies rocks (630 °C, 1.7 to 2.4 MPa) from the Münchberg Complex, Germany. Metasediments intercalated with calc-silicate bands contain high density CO_2-N_2 fluid inclusions whereas in adjacent eclogites and non-calcareous metasediments moderately saline brines along with N_2-rich inclusions were trapped. Local fluid production and/or buffering was suggested with only short-distance fluid transport. Analogous features were described from eclogites of the Orlica-Sniezik dome, Poland (Klemd et al. 1995), even though fluid inclusions consisting of brine and CO_2±N_2±CH_4 inclusions were reset during uplift at amphibolite facies conditions and immiscibility at eclogite facies was therefore not explicitly proven. Andersen et al. (1993) reported CO_2-N_2 fluid inclusions occurring with conjugate brines (30 wt% NaCl) from eclogite of the Norwegian Caledonides representing immiscible fluids at peak metamorphic conditions. Trapping of conjugate aqueous brine and CO_2+N_2 fluids occurred in felsic extension veins formed at peak eclogite conditions of 700 °C, 1.6 to 1.9 GPa at the Bergen Arcs nappe complex, Norway (Andersen et al. 1991). Immiscible fluids of the brine-CO_2+N_2 system were also trapped at later stages of the cooling and uplift history of these rocks. Herms (2002) reported fluid immiscibility in eclogite facies metabasites and metapelites from the Usagaran Belt, Tanzania. Although the primary fluid inclusions, brines and CO_2±N_2-rich

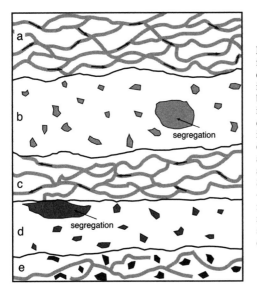

Figure 26. Schematic diagram of a model for the fluid distribution in cm-scale banded eclogites. Layers (a) and (c) have an interconnected fluid flow lining the grain boundaries (dihedral angle is < 60°); arrows indicate direction of fluid flow. Layers (b) and d) show isolated fluid pockets (dihedral angel is > 60°). Layer (e) contains fluid inclusions of immiscible fluids, one with a low (brine) and one with a high ($CO_2\pm N_2$-rich) dihedral angle, in the polycrystalline eclogite layer. Segregations consist of zoisite + clinozoisite + rutile formed either from ponding of nonwetting fluids or from hydrofracturing in response to devolatilization reactions. Vertical scale is about 2 cm. Modified after Selverstone et al. (1992).

vapors partially adjusted their densities during uplift at the eclogite-granulite facies boundary, their textures suggest trapping at the eclogite stage. Variable N_2 contents of the CO_2-rich inclusions on a mm-scale indicate local fluid inhomogeneity and point to an internal fluid source. Fu et al. (2003a,b) described fluid inclusions in a variety of eclogites from the Dabie-Sulu terranes, China. High salinity brines were trapped along with N_2, CO_2, or CH_4-rich vapors. Low salinity, water-rich fluids were interpreted as remnants of meteoric water which had modified compositions during pre- and syn-peak high *P* metamorphism. CH_4 was interpreted to have formed early under the influence of CO_2-rich aqueous fluids, which were reduced during serpentinization of mantle-derived peridotites prior to or during subduction. The large variety of fluid inclusions were believed to represent the large diversity of initial fluid compositions in the precursors implying very limited fluid-rock interaction and fluid exchange during syn- and post-peak high pressure metamorphism. Fluid inclusion studies of drill cores and UHP rocks form the Chinese continental drilling project (Dabie-Sulu terranes) revealed immiscible brines and CO_2, N_2, and/or CH_4-rich fluids from peak metamorphism through various stages of retrogression (Zhang et al. 2005; Xiao et al. 2006). Again, a common feature is the diversity in fluid inclusion populations and compositions from different vertical depths suggesting a more or less closed fluid system without large-scale migration during UHP metamorphism. Zhang et al. (2005) and Xiao et al. (2006) however, interpreted the presence of inclusions with low and medium salinity as due to influx from an external source during retrogression. Janák et al. (1996) reported fluid immiscibility between polyphase brines and gas-rich $H_2O+N_2+CH_4$ fluids during the retrograde, amphibolite facies stage of garnet-clinopyroxenite amphibolites from the Tatra Mountains, Carpathians, where N_2-rich inclusions represent relics of the primary, eclogite facies-related N_2, whereas brines originated by leaching and re-entrapment of saline inclusions from tonalitic-trondhjemitic layers of the amphibolite enclosing the eclogite relics. Immiscible H_2O-rich and CH_4-rich fluids were trapped in jadeite from the Myanmar jadeitite at about 300 to 400 °C, and stable isotope data indicate an abiogenic mechanism for CH_4 formation (Shi et al. 2005).

It is possible that different brine and vapor inclusions in eclogites, due to very limited fluid transport, may represent remnants of precursor fluids and were thus not actually trapped at distinct solvus boundaries. There is, however, no doubt that fluid immiscibility in eclogite facies rocks may occur. Fluids of moderate to high salinity from recycled seawater or inherited from pore space may be present during various stages of prograde high-*P* metamorphism (e.g., Philippot and Selverstone 1991; Scambelluri et al. 1997, 1998; Scambelluri and Philippot 2001; Touret 2001; Yardley and Graham 2002; Touret and Frezzotti 2003). Local reactions of minerals with brines may produce CO_2, CH_4, N_2 or a combination of them, driving the fluid into immiscibility. Because solvus boundaries and activities of fluid components at high pressure are not available, a link to specific mineral equilibria is not yet possible.

Granulites. Van den Berg and Huizenga (2001) reported coexisting high salinity brine and CO_2 (±CH_4)-rich inclusions trapped at peak metamorphic conditions (800-850 °C, 750-850 MPa) from granulites of the Limpopo Belt, South Africa. The brine originated from the sedimentary stage of the rock and CO_2 evolved from devolatilization reactions in the greenstone belt lithologies. The peak metamorphic mineral assemblage in mafic granulites was in equilibrium with fluids of low a_{H_2O} (0.2-0.3) in accordance with fluid compositions. During uplift, influx of low salinity, low CO_2 water-rich fluids at 600 °C, 500-600 MPa produced hydration reactions. Salinity and CO_2-contents concomitantly increased driving the retrograde fluid system again to immiscibility. Similarly, high salinity fluid inclusions coexisting with CO_2, CO_2+N_2, or CO_2-CH_4-N_2-rich inclusions at or near-peak granulite facies metamorphism were described from high-grade metapelites and pyroxene granulites from Tanzania (Herms and Schenk 1998) again with development of various immiscibility patterns during retrogression. Granulite-facies metapelites from Calabria, Southern Italy did not preserve peak metamorphic fluid inclusions (Herms and Schenk 1992). Early secondary fluid inclusions of CO_2-rich composition, partly with minor N_2, were trapped or re-equilibrated together with brines during isothermal uplift. In many cases, fluid modeling and fluid inclusion observations have shown to give widely different results leading to the idea that CO_2 and N_2- rich inclusions and brines are late and not related to peak metamorphic processes. Case studies, for example from the Adirondack granulites and from the Norwegian Caledonides, were reported by Lamb (1988), Lamb et al. (1991), Elvevold and Andersen (1993), Lamb and Morrison (1997).

Migmatites. Selective H_2O removal into the melt and concentration of the CO_2 and N_2 species is the main mechanism invoked to explain the common occurrence of CO_2-rich inclusions in migmatites (e.g., Touret 1971, 2001; Hollister 1988; Fitzsimmons and Mattey 1994; Frezzotti et al. 2004). *In-situ* crystallization of leucosomes during cooling releases H_2O of variable salinity so that brine and CO_2-rich fluid inclusions may occur in close spatial arrangement but probably do not represent true coexisting fluids. Immiscibility between CO_2(±CH_4 ±N_2)- and H_2O-dominated fluids occurred during dehydration melting and devolatilization of high-grade metapelites of the Tatra Mountains (Janák et al. 1999). Immiscible nitrogen-bearing fluids, $CaCl_2$-dominated brines and carbonate liquids in migmatites from the Tatra Mountains, Western Carpathians were interpreted to have been inherited from a high pressure stage pre-dating the migmatization (Hurai et al. 2000).

Amphibolite facies rocks. In a classical study, Sisson and Hollister (1990) reported CO_2 (±N_2±CH_4)-rich fluid inclusions from regional metamorphosed graphite schists and pelitic and carbonate rocks, south-central Maine. CO_2 was produced by decarbonation reactions, which were triggered by influx of water-rich fluids. Even though high salinity fluid inclusions were lacking and immiscibility therefore not proven the authors speculated that it must have occurred. In graphite-bearing schists, however, H_2O-CH_4 immiscibility was documented. Huizenga and Touret (1999) described fluid immiscibility in syntectonic quartz veins from shear zones in the Harare-Shamva-Bindura greenstone belt, Zimbabwe. H_2O-CO_2±Hl and graphite-bearing H_2O-CO_2-CH_4±Hl inclusions were trapped on the solvus at 600-650 °C and 400-600 MPa. Carbonic and aqueous-rich inclusions occurred within the same trails. Immiscibility was interpreted as resulting from a homogeneous, carbon-saturated, water-rich ($X_{H_2O} \geq 0.9$) fluid during cooling. *P-T* conditions and fluid compositions were used to calculate f_{O_2} of the fluid at 0.7-1.0 $\log_{10}$ units below the QFM buffer. Immiscibility of H_2O-CH_4-N_2-NaCl fluids were observed in graphitic lithologies of metasedimentary units from Chavez, Portugal (Guedes et al. 2002). Conjugate fluids were trapped near peak metamorphism at ~500 °C and 300-350 MPa. Fluids were interpreted as resulting from dehydration reactions plus interaction with the C-rich host rocks that provided CH_4 and N_2. Kolb et al. (2000) reported fluid immiscibility between brines and CO_2 ($\pm CH_4 \pm N_2$)-rich fluids in mid-crustal mylonites form southern Zimbabwe. Fluids were trapped at 600 to 700 °C. Immiscibility was interpreted to result from catastrophic pressure drops during seismic slip events and the fluids that infiltrated the thrust zone were generated by devolatilization of underlying granite-greenstone terrains of the Zimbabwe craton.

Very low and low grade metamorphic rocks. Immiscible CO_2-rich (+ trapped graphite) and brine-rich (+ trapped carbonate) fluid inclusions were described in pillow breccia fragments from the 3.8-3.7 Ga Isua Greenstone belt (Greenland) (Heijlen et al. 2006). Immiscible fluids were trapped at peak conditions of ~460 °C and 400 MPa and graphite and carbonate were precipitated cogenetically from physically separated endmember fluids. Recrystallization obliterated all fluid inclusion evidence of early processes and the fluids do not represent remnants of an early seafloor hydrothermal system. Further retrograde immiscibility was caused by influx of highly reducing fluids at 200-300 °C and 50-200 MPa. Kisch and van den Kerkhof (1991) reported immiscible water-rich fluids and CH_4 ($\pm CO_2$) vapor from very low-grade veins in the Valley and Ridge province and the anthracite fields of the Pennsylvania Appalachians. Conjugate fluids were trapped at 195-245 °C, 80-130 MPa, and at 180-230 °C, 220-250 MPa. Fluids were internally derived and the *P-T* conditions coincide with conodont alteration indices, illite crystallinities, the anthracite rank, slate cleavage development, and the stability filed of pyrophyllite present in the rocks. Superdense CO_2 ($\pm CH_4 \pm N_2$) inclusions that exsolved from an aqueous bulk fluid at 180-240 °C and 160-350 MPa, possibly up to 450 MPa, were reported from quartz-stibnite veins hosted in low grade Variscan basement of the Western Carpathians (Urban et al. 2006). *P-T* parameters are believed to represent those during vein opening and decompression of internally derived fluids at an environment with a low thermal gradient of about 12 °C/km.

Examples for fluid immiscibility in late veins formed at low and very low retrograde conditions during uplift and exhumation of higher grade metamorphic rocks of different lithologies are numerous. Immiscible fluids are mainly represented by fluid inclusions of low to medium saline aqueous fluids along with CO_2 and/or CH_4-rich vapors. Fluids may originate from different sources: dehydration and decarbonation from underlying metasediments, reaction with graphite or by influx of meteoric water during the last stages of uplift and immiscibility is generally believed to result from decompression. Case studies include rocks from the Southern Alps, New Zealand (Craw 1988, 1997; Craw and Norris 1993), the Central Alps, Switzerland (Mullis et al. 1994), the Annapurna Himal, Nepal (Craw 1990), North Victoria Land, Antarctica (Craw and Cook 1995), the Eastern and Northwestern European Alps, Austria, Italy (Craw et. al. 1993), the Austrian Modanubian Zone (Jawecki 1996), the Northwest Italian Alps (Diamond 1990), the Mount Isa fold Belt; NW Queensland (Xu and Pollard 1999), the Eastern Himalayan syntaxis (Craw et al. 2005) and many more. A summary of late fluid immiscibility processes in Archean lode-gold deposits worldwide is given by Hagemann and Brown (1996). Fluid immiscibility is also produced when basinal fluids of varying salinity infiltrate cooled high-grade basement rocks where deep wall-rock hydration reactions at 200 to 300 °C may lead to salt saturation of brine and immiscible H_2O-CO_2 vapor (e.g., Gleeson and Yardley 2002; Gleeson et al. 2003).

CONCLUDING REMARKS

There is no doubt that fluid immiscibility plays a significant petrologic role in metamorphic processes. The evidence obtained from the fluid inclusion record in rocks is overwhelming. However, it seems as if the control of fluid immiscibility on mineral compositions and mineral reactions (and vice versa) is often not sufficiently addressed, particularly in regional and subduction-related metamorphic rocks. For example, countless petrological studies and thermodynamic calculations of mineral equilibria in medium and high grade rocks revealed that assemblages with OH-bearing solids formed under a fluid regime of reduced water activities, the physicochemical meaning of which often remains unclear. Reduced water activities may come about through high salinity brines or through immiscible fluids or a combination of both where, generally speaking, activities of phase components at either side of the solvus may be lowered. One may speculate that the latter case is responsible for many of the calculated low a_{H_2O}. A sound analysis of mineral and mineral-fluid equilibria would require equations of state for complex immiscible fluids with information about the activities of phase components along the solvus boundaries and their P-T dependence. These are, aside from some restricted information for simple H_2O-CO_2-NaCl mixtures, not available for most metamorphic conditions, and are urgently needed.

Another important point is the significance of geochemical tracers such as trace elements and stable isotope systems during fluid-rock interaction when an immiscible fluid system operates. Unlike in hydrothermal environments, trace element and isotope fractionation factors between complex immiscible fluids are largely unknown at most metamorphic conditions. These are, however, a prerequisite for understanding mass distribution and transport in metamorphic rocks because immiscible fluids travel independently and imprint different geochemical signatures on rocks along different pathways. If available, modeling of two-fluid flow systems in the critical state coupled with mineral reactions and trace element or isotopic exchange based on field studies could be tackled, possibly by applying and expanding models currently used in groundwater hydrology.

The geophysical and petrophysical consequences of reaction-induced fluid immiscibility in metamorphic rocks are largely unexplored. Aside from some speculations on generating seismic reflections, the effects of immiscible fluids on parameters such as electrical conductivity of crustal rocks, transient parameters such as porosity and permeability, and on processes such as dehydration/devolatilization embrittlement and dilatancy hardening during rock deformation, just to mention a few, are simply unknown. It may well be that a mindful approach that combines fluid immiscibility processes with petrophysical properties of rocks would provide new insights into the dynamics of the Earth's crust.

ACKNOWLEDGMENTS

This review benefited from critical remarks by B.W.D. Yardley (University of Leeds) and A. Liebscher (Technische Universität Berlin). Thanks go also to M. Gottschalk (GeoForschungsZentrum Potsdam) for fruitful discussions on immiscible fluids, and to C. Heinrich, A. Liebscher and J. Rosso for their editorial efforts.

REFERENCES

Andersen T, Austrheim A, Burke EA, Elvevold S (1993) N₂ and CO₂ in deep crustal fluids: evidence from the Caledonides of Norway. Chem Geol 108:113-132

Andersen T, Austrheim H, Burke EAJ (1991) Mineral-fluid-melt interactions in high-pressure shear zones in the Bergen Arcs nappe complex, Caledonides of W. Norway: Implications for the fluid regime in Caledonian eclogite-facies metamorphism. Lithos 27:187-204

Ayers JC, Watson EB (1991) Solubility of apatite, monazite, zircon and rutile in supercritical aqueous fluids with implications for subduction zone geochemistry. Phil Trans R Soc Lond A 335:365-375

Bakker RJ, Jansen BH (1990) Preferential water leakage from fluid inclusions by means of mobile dislocations. Nature 345:58-60

Bakker RJ, Jansen BH (1991) Experimental post-entrapment water loss from synthetic CO₂-H₂O inclusions in natural quartz. Geochim Cosmochim Acta 55:2215-2230

Balashov VN, Yardley BWD (1998) Modeling metamorphic fluid flow with reaction-compaction-permeability feedbacks. Am J Sci 298:441-470

Baumgartner L, Valley JW (2001) Stable isotope transport and contact-metamorphic fluid flow. Rev Mineral Geochem 43:415-467

Bodnar RJ, Burnham CW, Sterner SM (1985) Synthetic fluid inclusions in natural quartz: III. Determination of phase equilibrium properties in the system NaCl-H_2O to 1000 °C and 1500 bars. Geochim Cosmochim Acta 49:1861-1873

Bowers TS, Helgeson HC (1983a) Calculation of the thermodynamic and geochemical consequences of nonideal mixing in the system H_2O-CO_2-NaCl on phase relations in geological systems: Equation of state for H_2O-CO_2-NaCl fluids at high pressures and temperatures. Geochim Cosmochim Acta 47:1247-1275

Bowers TS, Helgeson HC (1983b) Calculation of the thermodynamic and geochemical consequences of nonideal mixing in the system H_2O-CO_2-NaCl on phase relations in geological systems: Metamorphic equilibria at high pressures and temperatures. Am Mineral 68:1059-1075

Bowman JR, Willett SD, Cook SJ (1994) Oxygen isotopic transport and exchange during fluid flow: one-dimensional models and applications. Am J Sci 294:1-55

Buick IS, Cartwright I (2002) Fractured-controlled fluid flow and metasomatism in the contact aureole of the Marulan Batholith (New South Wales, Australia). Contrib Mineral Petrol 143:733-749

Cartwright I (1997) Permeability generation and resetting of tracers during metamorphic fluid flow: implications for advection-dispersion models. Contrib Mineral Petrol 129:198-208

Cartwright I, Buick IS (1996) Determining the direction of contact metamorphic fluid flow: an assessment of mineralogical and stable isotope criteria. J Metamorph Geol 14:289-305

Connolly JAD, Cesare B (1993) C-O-H-S fluid compositions and oxygen fugacity in graphitic metapelites. J Metamorph Geol 11:379-388

Cook SJ, Bowman JR, Forster CB (1997) Contact metamorphism surrounding the Alta Stock: finite element model simulation of heat- and $^{18}O/^{16}O$ mass-transport during prograde metamorphism. Am J Sci 297:1-55

Craw D (1988) Shallow-level metamorphic fluids in a high uplift rate metamorphic belt: Alpine Schist, New Zealand. J Metamorph Geol 6:1-16

Craw D (1990) Fluid evolution during uplift of the Annapurna Himal, Central Nepal. Lithos 24:137-150

Craw D (1997) Fluid inclusion evidence for geothermal structure beneath the Southern Alps, New Zealand. New Zealand J Geol Geophys 40:43-52

Craw D, Cook YA (1995) Retrogressive fluids and vein formation during uplift of the Priestley metamorphic complex, North Victoria Land, Antarctica. Antarctic Sci 7:283-291

Craw D, Koons PO, Zeitler PK, Kidd WS (2005) Fluid evolution and thermal structure in the rapidly exhuming gneiss complex of the Namche Barwa-Gyala Peri, eastern Himalayan syntaxis. J Metamorph Geol 23:829-845

Craw D, Norris J (1993) Grain boundary migration of water and carbon dioxide during uplift of garnet-zone Alpine schist, New Zealand. J Metamorph Geol 11:371-378

Craw D, Teagle DAH, Belocky R (1993) Fluid immiscibility in late Alpine gold-bearing veins, Eastern and Northwestern European Alps. Mineral Deposita 28:28-36

Crawford ML, Hollister LS (1986) Metamorphic fluids: the evidence from fluid inclusions. *In:* Fluid-rock interactions during metamorphism. Walther JV, Wood BJ (eds) Springer, p 1-35

Crawford ML, Kraus DW, Hollister LS (1979) Petrologic and fluid inclusion study of calc-silicate rocks, Prince Rupert, British Columbia. Am J Sci 279:1135-1159

Cui X, Nabelek PI, Liu M (2001) Heat and fluid flow in contact metamorphic aureoles with layered and transient permeability, with application to the Notch Peak aureole, Utah. J Geophys Res 106-B4:6477-6491

Cui X, Nabelek PI, Liu M (2003) Reactive flow of mixed CO_2-H_2O fluid and progress of calc-silicate reactions in contact metamorphic aureoles: insight from two-dimensional numerical modelling. J Metamorph Geol 21:663-684

Diamond LW (1990) Fluid inclusion evidence for P-V-T-X evolution of hydrothermal solutions in late-alpine gold-quartz veins at Brusson, Val d'Ayas, Northwest Italian Alps. Am J Sci 290:912-958

Diamond LW (2003) Introduction to gas-bearing, aqueous fluid inclusions. *In:* Fluid inclusions: Analysis and Interpretation. Short Course Series Volume 32. Samson I, Anderson A, Marshall D (eds) Mineralogical Association of Canada, p 101-158

Dipple GM, Ferry JM (1996) The effect of thermal history on the development of mineral assemblages during infiltration driven contact metamorphism. Contrib Mineral Petrol 124:334-345

Duan Z, Møller N, Weare JH (1995) Equations of state for the NaCl-H_2O-CO_2 system: Prediction of phase equilibria and volumetric properties. Geochim Cosmochim Acta 59:2869-2882

Duan Z, Møller N, Weare JH (2003) Equation of state for the NaCl-H_2O-CH_4 system and the NaCl-H_2O-CO_2-CH_4 system: phase equilibria and volumetric properties above 573 K. Geochim Cosmochim Acta 67:671-680

Elvevold S, Andersen T (1993) Fluid evolution during metamorphism at increasing pressure: carbonic- and nitrogen-bearing fluid inclusions in granulites from Oksfjord, north Norwegian Caledonides. Contrib Mineral Petrol 114:236-246

Fernandez-Caliani JC, Casquet C, Galán E (1996) Complex multiphase fluid inclusions in wollastonite from the Mérida contact-metamorphic deposit, Spain: evidence for rock/HCl-rich fluid interaction. Eur J Mineral 8:1015-1026

Ferry JM, Gerdes ML (1998) Chemically reactive fluid flow during metamorphism. Annu Rev Earth Planet Sci 26:255-287

Ferry JM, Sorensen SS, Rumble III D (1998) Structurally controlled fluid flow during contact metamorphism in the Ritter range pendant, California, USA. Contrib Mineral Petrol 130:358-378

Ferry JM, Wing BA, Penniston-Dorland SC, Rumble III D (2002) The direction of fluid flow during contact metamorphism of siliceous carbonate rocks: new data for the Monzoni and Predazzo aureoles, northern Italy, and a global review. Contrib Mineral Petrol 142:679-699

Ferry JM, Wing BA, Rumble III D (2001) Formation of wollastonite by chemically reactive flow during contact metamorphism, Mt. Morrison Pendant, Sierra Nevada, California, USA. J Petrol 42:1705-1728

Fitzsimmons ICW, Mattey DP (1994) Carbon isotope constraints on volatile mixing and melt transport in granulite-facies migmatites. Earth Planet Sci Lett 134:319-328

Frantz JD, Popp RK, Hoering TC (1992) The compositional limits of fluid immiscibility in the system H_2O-CO_2-NaCl as determined with the use of synthetic fluid inclusions in conjunction with mass spectrometry. Chem Geol 98:237-255

Franz L, Harlov DE (1998) High-grade K-feldspar veining in granulites from the Ivrea-Verbano zone, northern Italy: fluid flow in the lower crust and implications for granulite genesis. J Geol 106:455-472

Frey M, Bucher K, Frank E, Mullis J (1980) Alpine metamorphism along the Geotraverse Basel-Chiasso – a review. Eclogae Geol Helv 73:527-546

Frezzotti ML, Cesare B, Scambelluri M (2004) Fluids at extreme P-T metamorphic conditions: the message from high-grade rocks. Per Mineral 73:209-219

Fu B, Touret JLR, Zheng YF (2003a) Remnants of premetamorphic fluid and oxygen isotopic signatures in eclogites and garnet clinopyroxenite from the Dabie-Sulu terranes, eastern China. J Metamorph Geol 21: 561-578

Fu B, Touret JLR, Zheng YF, Jahn BM (2003b) fluid inclusions in granulites, granulitized eclogites and garnet clinopyroxenites from the Dabie-Sulu terranes, Eastern China. Lithos 70:293-319

Gehrig M (1980) Phasengleichgewichte und PVT-Daten ternärer Mischungen aus Wasser, Kohlendioxid und Natriumchlorid bis 3 Kbar und 550 °C. Hochschulverlag Freiburg, Germany

Gehrig M, Lentz H, Franck, EU (1986) Concentrated aqueous sodium chloride solutions from 200 to 600 °C and to 3000 bar. Ber Bunsenges Phys Chem 87:597-600

Gerdes ML, Baumgartner LP, Person M, Rumble III D (1995) One and two dimensional models of fluid flow and stable isotope exchange at an outcrop in the Adamello contact aureole, Southern Alps, Italy. Am Mineral 80:1004-1019

Gibert F, Guillaume D, Laporte D (1998) Importance of fluid immiscibility in the H_2O-CO_2-NaCl –system and selective CO_2 entrapment in granulites: Experimental phase diagram at 5-7 kbar, 900 degrees C and wetting textures. Eur J Mineral 10:1109-1123

Gleeson SA, Yardley BWD (2002) Veins and mineralization in basement rocks: the role of penetration of sedimantary fluids. *In:* Water-rock interaction. Stober I, Bucher K (eds) Kluwer Academic Publishers p 189-205

Gleeson SA, Yardley BWD, Munz IA, Boyce AJ (2003) Infiltration of basinal fluids into high-grade basement, South Norway: sources and behaviour of waters and brines. Geofluids 3:33-48

Glushko VP, Gurvich LV, Bergman GA, Veitz IV, Medvedev VA, Khachkuruzov GA, Yungman VS (1976-1982) Thermodynamic Properties of Individual Substances. Glushko VP (ed) Nauka V 1-4

Goldschmidt VM (1911) Die Kontaktmetamorphose im Kristianagebiet. Skrifter fra Norges Vitenskaps Akademi Oslo, Matematisk Naturvitenskapelig Klasse 11

Gómez-Pugnaire MT, Franz G, Sánchez-Vizcaino VL (1994) Retrograde formation of NaCl-scapolite in high pressure metaevaporites from the Cordilleras Béticas (Spain). Contrib Mineral Petrol 116:448-461

Gottschalk M (1997) Internally consistent thermodynamic data set for rock forming minerals in the system SiO_2-TiO_2-Al_2O_3-Fe_2O_3-CaO-MgO-FeO-K_2O-Na_2O-H_2O-CO_2: an alternative approach. Eur J Mineral 9: 175-223

Gottschalk M (2007) Equations of state for complex fluids. Rev Mineral Geochem 65:49-97

Greenwood HJ (1975) Buffering of pore fluids by metamorphic reactions. Am J Sci 275:573-593

Guedes A, Noronha F, Boiron MC, Banks DA (2002) Evolution of fluids associated with metasedimentary sequences from Chaves (North Portugal). Chem Geol 190:273-289

Hagemann SG, Brown PE (1996) Geobarometry in Archean lode-gold deposits. Eur J Mineral 8:937-960

Hansen EC, Newton RC, Janardhan AS, Lindenberg S (1995) Differentiation of Late Archean crust in the Eastern Dharwar Craton, Krishnagiri-Salem area, South India. J Geol 103:629-651

Harlov DE, Hansen EC, Bigler C (1998) Petrologic evidence for K-feldspar metasomatism in granulite facies rocks. Chem Geol 151:373-386

Harlov DE, Johansson L, Van den Kerkhof A, Förster HJ (2006) The role of advective fluid flow and diffusion during localized, solid-state dehydration: Söndrum Stenhuggeriet, Halmstad, SW Sweden. J Petrol 47: 3-33

Heijlen W, Appel PU, Frezzotti ML, Horsewell A, Touret JRL (2006) Metamorphic fluid flow in the northeastern part of the 3.8-3.7 Ga Isua Greenstone Belt (SW Greenland): A re-evaluation of fluid inclusion evidence for Early Archean seafloor-hydrothermal systems. Geochim Cosmochim Acta 70:3075-3095

Heinrich CA (2007) Fluid – fluid interactions in magmatic-hydrothermal ore formation. Rev Mineral Geochem 65:363-387

Heinrich W (1993) Fluid infiltration through metachert layers at the contact aureole of the Bufa del Diente intrusion, northeast Mexico: Implications for wollastonite formation and fluid immiscibility. Am Mineral 78:804-818

Heinrich W, Churakov SS, Gottschalk M (2004) Mineral-fluid equilibria in the system CaO-MgO-SiO$_2$-H$_2$O-CO$_2$-NaCl and the record of reactive fluid flow in contact metamorphic aureoles. Contrib Mineral Petrol 148:131-149

Heinrich W, Gottschalk M (1994) Fluid flow patterns and infiltration isograds in melilite marbles from the Bufa del Diente contact metamorphic aureole, north-east Mexico. J Metamorph Geol 12:345-359

Heinrich W, Gottschalk M (1995) Metamorphic reactions between fluid inclusions and mineral hosts I: Progress of the reaction calcite + quartz ↔ wollastonite + CO$_2$ in natural wollastonite-hosted fluid inclusions. Contrib Mineral Petrol 122:51-61

Heinrich W, Hoffbauer R, Hubberten HW (1995) Contrasting fluid flow pattern at the Bufa del Diente contact metamorphic aureole, north-east Mexico: evidence from stable isotopes. Contrib Mineral Petrol 119:362-376

Helgeson HC, Delany JM, Nesbitt HW, Bird DK (1978) Summary and critique of the thermodynamic properties of rock-forming minerals. Am J Sci 278:1-229

Herms P (2002) Fluids in 2 Ga old subduction zone – deduced from eclogite-facies rocks of the Usagaran belt, Tanzania. Eur J Mineral 14:361-373

Herms P, Schenk V (1992) Fluid inclusions in granulite-facies metapelites of the Hercynian ancient lower crust of the Serre, Calabria, Southern Italy. Contrib Mineral Petrol 112:393-404

Herms P, Schenk V (1998) fluid inclusions in high pressure granulites of the Pan-African belt in Tanzania (Uluguru Mus): a record of prograde to retrograde fluid evolution. Contrib Mineral Petrol 130:199-212

Hollister (1990) Enrichment of CO$_2$ in fluid inclusions in quartz by removal of H$_2$O during crystal-plastic deformation. J Struct Geol 12:895-901

Hollister LS (1988) On the origin of CO$_2$-rich inclusions in migmatites. J Metamorph Geol 6:467-474

Holness MB (1992) Equilibrium dihedral angles in the system quartz-CO$_2$-H$_2$O-NaCl at 800 °C and 1-15 kbar: the effects of pressure and fluid composition on the permeability of quartzites. Earth Planet Sci Lett 114: 171-184

Holness MB, Graham CM (1991) Equilibrium dihedral angles in the system CO$_2$-H$_2$O-NaCl-calcite, and implications for fluid flow during metamorphism. Contrib Mineral Petrol 108:368-383

Huizenga JM, Touret JLR (1999) Fluid inclusions in shear zones: the case of the Umwindsi shear zone in the Harare-Shamva-Bindura greenstone belt, NE Zimbabwe. Eur J Mineral 11:1079-1090

Hurai V, Horn EE (1992) A boundary layer-induced immiscibility in naturally re-equilibrated H$_2$O-CO$_2$-NaCl inclusions from metamorphic quartz (Western Carpathians, Czechoslovakia). Contrib Mineral Petrol 112: 414-427

Hurai V, Janák M, Ludhová L, Horn EE, Thomas R, Majzlan J (2000) Nitrogen-bearing fluids, brines and carbonate liquids in Variscan migmatites of the Tatra Mountains, Western Carpathians - Heritage of high pressure metamorphism. Eur J Mineral 12:1283-1300

Jamtveit B, Andersen T (1993) Contact metamorphism of layered shale-carbonate sequences in the Oslo Rift: III. Nature of skarn-forming fluids. Econ Geol 88:1830-1849

Jamtveit B, Bucher-Nurminen K, Stijfhoorn DE (1992a) Contact metamorphism of layered shale-carbonate sequences in the Oslo Rift: I. Buffering, infiltration, and the mechanisms of mass transport. J Petrol 33: 377-422

Jamtveit B, Grorud HF, Bucher-Nurminen K (1992b) Contact metamorphism of layered carbonate-shale sequences in the Oslo Rift. II. Migration of isotopic and reaction fronts around cooling plutons. Earth Planet Sci Lett 114:131-148

Janák M, Hurai V, Ludhová L, O'Brien PJ, Horn EE (1999) Dehydration melting and devolatilization during exhumation of high-grade metapelites: The Tatra Mountains, Western Carpathians. J Metamorph Geol 17: 379-395

Janák M, O'Brien PJ, Hurai V, Reutel C (1996) metamorphic evolution and fluid composition of garnet-clinopyroxene amphibolites from the Tatra Mountains, Western Carpathians. Lithos 39:57-79

Jawecki C (1996) fluid regime in the Austrian Moldanubian Zone as indicated by fluid inclusions. Mineral Petrol 58:235-252

Johnson EL (1991) Experimentally determined limits for H_2O-CO_2-NaCl immiscibility in granulites. Geology 19:925-928

Johnson EL, Hollister LS (1995) Syndeformational fluid trapping in quartz: determining the pressure-temperature conditions of deformation from fluid inclusions and the formation of pure CO_2 fluid inclusions during grain-boundary migration. J Metamorphic Geol 13:239-249

Joyce DB, Holloway JR (1993) An experimental determination of the thermodynamic properties of H_2O-CO_2-NaCl fluids at high pressures and temperatures. Geochim Cosmochim Acta 57:733-746

Kisch HJ, van den Kerkhof AM (1991) CH_4-rich inclusions from quartz veins in the Valley-and Ridge province and the anthracite fields of the Pennsylvania Appalachians. Am Mineral 76:230-240

Klemd R, Bröcker M, Schramm J (1995) Characterization of amphibolite-facies fluids of Variscan eclogites from the Orlica-Snieznik dome (Sudetes, SW Poland) Chem Geol 119:101-113

Klemd R, van den Kerkhof AM, Horn EE (1992) High density CO_2-N_2 inclusions in eclogite facies metasediments of the Münchberg gneiss complex, SE Germany. Contrib Mineral Petrol 111:409-419

Kolb J, Kisters AFM, Hoernes S, Meyer FM (2000) The origin of fluids and nature of fluid-rock interaction in mid-crustal auriferous mylonites of the Renco mine, southern Zimbabwe. Mineral Deposita 35:109-125

Kotel'nikov AR, Kotel'nikova ZA (1990) The phase state of the H_2O-CO_2-NaCl system examined from synthetic fluid inclusions in quartz. Geokhimiya 4:526-537

Krader T, Franck EU (1987) The ternary systems H_2O-CH_4-NaCl and H_2O-CH_4-$CaCl_2$ to 800 K and 250 MPa. Ber Bunsenges Phys Chem 91:627-634

Kretz (1983) Symbols for rock-forming minerals. Am Mineral 68:277-279

Labotka TC (1991) Chemical and physical properties of fluids. Rev Mineral Geochem 26:43-104

Labotka TC, Nabelek PI, Papike JJ (1988) Fluid infiltration through the Big Horse Limestone in the Notch Peak contact-metamorphic aureole, Utah. Am Mineral 73:1302-1324

Lamb WM (1988) CO_2-rich fluid inclusions in granulites: evidence of entrapment after the peak metamorphism. Mem Geol Soc India 11:101-115

Lamb WM, Brown PE, Valley JW (1991) Fluid inclusions in Adirondack granulites: implications for the retrograde *P-T* path. Contrib Mineral Petrol 107:472-483

Lamb WM, McShane CJ, Popp RK (2002) Phase relations in the CH_4-H_2O-NaCl system at 2 kbar, 300 to 600 °C as determined using synthetic fluid inclusions. Geochim Cosmochim Acta 66:3971-3986

Lamb WM, Morrison J (1997) Retrograde fluids in the Archean Shawmere anorthosite, Kapuskasing Structural Zone, Ontario, Canada. Contrib Mineral Petrol 129:105-119

Lamb WM, Popp RK, Boockoff LA (1996) The determination of phase relations in the CH_4-H_2O-NaCl system at 1 kbar, 400 to 600 °C using synthetic fluid inclusions. Geochim Cosmochim Acta 60:1885-1897

Liebscher A (2007) Experimental studies in model fluid systems. Rev Mineral Geochem 65:15-47

Markl G, Bucher K (1998) Composition of fluids in the lower crust inferred from metamorphic salt in lower crust. Nature 391:781-783

Meen JK, Eggler DH, Ayers JC (1989) Experimental evidence for very low solubility of rare earth elements in CO_2-rich fluids at mantle conditions. Nature 240:301-303

Mercolli I (1982) Le inclusione fluide nei nodulo de quarzo dei marmi dolomitici della regione del Campolungo (Ticino). Schweiz Mineral Petrogr Mitt 62:245-312

Milsch H, Heinrich W, Dresen G (2003): Reaction-induced fluid flow in synthetic quartz-bearing marbles. Contrib Mineral Petrol 146:286-296

Mullis J, Dubessy J, Poty B, O'Neil J (1994) Fluid regimes during late stages of a continental collision: Physical, chemical, and stable isotope measurements of fluid inclusions in fissure quartz from a geotraverse through the Central Alps, Switzerland. Geochim Cosmochim Acta 58:2239-2267

Nabelek PI (2002) Calc-silicate reactions and bedding controlled isotopic exchange in the Notch Peak aureole, Utah: implications for differential fluid fluxes with metamorphic grade. J Metamorphic Geol 20:429-440

Nabelek PI, Labotka TC (1993) Implications of geochemical fronts in the Notch Peak contact metamorphic aureole, Utah, USA. Earth Planet Sci Lett 119:539-559

Nabelek PI, Labotka TC, Ross-Labotka C (1992) Stable isotope evidence for the role of diffusion, infiltration, and local structure on contact metamorphism of calc-silicate rocks at Notch Peak, Utah. J Petrol 33:557-583

Naumov VB, Khakimov AKH, Khodakovskiy IL (1974) Solubility of carbon dioxide in concentrated chloride solutions at high temperatures and pressures. Geochem Int 11:31-41

Newton RC (1992) Charnockitic alteration: evidence for CO_2 infiltration in granulite facies metamorphism. J Metamorphic Geol 10:383-400

Olsen SN (1988) High density CO_2 inclusions in the Colorado Front Range. Contrib Mineral Petrol 100:226-235

Pattison DRM (2006) The fate of graphite in prograde metamorphism of pelites: an example from the Ballachulish aureole, Scotland. Lithos 88:85-99

Perchuk LL, Gerya TV (1993) Fluid control of charnockitization. Chem Geol 108:175-186

Philippot P, Selverstone J (1991) Trace element rich brines in eclogitic veins: implications for fluid composition and transport during subduction. Contrib Mineral Petrol 106:417-430

Pöter B, Gottschalk M, Heinrich W (2004) Experimental determination of the ammonium partitioning among muscovite, K-Feldspar, and aqueous chloride solutions. Lithos 74:67-90

Romer RL, Heinrich W (1998) Transport of Pb and Sr in leaky aquifers of the Bufa del Diente contact metamorphic aureole, north-east Mexico. Contrib Mineral Petrol 131:155-170

Romer RL, Heinrich W, Schröder-Smeibidl B, Meixner A, Fischer CO, Schulz C (2005) Elemental dispersion and stable isotope fractionation during reactive fluid-flow and fluid immiscibility in the Bufa del Diente aureole, NE-Mexico: evidence from radiographies and Li, B, Sr, Nd, and Pb isotope systematics. Contrib Mineral Petrol 149:400-429

Scambelluri M, Pennacchioni G, Philippot P (1998) Salt-rich aqueous fluids formed during eclogitization of metabasites in the Alpine continental crust. Lithos 43:151-161

Scambelluri M, Philippot P (2001) Deep fluids in subduction zones. Lithos 55:213-227

Scambelluri M, Piccardo GB, Philippot P, Robbiano A, Negretti L (1997) High salinity fluid inclusions formed from recycled seawater in deeply subducted alpine serpentinite. Earth Planet Sci Lett 148:485-499

Schmidt C, Bodnar RJ (2000) Synthetic fluid inclusions XIV. PVTX properties in the system H_2O-CO_2-NaCl at elevated temperatures, pressures, and salinities. Geochim Cosmochim Acta 64:3853-3869

Schmidt C, Rosso KM, Bodnar RJ (1995) Synthetic fluid inclusions XIII. Experimental determination of PVT properties in the system H_2O + 40 wt% NaCl + 5 mol% CO_2 at elevated temperature and pressure. Geochim Cosmochim Acta 59:3953-3959

Selverstone J, Franz G, Thomas S, Getty S (1992) fluid variability in 2 GPa eclogites as an indicator of fluid behavior during subduction. Contrib Mineral Petrol 112:341-357

Shi GU. Tropper P, Cui W, Tan J, Wang C (2005) Methane (CH_4)-bearing fluid inclusions in the Myanmar jadeitite. Geochem J 39:503-516

Shinohara H (1989) Exsolution of immiscible vapor and liquid phases from a crystallizing silicate melt: Implications for chlorine and metal transport. Geochim Cosmochim Acta 58:5215-5221

Shinohara H, Iiyama JT, Matsuo S (1989) Partition of chlorine compounds between silicate melts and hydrothermal solutions. I. Partition of NaCl-KCl. Geochim Cosmochim Acta 53:2617-2630

Shmulovich KI, Graham CM (1999) An experimental study of phase equilibria in the system H_2O-CO_2-NaCl at 800 °C and 9 kbar. Contrib Mineral Petrol 136:247-257

Shmulovich KI, Graham CM (2004) An experimental study of phase equilibria in the systems H_2O-CO_2-$CaCl_2$ and H_2O-CO_2-NaCl at high pressures and temperatures (500-800 °C, 0.5-0.9 GPa): geological and geophysical applications. Contrib Mineral Petrol 146:450-462

Shmulovich KI, Landwehr D, Simon K, Heinrich W (1999) Stable isotope fractionation between liquid and vapour in water-salt systems. Chem Geol 157:343-354

Shmulovich KI, Plyasunova NV (1993) Phase equilibria in ternary systems formed by H_2O and CO_2 with $CaCl_2$ or NaCl at high T and P. Geochem Int 30:53-71

Shmulovich KI, Tkachenko SI, Plyasunova NV (1995) Phase equilibria in fluid systems at high pressures and temperatures. *In:* Fluids in the Crust. Shmulovich KI, Yardley BWD, Gonchar GG (eds) Chapman and Hall, p 193-214

Sisson VB, Crawford ML, Thompson PH (1981) CO_2-brine immiscibility at high temperatures: Evidence from calcareous metasedimentary rocks. Contrib Mineral Petrol 78:371-378

Sisson VB, Hollister LS (1990) A fluid-inclusion study of metamorphosed pelitic and carbonate rocks, south-central Maine. Am Mineral 75:59-70

Skippen G, Trommsdorff V (1986) The influence of NaCl and KCl on phase relations in metamorphosed carbonate rocks. Am J Sci 286:81-104

Sourirajan S, Kennedy GC (1962) The system H_2O-NaCl at elevated pressures and temperatures. Am J Sci 260: 115-141

Stähle HJ, Raith M, Hoernes S, Delfs A (1987) Element mobility during incipient granulite formation at Kabbaldurga, Southern India. J Petrol 28:803-834

Takenouchi S, Kennedy G (1965) the solubility of carbon dioxide in NaCl solutions at high temperatures and pressures. Am J Sci 263:445-454

Touret (1981) Fluid inclusions in high grade metamorphic rocks. *In:* Fluid Inclusions: Applications to Petrology. Short Course Volume 6. Hollister LS, Crawford ML (eds) Mineralogical Association of Canada, p 182-208

Touret JLR (1971) Le facies granulite en Norvege meridionale. II. Les inclusions fluides. Lithos 4:423-436

Touret JLR (1977) The significance of fluid inclusions in metamorphic rocks. *In:* Thermodynamics in Geology. Fraser DG (ed) Reidel, p 203-227

Touret JLR (1985) Fluid regime in southern Norway: the record of fluid inclusions. *In:* The Deep Proterozoic Crust in the North Atlantic Provinces. Tobi AC, Touret JLR (eds) Reidel, p 517-549

Touret JLR (1995) Brines in granulites: the other fluid. Bol Soc Esp Mineral 18:250-251

Touret JLR (2001) Fluids in metamorphic rocks. Lithos 55:1-25

Touret JLR, Huizenga JM (1999) Precambrian intraplate magmatism: high temperature, low pressure crustal granulites. J Afr Earth Sci 28:367-382

Touret JLR, Olsen SN (1985) Fluid inclusions in migmatites. *In:* Migmatites. Ashworth JR (ed) Blackie and Son, p 265-288

Touret JRL, Frezzotti ML (2003) Fluid inclusions in high pressure and ultrahigh pressure rocks. *In:* Ultrahigh pressure Metamorphism. EMU Notes in Mineralogy Vol 5. Carswell DA, Compagnoni R (eds), Eötvös University Press, p 467-487

Trommsdorff V, Skippen G (1986) Vapour loss ("boiling") as a mechanism for fluid evolution in metamorphic rocks. Contrib Mineral Petrol 94:317-322

Trommsdorff V, Skippen G (1987) Metasomatism involving fluids in CO_2-H_2O-NaCl. *In:* Chemical Transport in Metasomatic Processes. Proc NATO Advanced Study Inst. Helgeson HC (ed) Reidel, p 133-152

Trommsdorff V, Skippen G, Ulmer P (1985) Halite and sylvite as solid inclusions in high-grade metamorphic rocks. Contrib Mineral Petrol 89:24-29

Urbán M, Thomas R, Hurai V, Konečný P, Chovan M (2006) Superdense CO_2 inclusions in Cretaceous quartz-stibnite veins hosted in low-grade Variscan basement of the Western Carpathians, Slovakia. Mineral Deposita 40:867-873

Van den Berg R, Huizenga JM (2001) Fluids in granulites of the southern Marginal Zone of the Limpopo Belt, south Africa. Contrib Mineral Petrol 141:529-545

Van Reenen DD, Hollister LS (1988) Fluid inclusions in hydrated granulite facies rocks, southern marginal zone of the Limpopo Belt, South Africa. Geochim Cosmochim Acta 52:1057-1064

Walther JV (1983) Description and interpretation of metasomatic phase relations at high pressures and temperatures; 2. Metasomatic relations between quartz and dolomite at Campolungo, Switzerland. Am J Sci 283A:459-485

Warner M (2004) Free water and seismic reflectivity in the lower continental curst. J Geophys Eng 1:88-101

Watson EB, Brenan JM (1987) Fluid in the lithosphere. I. Experimentally determined wetting characteristics of CO_2-H_2O fluids and their implications for fluid transport, host-rock physical properties, and fluid inclusion formation. Earth Planet Sci Lett 85:497-515

Webster J, Mandeville C (2007) Fluid immiscibility in volcanic environments. Rev Mineral Geochem 65:313-362

Whitney (1992) Origin of CO_2-rich fluid inclusions in leucosomes from the Skagit migmatites, North Cascades, Washington, USA. J Metamorphic Geol 10:715-725

Williams-Jones AE (1982) Patapedia: an Appalachian calc-silicate hosted copper prospect of porphyry affinity. Can J Sci 19:438-455

Williams-Jones AE, Ferreira DR (1989) Thermal metamorphism and H_2O-CO_2-NaCl immiscibility at Patapedia, Quebec: Evidence from fluid inclusions. Contrib Mineral Petrol 102:247-254

Xiao YL, Zhang ZM, Hoefs J, van den Kerkhof A (2006) Ultrahigh-pressure metamorphic rocks from the Chinese Continental Scientific Drilling Project: II. Oxygen isotope and fluid inclusion distributions through vertical sections. Contrib Mineral Petrol 152:443-458

Xu G, Pollard PJ (1999) Origin of CO_2-rich fluid inclusions in synorogenic veins from the Eastern Mount Isa Belt, NM Queensland, and their implications for mineralization. Mineral Deposita 34:395-404

Yardley BWD, Bottrell SH (1988) Immiscible fluids in metamorphism: implications of two-phase fluid flow for reaction history. Geology 16:199-202

Yardley BWD, Graham JT (2002) The origins of salinity in metamorphic fluids. Geofluids 2:249-256

Yardley BWD, Lloyd GE (1989) An application of cathodoluminescence microscopy to the study of textures and reactions in high-grade marbles from Connemara, Ireland. Geol Mag 126:333-337

Yardley BWD, Valley JW (1997) The petrological case for a dry lower crust. J Geophys Res 102:12173-12185

Zatsepin SV, Crampin S (1997) Modelling the compliance of crustal rocks: I – Response of shear-wave splitting to differential stress. Geophys J Int 129:477-494

Zhang YG, Frantz JD (1989) Experimental determination of the compositional limits of immiscibility in the system $CaCl_2$-H_2O-CO_2 at high temperatures and pressures using fluid inclusions. Chem Geol 74:289-308

Zhang YG, Frantz JD (1992) Hydrothermal reactions involving equilibrium between minerals and mixed volatiles. 2. Investigations of fluid properties in the CO_2-CH_4-H_2O system using synthetic fluid inclusions. Chem Geol 100:51-72

Zhang ZM, Shen K, Xiao YL, van den Kerkhof A, Hoefs J, Liou JG (2005) Fluid composition and evolution attending UHP metamorphism: study of fluid inclusions from drill cores, southern Sulu Belt, eastern China. Int Geol Rev 47:297-309